INVESTIGATIVE
IMMUNOTOXICOLOGY

INVESTIGATIVE IMMUNOTOXICOLOGY

Edited by

Helen Tryphonas
Michel Fournier
Barry R. Blakley
Judit E. G. Smits
Pauline Brousseau

CRC Press
Taylor & Francis Group
Boca Raton London New York

CRC Press is an imprint of the
Taylor & Francis Group, an **informa** business

CRC Press
Taylor & Francis Group
6000 Broken Sound Parkway NW, Suite 300
Boca Raton, FL 33487-2742

First issued in paperback 2019

ISBN-13: 978-0-415-30854-0 (hbk)
ISBN-13: 978-0-367-39329-8 (pbk)

Library of Congress Cataloging-in-Publication Data

Investigative immunotoxicology / edited by Helen Tryphonas ... [et al.].
    p. cm.
Includes bibliographical references and index.
ISBN 0-415-30854-2 (alk. paper)
    1. Immunotoxicology—Research—Methodology. 2. Experimental immunology. I. Tryphonas, Helen.
    [DNLM: 1. Immune System—drug effects. 2. Disease Models, Animal. 3. Environmental Monitoring—methods. 4. Immunologic Diseases—chemically induced. QW504 [629 2004]
RC582.17.1585 2004
616.97'071—dc22
                                   2004057045

Library of Congress Card Number 2004057045

Visit the Taylor & Francis Web site at
http://www.taylorandfrancis.com

and the CRC Press Web site at
http://www.crcpress.com

Dedication

To our families for their support and understanding

Preface

Immunotoxicology, which is defined as the study of immunomodulatory effects of chemicals on the immune system, is an evolving discipline and is rapidly becoming an important component of toxicologic studies. This book provides a critical evaluation of proposed experimental animal models and approaches, and discusses the contribution immunotoxicity can make to the overall assessment of chemical-induced adverse health effects in humans and on the ecosystem. The apparent lack of much needed published literature on immunotoxicological applications for many species in the ecosystem has prompted us to pay special attention to this aspect of the science. This information is organized as follows.

Section I is an account of general concepts of the immune system which may impact on the immunotoxicologic evaluation of chemicals and on approaches to risk assessment. Interactions of the immune system with the neuro-endocrine system are also presented.

Section II focuses on a comprehensive review of approaches and models used for testing chemical-induced immunotoxic effects in a variety of invertebrates as well as terrestrial wildlife species and marine mammals. The predictive value of selected immune markers across species is discussed and recommendations for the selection of sentinel species in ecoimmunotoxicology are presented.

Section III is a critical evaluation of the use of rodent species as well as non-human primates as experimental animal models for the study of chemical-induced immunotoxicity. Practical considerations in enhanced histopathology of the immune system of these species are also presented and discussed. This section also includes a comprehensive discussion on feasible approaches to human immunotoxicology.

Section IV presents several animal models and discusses recent developments in chemical-induced allergenicity and allergic contact dermatitis. Autoimmunity is discussed in Section V.

Section VI presents several emerging techniques for *in vivo* and *in vitro* approaches in immuno-toxicology including the contribution of transgenic rodent models and genomics to immunotoxicology.

Finally, the role of immunotoxicology in risk assessement is discussed in Section VII. The following are included in this section: selection of appropriate biomarkers of effect, the extrapolation of experimental animal data to human immunotoxicity guidelines, and initiatives toward standardization of methodologies across laboratories at the national and international levels. This section also includes approaches to statistical analysis of immunotoxicity data in support of risk assessment.

This book is an excellent reference source for biologists, toxicologists, and those actively involved in immunotoxicology. It should be especially helpful to those specializing in immunotoxicology related to ecosystem health and those anticipating engaging in the field of immunotoxicology. The book also contains a wealth of information that can be useful to regulatory agencies at the international level.

Acknowledgment

The support and encouragement received during the preparation of this book from Dr. Rekha Mehta, Chief, Toxicology Research Division, Food Directorate, Health Canada, is greatly appreciated. In addition, the authors wish to thank Dr. Rudi Mueller for his contribution to the art design used on the cover.

The Editors

Helen Tryphonas, Ph.D., received her Master of Science (M.Sc.) degree in microbiology with a major in immunology from the University of Saskatchewan (1972) and a Doctor of Philosophy (Ph.D.) degree in environmental toxicology/immunotoxicology from the Université du Québec à Montréal, Quebec, Canada (1997). Upon completion of her M.Sc. degree, Dr. Tryphonas accepted a two-year appointment with the Western College of Veterinary Medicine as a teaching assistant, and in 1974 she joined the Food Directorate of the Food Safety and Health Products Branch of Health Canada. During her 28 years as a senior scientist with Health Canada, Dr. Tryphonas developed the immunotoxicology program of the Food Directorate, directed research projects in immunotoxicology, and carried out significant collaborative studies at the national and international levels. Presently, she is a senior member of the CPT Consultants in Pathology and Toxicology Inc.

Dr. Tryphonas' research contributions are in the areas of regulatory immunotoxicology of environmental contaminants, food additives and fungal toxins, and allergenicity of foods and GM proteins. She is an adjunct professor of the UQAM-TOXEN and has served in several local, national, and international committees and scientific panels as a consultant in immunotoxicology and allergenicity of foods. She is a member of the Canadian Network of Toxicology Centres, and a member of the editorial review board of *Environmental Health Perspectives*.

Dr. Tryphonas has authored or co-authored over 110 scientific papers including publications in peer-reviewed journals, books and book reviews, and presentations at scientific meetings. Recently, she was the recipient of the Food Directorate's Award for Creativity and Innovation in recognition of her contribution to building laboratory research capacity in support of the detection and evaluation of the safety, nutritional quality, and health effects of GM foods, and the recipient of a certificate of appreciation from the City of Ottawa in recognition of her contributions to the community.

Michel Fournier, Ph.D., has a B.Sc. in cellular and molecular biology (1976) from the Université du Québec à Montréal, an M.Sc. in microbiology-immunology (1978) from the Université de Montréal, and a Ph.D. in experimental medicine from McGill University. Dr. Fournier has been a professor in environmental immunotoxicology since 1980 in the Biological Sciences Department of the Université du Québec à Montréal and was a member of the environmental toxicology research laboratory (TOXEN). In 1998, he became professor in INRS-Institut Armand-Frappier. He holds a Canada Research Chair in Environmental Immunotoxicology. Dr. Fournier is also the Director of St. Lawrence Ecotoxicology Network. The research projects on which he has worked since his graduate studies involve the effects of environmental substances on the immune system.

Barry R. Blakley, D.V.M., Ph.D., is a professor in the Department of Veterinary Biomedical Sciences at the University of Saskatchewan. Dr. Blakley received his B.Sc. in chemistry in 1971 and his D.V.M. in 1975 from the University of Saskatchewan. He completed his M.Sc. in 1977 at the University of Saskatchewan and his Ph.D. at the University of Cincinnati in 1980 in immunotoxicology. In the same year he joined the Department of Veterinary Physiological Sciences as an Associate Professor. He was promoted to Professor in 1986 and was appointed as Chair in the Department of Veterinary Biomedical Sciences in 2003.

Dr. Blakley has also participated as an executive member of the Toxicology Group at the University of Sasktachewan. He has served in the capacity of Chair or Academic Coordinator of the Toxicology Graduate Program for many years. Dr. Blakley is also the supervisor of the Diagnostic Toxicology Laboratory associated with Prairie Diagnostic Services. This laboratory provides analytical support and consultation for veterinarians in Western Canada.

Dr. Blakley has published more than 100 manuscripts and book chapters in immunotoxicology, nutritional toxicology, and veterinary toxicology. His current research interests are focused on developmental immunotoxicology associated with various environmental contaminants.

Judit E.G. Smits, D.V.M., Ph.D., is a professor in the Department of Veterinary Pathology with a cross appointment to the Toxicology Centre at the University of Saskatchewan. She practiced veterinary medicine for 7 years before returning to academia to combine a new discipline, environmental toxicology, with her experience in animal health. Dr. Smits developed an interest in immunotoxicology while investigating the biological impacts of environmental contaminants on wildlife. She discovered the immune system to be one of the most sensitive endpoints. Her focus for the past 8 years has been on wild birds that breed and raise their young in contaminated areas. Study of the immune response and immunotoxicology are integral components of her search for mechanisms through which toxicity is expressed in wildlife.

Dr. Smits has research projects in environmental toxicology in Spain as well as numerous locations across Canada. She is active in the Society of Toxicology of Canada and the Saskatoon Environmental Advisory Committee, and holds memberships in the Canadian and Saskatchewan Veterinary Medical Associations, the Wildlife Disease Association, and the Society of Environmental Toxicology and Chemistry.

Dr. Smits has authored or co-authored over 60 peer reviewed scientific papers in toxicology, immunotoxicology, wildlife disease, and environmental health.

Pauline Brousseau, Ph.D., obtained her B.Sc. in cellular and molecular biology from UQAM and her M.Sc. in microbiology and immunology from Université de Montréal. In 1983 Dr. Brousseau obtained her Ph.D. in experimental medicine from McGill University. She was then hired by Bio-Research Ltd., and was in charge of the Department of Immunopharmacology/Immunotoxicology to develop and validate methods in support of pre-clinical toxicological studies.

After 10 years in the private sector, Dr. Brousseau returned to fundamental research at UQAM. Subsequently, she moved to INRS-Institut Armand-Frappier. During this period, her research activities were directed mainly toward environmental immunotoxicology. During this time, she was responsible for the Canadian effort of harmonization of the immunological methods, a program of the Canadian Network of Toxicology Centres involving three Canadian universities. In February 2001, she joined Biophage Inc. as the director of the Division of Contractual Research (CRO), specialized in immunology and was involved in the sector of pre-clinical and clinical studies. She devoted much energy to set up and validate immunological methods within a framework of GLPs. She was instrumental in setting up the first Canadian laboratory for the detection of workers sensitized to beryllium.

Dr. Brousseau has authored or co-authored 50 scientific papers, books, and book chapters and was involved in over 100 scientific communications.

Contributors List

Douglas L. Arnold, Ph.D.
Frederick G. Banting Research Center
Ottawa, Ontario, Canada

Michel Auffret, Ph.D.
UBO Institut Universitaire Européen de la Mer
Plouzane, France

Pierre Ayotte, Ph.D.
Université Laval
Sainte Foy, Québec, Canada

Jacques Bernier, Ph.D.
INRS-Institut Armand-Frappier
Pointe-Claire, Québec, Canada

Pierluigi E. Bigazzi, M.D.
University of Connecticut Health Center
Farmington, Connecticut

Barry R. Blakley, D.V.M., Ph.D.
University of Saskatchewan
Saskatoon, Saskatchewan, Canada

Patricia M. Blakley, M.D., Ph.D.
Kinsmen Children's Centre
Saskatoon, Saskatchewan, Canada

Laura Blanciforti
National Institute for Occupational Safety
 and Health
Morgantown, West Virginia

Herman J. Boermans, Ph.D.
University of Guelph
Guelph, Ontario, Canada

Genevieve S. Bondy, Ph.D.
Health Canada
Ottawa, Ontario, Canada

Pauline Brousseau, Ph.D.
CRO Division
Montréal, Québec, Canada

Don Caldwell, D.V.M.
Health Canada
Ottawa, Ontario, Canada

Daniel G. Cyr, Ph.D.
INRS-Institut Armand-Frappier
Pointe-Claire, Québec, Canada

Michele D'Elia, M.Sc.
INRS-Institut Armand-Frappier
Pointe-Claire, Québec, Canada

Frédéric Dallaire, Ph.D.
Unité de recherche en santé publique
 CHUQ-CHUL
Sainte Foy, Québec, Canada

Claire Dautremepuits, Ph.D.
INRS-Institut Armand-Frappier
Pointe-Claire, Québec, Canada

Rebecca J. Dearman, Ph.D.
Syngenta Central Toxicology Laboratory
Cheshire, England

Sylvain De Guise, D.M.V., Ph.D.
University of Connecticut
Storrs, Connecticut

Éric Dewailly, Ph.D.
Unité de recherche en santé publique
 CHUQ-CHUL
Sainte Foy, Québec, Canada

Jessica E. Duffy
New York University School of Medicine
Tuxedo, New York

Lloyd C. Fitzpatrick, Ph.D.
University of North Texas
Denton, Texas

Alain Fournier, Ph.D.
INRS-Institut Armand-Frappier
Pointe-Claire, Québec, Canada

Michel Fournier, Ph.D.
INRS-Institut Armand-Frappier
Pointe-Claire, Québec, Canada

Dori R. Germolec, Ph.D.
National Institute of Environmental Health
 Sciences
Research Triangle Park, North Carolina

Santokh Gill, Ph.D.
Frederick G. Banting Research Center
Ottawa, Ontario, Canada

Arthur J. Goven, Ph.D.
University of North Texas
Denton, Texas

Sami Haddad, Ph.D.
INRS-Institut Armand-Frappier
Pointe-Claire, Quebec, Canada

Fiona A. Harding, Ph.D.
Genencor International Inc.
Palo Alto, California

Stephen Hayward
Frederick G. Banting Research Center
Ottawa, Ontario, Canada

Ricki M. Helm, Ph.D.
University of Arkansas
Little Rock, Arkansas

Kevin S. Henry, Ph.D.
The Dow Chemical Company
Midland, Michigan

Michael P. Holsapple, Ph.D.
ILSI Heath and Environmental Sciences
  Institute
Washington, D.C.

Robert V. House, Ph.D.
DVC
Frederick, Maryland

David Janz, Ph.D.
University of Saskatchewan
Saskatoon, Saskatchewan, Canada

Michael Kashon
National Institute for Occupational Safety
  and Health
Morgantown, West Virginia

Meghan Kavanagh
Frederick G. Banting Research Center
Ottawa, Ontario, Canada

Ian Kimber, Ph.D.
Syngenta Central Toxicology Laboratory
Cheshire, England

Edward Kouassi, Ph.D.
Guy-Bernier Research Center
Montreal, Quebec, Canada

Robert W. Luebke, Ph.D.
U.S. Environmental Protection Agency
Research Triangle Park, North Carolina

Michael I. Luster, Ph.D.
National Institute for Occupational Safety
  and Health
Morgantown, West Virginia

Donald P. Naki, Ph.D.
Genencor International Inc.
Palo Alto, California

Jean-Marc Nicolas, Ph.D.
INRS-Institut Armand-Frappier
Montreal, Quebec, Canada

Christine G. Parks*
National Institute of Environmental Health
  Sciences
Research Triangle Park, North Carolina
*Currently with NIOSH

James J. Pestka, Ph.D.
Michigan State University
East Lansing, Michigan

Stéphane Pillet, Ph.D.
INRS-Institut Armand-Frappier
Montreal, Quebec, Canada

Olga Pulido, M.D.
Frederick G. Banting Research Center
Ottawa, Ontario, Canada

Louise A. Rollins-Smith, Ph.D.
Vanderbilt University
Nashville, Tennessee

Andrew A. Rooney, Ph.D.
U.S. Environmental Protection Agency
Research Triangle Park, North Carolina

Harri Salo, Ph.D.
INRS-Institut Armand-Frappier
Pointe-Claire, Quebec, Canada

Katherine Sarlo, Ph.D.
The Procter & Gamble Company
Cincinnati, Ohio

Judit E.G. Smits, D.V.M., Ph.D.
University of Saskatchewan
Saskatoon, Saskatchewan, Canada

Helen Tryphonas, Ph.D.
CPT Consultants in Pathology and
  Toxicology Inc.
Nepean, Ontario, Canada

Barney J. Venables, Ph.D.
University of North Texas
Denton, Texas

David L. Wong, Ph.D.
Genencor International Inc.
Palo Alto, California

Michael R. Woolhiser, Ph.D.
The Dow Chemical Company
Midland, Michigan

Judith T. Zelikoff, Ph.D.
New York University School of Medicine
Tuxedo, New York

Table of Contents

SECTION I

The Immune System

CHAPTER 1

Structural and Functional Complexity of the Immune System and Its Relationship to Immunotoxicology

Barry R. Blakley and Edward Kouassi

CONTENTS

1.1 INTRODUCTION

Understanding of the immune system and its varied functions has advanced considerably in recent years. Underlying physiological mechanisms and the complex multicellular interactions that are required to maintain homeostasis provide endless research opportunities. In the past, the immune system was considered to be an isolated organ system with limited physiological interaction with other body systems. Recent studies have clearly demonstrated that this is not the case. The immune system is far more complex than anyone ever anticipated. Its interactions with many other systems, including the endocrine and nervous systems, for example, have made it difficult to determine the full relevance and mechanisms of immune alteration. Immune function appears to be influenced by stress, neuromediators, hormones, and various psychological factors. In addition to immuno-modulatory disturbances, effects on behavior, reproduction, or other disease states, may be under a degree of immune regulation. Such associations have led to the establishment of several new subdisciplines, including immunoteratology and psychoneuroimmunology (Madden and Felten, 1995). Nutritional status and aging have also been identified as important factors with respect to immune function.

The discipline of immunotoxicology has benefited from recent advancements in the field of immunology. For more than 30 years, toxicologists have been aware of immune alteration associated with a variety of immunotoxicants including dioxins (Kerkvliet et al., 1985), mycotoxins (Tomar et al., 1988), metals (Blakley and Tomar, 1986), and pesticides (Lee et al., 2001). Extensive documentation describing the diversity of immunomodulating agents has been reviewed by Luster et al. (1989) and Holladay (1999). Many studies in the past have emphasized immunosuppression or a specific immune function such as humoral immunity as the exclusive investigative endpoint. The regulatory influence of the nervous system or the endocrine system was not considered. In order to fully assess the overall immunocompetence in the host, it is critical to understand the impact of other systems involved in homeostasis and immune regulation.

1.2 STRUCTURAL AND FUNCTIONAL COMPONENTS OF THE IMMUNE SYSTEM

The immune system has the unique ability to respond rapidly with highly specific protective responses when challenged with a foreign substance. At the same time the immune system must be tolerant of self-antigens. The ability of the immune system to discriminate between antigenic determinants expressed on foreign substances and determinants expressed by host tissues is an area of intense interest and research. The concept of immune tolerance, which involves the elimination or inactivation of cells that recognize self-antigens, is fundamental to this discriminatory phenomenon.

The immune system contains a range of distinct cell types located in a variety of tissues, including the thymus and bone marrow centrally, and the spleen, lymph nodes, and various other lymphoid tissues throughout the body. The lymphocyte plays a central role in the immune system. Lymphocyte subpopulations interact with various cells to initiate an immune response. Lymphocytes are highly specialized cells. They have the ability to respond to specific determinants on the antigen through specific receptors on the surface of the lymphocyte. Receptor specificity and functional heterogeneity allow for the lymphocyte subpopulations to respond to virtually any antigen.

There are two classes of lymphocytes including the B-lymphocyte and the T-lymphocyte. The B-lymphocytes, which are precursors of the antibody secreting cells, are derived from haematopoietic stem cells located in the fetal liver or in the bone marrow. B-lymphocytes may be activated directly through cross-linkage of the membrane immunoglobulin with the antigen or indirectly by interaction with T-lymphocytes. The B-lymphocytes may differentiate into a variety of specific antibody-secreting cells. This process is controlled in part by T-lymphocytes and a variety of interleukins including IL-2, IL-4, IL-5, and IL-6. The B-lymphocytes may also differentiate into many types of memory cells. This immunoglobulin class–switching process occurs in the spleen or in the lymph nodes.

The T-lymphocytes are the second major class of lymphocytes that are thymus derived. The T-lymphocytes are divided into several subpopulations based on the cell surface receptors that they express. These cells are involved in a variety of regulatory and effector functions. The precursors for T-lymphocytes in the thymus do not express specific characteristics for CD_3, CD_4, or CD_8 subpopulations. The thymocytes ultimately differentiate to form the specific lymphocyte populations that are distinguished based on cell surface–receptor differences. This process is critical for the development of self-tolerance and other regulatory functions of the immune system.

There are two subpopulations of T-lymphocytes with special significance to immunotoxicology. The CD_4 lymphocytes play an important regulatory role. The production of this subpopulation of lymphocytes is influenced by various cytokines. Two CD_4 subsets are produced including the Th_2 subset, which is classified as the T-helper lymphocyte. This lymphocyte enhances the ability of the B-lymphocyte to produce antibodies. This cell type produces a variety of interleukins that have the ability to modulate specific immune responses. Another subset, the Th_1 lymphocytes induces cellular responses that enhance microbiocidal activity. Interferon-γ and lymphotoxin are cytokines secreted by the Th_1 lymphocytes.

The CD_8 subpopulation of T-lymphocytes become cytotoxic T-lymphocytes. These cells are actively involved in the lysis of undesirable cells such as tumor cells. Suppressor T-lymphocytes, which have a major regulatory function are a group of lymphocytes with varying phenotypes. These cells have the ability to produce many different cytokines with the ability to block or enhance the activity of other cytokines or cell types.

Similar to the B-lymphocytes, the T-lymphocyte recognizes antigens on the surface of cells. The T-lymphocyte interacts with these antigen-presenting cells to produce a response. Each T-lymphocyte subpopulation recognizes and responds to a different peptide complex, thus generating a unique cytokine or immune response.

In addition to the cell-mediated (T-lymphocyte) and humoral (B-lymphocyte) immune responses that have a considerable degree of specificity, there are also nonspecific immune responses that are the first line of defense against invading pathogens. In many instances, these cell types mediate effector function as well. There are primarily two types of cells associated with nonspecific immunity. These include the natural killer (NK) cells and various phagocytic cells. The NK cells are closely related to T-lymphocytes, but they lack typical T-lymphocyte receptors. They express receptors that permit the destruction of cancer cells or virally infected cells. NK cells are also involved in antibody dependent cellular cytotoxicity.

Monocytes, various macrophages, and granulocytes are found in most tissues. They have a distinct phagocytic function to control infectious disease. They can be activated in many instances by cytokines to become more efficient killer cells. Several of these cell types, in particular the monocytes and the macrophages, produce reactive oxygen species, nitric oxide, or various cytokines such as interleukin-1, interleukin-6, or tumor necrosis factor to enhance killing efficiency. The monocytes and macrophages also act as antigen-presenting cells in specific immune responses. Mast cells, basophils, and the complement system also play important roles in the inflammatory processes involving nonspecific immune function.

In summary, a number of cell types and cell subpopulations involved in humoral, cell-mediated, and nonspecific immune responses act in concert with each other. Through cytokines or by direct cell–cell interaction, each component of the immune system regulates and influences the responses of many cell types throughout the immune system.

1.3 IMMUNE SYSTEM REGULATION

The immune system is functionally and structurally complex. In order to maintain optimal function and organ homeostasis, it is critical that all cells within the immune system communicate and respond in a coordinated fashion. Many functions of the immune system are mediated by the production of cytokines. These cytokines, which affect a variety of cell types, are produced by several different cells. The cytokines, in general, are extremely potent and bind strongly to specific receptors. The effects are often highly localized, potentially within a specific cell. Many of the cytokines are protein products produced by T-lymphocytes. They often regulate T-lymphocyte function, and are produced in response to the activation of the other cell types. Examples of cytokines include the interleukins, interferons, tumor necrosis factor, lymphotoxin, transforming growth factor, migration inhibitory factor, and so on. Cytokines are not only produced by immune cells such as NK cells, macrophages, T-lymphocytes, or B-lymphocytes, but cell types from other organ systems also produce cytokines under certain conditions. Fibroblasts, endothelial cells, platelets, hepatocytes, pituitary cells, and chondrocytes are examples. Any alteration in cell function or cell number may result in abnormal cytokine production and a loss of regulatory control. It is important to note that the alteration may be associated directly with the immune system or indirectly through other organ systems. Consequently, to fully understand the relevance and mechanisms of compromised immune function or regulation, it is critical to extend the evaluation process beyond immune system function and integrity.

1.4 THE NEURO–ENDOCRINE–IMMUNE CONNECTION

Accumulating evidence indicates that the nervous, endocrine, and immune systems are connected through pathways involving a number of soluble mediators and their specific receptors. The immune system signals the central nervous system (CNS) through cytokines and chemokines (Benveniste, 1998). In turn, the CNS signals the immune system through two primary pathways: (1) the neuroendocrine system via production of glucocorticoids (Webster et al., 2002); and (2) the sympathetic nervous system via noradrenergic innervation of lymphoid organs, release of norepinephrine, and stimulation of beta 2-adrenergic receptors selectively expressed on B-lymphocytes, and CD4+ Th_1 cells (Kohm and Sanders, 2001). Other neurotransmitters, including serotonin, may play an important role in neuro–immune interactions, especially at inflammation sites (Serafeim and Gordon, 2001; Abdouh et al., 2001, 2004). The nervous and immune systems not only have bidirectional regulatory influences, but also share several molecular similarities. These include the fact that agrin, an aggregating protein crucial for formation of the neuromuscular junction, is also expressed in lymphocytes and is important in reorganization of membrane lipid microdomains and setting the threshold for T-cell signaling (Khan et al., 2001). The neuronal repellent Slit regulates leukocyte migration (Wu et al., 2001), and the neuronal receptor neuropilin-1 is expressed by human dendritic cells and T-lymphocytes, and it is essential for the initiation of the primary immune response (Tordjman et al., 2002). The analogies and cross-talk among the nervous, endocrine, and immune systems may contribute to the maintenance of homeostasis in the body.

Despite recent advances, the understanding of mechanisms underlying neuroendocrine–immune interactions is still embryonic. As a consequence, little is known about the potential influence of environmental toxicants on these interactions. The term neuro-immunotoxicology has been coined to indicate specifically the emerging research area dealing with the detrimental effects of drugs and environmental toxicants on the neuroendocrine–immune network (Lawrence and Harry, 2000). It is suspected that perturbation of one system in this network may affect the others, and more investigations that bridge the analyses of immunological, neurobiological, and endocrinological factors are needed (Lawrence and Kim, 2000).

1.5 CONSEQUENCES OF IMMUNE ALTERATION

Following exposure to an immunomodulating agent, the host may attempt to restore normal immune function or to eliminate the immune alteration or imbalance using compensatory mechanisms. Both the primary immune alteration and the compensatory response have the potential to produce disease. The immune dysfunction may result in a loss of self-tolerance or increased susceptibility to other immune-mediated diseases. Possible consequences of primary or secondary immune imbalance subsequent to exposure to immunotoxicants related to infectious disease may include increased susceptibility to infection, increased carrier states, prolonged recovery periods, reduced responses to vaccinations or antibiotic therapy, and impaired immunosurveillance mechanisms. Consequences related to loss of self-tolerance or the regulatory or compensatory mechanisms may include hypersensitivity or autoimmunity (Dean et al., 1994; Zelikoff et al., 1994). The impact may be extended further if interactions and feedback mechanisms associated with the endocrine and central nervous systems are considered (Madden and Felten, 1995; Zelikoff et al., 1994; Husband, 1993).

The underlying immunological, cellular, or biochemical processes associated with immune imbalances or the accompanying compensatory mechanisms are poorly defined in many instances. Disruption of homeostasis that occurs upon exposure to immunotoxicants may alter physiological or biochemical function. Secondary compensatory responses that are attempts to restore normal ordered processes may, in fact, cause further complications.

Table 1.1 Immune Alterations or Compensatory Responses Induced Following Exposure to Immunotoxicants That May Produce Immune-Mediated Disease

Loss of cell population
Induction of cell population
Impaired cell function
Impaired cell differentiation
Altered cell metabolism
Membrane damage
Loss of membrane receptors
Altered cell-surface antigen expression
Expression of sequestered (hidden) antigens
Altered cytokine production
Impaired protein synthesis
Impaired transport or secretion of cytokines
Altered apoptosis
Altered cell oxidation — reduction status
Hapten formation

Table 1.1 provides examples of immune alterations or compensatory responses that may eventually lead to immune-mediated disease. Many of the alterations as compensatory responses cited in Table 1.1 are interrelated. The loss of a cell population or impaired cell function may also reduce cytokine production. Cytokines function as specific chemical mediators that ultimately regulate growth and differentiation of cells. Cell function and cell–cell interactions that require membrane integrity or specific receptor recognition may also be controlled by cytokines (reviewed by Kimber, 1994). The immune alteration or immune deficits as perceived by the host may imply a loss of cells or impaired cellular function. Ultimately, these deficits induce further adaptive or compensatory responses in the host, which in most instances, reduce the impact of the deficit on all physiological systems. From an immunological perspective, there is generally a balance between cell-mediated immunity, humoral immunity, and nonspecific immunity. For example, if cell-mediated immunity is suppressed by an immunotoxicant, the host may augment humoral or nonspecific immune responses to compensate for the suppression. The augmentation may involve increased cell production (cellular proliferation or clonal expression) or altered cytokine production by specific cells. This immunoenhancement would appear to be beneficial to the host, although the enhanced responses may induce further immunological alterations that may suppress desirable responses, potentially triggering an autoimmune or hypersensitive predisposition. Initially, toxic damage to the immune system disrupts the balance associated with normal immune function. Following the period of adjustment or compensation by the host, the new balance that is created may result in toxic damage to the host by the immune system.

Given the diversity of cell types and cell functions, it is possible to speculate on the development of many different compensatory responses. These responses may occur at the cellular, membrane, or biochemical level, or in combination. Alterations specifically involving cells with regulatory function such as T-lymphocytes or macrophages or their regulatory cell products, particularly cytokines, have considerable potential to be associated with immune-mediated disease. The disease may develop directly from exposure to the immunotoxicant or from the subsequent imbalance created by the compensatory responses. In the vast majority of instances, the compensatory responses trigger no adverse effects. The compensated immune function will have little impact on health, which is the intended purpose of such responses. Adverse health effects may occur if the cell type(s) involved in the compensatory responses are critical for long-term immunological homeostasis or maintenance of health. Cells involved in regulatory functions or their cell products are prime candidates.

1.6 COMPENSATION ASSOCIATED WITH IMMUNOTOXICANTS

In classical immunotoxicological studies normal endpoints for evaluation have included humoral immunity, lymphocyte blastogenesis, macrophage function, and phenotypic expression of cell-surface antigens. To provide a more realistic assessment of immune dysfunction, using a battery of tests has been recommended (Luster et al., 1988). Proper selection of tests within the battery should enable the investigator to assess subtle imbalances and compensatory mechanisms with greater reliability and predictability to improve risk assessment determinations (Luster et al., 1992, 1993). Using a multitest approach, it is possible to assess overall immunocompetence or immune imbalance rather than immunosuppression in isolation. The presence of an immune imbalance may be more significant from a long-term health perspective as compared to simply immunosuppression.

Few studies have utilized sufficient immunological endpoints in long-term studies to adequately address the impact of immunological imbalance and subsequent compensatory responses. In one study (Blakley and Tomar, 1986), cadmium suppressed T-lymphocyte–dependent humoral immunity in mice. Corresponding T-lymphocyte–independent humoral immunity was enhanced. It was suggested that the enhanced T-lymphocyte–independent antibody responses were compensating for the T-lymphocyte–dependent immunosuppression. The nature of the compensatory mechanism or the long-term implications were not investigated further.

The compensatory responses, in many instances, may not be so straightforward. Suppression of one response such as antibody production may not always lead to an increase in lymphocyte blastogenesis or macrophage function. One example to illustrate the complexity was observed in a subacute study in rats exposed to chlorpyrifos, an organophosphate (Blakley, 1999). The T-lymphocyte–dependent antibody response and the T-lymphocyte blastogenesis were suppressed, suggesting impaired T-lymphocyte function. If the antibody response was expressed per spleen rather than per viable spleen cell, no suppression was observed. This indicated that the chlorpyrifos exposure impaired spleen cell (lymphocyte) function, but the impact of the suppression was overcome by producing more subfunctional cells. Other interpretations are also possible to explain these observations, but were not investigated. In the same study, the expression of CD_5 and CD_8 cell-surface antigens on lymphocytes was enhanced. The presence of both enhanced and suppressed immune responses following exposure to chlorpyrifos suggests that the immune alterations are not just generalized cytotoxicity. More complex immunomodulation involving immunocompensation may be present, although the extent, exact nature, and environmental or health significance of the compensatory activity are unknown. In a similar study, the organophosphate insecticide malathion increased T-lymphocyte–dependent humoral immune responses following low-level exposure over an extended period of time (Johnson et al., 2002). In contrast, acute studies of higher dosages of malathion have suppressed humoral immunity (Casale et al., 1983). It has been suggested that the compensatory ability of the immune system may account for the disparate results observed in these malathion studies.

1.7 SUMMARY

The immune system is essential for long-term survival. Its complexity is only partially understood. Both specific (humoral and cell-mediated immunity) and nonspecific (NK cells and phagocytes) immune functions play crucial roles in homeostasis and immune regulation. Numerous cell types and subpopulations of cells interact directly via cell–cell contact or indirectly via the secretion of soluble factors (cytokines) to regulate immunological and physiological processes within the immune system and other organ systems including the endocrine and nervous systems. Disruption of these immunological responses and feedback systems by immunotoxicants may result in direct adverse effects. In some instances, compensatory responses within the immune system will occur to restore normal function. These compensatory responses have the potential to create further imbalances that may result in immune-mediated disease problems such as autoimmunity or hypersensitivity. The full impact of these compensatory responses to correct immune imbalance following exposure to immunotoxicants has been investigated to a limited extent.

ACKNOWLEDGMENTS

This work was supported in part by a grant from the Canadian Institutes of Health Research (MOP-67183).

REFERENCES

Abdouh, M., et al., Transcriptional mechanisms for induction of 5-HT1A receptor mRNA and protein in activated B and T lymphocytes, *J. Biol. Chem.*, 276, 4382–4388, 2001.

Abdouh, M., et al. 5-HT1A-mediated promotion of mitogen-activated T and B cell survival and proliferation is associated with increased translocation of NF-kB to the nucleus, *Brain Behav. Immun.*, 18, 24–34, 2004.

Benveniste, E.N., Cytokine actions in the central nervous system, *Cytokine Growth Factor Rev.*, 9, 259–275, 1998.

Blakley, B.R., and Tomar, R.S., The effect of cadmium on antibody response to antigens with different cellular requirements, *Intl. J. Immunopharmacol.*, 8, 1009–1015, 1986.

Blakley, B.R., et al., Effect of chlorpyrifos on immune function in rats, *Vet. Hum. Toxicol.*, 41, 140–144, 1999.

Casale, G.P., Cohen, S.D., and Dicapua, S.D., The effects of organophosphate-induced cholinergic stimulation on the antibody response to sheep erythrocytes in inbred mice, *Toxicol. Appl. Pharmacol.*, 68, 198–205, 1983.

Dean, J.H., et al., Immune system: evaluation of injury, in *Principles and Methods of Toxicology*, 3rd ed., Hayes, A.W., Ed., Raven Press, New York, 1994, chap. 30.

Holladay, S.D., Prenatal immunotoxicant exposure and postnatal autoimmune disease, *Environ. Health Perspect.*, 107 (Suppl. 5), 687–691, 1999.

Husband, A.J., Role of central nervous system and behavior in the immune response, *Vaccine*, 11, 805–816, 1993.

Johnson, V.J., et al., Increased T-lymphocyte dependent antibody production in female SJL/J mice following exposure to commercial grade malathion, *Toxicology*, 170, 119–129, 2002.

Kerkvliet, N.I., Brauner, J.A., and Matlock, J.P., Humoral immunotoxicity of polychlorinated diphenyl ethers, phenoxyphenols, dioxins and furans present as contaminants of technical grade pentachlorophenol, *Toxicology*, 36, 307–324, 1985.

Khan, A.A., et al., Physiological regulation of the immunological synapse by agrin, *Science*, 292, 1681–1686, 2001.

Kimber, I., Cytokines and regulation of allergic sensitization to chemicals, *Toxicology*, 93, 1–11, 1994.

Kohm, A.P., and Sanders, V.M., Norepinephrine and beta 2-adrenergic receptor stimulation regulate CD4+ T and B lymphocyte function in vitro and in vivo, *Pharmacol. Rev.*, 53, 487–525, 2001.

Lawrence, D.A., and Harry, G.J., Environmental stressors and neuroimmunotoxicological processes, *Brain Behav. Immun.*, 14, 231–238, 2000.

Lawrence, D.A., and Kim, D., Central/peripheral nervous system and immune responses, *Toxicology*, 142, 189–201, 2000.

Lee, K., Johnson, V.J., and Blakley, B.R., The effect of exposure to a commercial 2,4-D formulation during gestation on the immune response in CD-1 mice, *Toxicology*, 165, 39–49, 2001.

Luster, M.I., et al., Development of a testing battery to assess chemical-induced immunotoxicology: national toxicology program's guidelines for immunotoxicity evaluation in mice, *Fundam. Appl. Toxicol.*, 10, 2–19, 1988.

Luster, M.I., et al., Perturbations of the immune system by xenobiotics, *Environ. Health Perspect.*, 81, 157–162, 1989.

Luster, M.I., et al., Risk assessment in immunotoxicology: I. Sensitivity and predictability of tests, *Fundam. Appl. Toxicol.*, 18, 200–210, 1992.

Luster, M.I., et al., Risk assessment in immunotoxicology: relationships between immune and host resistance tests, *Fundam. Appl. Toxicol.*, 21, 71–82, 1993.

Madden, K.S., and Felten, D.L., Experimental basis for neural-immune interactions, *Physiol. Rev.*, 75, 77–106, 1995.

Serafeim, A., and Gordon, J., The immune system gets nervous, *Curr. Opin. Pharmacol.*, 1, 398–403, 2001.

Tomar, R.S., Blakley, B.R., and DeCoteau, W.E., Antibody producing ability of mouse spleen cells after subacute dietary exposure to T–2 toxin, *Intl. J. Immunopharmacol.*, 10, 145–151, 1988.

Tordjman, R., et al., A neuronal receptor, neuropilin-1, is essential for the initiation of the primary immune response, *Natl. Immunol.*, 3, 477–482, 2002.

Webster, J.I., Tonelli, L., and Sternberg, E.M., Neuroendocrine regulation of immunity, *Annu. Rev. Immunol.*, 20, 125–163, 2002.

Wu, J.Y., et al., The neuronal repellent Slit inhibits leukocyte chemotaxis induced by chemotactic factors, *Nature*, 410, 948–952, 2001.

Zelikoff, J.T., et al., Immunomodulation by metals, *Fundam. Appl. Toxicol.*, 22, 1–7, 1994.

Approaches and Models Relevant to the Assessment of Chemical-Induced Immunotoxic Effects in Species of the Ecosystem

CHAPTER **2**

Ecoimmunotoxicology: State of the Science

Michael R. Woolhiser, Kevin S. Henry, and Michael P. Holsapple

CONTENTS

2.1 INTRODUCTION: WHAT IS ECOIMMUNOTOXICOLOGY?

With the relatively recent advent of mammalian immunotoxicology as a subdiscipline of toxicology over the past few decades, the evaluation of the immune system is increasingly recognized as a critical target in the process of xenobiotic risk assessment. Unintended immune suppression has the potential to increase the susceptibility of individuals to infectious organisms. As described below, a number of approaches have evolved to assess the potential for this type of immunotoxicity in traditional laboratory animals, most notably rats and mice. Likewise, compromised immune suppression in gulls, frogs, flounder, or other animals can result in analogous challenges. Thus, there is growing support that evaluation of immunotoxicity potential using wildlife may help to fill a gap in current environmental risk assessment. For the purposes of this chapter, we suggest that "ecoimmunotoxicology" may be largely defined in the context of nontraditional laboratory animal models, and can include terrestrial mammalian and invertebrate species, birds, amphibians, reptiles, fish, or other aquatic species. The peer-reviewed literature has numerous examples of models that appear to serve as sensitive markers for environmental stress; thus, some scientists feel that the immune status of wildlife itself is challenged and needs to be assessed. Aquatic environments in particular readily sustain many disease-causing microorganisms, and decreased host resistance due to unintended immune suppression may be of particular concern in these systems.

While considering the various immunotoxicological endpoints discussed herein, it is important to keep in mind a principal difference between ecological risk assessment and human health risk assessment: Human health risk assessment is generally intended to offer conservative protection of the individual person, while ecological risk assessment is often directed toward protecting wildlife at population or ecosystem levels of organization (Suter, 1993). Classically trained mammalian toxicologists may concentrate on lower levels of biological organization, and these endpoints can be useful as biomarkers of exposure or effect. Such biomarkers may include distinct and predictable physiological, biochemical, histological, or immunological responses. These responses may not, however, be expressed in the field as ecologically relevant effects if they do not impact environmental populations and the resultant ecosystem function. Beyond concerns for wildlife, it has been suggested that immunological data sets derived from environmental evaluation can serve as sentinels for human risk stemming from chemical exposures (Luebke et al., 1997; Fox, 2001). While such data may be useful as part of a weight-of-evidence risk assessment, as described below, a number of confounding factors makes this objective vulnerable to over interpretation beyond the state of the science.

2.2 BACKGROUND ON CURRENT IMMUNOTOXICOLOGY GUIDELINES

Immunotoxicity is most frequently discussed in the context of immune suppression (U.S. Environmental Protection Agency, 1998). Conversely, it must be recognized that a hyperactive immune system is also undesirable and may present itself as allergy or autoimmunity. This aspect of immunotoxicology is oftentimes referred to as a "continuum," and there is no doubt that damage to the immune system in either direction may be associated with adverse health effects. However, a consideration of the impact of allergy and autoimmunity in the context of ecoimmunotoxicology is largely beyond the scope of this current discussion. While an extensive list of immunological endpoints using multiple laboratory species have contributed to understanding immunotoxicity potential, a finite number of models have been sufficiently characterized to be heavily relied on in support of human risk assessment. Over the past several decades, a number of specific laboratory models using rodents have distinguished themselves as valuable predictors of immunotoxicity potential. Compilations of historical data sets have allowed concordance comparisons using data from laboratory mice, and have highlighted that no single test appears to be predictive of immunotoxicity potential (Luster et al., 1992). Using a liberal definition of immunotoxicity classification, measurements such as the primary antibody response to sheep red blood cells (SRBC), or other so-called T-dependent antigens, in tandem with thymus weight serve as effective, "Tier One" indications of immunotoxicity. Several endpoints typically evaluated in general, mammalian toxicology studies, and thought to be indicators of immunotoxicity potential (e.g., leukocyte counts, thymus weights, and spleen cellularity), are considerably less reliable and are not sufficient in the absence of a functional immune test to predict immunotoxicity (Luster et al., 1992). Measurements of lymphocyte surface markers via flow cytometry or assessment of innate immunity via natural killer (NK) cell activity have been shown to be effective at elucidating the potential mechanisms of action for immune suppressive effects (Luster et al., 1992), and thus, have been outlined as additional methods to assess immunotoxicity potential. Such efforts have become the cornerstone for experimental evaluation of chemical immunotoxicity potential in the United States. For example, the most recent version of the Environmental Protection Agency (EPA) immunotoxicity guidelines (U.S. EPA, 1998) is based on measurement of the antibody response to SRBC, or another well-characterized T-dependent antigen. Depending on the outcome of that parameter, additional tests may be conducted on a case-by-case basis. If the antibody response is suppressed, then the measurement of lymphocyte surface markers may be required. Conversely, if the antibody response is unaffected, then the NK cell assay may be required (U.S. EPA, 1998). While retrospective analysis suggests that these endpoints serve as reasonable predictors of immunotoxicity potential, one should not rely solely on experimental results using laboratory animals in instances where convincing human data are available.

The evaluation of immunotoxicological endpoints in standardized regulatory guidelines does not extend beyond mammalian toxicology to any significant degree; there are no endpoints that are specifically assessed for immunotoxicity potential in the ecotoxicology regulatory framework. Traditional ecotoxicology guidelines (e.g., Organization of Economic Cooperation and Development; Federal Insecticide, Fungicide, and Rodenticide Act; Toxic Substances Control Act) address endpoints such as mortality, growth, development, or reproduction. More recent developments within the ecotoxicology regulatory environment may incorporate biochemical and histological endpoints into the suite of testing for some classes of compounds, although these endpoints may be more prominent in higher-tiered testing than in initial screening.

2.3 BACKGROUND ON THE IMMUNE SYSTEM IN CONTEXT OF ALTERNATIVE, NONTRADITIONAL LABORATORY SPECIES

2.3.1 Mammalian Immune System

The ultimate purpose of the immune system is to protect against infectious diseases, caused by parasites, viruses, bacteria, or other microorganisms, in addition to cancerous cells. Protection against these foreign entities involves immediate, nonspecific immune responsiveness in the form of innate immunity, as well as orchestrated, coordinated responses that are acquired, and provide long-term memory to rapidly defend against repeated assault. As an initial biological response, innate immunity requires a lesser degree of cellular and molecular coordination, and involves more direct recognition of foreign cells and particles to allow for phagocytosis and lysis. Key cells involved in innate immunity in humans and other mammalian species include macrophages, neutrophils, and NK cells. In humans and other mammalian species, the primary cell types of the adaptive immune response include B cells, T cells, and antigen-presenting cells. Through molecule recognition, cellular interactions and soluble mediators, these cells orchestrate the expansion of explicit cell types and antibodies that are specific for offending, antigenic stimuli. Other mammalian hematopoietic cells involved with adaptive immune responses include mast cells and basophils, which play the primary effector roles in immediate allergic reactions, and eosinophils, which aid in parasitic infection following their activation by lymphocytes. Cells of the mammalian immune system originate in the bone marrow and must migrate to lymphoid tissues (e.g., spleen, thymus, lymph nodes) to mature and differentiate before protecting the body. Details regarding these concepts of the immune system were outlined in Chapter 1, and are well described elsewhere in immunology texts (e.g., Janeway et al., 2001).

2.3.2 Avian Immune System

Many of the above cell types and functions are conserved across vertebrate species. Avian immunology has been largely described through research using domesticated poultry and has been most recently reviewed by Fairbrother et al. (2004). Similar to humans and laboratory rodents, avian immune responses are divided into innate and adaptive immune function. Most, if not all, cell types and homologous signaling molecules (e.g., cytokines) appear to be represented in birds. Characterization of T-cell subsets are beginning to be elucidated based on cytokine responsiveness (Gobel et al., 2003). Stem cells originate from bone marrow and distribute to lymphoid organs for maturation. As in humans, birds also experience T-cell maturation in the thymus (Pickel et al., 1993).

Most avian species appear to lack well-defined lymph nodes but have most of the mucosa-associated lymphoid tissues (MALT), including tonsils, Peyer's patches, and bronchial-, nasal-, and gut-associated lymphoid tissues, necessary to protect against environmental pathogens and antigens. In the case of B-cell development, cell differentiation and clonal expansion primarily occurs in a thymus-like lymphoid gland termed the bursa of Fabricius (McCormack et al., 1991; Koskela et al., 2003). Avian species are capable of producing IgM, IgG, and IgA with repertoire heterogeneity developed via immunoglobulin gene conversion, rather than the VDJ gene rearrangement observed in most humans and laboratory animals (McCormack et al., 1993).

2.3.3   Aquatic Vertebrate Immunology

Similar to avian species, fish have an immune system that has many comparable features to that of mammalian vertebrates. Generally speaking, fish have hematopoietic tissues and a majority of the circulating white blood cell (WBC) lineages observed in humans, including lymphocytes (T and B cells), granulocytes, and monocytes/macrophages (Anderson and Zeeman, 1995). In contrast to the vertebrate species considered thus far in this chapter, stem cells of fish primarily originate from the anterior kidney. These myeloid and lymphoid progenitors then mature and reside in the spleen, thymus, kidney, and several other immunopoietic tissues. Fish demonstrate both innate and adaptive immune responses, and produce antibodies that consist of tetrameric IgM and IgG, which demonstrate an antigenic repertoire resulting from genetic segmentation and somatic mutations (Ghaffari and Lobb, 1989; Anderson and Zeeman, 1995). Despite the many similarities in immune system structure between mammals and fish, the specific immune response and immunological tissue differentiation of fish are less evolved than in mammals. The same appears to hold true for amphibian species. Research on immune responses in the African clawed frog, *Xenopus laevis*, suggests a high degree of similarity with mammalian immune system components, including macrophages, neutrophils, lymphocytes with T-cell and immunoglobulin receptors, complements, NK cells, and cytokines (Carey, et al., 1999). To date, only minor differences in tissue organization have been noted between amphibian and mammalian immune systems. At least some frogs also produce antibodies of the isotypes IgX and IgY, molecules essentially equivalent to mammalian IgA and IgG, respectively (Rollins-Smith, 1998).

2.3.4   Invertebrate Immunology

One of the most prevalent differences in immune system complexity as one travels "down" the evolutionary ladder involves the ability of immune cells to coordinate responses that confer lifelong protection via adaptive immunity. The degree of organization, complexity, and memory of the immune system declines in lower vertebrate and invertebrate species. Invertebrate species such as crustaceans, mollusks, sponges, oligochaetes, and earthworms have an immune system that primarily relies on nonspecific, innate responses (Ratcliffe, 1985; Galloway and Depledge, 2001). In general, invertebrates have little, to no, cellular cooperation during defense responses and rely on hemocytes or coelomocytes (equivalent to WBCs) as the primary cell types involved with innate immunity. These cells essentially circulate freely in the blood compartment, and phagocytose foreign cells or particles in a manner similar to that which occurs in mammals. Hematopoietic tissue organization varies among invertebrate species, but is limited from having lymph nodes only near the mid-gut and pharynx of urochordates, to a small number of loosely grouped cells in other species. Soluble mediators of immune function in lower species exclude antibodies and are largely limited to molecules such as enzymes (e.g., prophenoloxidase, an enzyme cascade that is analogous to the mammalian complement system) (Soderhall, et al., 1997), lectins eicosanoids and cytokine-like molecules, particularly correlates of the interleukins IL-1 and IL-6, and tumor necrosis factor (Beck and Habicht, 1996; Beschin et al., 2004).

2.4 ASSESSMENT OF IMMUNE ENDPOINTS IN WILDLIFE SPECIES

The high degree of conservation in immunological organization across species has aided the development of models that can be used to assess the immunotoxicity potential of environmental challenges and influences. Petrochemicals, metals, pesticides, and industrial chemicals are some of the anthropogenic substances most frequently considered to be environmental xenobiotics of concern. A vast number of effects has been reported across numerous vertebrate and invertebrate species following exposures to such chemicals. The majority of the ecoimmunotoxicological endpoints considered are consistent with those parameters and assays used in traditional laboratory rodents. In addition to basic evaluation of organ mass and morphology, functional endpoints that rely on responses to specific antigens such as SRBC or to nonspecific mitogens (e.g., substances capable of triggering lymphoproliferation), for example, have been developed using nontraditional laboratory species (Fowles et al., 1993; Trust et al., 1994; Fairbrother et al., 2004). However, some researchers have suggested that wildlife endpoints may not be as readily interpreted as standard laboratory immunotoxicity results. For example, Smith and Hunt (2004) suggest spleen mass is too easily affected by normal behaviors and environmental conditions, thus challenging the premise that something as fundamental as spleen size can be readily interpreted as an indication of immunotoxicity.

Such observations illustrate the fact that one must be intimately familiar with the environmental conditions as well as the experimental model in order to properly understand and interpret the results of wildlife immunotoxicity studies. Any association between the potential effects of anthropogenic chemical exposures and immunotoxicity mechanisms must be preceded by a thorough understanding of normal immune responsiveness, variability, and mechanisms. As a case in point, in order to establish baseline macrophage data in fish, Zelikoff et al. (1991) examined morphology, phagocytic activity, migration, and superoxide (O^{2-}) production using macrophages from naive, lipopolysaccharide (LPS) or *Aeromonas salmonicidae*–immunized trout (*Oncorhynchus mykiss*). Subsequently, these endpoints were used to demonstrate altered phagocytosis and free radical production in trout exposed to relevant contamination levels of cadmium (Zelikoff et al., 1995) while *in vitro* treatment of macrophages with nickel sulfate ($NiSO_4$) did not affect phagocytic activity under conditions that reduced unstimulated migration and reactive oxygen species production (Bowser et al., 1994).

Evaluation of immunotoxicity using wildlife species requires more than mere characterization of best-understood and readily obtainable immune parameters. Beyond establishing exposure regimens that should capture the range of ecological exposure scenarios, one needs to develop immunization conditions to evaluate functional responsiveness. Experimental models of viral infection using *Xenopus* have further elucidated both the cellular and humoral adaptive immune responses involved in immune defense against iridovirus, one of the primary pathogens associated with declines in ectothermic vertebrate species such as frogs (Gantress et al., 2003). More recent models of fish immunotoxicity testing have stressed the use of antigen challenge models to evaluate immune status, and have helped to broaden the range of functional endpoints via evaluation of specific antibodies and cell-mediated cytotoxicity following immunization and/or pathogen challenge (Harrahy et al., 2001; Alcorn et al., 2002; Kollner et al., 2002). Development of a functional wildlife model may frequently involve modification of established procedures to overcome diverse subtleties in immunological components. For example, due to high nonspecific binding by toad sera antibodies, an ELISA method required microtiter plates to be coated with complexed amphibian antibody-SRBC before addition of detection antibodies to measure specific immunoglobulins (Rosenberg et al., 2002). Once developed, the complete model was useful to demonstrate an elevated humoral response due to lead (acetate) exposure. Not to be overlooked, availability of immune reagents for wildlife species (a factor that once restricted immunological testing using rats) adds limitations to the breadth of experimental questions as well as the ease, reproducibility, and widespread conduct of wildlife immunotoxicology.

Using amphibian species in field studies, investigators have been able to measure changes in B-cell responses to lipopolysaccharide (LPS) and alterations in white pulp (lymphocyte zones) of spleens (Linzey et al., 2003). While definitive association to particular pesticides or heavy metal soil residuals could not be made, p,p'-dichlorodiphenyldichloroethylene (DDE) was determined to be the most abundantly measured bioactive anthropogenic molecule. In general, Bermuda toads (*Bufo marinus*) demonstrated a decrease in white pulp among locations demonstrating the greatest concentrations of DDE. Splenocytes evaluated from one particular site exhibiting high cadmium levels demonstrated a reduced mitogenic response towards LPS *in vitro* (Linzey et al., 2003). Impaired cell-mediated immunity as measured via intradermal response to phytohemagglutinin (PHA) has been reported using avian species. Through measurement of skin-swelling responses following intradermal injection of PHA, an association between selenium concentrations in the liver of eider ducks and reduced cell-mediated immunity was reported (Wayland et al., 2002). Highlighted by this paper is the possibility of potential confounders when conducting research using field species. Not only were elevated residues of multiple trace metals detected systemically in the eider ducks (*S. mollissima borealis*) but capture-induced stress was confirmed by demonstrating increases in corticosterone concentrations, an accepted immunosuppressive mechanism.

In an extensive avian environmental immunotoxicology study, hematology, histopathology, *in vitro* lymphocyte proliferation (concanavalin A [Con A], PHA, LPS, and pokeweed mitogen [PWM]), leukocyte responsiveness, and phagocytic activity were evaluated as immunological parameters using feral tree swallows in apple orchards following pesticide exposure from nesting through day 16 of development (Bishop et al., 1998). Following various orchard treatments, which consisted of up to three pesticide mixtures, only reduced bursal mass size, cortex cell proliferation, and delayed thymic involution were associated with spray frequency, while remaining unconfounded by such things as date of nest collection, seasonal temperatures, or parasitic infection. Although confounded by collection date, lymphocytes may have demonstrated increased blastogenesis among spray groups, particularly in response to pokeweed mitogen stimulation.

Wildlife species can be, and frequently are, taken into captive, laboratory conditions for purposes of study. However, as with any other biological system or animal model, validation is necessary for proper interpretation of study results. In the absence of sufficient validation, the direct relevance to possible field effects is often difficult to establish, and extrapolations accompanied by additional uncertainty may be required (Grinwis et al., 2000). Measurement of the humoral immune response using Chinook salmon (*Oncorhynchus tshawytscha*) has been reported to assess the hazard of polychlorinated biphenyl (PCB) pesticide exposure (Arkoosh et al., 1994). Chinook salmon injected intraperitoneally (i.p.) with the PCB Aroclor 1254, or the polycyclic aromatic hydrocarbon, dimethylbenz[a]anthracene (DMBA), at doses resulting in body burdens greater than those typically observed in the environment, demonstrated selective suppression of the B-cell–mediated, T-independent antibody response using trinitrophenyl (TNP)-LPS. In contrast, another aggressive exposure regimen via gavage (up to 500 mg/kg) using adult ducks and Aroclor 1254 did not result in altered antibody titers (SRBC), NK cell activity, or lymphocyte blastogenesis (Fowles et al., 1997). An absence of immunological impairment as a result of Aroclor 1254 administration is consistent with that observed following oral exposure of juvenile salmon using four levels of the PCB congener mixture (Powell et al., 2003). Mortalities and acquired immunity to pathogenic *Listonella anguillarum* bacteria among treated, vaccinated, control groups were found to be unaffected following high levels of dietary PCB administration.

Evaluation of a captive, wild mouse strain resulted in unexpected inconsistencies following slight modifications in dosing regimens that represented more relevant exposure conditions. Offspring born to dams treated with a single, i.p. dose of the PCB Aroclor 1254 (300 mg/kg) before conception demonstrated effects on the spleen, WBC, lymphocyte, and monocyte numbers, and PHA mitogen responsiveness (Wu et al., 1999). However, a subsequent study using relevant, subchronic dosing via diets containing 25 ppm PCB, only observed thymocyte atrophy accompanied by a heightened proliferative response by thymocytes to Con A and a normal, specific humoral response to keyhole limpet hemocyanin (KLH), another T-dependent antigen (Segre et al., 2002). While the test material intake was not reported for this work, approximating food consumption for a 30-g *Peromyscus leucopus* mouse to be 4.5 g (Duffy et al., 1997) results in a theoretical PCB exposure near 3.75 mg/kg/day. These results illustrate the fact that dosing regime and methodology must be carefully considered in order to avoid overestimation of the effects of environmental contaminants.

Parasitic lungworms have been shown to increase rates of migration, maturation, and reproduction within leopard frogs (*Rana pipiens*) following a 3-week exposure using sublethal concentrations of pesticide mixtures (Gendron et al., 2003). These results were accompanied by a reduction in subsequent lymphocyte proliferation; specifically, responsiveness and proliferation were impaired (Christin et al., 2003). Phagocytic ability and spleen cell numbers also appeared to be impaired 21 days into a subsequent parasitic challenge. These associations may add to the ongoing debate regarding the role of anthropogenic agents in disease-related reductions in amphibian populations, although a number of nonchemical factors have been identified as possible causes of population declines (e.g., UV radiation, climate change, introduction of novel species) (Christin et al., 2003). In this case, field evidence has not associated environmental levels of these pesticides with direct acute or chronic toxicity, nor with disease-related die-offs.

Even species such as polar bears and harbor seals have been evaluated for associations between their immunological status and anthropogenic contamination of the environment. Dietary organochlorine exposure of polar bears (*Ursus maritimus*), as determined via PCB and HCB (hexachlorobenzene) measurement of plasma samples, was associated with decreasing total IgG levels (Bernhoft et al., 2000). Following a 1-year acclimation to captivity, 22 harbor seal pups (*Phoca vitulina*) were fed herring collected from two distinct locations containing appreciably different levels of polychlorinated biphenyl equivalents (Van Loveren et al., 2000). Comparisons of seals among the two treatments suggested an association between higher PCB administration and reduced NK cell activity, retarded mitogen-induced blastogenesis, reduced mixed lymphocyte response, and diminished delayed-type hypersensitivity response to ovalbumin.

One approach to evaluate immune status in the field involves the use of comparative or control animals from ecologically matched reference sites. Cotton rats (*Sigmodon hispidus*) living on petrochemical wasteland farms were reported to demonstrate reduced spleen size compared to rats evaluated from five regionally matched sites (Wilson et al., 2003). During summer months, increases in hemoglobin, hematocrit, and platelet counts, and decreases in leukocytes, were reported when exposure might be expected to be higher. In spite of these observations, no treatment-related functional effects were demonstrated; no differences were noted for *in vitro* lymphocyte stimulation using Con A, PWM, or IL-2, and killer cell lytic activity was comparable between sites. Cell-mediated hypersensitivity reactions in response to intradermal injection of PHA showed no differences. Using a similar design, American alligators (*Alligator mississippiensis*) from regional lakes were compared for thymus and spleen pathology in association with agricultural pesticide exposure. Investigators noted reduced medullary:cortex thymic ratios and decreased lymphocyte sheath width associated with lake contamination compared to two reference lakes in which contamination was not detected (Rooney et al., 2003). Similarly, Grasman and Fox (2001) reported findings from 1997–1999 which used PHA skin responses and SRBC antibody titers in an attempt to associate immune responsiveness with PCB concentration in eggs and plasma collected from Caspian terns (*Sterna caspia*) residing on two islands of Lake Huron (North America) having distinctly different levels of contamination. T-lymphocyte function (PHA test) was reduced by 42% in birds associated with higher PCB contamination, but combined IgM and IgG SRBC antibody titers were up to 2.5-fold greater from the same location. Perhaps a more sophisticated means by which to create matched reference sites is to design enclosures to allow distinct field exposures (Fairbrother et al., 1998). Gray-tailed voles (*Microtus canicaudus*) have been evaluated in this manner to determine biomarker sensitivities (including neutrophil and lymphocyte counts) for spray exposure to an organophosphorous insecticide (azinphos-methyl). Enclosures (45 × 45 meters) constructed of galvanized sheet metal were built to isolate populations of voles using six heterosexual pairs in each enclosure.

Another interesting field design strategy involved a block design in which nestlings from multiple nests were assigned to particular treatment groups to minimize brood effects (Fair and Myers, 2002). Thus, every bird from a given nest was treated differently with a combination of lead shot (mealworm gavage) and/or antigen (killed Newcastle disease virus or SRBC). Ultimately, lead shot administration did not appear to alter humoral-mediated antibody response, while the cell-mediated PHA response was significantly lower in the high lead–treated nestlings. This type of experimental design can be effectively used when exposure conditions in the field can be controlled for individual animals.

2.5 ECOIMMUNOTOXICOLOGY EXPERIMENTAL DESIGN

2.5.1 Advantages

As described above and in other sections of this book, measuring, and even demonstrating effects on immune system parameters have become increasingly more routine. While many might suspect that immunotoxicological data stemming from wildlife species are rare, there are many hundreds of peer-reviewed citations which associate an anthropogenic, environmental contaminant with some immunological impairment. That which has been presented in this chapter is only intended to provide a representation of studies and models to outline the current state of ecoimmunotoxicology investigation. The advantages, disadvantages, and ultimate interpretation of these experiments should be considered to appreciate the current utility of ecoimmunotoxicology in safety and risk assessment schemes.

For the time being, the use of environmental species (for example, amphibians, earthworms, and fish) in regulatory testing for environmental risk assessment is perceived as considerably less controversial than traditional mammalian species in the context of animal usage and welfare. The degree to which these tests can be modified or expanded to include immunotoxicology endpoints has yet to be established. It also appears to be more economical to purchase and maintain species such as fish and earthworms in comparison to traditional laboratory rodents (Zelikoff, 1998). Certainly, the capture and maintenance of feral mammalian and avian species for experiment and exposure(s) under laboratory conditions would eliminate this economical benefit. Although there are few costs involved in the maintenance of species in field studies, resources associated with collection of organisms should not be overlooked. Ecoimmunotoxicology studies, especially when conducted in the field (Bishop et al., 1998) or using fish (Zelikoff, 1998), can have the advantage of more relevant, "natural" exposures. In some instances however, investigators captured feral species but administered anthropogenic substances by less relevant route(s) as a means to develop an experimental model or to demonstrate proof in theory (Arkoosh et al., 1994; Wu et al., 1999). The observation of environmental effects as early biomarkers of risk may be another advantage of ecoimmunotoxicity testing. For example, if a decline in species populations is noted, properly designed studies using validated methodologies and endpoints, coupled with knowledge regarding physiological bio-mechanisms of these species, may provide insight into potential mechanism(s) of action. Results interpreted in the context of well-understood endpoints for the species in question may aid in the elucidation of environmental effects, serve as a trigger to further investigate mechanism and species differences not recognized during initial safety testing, or help conduct definitive guideline studies to clarify immunotoxicity potential if undetermined in humans.

The weight-of-evidence suggests that there may be a number of chemicals which interfere with laboratory rodent immune status (i.e., the "gold standard" for immunotoxicology safety testing) and alter one or more homologous endpoints in environmental species. One caveat with this generalization involves diversity of responses. As illustrated in the brief description of experimental models presented in this chapter, immunotoxicity effects for a given substance (or class of materials) can vary widely based on experimental influences and species differences. As an example, Chinook salmon treated with the PCB Aroclor 1254 by i.p. injection demonstrated altered B-cell mediated antibody responses to TNP-LPS (Arkoosh et al., 1994) while SRBC antibody titers (and lymphocyte blastogenesis) in Mallard ducks treated orally with 500 mg/kg Aroclor 1254 were unaffected (Fowles et al., 1997). Studies using rodents do demonstrate some evidence of immunotoxicity by Aroclor 1254 under various study conditions. Specifically, Aroclor 1254 has elicited reduced thymus weights, NK cell activity, and T-cell blastogenesis, but at doses (above 10 mg/kg) which concomitantly demonstrate overt toxicity (Smialowicz et al., 1989). Also, i.p. administration of Aroclor 1254 (300 mg/kg) led to immunotoxicity effects in mouse offspring (*Peromyscus leucopus*) while evaluations using relevant exposure scenarios via the diet did not support these findings very well (Wu et al., 1999; Segre et al., 2002). Consistent with DMBA-mediated immunotoxicity targeted towards B-cell responsiveness in salmon (Arkoosh et al., 1994), B6C3F1 mice injected with a (similar) dose have exhibited reduced B-cell reactivity to TNP-LPS (Ward et al., 1984). In contrast, Ward et al. also demonstrated an effect on the response to a T-dependent antigen, an endpoint that was not altered when evaluated in salmon by Arkoosh et al. (1994). Considering a heavy metal contaminant, an association between lead acetate exposure (6 weeks) and elevated humoral, SRBC antibody titers by toads (Rosenberg, Salibian and Fink, 2002) are consistent with that noted in studies using rodents; dams administered lead acetate during pregnancy provided offspring which exhibited an apparent shift in humoral, immunological responsiveness as evidenced by heightened IgE antibody levels (Miller et al., 1998).

While the weight-of-evidence and general agreement among traditional laboratory studies and alternative environmental species supports the use of ecoimmunotoxicology, evidence of immunotoxicity potential by substances such as pesticides at environmentally relevant concentrations appears to be sparse, and sufficiently contradictory experimental data persist such that meaningful risk assessment remains unsupported. For example, in spite of the experimental evidence in animals which reports immune suppression due to organophosphates (OP), reports suggest a lack of conclusive support that human exposure to OP adversely alters immune function (Pruett 1994; Galloway and Handy, 2003). Even in instances where OP metabolites have been measured in the urine of human subjects (children) following environmental exposure, no adverse symptoms were reported suggesting a lack of observable immunotoxicity (Lu et al., 2001). Similar observations have been made in invertebrate species where experimental immunotoxicity results due to anthropogenic materials in the laboratory are not supported by profound incidences of disease or mortality in the environment (Galloway and Depledge, 2001).

2.5.2 Complications

While there are some appealing advantages to pursuing ecoimmunotoxicology models, there are a number of confounders that would have to be addressed before this approach could be widely accepted for xenobiotic safety testing. When attempting to evaluate immune parameters in a field study, one presumes that measurements of organ mass and cellularity readily serve as indicators of immune system status. As alluded to earlier, however, when evaluating a dynamic, responsive parameter like the immune system using field studies, the premise that there is a positive relationship between immunocompetence and immune component size should be critically considered. Smith and Hunt (2004) highlight a host of complicating factors including migration, breeding, parasite load, and sex when using avian spleen mass as a primary indication of immune strength. For example, the spleen of birds demonstrates significant variation in mass and histology in accordance with breeding in early summer (enlargement) and winter (atrophy) (Silverin et al., 1999). The reason for this observation is unknown, but as posed by Silverin, the growth of the spleen is consistent with developing immune responses towards numerous antigen and microbial attacks encountered during warmer seasons. As a case in point, Brown and Bomberger Brown (2002) reported an increase in spleen mass as a result of parasite load when comparing groups from untreated nests to those of fumigated nests. Researchers untrained in the field of avian physiology could interpret these results as immunotoxicity due to the reduction in spleen mass associated with the chemical fumigation of nests, for example.

Additionally, immunohistopathology of lymphoid organs may provide another level of detail and insight, but recent interlaboratory comparisons with traditionally used rodents suggest limitations (Germolec et al., 2004). A common set of lymphoid tissues were examined by four toxicologic pathologists from experiments evaluating both immunotoxicants (positive controls and test chemicals) as well as negative controls. In short, the level of agreement decreased among pathologists as lesion severity decreased; this scenario was consistent with test chemical effects. Pathological changes following treatments with strong immunosuppressive, positive control materials were easily interpreted by all laboratories. Aside from severity, the ability to detect pathological changes was associated with a pathologist's experience in reading for immunohistopathology effects. Furthermore, since different immune tissues have unique anatomies, and are comprised of different cell populations, ability to observe changes is also influenced by the lymphoid tissue that is affected by chemical exposure. Specific cell effects (e.g., B cell versus T cell) may be detected differently in each respective tissue and/or compartment. Taken together, histopathological evaluation of immune organs and tissues cannot be assumed to be routine or simple (Germolec et al., 2004).

While "static" immune endpoints can be readily studied, these parameters have a number of drawbacks. As noted earlier in this chapter, concordance comparisons using laboratory rodent data indicate that "static" immune endpoints (e.g., leukocyte counts, lymphoid organ weight) appear to be less reliable at identifying immunotoxicity potential than measurements of functional, T-dependent antibody responses (Luster et al., 1992). If this conclusion holds true for ecoimmunotoxicology models, conducting immune function tests that require experimentation to immunize or challenge feral animals over repeated time intervals becomes more critical, but can limit field investigations in the absence of test animal containment (Fairbrother, 2004). Evaluation of innate immune responses, or development of in vitro and in vivo stimulation assays using lymphocyte mitogens, may partially address functional responsiveness. As these parameters do not necessarily assess a coordinated immune response, development of an in vitro assay such as the Mishell–Dutton assay (Mishell and Dutton, 1967), which immunizes and elicits an adaptive immune response entirely in vitro following lymphoid tissue processing (Holsapple, 1995), might help alleviate this limitation. Evidence of this approach appears to be rather limited, but has been demonstrated using catfish and rainbow trout (Miller and Clem, 1984; Anderson et al., 1986).

Aside from limitations due to immune analysis concerns, factors that may confound the design of an ecological study should be considered. Definitive characterization of chemical exposures is one complication that compels investigators to conduct test material exposures to feral animals under captive conditions. For example, it is not uncommon for pesticides to be applied as mixtures, and potential exposure to such an application would make it difficult to identify a primary immunotoxicant (Bishop et al., 1998). Investigators have reported alterations in spleen pathology (white pulp) and B-cell response in toads as a result of environmental pollutants among 15 sites in Bermuda (Linzey et al., 2003). Soil analyses detected the presence of multiple pesticides (or metabolites) at most sites, while cadmium, chromium, copper, and zinc were detected in the livers of these amphibians. Confounding the analysis and characterization of chemical exposures further is the fact that feral species may migrate between sites of differing chemical identities and exposure levels, making estimation of actual exposure very difficult (Luebke et al., 1997). In addition, interpretation of effects demonstrated following exposure to individual environmental pollutants within a controlled laboratory setting, under a controlled set of conditions, will not represent the multitude of environmental influences (e.g., food availability, competition predation, chemical mixtures, parasites) that are commonly encountered by feral species (Fairbrother et al., 2004).

In the context of exposure characterization, an additional aspect that must be considered for interpretation of ecoimmunotoxicological research involves relevant dose. Traditional toxicology has often relied on the extrapolation of effects from studies with unrealistically high doses, a method that is overly conservative in many cases. As illustrated in many of the papers discussed herein, investigators need to select dose levels that are well above realistic exposure conditions so that a measurable response can be elicited; these doses may have very little to do with actual environmental conditions. The full range of exposure response, including concentrations above environmentally observable levels, may be useful for interpretation of the dose–response curve. However, in order to accurately predict field ecoimmunotoxicological effects, dose levels must be selected over environmentally realistic ranges, many of which may be well below no-effect levels. Only through this strategy can relevant no observed effect concentration (NOEC)/no observable effect level (NOEL) values be identified and considered with respect to realistic environmental exposure conditions.

Seasonality and weather are primary confounders that cut across multiple environmental influences. As reviewed by Galloway and Depledge (2001), invertebrate hemocytes demonstrate fluctuations in numbers and motility in response to seasonal conditions that need to be considered. Temperature and light cycle are two seasonal factors that affect immune function by fish. The humoral and cellular components of an immune response were demonstrated to be affected in fish reared at either 8°C or 12°C through an entire life cycle (Alcorn et al., 2002). Interestingly, Sockeye salmon (*Oncorhynchus nerka*) raised at 8°C had a greater expression of nonspecific immune cells (macrophages), while fish reared at 12°C demonstrated elevated blood lymphocyte percentages and a greater T-dependent antibody response to *Renibacterium salmoninarum*. The health evaluation of tree swallows (*Tachycineta bicolor*) by Bishop et al. (1998) took a number of seasonality confounders into account to critically interpret multiple study endpoints. Bishop et al. (1998) noted that changes in air temperature over the course of sampling periods may have been responsible for temperature stress, which can account for reductions in humoral and cell-mediated immune responses.

Partially influenced by seasonality, parasite exposure can also dramatically influence parameters measured as part of an immunotoxicity analysis. Local inflammation noted in some tree swallows evaluated by Bishop et al. (1998) coincided with the presence of trematode infections; such parasites could be responsible for observed changes in hematological parameters (e.g., increased WBC). As mentioned above, increased spleen mass is consistent with added antigen and microbial challenges during summer months (Silverin et al., 1999), and thus fumigated areas may actually have reduced parasite populations that may translate to diminished immune parameters. As it is generally accepted that appreciable stress can have deleterious effects on immune status, such considerations should influence design and interpretation of an ecoimmunotoxicity experiment. In addition to natural sources of stress such as weather and nutritional status, stress through capture and/or confinement should be recognized. Evidence of capture-induced stress can be demonstrated in eider ducks via measurement of corticosterone (Wayland et al., 2002), a well-known immunosuppressive molecule. The ability of hemocytes to produce corticoid-like molecules suggests that invertebrates may also be susceptible to stress-related influences on the immune system (Galloway and Depledge, 2001).

Although a high degree of conservation is observed in immunological systems across wildlife species and mammals, the immune systems of wildlife species have sufficient differences in immune anatomy to confound data interpretation and risk assessment extrapolations beyond the specific species under investigation. An extreme example might include invertebrates as individual cellular homology and function involved with phagocytosis is quite similar with other taxa but the entire repertoire of their coordinated immune response, which is primarily innate, is too dissimilar for extensive, *in vivo* interpretations of mammalian risk (Galloway and Depledge, 2001). Aside from obvious limitations involving unique immunological architecture, investigators need to consider biomechanisms of absorption, distribution, metabolism, and elimination (ADME) that may also play a large role in species-specific bioactivation and chemical sensitivity. In some circumstances, wildlife populations can be more sensitive to environmental stressors than humans because exposures to nonhuman organisms may be sustained for greater periods. For example, fish may be constantly exposed to water-borne contaminants, when present, through respiration across their gills and through ingestion of food items. It is also likely that some wildlife are more sensitive to various compounds than humans as many of the highly developed detoxicification mechanisms present in humans are less sophisticated, or simply absent in nonhuman species (deFur et al., 1999). Characterizing ADME kinetics and profiles would be necessary to even consider extrapolation among species. Considering differences in xenobiotic uptake, metabolism, and immune responses across taxa, it seems more than difficult to extrapolate immunotoxicity data beyond closely related organisms. Field studies using feral animals should be considered the most relevant means by which to demonstrate a direct association between environmental influences or contaminants, and an immune-mediated effect. Although field studies have the advantage of low, relevant, and "natural" exposures, the lack of controls and complex experimental design, added to variability, makes interpretation of these results rather difficult. As discussed above, there are at least 20,000 species of fish for which there appears to be a variety of immune system responses according to phylogeny (Zelikoff, 1998). Diversity of such populations becomes even greater when one considers environmental influences (e.g., temperature, stress) and that species adapt to their immediate environment.

2.6 SUMMARY: STATE OF THE SCIENCE

Protection and integrity of the environment are critical elements in ecological risk assessment and must be placed on par with human safety assessment. So how can ecoimmunotoxicological techniques, many of which have been derived from mammalian immunotoxicity protocols, be effectively used? With a growing abundance of data, it appears that ecoimmunotoxicology methods continue to develop and can provide useful data that may be leveraged to evaluate immunotoxicity potential in wildlife species. The value of such research is the identification of potential, unintended harm to the environment expressed through subtle immunological endpoints. When taking such confounders as field exposure(s) and other environmental stressors into consideration, it becomes exceedingly difficult to identify definitive associations, but such studies can be considered in tandem with controlled, laboratory models to assess environmental risk using a weight-of-evidence approach. In the case of "lower" species, convincing identification of harm to immunological status may be fraught with numerous differences and uncertainties when considered in relation to effects on mammalian immune status. Evidence to support immunotoxicity based on ecotoxicological models should be evaluated on a case-by-case basis with hypotheses and causality determined, possibly using a tiered approach. As an initial screen, cellular endpoints could be used to screen immune function prior to defining associations of xenobiotics with effects through controlled experiments that establish definitive exposure. If cellular endpoints trigger further investigation, then evaluation of host resistance and environmental impact can be subsequently considered. Associations with mammalian immunotoxicology potential may be appropriate in the context of a thorough understanding of species physiology.

As discussed above, the goal of ecological risk assessment is generally not protection of the individual, but rather protection of populations and ecosystems. As such, it is critical that the application of immunological endpoints to wildlife models be accompanied by a practical understanding of the relevance of these endpoints, not only to single organisms, but to higher, sustainable levels of biological organization. The NOEL values from these endpoints should be considered with respect to realistic environmental exposure conditions. Currently, no provisions in any regulatory body's guidance refer to evaluation of functional immune responses using a model perceived as an alternative or environmental species. If properly validated and designed, one could envision that the addition of functional immune endpoints using guideline ecotoxicology models (e.g., wild mice, quail) may serve to reduce experimentation and animal usage for xenobiotic safety testing. The large amount of resources likely required for validation of these endpoints and collection of the additional data is also a consideration. To date, the issues discussed here make definitive interpretation of ecoimmunotoxicological data in the regulatory industrial arenas very difficult, and preclude the widespread acceptance of results collected in this emerging field. An important consideration is that the science of mammalian immunotoxicology is relatively less developed than many other fields of toxicology, and the current knowledge regarding ecoimmunotoxicology lies considerably further behind mammalian immunotoxicology. The current state of the science of ecoimmunotoxicology must be considered in the context of the evolution of traditional immunotoxicology over the past several decades. The development of rodent immunotoxiology guidelines has relied on approaches that are based on models that are well established. The series of papers by Luster and co-workers (1988, 1992, 1993) described and evaluated elegant concordance analyses involving immunotoxicity studies in rodents with over 50 prototypical immunotoxicants in at least a dozen different endpoints, including host resistance assays. Only subsequent to these hallmark papers was the EPA able to draft appropriate immunotoxicity guidelines that were distributed in 1998 (U.S. EPA, 1998). The focused effort to advance and embrace immunotoxicology as regulatory guidance has spanned at least 2 decades and to date, these guidelines are still considered drafts and have yet to be finalized into law. As such, it is reasonable to conclude that the integration of ecoimmunotoxicology in a regulatory guideline context can only take place after individual models and parameters have been similarly characterized. Ultimately, an understanding of immune systems across species and some consensus regarding the interpretation of novel data sets are needed to assess the role of ecoimmunotoxicology in current risk assessment strategies.

REFERENCES

Alcorn, S.W., Murra, A.L., and Pascho, R.J., Effects of rearing temperature on immune functions in Sockeye salmon (Oncorhynchus nerka), *Fish Shellfish Immunol.*, 12, 303–334, 2002.

Anderson, D.P., Dixon, O.W., and Lizzio, E.F., Immunization and culture of rainbow trout organ sections in vitro, *Vet. Hum. Toxicol.*, 12(1–4), 203–211, 1986.

Anderson, D.P., and Zeeman, M.G., Immunotoxicology in fish, in *Fundamentals of Aquatic Toxicology*, Rand, G.M., Ed., Taylor & Francis, New York, 1995.

Arkoosh, M.R., et al., Suppression of B-cell mediated immunity in juvenile chinook salmon (Oncorhynchus tshawytscha) after exposure to either a polycyclic aromatic hydrocarbon or to polychlorinated biphenyls, *Immunopharmacol. Immunotoxicol.*, 16, 293–314, 1994.

Beck, G., and Habicht, G.S., Immunity and the invertebrates, *Sci. Am.*, 275, 60–63, 66, 1996.

Bernhoft, A., et al., Possible immunotoxic effects of organochlorines in polar bears (Ursus maritiums) at Svalbard, *J. Toxicol. Environm. Health*, Part A, 59, 561–574, 2000.

Beschin, A., et al., Functional convergence of invertebrate and vertebrate cytokine-like molecules based on a similiar lectin-like activity, *Prog. Mol. Subcell. Biol.*, 34, 145–163, 2004.

Bishop, C.A., et al., Health of tree swallows (Tachycineta bicolor) nesting in pesticide-sprayed apple orchards in Ontario, Canada. I. Immunological parameters, *J. Toxicol. Environ. Health, Part A*, 55, 531–559, 1998.

Bowser, D., Frenkel, K., and Zelikoff, J.T., Effects of in vitro nickel exposure on the macrophage-mediated immune functions of rainbow trout (Oncorhynchus mykiss), *Bull. Environ. Contam. Toxicol.*, 52, 367–373, 1994.

Brown, C.R., and Bomberger Brown, M., Spleen volume varies with colony size and parasite load in a colonial bird, *Proc. R. Soc. Lond. B. Biol. Sci.*, 269(1498), 1367–1373, 2002.

Carey, C., Cohen, N., and Rollins-Smith, L., Amphibian declines: an immunological perspective, *Dev. Comp. Immunol.*, 23, 459–472, 1999.

Christin, M.S., et al., Effects of agricultural pesticides on the immune system of Rana pipiens and on its resistance to parasitic infection, *Vet. Immunol. Immunopathol.*, 22, 1127–1133, 2003.

deFur, P.L., et al. Endocrine disruption in invertebrates: endocrinology, testing, and assessment. SETAC Press, Pensacola, FL, 1999.

Duffy, P.H., et al., Age and temperature related changes in behavioral and physiological performance in the Peromyscus leucopus mouse, *Mechan. Ageing Dev.*, 95, 43–61, 1997.

Fair, J.M., and Myers, O.B., The ecological and physiological costs of lead shot and immunological challenge to developing western bluebirds, *Ecotoxicology*, 11, 199–208, 2002.

Fairbrother, A., et al., A novel nonmetric multivariate approach to the evaluation of biomarkers in terresterial field studies, *Ecotoxicology*, 7, 1–10, 1998.

Fairbrother, A., Smits, J., and Grasman, K.A., Avian immunotoxicology, *J. Toxicol. Environ. Health B Crit. Rev.*, 7, 105–137, 2004.

Fowles, J., et al., Glucocorticoid effects upon natural and humoral immunity in mallards, *Dev. Comp. Immunol.*, 17, 165–177, 1993.

Fowles, J.R., et al., Effects of Aroclor 1254 on the thyroid gland, immune function, and hepatic cytochrome P450 activity in mallards, *Environ. Res.*, 75, 119–129, 1997.

Fox, G.A., Wildlife as sentinels of human health effects in the Great Lakes–St. Lawrence basin, *Environ. Health. Perspect.*, 109(Suppl 6), 853–861, 2001.

Galloway, T., and Handy, R., Immunotoxicity of organophosphorous pesticides, *Ecotoxicology*, 12, 345–363, 2003.

Galloway, T.S., and Depledge, M.H., Immunotoxicity in invertebrates: measurement and ecotoxicological relevance, *Ecotoxicology*, 10, 5–23, 2001.

Gantress, J., et al. Development and characterization of a model system to study amphibian immune responses to iridoviruses, *Virology*, 311, 254–262, 2003.

Gendron, A.D., et al., Exposure of leopard frogs to a pesticide mixture affects life history characteristics of the lungworm Rhabdias ranae, *Oecologia*, 135, 469–476, 2003.

Germolec, D.R., et al., Extended histopathology in immunotoxicity testing: interlaboratory validation studies, *Toxicol. Sci.*, 78, 107–115, 2004.

Ghaffari, S.H., and Lobb, C.J., Nucleotide sequence of channel catfish heavy chain cDNA and genomic blot analyses, *J. Immunol.*, 143, 2730–2739, 1989.

Gobel, T.W., et al., IL–18 stimulates the proliferation and IFN-gamma release of CD4+ T cells in the chicken: conservation of a Th1-like system in a non-mammalian species, *J. Immunol.*, 171, 1809–1815, 2003.

Grasman, K.A., and Fox, G.A., Associations between altered immune function and organochloride contamination in young Caspian terns (Sterna caspia) from Lake Huron, 1997–1999, *Ecotoxicology*, 10, 101–114, 2001.

Grinwis, G.C.M., et al., Toxicology of environmental chemicals in the flounder (Platictyhs flesus) with emphasis on the immune system: field, semi-field (mesocosm) and laboratory studies, *Toxicol. Lett.*, 112–113, 289–301, 2000.

Harrahy, L.N., Schreck, C.B., and Maule, A.G., Antibody-producing cells correlated to body weight in juvenile Chinook salmon (Oncorhynchus tshawytscha) acclimated to optimal and elevated temperatures, *Fish Shellfish Immunol.*, 11, 653–659, 2001.

Holsapple, M.P., The plaque-forming cell (PFC) response in immunotoxicology: an approach to monitoring the primary effector function of B lymphocytes, in *Methods in Immunotoxicology*, vol. 1, Burleson, G.R., Dean, J.H., and Munson, A.E., Eds., Wiley-Liss, New York, 1995, pp. 71–108.

Janeway, C.A., et al., *Immunobiology: The Immune System in Health and Disease*, Garland Publishing, New York, 2001.

Kollner, B., et al., Evaluation of immune functions of rainbow trout (Oncorhynchus mykiss): how can environmental influences be detected? *Toxicol. Lett.*, 131, 83–95, 2002.

Koskela, K., et al., Insight into lymphoid development by gene expression profiling of avian B cells, *Immunogenetics*, 55, 412–422, 2003.

Linzey, D., et al., Role of environmental pollutants on immune functions, parasitic infections and limb malformations in marine toads and whistling frogs from Bermuda. *Int. J. Environ. Health Res.*, 13, 125–148, 2003.

Lu, C., et al., Biological monitoring survey of organophosphorus pesticide exposure among preschool children in the Seattle metropolitan area, *Environ. Health Perspect.*, 108, 299–303, 2001.

Luebke, R.W., et al., Symposium overview. Aquatic pollution-induced immunotoxicity in wildlife species, *Fundam. Appl. Toxicol.*, 37, 1–15, 1997.

Luster, M.I., et al., Development of a testing battery to assess chemical-induced immunotoxicity: national toxicology program's guidelines for immunotoxicity evaluation in mice, *Fundam. Appl. Toxicol.*, 10, 2–19, 1988.

Luster, M.I., et al., Risk assessment in immunotoxiclogy. II. Relationships between immune and host resistance tests, *Fundam. Appl. Toxicol.,* 21, 71–82, 1993.

Luster, M.I., et al., Risk assessment in immunotoxicology. I. Sensitivity and predictability of immune tests, *Fundam. Appl. Toxicol.,* 18, 200–210, 1992.

McCormack, W.T., Tjoelker, L.W., and Thompson, C.B., Avian B-cell development: generation of an immunoglobulin repertoire by gene conversion, *Ann. Rev. Immunol.,* 9, 219–241, 1991.

McCormack, W.T., Tjoelker, L.W., and Thompson, C.B., Immunoglobulin gene diversification by gene conversion, *Prog. Nucleic Acid Res. Mol. Biol.,* 45, 27–45, 1993.

Miller, N.W., and Clem, L.W., Microsystem for in vitro primary and secondary immunization of channel catfish (Ictalurus punctatus) leukocytes with hapten-carrier conjugates, *J. Immunol. Methods,* 72, 367–379, 1984.

Miller, T.E., et al., Developmental exposure to lead causes persistent immunotoxicity in Fischer 344 rats, *Toxicol. Sci.,* 42, 129–135, 1998.

Mishell, R.I., and Dutton, R.W., Immunization of dissociated spleen cell cultures from normal mice, *J. Exp. Med.,* 126, 423–442, 1967.

Pickel, J.M., et al., Differential regulation of V(D)J recombination during development of avian B and T cells, *Int. Immunol,* 5, 919–927, 1993.

Powell, D.B., et al., Immunocompetence of juvenile Chinook salmon against Listonella anguillarum following dietary exposure to Aroclor 1254, *Vet. Immunol. Immunopathol.,* 22, 285–295, 2003.

Pruett, S.B., Immunotoxicity of agrochemicals. An overview of currently available information, *Toxicol. Ecotoxicol. News,* 1, 49–54, 1994.

Ratcliffe, N.A., Invertebrate immunology: a primer for the non-specialist, *Immunol. Lett.,* 10, 253–270, 1985.

Rollins-Smith, L.A., Metamorphosis and the amphibian immune system, *Immunol. Rev.,* 166, 221–230, 1998.

Rooney, A.A., Bermudez, D.S., and Guillette Jr., L.J., Altered histology of the thymus and spleen in contaminant-exposed juvenile American alligators, *J. Morphol.,* 256, 349–359, 2003.

Rosenberg, C.E., Salibian, A., and Fink, N.E., An enzyme-linked immunosorbent assay for measuring anti-sheep red blood cells antibodies in lead-exposed toads, *J. Pharmacol. Toxicol. Methods,* 47, 121–128, 2002.

Segre, M., et al., Immunological and physiological effects of chronic exposure of Peromyscus leucopus to Aroclor 1254 at a concentration similar to that found at contaminated sites, *Toxicology,* 174, 163–172, 2002.

Silverin, B., et al., Seasonal changes in mass and histology of the spleen in willow tits Parus montanus, *J. Avian Biol.,* 30, 255–262, 1999.

Smialowicz, R.J., et al., Evaluation of the immunotoxicity of low level PCB exposure in the rat, *Toxicology,* 56, 197–211, 1989.

Smith, K.G., and Hunt, J.L., On the use of spleen mass as a measure of avian immune system strength, *Oecologia,* 138, 28–31, 2004.

Soderhall, K., Cerenius, L., and Johansson, M.W., The prophenoloxidase activating system in invertebrates, in *New Directions in Invertebrate Immunology,* Soderhall, K., Iwanaga, S., and Vasta, G.R., Eds., SOS Publications, Fairhaven, NJ, 1997, pp. 229–253.

Suter, G.W.I., *Ecological Risk Assessment,* Lewis Publishers, Chelsea, MI, 1993.

Trust, K. A., et al., Cyclophosphamide effects on immune function of European starlings, *J. Wildl. Dis.,* 30, 328–334, 1994.

U.S. Environmental Protection Agency, Health Effects Test Guidelines, Immunotoxicity, Office of Prevention, Pesticides and Toxic Substances, Environmental Protection Agency, Washington, DC, August 1998 (OPPTS 870.7800).

Van Loveren, H., et al., Contaminant-induced immunosuppression and mass mortalities among harbor seals, *Toxicol. Lett.,* 112–113, 319–324, 2000.

Ward, E.C., et al., Immunosuppression following 7,12-dimethylbenz[a]anthracene exposure in B6C3F1 mice. I. Effects on humoral immunity and host resistance, *Toxicol. Appl. Pharmacol.,* 75, 299–308, 1984.

Wayland, M., et al., Immune function, stress response, and body condition in arctic-breeding common eiders in relation to cadmium, mercury, and selenium concentrations, *Environ. Res.,* 90, 47–60, 2002.

Wilson, J., et al., Ecotoxicological risks associated with land treatment of petrochemical wastes. III. Immune function and hematology of cotton rats, *J. Toxicol. Environ. Health A,* 66, 345–363, 2003.

Wu, P.J., et al., Immunological, hematological, and biochemical responses in immature white-footed mice following maternal Aroclor 1254 exposure: a possible bioindicator, *Arch. Environ. Contam. Toxicol.,* 36, 469–476, 1999.

Zelikoff, J.T., Biomarkers of immunotoxicity in fish and other non-mammalian sentinel species: predictive value for mammals? *Toxicology,* 129, 63–71, 1998.

Zelikoff, J.T., et al., Immunotoxicity of low level cadmium exposure in fish: an alternative animal model for immunotoxicological studies, *J. Toxicol. Environ. Health,* 45, 235–248, 1995.

Zelikoff, J.T., et al., Development of fish peritoneal macrophages as a model for higher vertebrates in immunotoxicological studies. I. Characterization of trout macrophage morphological, functional, and biochemical properties, *Fundam. Appl. Toxicol.,* 16, 576–589, 1991.

CHAPTER 3

Bivalves as Models for Marine Immunotoxicology

Michel Auffret

CONTENTS

3.1 INTRODUCTION

Marine bivalve mollusks comprise a zoological group that offers interesting ecopathological models for invertebrate immunotoxicologists. First, either feral or reared populations naturally sustain heavy microbial and parasitic pressure that generates a constant stimulation of their defense system. Thus, additional mortality generated by pathogens play a major role in regulating populations (Renault, 1996). Second, their prolific populations inhabit coastal waters where environmental conditions are highly variable, making most physicochemical factors potential stressors. Consequently, any alteration of their internal defenses, especially stress-induced immunosuppression, appears potentially critical when considering bivalve population dynamics. Extended chemical contamination in coastal ecosystems of industrialized countries is a relatively recent environmental parameter that constitutes a major source of immunosuppressive factors. Most bivalve mollusks, as filter feeders inhabiting estuarine, brackish waters, are highly exposed to contaminant accumulation. Consequently, large bioconcentration factors have been recorded for metals and organics in oysters and mussels from contaminated areas (Phillips, 1995).

3.2 THE IMMUNE SYSTEM OF BIVALVE MOLLUSKS

It is now well established that circulating hemolymph cells — or hemocytes — constitute a major medium for immune defense of bivalve mollusks against microbial and parasitic attacks. These ubiquitous cells are found in a semiclosed system including a heart, vessels, and variably sized sinuses localized in the major organs. It is remarkable that an as yet unknown and certainly highly variable proportion of hemocytes does not appear in the circulatory flux. Resident cells nested in interstitial tissues may, however, return to circulation as suggested by observations described in the next section. The morphology, cytochemistry, and ultrastructure of bivalve hemocytes are known in numerous species (Cheng, 1981; Fisher, 1986; Auffret, 1988). The cell types as identified and described in stained monolayers (Auffret and Oubella, 1995) have been classified in two major types: granular and nongranular cells. The formers, which are most often called granulocytes, have been found in all groups, except in scallops. Several sub-populations of granulocytes may be distinguished microscopically by the staining affinity of their cytoplasmic granules. The hyalinocytes are the major nongranular hemocytes, that is, cells lacking cytoplasmic granules. They have a lucent, frequently vacuolized cytoplasm and a large, central nucleus.

In any experimental approach to immunotoxicology, identifying the cell types is a routine and necessary step, either for establishing differential counts or for studying their respective functions. The application of flow cytometry to cell identification and enumeration has considerably enhanced the possibilities of investigating invertebrate hemolymph cells. However, pioneering studies have indicated that bivalve hemocyte populations show a great deal of natural variability in their size and structure (Ford et al., 1994). This feature presents an obstacle in any attempt to differentiate between cell types by using solely structural parameters in the light side scatter obtained by flow cytometry. Unfortunately, reagents for the phenotypic discrimination of hemocytes are not commercially available. Indeed, the panel of specific antibodies intended to study non-neoplastic hemocytes in bivalve mollusks is scarce. Dyrynda et al. (1997) raised monoclonal antibodies against hemocytes subpopulations in mussels. Recently, Xue and Renault (2001) were able to label oyster granulocytes by immunohistochemistry. As an alternative approach to the use of immunological reagents, plant lectins have been successfully applied to label, and therefore, to distinguish, the cell types in mussels (Renwrantz et al., 1985; Pipe, 1990).

In bivalve mollusks as in other invertebrates, the mechanisms of antimicrobial internal defense mechanisms are entirely innate in nature and rely mainly on phagocytosis. In this universally distributed biological function, cells are able to engulf, kill, and destroy foreign particles. Alternately, large aggressors such as metazoans are encapsulated. Even before this, within minutes following hemolymph withdrawal, hemocytes from bivalve mollusks exhibit an impressive capacity to spontaneously aggregate in large clots. This conspicuous "stress" response is most likely a true component of the immune function of hemolymph cells. Whether this corresponds to an activation of immunocompetent cells to as yet unknown signals or to a basic involvement in casual, blood-vessel breach repair remains to be elucidated. It is worth noting here that the success of the phagocytic process is strictly dependent on several elementary mechanisms, including recognition, locomotion (chemotaxis), and membrane adhesion (Verhoef and Visser, 1993). Membrane mechanisms are essential for cell recognition; however, these have been only partially elucidated in bivalves. Hydrophobicity and electrical charge interactions are probably basic processes, but the involvement of surface glycoproteins could explain the observed opsonizing and agglutinating properties in hemolymph (Renwrantz and Stahmer, 1983; Olafsen, 1986).

To complete these cellular mechanisms, bivalve mollusks possess a humoral immunity based on peptides and proteins with cytotoxic properties (Roch, 1999). Among these, the first to be described was the lysozyme. However, recent work in this area, largely stimulated by the extensive knowledge available for insects and other arthropods, allowed the description of additional proteins including bactericidins and serine proteases. At present, the molecular biology of antimicrobial peptides has unraveled several bivalve gene sequences and translation data, especially in mussels (Mitta et al., 2000).

3.3 EVIDENCE FOR IMMUNOTOXICITY OF CHEMICALS IN BIVALVE MOLLUSKS: AN OVERVIEW

In bivalve mollusks, exposure to xenobiotics most often induces immunological alterations at two levels of organization. An elementary level would correspond to cellular and tissue damage. Alterations at a systemic level would lead to immunosuppression. It is conceivable that both types of effects can be observed after chemical exposure, most likely in a sequential manner, since cellular damage is likely to lead to functional impairment. However, immunosuppression may occur without real structural alterations when cell signaling pathways are disrupted. An example is given in vertebrates by the effects of the fixation of halogenated aromatic hydrocarbons on Ah receptors (Hahn, 1998). The generation of reactive oxygen species (ROS), which generate severe secondary toxic effects, are among the intracellular responses triggered by this ligand–receptor interaction. At a higher organizational level of the immune system, when the xenobiotic interaction with the Ah receptor disrupts cellular responses such as the activation of strains with cytotoxic potential, immunosuppression is directly induced (Harper et al., 1993). In organisms lacking any specific immunity, as is the case in bivalve mollusks, the approach to describing immunotoxicologic effects usually includes two classes of biological responses. One identifies basic structural changes in hemolymph, with special attention to the composition of its two components, serum and cell fraction. The other, based on *in vitro* assays, intends to characterize functional impairment of the immune function and assesses the immunocompetence of hemocytes.

In bivalve mollusks, evidence for immunotoxicity induced by environmental stressors has existed for over ten years (Anderson, 1988, 1993). As in most invertebrates, changes in the number of circulating hemolymph cells, whether absolute numbers (total cell counts) or relative numbers of different cell types (differential cell counts), have been observed to vary extensively with biological cycles (season) and biotic factors, including microbial or parasitic attacks (Oubella et al., 1993). These observations assist in the unraveling of physiological changes induced in the body by abiotic, environmental factors (Auffret and Oubella, 1994). Experimental exposure to sublethal concentrations of xenobiotics (heavy metals and organics) induced in mussels (Coles et al., 1995; Pipe et al., 1999) and in oysters (Auffret et al., 2002), a dramatic increase of total numbers of circulating hemocytes. The degree of change in total numbers of circulating hemocytes was dependent on the intensity of stress, either in terms of dose or duration (Figure 3.1). The exact mechanisms governing such changes in cell numbers are not known. It is conceivable that an alteration of cellular adherence following contaminant accumulation in the body can generate a transfer of resting hemocytes from tissues toward hemolymph. Indeed, George et al. (1983) observed in oysters a decrease in so-called "tissular" hemocytes when the internal concentration of metals increased.

Exposure of oyster hemocytes *in vitro* to either heavy metals or organics altered their clotting capacity, suggesting a potential disruption of membrane adhesion mechanisms (Auffret and Oubella, 1997). Thus, resident cells normally located in organs and closely associated to tissue structures would return to the circulatory flux. This toxic response indirectly makes the quantification of cell numbers in hemolymph an appropriate biomarker of exposure to environmental contaminants (Auffret et al., in press). Surprisingly, in comparison to total cell counts, less information is available about changes in hemocyte subpopulations. As discussed above, this is probably due to the relative difficulty in establishing differential counts in these organisms. However, experimental contamination of mussels has resulted in conspicuous changes in the proportion of granular and agranular cells (McCormick-Ray, 1987; Pipe et al., 1999). In the case of contamination by phenol, the occurrence

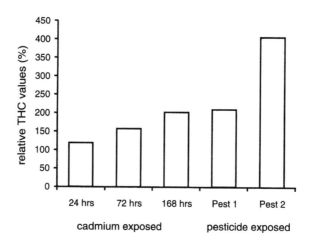

Figure 3.1   Total hemocyte counts were severely altered when Pacific oysters, *Crassostrea gigas*, were exposed in the laboratory either for 7 days to 0.5 ppm cadmium (left bars, n = 20 individuals) or 6 days to increasing concentrations of a pesticide cocktail (right bars, n = 12 individuals). Pest 1 = peak value measured in a French estuary; Pest 2 = Pest 1 × 10. The bars represent relative values compared to initial measurement. (From Auffret M. and R. Oubella, unpublished data.)

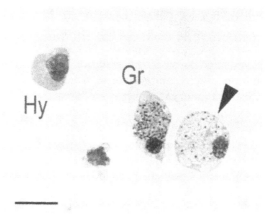

Figure 3.2   Exposure of Pacific oysters, *Crassostrea gigas*, for 7 days in the laboratory to 1 ppm copper, induced large, vacuolated cell (arrowhead) to appear, probably resulting from cell degranulation. Micrograph from a hemocyte monolayer prepared by cytospin and stained with May-Grünwald-Giemsa. Scale bar = 10 μm. Gr, basophilic granulocyte; Hy, hyalinocyte.

of cellular alteration in clam granulocytes suggested a selective cytolysis (Fries and Tripp, 1980). This hypothesis of a marked sensitivity of mature, differentiated cells was confirmed by the observation of increased numbers of small hemocytes in oysters exposed to hydrocarbons (Sami et al., 1992).

Cytolytic effects are among the most frequent toxic responses induced in cells exposed to chemical contaminants (Moore, 1985), and are regularly associated to membrane destabilization (Regoli, 1992). Specifically, lysosomal disruption will result in the release of several hydrolases in cytosol and even, in the extracellular fluid, with subsequent alteration of other cellular and tissue structures (Figure 3.2). Released products may be assayed in serum by common spectrophometric or microtitration methods.

As mentioned above, the mechanisms of antimicrobial internal defense mainly rely on phago-cytosis. In order to assess the immunocompetence of hemocytes in bivalve mollusks, functional assays have been developed where parameters describing the phagocytic process were established (Fisher and Tamplin, 1988). A phagocytic index may be calculated as the number of cells having engulfed foreign particles while the phagocytic capacity will correspond to the number of particles counted in phagocytic cells. This was initially obtained by microscopic observation of cell monolayers. However, the measurement of these complementary parameters has been notably improved since the pioneering work of Alvarez et al. (1989) by using fluorescein-labeled latex beads (1- to 2-μm diameter) in cell suspensions analyzed by flow cytometry (Brousseau et al., 2000). At present, many xenobiotics have been tested in the laboratory, either by exposing individuals or isolated hemocytes, in various species. Sublethal concentrations of heavy metals or organic molecules such as organotin compounds had a dramatic depressive effect on the *in vitro* phagocytosis (Cima et al., 1998; Bouchard et al., 1999; Brousseau, 2000; Auffret et al., 2002). Apparently contradictory results were obtained in mussel hemocytes exposed *in vitro* to cadmium (Olabarrieta et al., 2001). However, no dose–effect relationship was evident as the metal concentration increased, suggesting that cytotoxic alterations were induced. Recent results obtained by Sauvé et al. (2002) also indicated that low doses of some pollutants had a stimulating effect corresponding to a hormetic-like response. Other observations reporting a differential toxicity among members of chemical families (Bouchard et al., 1999) confirm that the phagocytic is a complex target for xenobiotics. Finally, in bivalve mollusks, as in higher organisms where nonspecific immunity was assessed, phagocytosis measurement was found to be an interesting, sensitive tool for immunotoxicologists.

Among early cellular processes involved in internal defense of bivalve mollusks, chemotaxis is a mechanism related to hemocyte migration, and in some cases, could be considered as a prerequisite to phagocytosis. This complex phenomenon was found, as expected, to be sensitive to poisons inhibiting cell movements (Cheng and Howland, 1982). The capacity of hemocytes to spontaneously aggregate and form clots is also probably related to phagocytosis and relies on membrane adhesion mechanisms. In vertebrates, an inhibition of receptors that trigger phagocyte adhesion properties has been shown to favor microbial attacks (Wright and Detmers, 1988). In mussels, Chen and Bayne (1994, 1995) demonstrated that various molecules able to interact with cell membranes including sugars, inhibited the *in vitro* aggregation of hemocytes. Alvarez and Friedl (1992) observed that exposure of American oysters to fungicides reduced the capacity of hemocytes to adhere *in vitro* to a support. By calculating an aggregation index from hemolymph cell suspension maintained *in vitro* and analyzed in a particle counter we have demonstrated in Pacific oysters that acute exposure of hemocytes to organic contaminants or to heavy metals could induce a dose–dependent inhibition of spontaneous aggregation (Figure 3.3). Considering the broad range of cytotoxic effects of xenobiotics, various other mechanisms of action could be suspected

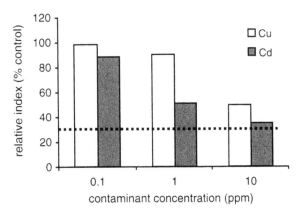

Figure 3.3 Heavy metal exposure inhibits the clotting activity of Pacific oyster, *Crassostrea gigas*, hemocytes as demonstrated by the decrease of the aggregation index measured *in vitro*. Mean values obtained from pools of hemocytes (n = 15 individuals) after 45-min incubation in medium supplemented with either copper or cadmium ions. The dashed line represents the level obtained with the chelator EDTA as a standard inhibitor. (From Auffret, M., and Oubella, R., *Comp. Biochem. Physiol.*, 118A, 705–712, 1997.)

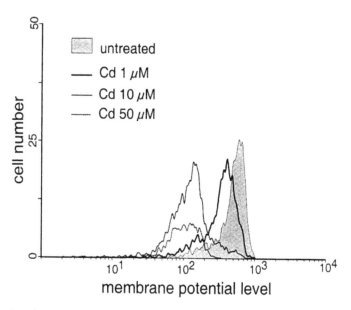

Figure 3.4   Alteration of granulocyte membrane potential assessed by flow cytometry through the fluorescence
due to the lipophilic probe $DIOC_6$ after a 1-week exposure of flat oysters, *Ostrea edulis*, in the
laboratory to sublethal concentrations of cadmium. (From Auffret, M., et al., *Mar. Environ. Res.*,
54, 585–589, 2002.)

to explain impairment of phagocytosis-related functions. The role of the cytoskeleton in controlling
cell movements is central, and we now have evidence that it is a target of chemicals. In mussel
hemocytes, the organization of filamentous proteins as actin was disturbed by exposure to heavy
metal (Fagotti et al., 1996) or organics (Gomez-Mendicute et al., 2002). However, it is known that
several xenobiotics disrupt cellular homeostasis by inhibiting mitochondrial function (Rice et al.,
1995) or by inactivating enzymatic systems (Viarengo, 1989; Fent and Bucheli, 1994). Thus, either
metabolic disorders in hemocytes or structural alterations reducing adherence and internalization
processes may be considered in explaining the impairment of the phagocytic function following
contamination.

Mitochondria are essential to provide energy resources for normal cell activity. Their membrane
potential is an electrophysiologic parameter linked to transmembranous ion fluxes. Its value may
vary with cellular physiological processes as differentiation or, in the case of immunocompetent
cells, activation in response to external stimuli (Shapiro et al., 1979). Such changes of membrane
potential can be indirectly observed by flow cytometry after incorporation of appropriate fluorescent
probes such as carbocyanines. In a recent study (Auffret et al., 2002), we observed a dose–dependent
reduction of hemocyte membrane potential in oysters contaminated with heavy metals (Figure 3.4).
Because this phenomenon has been described in vertebrate cells entering apoptosis (Hakem et al.,
1998), it is likely that metal ions had induced serious membrane alterations in oyster hemocytes.
Possible negative consequences in mitochondria could be a reduction of ion exchange through
intracellular compartments and impaired cell function. Thus, assessment of mitochondrial function
appears to be a valuable tool for detecting very early cellular alterations due to chemical stress and
should be further developed in bivalve hemocytes.

Among metabolic processes accompanying phagocytosis, those related to the intracytoplasmic
microbicidal (or killing) phase is a step of major concern since any alteration leads to immunosuppression (Roos, 1980). The production of ROS by phagocytic cells is a powerful mechanism
in neutralizing engulfed microorganisms. First described in vertebrate neutrophils, ROS production
has been reported in hemocytes from several bivalve species (Anderson, 1994). This mechanism
relies on a complex cascade of intracellular processes that may become a target for xenobiotics,

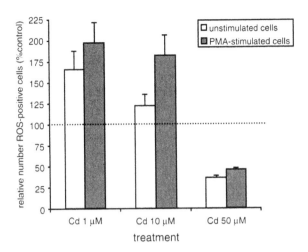

Figure 3.5 Alteration of the reactive oxygen species (ROS) level in hemocyte after a 1-week exposure of oysters, *Ostrea edulis*, in the laboratory to sublethal concentrations of cadmium. Phorbol myristate acetate (PMA) was applied as a membrane stimulant and ROS production measured by flow cytometry with the fluorescent probe di-hydrorhodamine 123. Data were expressed as relative values compared to untreated control individuals. Levels above the dashed line are indicative of cellular oxidative stress. (From Auffret, M., et al., *Mar. Environ. Res.*, 54, 585–589, 2002.)

and has been adopted as an assay to indirectly assess phagocytosis in bivalves. Indeed, various molecules including heavy metals or butyltins were able to inhibit ROS production in oysters (Larson et al., 1989; Anderson et al., 1992; Fisher et al., 1990). More recently, flow cytometry allowed the development of sensitive assays in bivalve hemocytes, using a combination of chemical membrane stimulation and oxygen radical detection by specific fluorescent probes. We have thus demonstrated that exposure to sublethal concentrations of cadmium could dramatically inhibit ROS production in oyster hemocytes (Auffret et al., 2002). Surprisingly, the level of oxygen radical was increased when lower concentrations were applied, suggesting that oxidative stress had been induced (Figure 3.5). Such an increase in ROS production by hemocytes had been previously reported in mussels exposed to organic contaminants or even polluted environments (Coles et al., 1994; Cajaraville et al., 1996). Responsible processes could be a preliminary lipid peroxidation or even free radical production by metal ions (Viarengo, 1989; Romeo and Gnassia-Barelli, 1997).

On the contrary, only sketchy information is available thus far on possible alterations of humoral defense mechanisms. As mentioned above, these are related to microbial killing during phagocytosis but not strictly dependent on it. Nevertheless, observations that the antibacterial response of mussel hemocytes could be influenced by temperature (Mitta et al., 2000; Hernroth, 2003) suggest that other abiotic factors including chemical contaminants could be potential immunosuppressors with respect to humoral defense. Organic molecules such as PCBs were observed to alter the transduction of bacterial signals leading to immune responses in mussels (Canesi et al., 2003). Consequently, further intracellular processes leading to microbial killing could be impaired, resulting in humoral immunosuppression.

This section has focused on evidence of immunotoxicity in bivalve mollusks. However, in view of the multiplicity and complexity of the mechanisms involved, the immune system response must not be oversimplified. In bivalves, as is the case in other groups, exposure to chemical stress does not systematically result in depressed immune function. Controlled acute contamination in the laboratory, where dose–effect relationships could be studied, has revealed several cases of stimulated hemocyte function, especially phagocytosis, at sublethal concentrations (Pipe et al., 1999; Sauvé et al., 2002). The assay applied, which quantified particle internalizations, allowed the investigators to conclude that a hormetic response occurred. However, in most cases, phagocytosis-related functions such as oxygen-dependent or -independent mechanisms were inconsistently affected, suggesting that the overall response of contamination was immunosuppression. In addition, Pipe et al. (1999) suggested that in the case of temporarily enhanced cell functions, the energetic cost would be at least detrimental for the organism.

3.4 EXPERIMENTAL DESIGN

3.4.1 *In Vitro* Exposure

Controlled experiments in the laboratory — where isolated hemocytes obtained from hemolymph samples were exposed *in vitro* to selected molecules — were used to assess toxic effects of contaminants on the immune system of bivalve mollusks. By reducing the complexity of the stress and by passing over all tissue barriers that separate xenobiotics in the environment from internal compartments, one is able to identify immunotoxic molecules and possible cellular targets. By using isolated cells, xenobiotics may be applied in a range of concentrations beyond those normally withstood by whole organisms. In addition, standardized experimental conditions allow comparing the toxicity of various contaminants or various forms within a xenobiotic family, considering the key role of biotransformation in organs involved in detoxication processes.

In the absence of cultured hemolymph cell lines, the hemocytes are usually separated from serum by centrifugation after addition of antiaggregant agents. The isolated hemocytes are resuspended in cold, buffered, isoosmotic incubation media supplemented with albumin, glucose, and antibiotics (Anderson et al., 1995; Auffret et al., 2002). Under these conditions, the cells can be exposed to xenobiotics up to 48 h in a cell incubator at 15°C.

Detecting immune alterations following chemical exposure requires a first tier to screen for possible structural damage to cells. The assessment of cell viability allowed Olabarrieta et al. (2001) to determine the LC_{50} of bivalve hemocytes for major aquatic pollutants. In oyster hemocytes, molecules such as the pesticide Paraquat induced an acute, dose-dependent toxicity as demonstrated by a severe depression of neutral red uptake (Figure 3.6).

By measuring phagocytosis by isolated hemocytes as a representative function of nonspecific immunity, the toxic potential of common pollutants such as heavy metals and butyltins could be

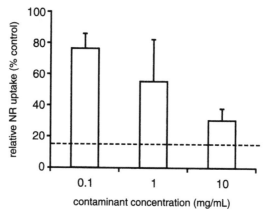

Figure 3.6 The neutral red uptake assay allows assessment of cell integrity. This function was impaired in Pacific oyster, *Crassostrea gigas,* hemocytes incubated *in vitro* in a concentration gradient of the pesticide Paraquat. The values were expressed as a percentage of the negative control (untreated cells). Vertical bars represent the standard error of the mean determined in triplicates. The dashed line gives the level obtained with glutaraldehyde as a positive control. (From Auffret M. and R. Oubella, unpublished data.)

ranked (Bouchard et al., 1999; Brousseau, 2000). Recently, Sauvé et al. (2002) demonstrated that marine bivalve species expressed different levels of sensitivity to heavy metals. If phagocytosis was regularly inhibited at high concentration, the response to low doses, somewhat more environmentally relevant, appeared to be more finely shaded. This observation underlined how important such species–species differences are when immunotoxicity is introduced for risk assessment in ecotoxicologic monitoring studies.

3.4.2   *In Vivo* Controlled Exposure Regimens

Most bivalve mollusk species are suitable aquatic animals for laboratory experiments because of their size, resistance to handling, feeding habits, and other biological characteristics (Auffret, 1995). Experimental protocols for *in vivo* exposure have most often been based on microcosms, which are plastic or glass tanks with either recirculating water or flow-through conditions. The latter protocol allows the control of a nominal contaminant concentration during exposure by renewing the water every day or more often if required.

By challenging the whole organism with pollutants, the disposition phase of foreign compounds in the body may be considered. In these filter-feeding invertebrates, tegument absorption, where the gills are largely involved, represents a prominent route of entry for bioavailable forms of contaminants. However, assimilation through the gut and especially the digestive gland, appears to play a major role. This is highlighted by the observation of several biotransformation systems in this organ (Petushok et al., 2002). After absorption, foreign compounds enter the hemolymph. The open structure of the vascular system in bivalve mollusks certainly influences the kinetics of this phase.

Short-term *in vitro* exposure to heavy metals or hydrocarbons at environmentally realistic levels affected the immune function of mussels (Coles et al., 1994, 1995; Pipe et al., 1999). The most responsive parameters were hemocyte counts and phagocytosis-related cellular events. In oysters, we recently observed that pesticides applied even at environmentally realistic levels induced immune alterations within days after initiation of exposure (Figure 3.7). Interestingly, when the concentration

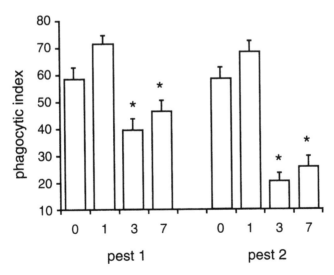

Figure 3.7   Evolution of the phagocytic index established for isolated hemocytes from Pacific oyster, *Crassostrea gigas*, exposed for 7 days in the laboratory to a mixture of pesticides at two concentrations (please refer to Figure 3.1). Data are the mean percentage (n = 10 individuals) of hemocytes containing phagocytosed yeast cells in monolayers observed under the microscope at day 0, 1, 3, or 7 after exposure started. Vertical bars represent the standard error of the mean. * Significantly different ($p < 0.05$) when compared to day 0. (From Auffret M. and R. Oubella, unpublished data.)

was increased, the effect was more pronounced but the delay remained unchanged. This clearly demonstrated that distribution processes in the body play a major role in the induction of immuno-toxic responses.

Controlled experiments with whole individuals in the laboratory allow modulation of chemical stress intensity by increasing the duration of exposure to several weeks and decreasing the con-centration of the chemicals so as to mimic conditions of long-term, chronic exposure. Relationships with bioaccumulation could then be established. In clams, a correlation between the contaminant body burden (organochlorine compounds and organic forms of heavy metals) and immunotoxic effects were described by Fournier et al. (2001, 2002). By exposing whole mussels to butyltin compounds, the no observable effect level (NOEL) was found near the threshold of usual chemical detection, that is, $1 \ ng/L^{-1}$ (Pelletier, 2004).

As underlined by Pipe et al. (1999), results of controlled experiments may help in establishing precise cause-and-effect relationships between pollution and immunotoxicity with possible effects on disease susceptibility in populations exposed to anthropogenic stress. However, *in vitro* exposure regimens have their disadvantages compared to other protocols. Among these is the ever-present difficulty in data extrapolation. Nevertheless, this approach is an alternative to whole animal testing and should be further developed. In this respect, establishing cell cultures in a bivalve species would greatly contribute to immunotoxicologic research in bivalves.

3.4.3 Microbial Infectivity Models

Experimental research, both in the laboratory and in the field, and observations in natural popula-tions of bivalve mollusks have demonstrated that environmental contamination is correlated to noninfectious diseases, resulting primarily from toxic effects on organs (Sparks, 1985). Information concerning possible links between pollution and infectious diseases is more limited (Chu and Hale, 1994). Furthermore, if one considers the fundamental role of the immune system in protecting the organism against such risk, it is obvious that establishing possible correlations between diseases and chemical contamination would help in exploring disruptions of the immune system at its highest level of complexity.

Most of the research undertaken to develop microbial infectivity models has been achieved through laboratory experiments. A bacterial clearance assay has been applied by Anderson et al. (1981) to clams exposed to sublethal concentrations of organic pollutants. The ability of hemolymph to clear cultured, marine bacteria from a strain isolated in seawater tanks and injected in an adductor muscle, was impaired and even in some cases completely abolished, suggesting that resistance to infection was significantly compromised. Pipe and Coles (1995) investigated the effect of heavy metals on the susceptibility of mussels to infection. Again, a marine strain found in bivalve environments was used, but in these experiments, the individuals were exposed to waterborne bacteria. After 7 days, several immune parameters were found altered in mussels, which indicated that chemical contamination had induced immunotoxic effects. Furthermore, an increased incidence of mortality was observed in mussels exposed to heavy metals and bacteria. The authors concluded that pre-exposure of bivalves to pollution could affect their immune system and as a consequence, increase the pathogenicity of opportunistic strains.

Although the literature on infectivity models in bivalve mollusks is scarce, the studies mentioned above clearly demonstrate that chemical contamination could be related to infectious disease outbreaks in natural populations of polluted coastal ecosystems. Much work is needed to explore such responses of other species in experimental conditions including challenge with both pathogenic and nonpathogenic strains. Nevertheless, it is worth noting that any result obtained from bacterial enumeration in hemolymph should take into account that in bivalves, this compartment is not considered as sterile. To bypass this problem, experimental designs including labeled bacteria or strains not found in the environment of the host could be considered.

3.4.4   Field Studies

Increasing concentrations of chemical contaminants in coastal waters is a subject of great concern since several of them are potent immunotoxicants for aquatic organisms. Even very low levels should be considered since biological effects of contamination may appear even before the threshold of detection by chemical analysis is reached. Such low concentrations may be found when repeated, reduced inputs occur in chronic pollution. In fact, this kind of pollution appears to be the most widespread in aquatic ecosystems. The complexity and sophistication of the immune system, even in lower animals, makes it one of the most sensitive physiological functions used to detect adverse effects of pollution. Among bivalve mollusks, oysters and mussels are the most frequent species used in field studies. Both have interesting biological and ecological characteristics: they are sedentary, filter feeders that potentially accumulate large quantities of xenobiotics (Zatta et al., 1992).

3.4.4.1   Site Comparison

Monitoring studies imply sampling series that most often extend over years. As many biological functions of bivalve mollusks undergo marked seasonal cycles, hemolymph parameters may also vary during the year. In comparing several hematological and functional parameters in oysters from two estuarine sites in the U.S., Oliver and Fisher (1995) revealed that both geographical and seasonal changes occurred naturally. In this respect, neither the total nor the differential hemocyte counts were comparable. Furthermore, in one site, both the phagocytic activity and ROS production by hemocytes were higher.

Pipe et al. (1995) sampled mussels several times a year in the Mediterranean Sea, from the Venice lagoon and the surroundings, an area having received particularly high inputs of chemical contaminants for several decades. Even if seasonal fluctuations were revealed in both biological and chemical data, several immune parameters showed a correlation with the level of contaminants measured in the tissues. Among those, changes in total cell counts and phagocytosis-related mechanisms appeared to be the most responsive. This essential process for bacterial clearance was inhibited when high metal levels were measured. In mussels from the Atlantic coast of Spain, seasonality has been demonstrated in the number of tissue-infiltrating hemocytes (Cajaraville et al., 1996). Furthermore, the pollution level of sampling sites was found to alter this pattern. These studies clearly indicated that seasonal cycles must be investigated in sentinel species and be considered when performing intersite comparisons of immunological parameters.

3.4.4.2   Caging Bivalves

Site comparison where natural beds of bivalves are monitored may be hampered when genetic differences among populations are suspected or more simply, when morphometric parameters due to age or population dynamics are observed. An alternative experimental design for field studies is to transfer subsamples from a single stock in the various sites to be monitored. For that purpose, transferred individuals have to be kept protected both from being scattered by hydrodynamic factors and from being invaded by native shells if present. The use of oyster-farming bags appears to be a judicious choice since the experimental species to be caged might not suffer from this kind of imprisonment. Pacific oysters, *Crassostrea gigas*, and to a lesser extent mussels *Mytilus edulis* or *Mytilus galloprovincialis*, are among the most convenient species for such caging experiments. In convenient sites, experimental bivalves could be maintained for weeks (short-term or acute exposure) to months (long-term or chronic exposure).

As part of a monitoring program, oysters (*Crassostrea gigas*) were transplanted in several locations of the Bay of Brest (Brittany, France) to demonstrate possible immunological alterations

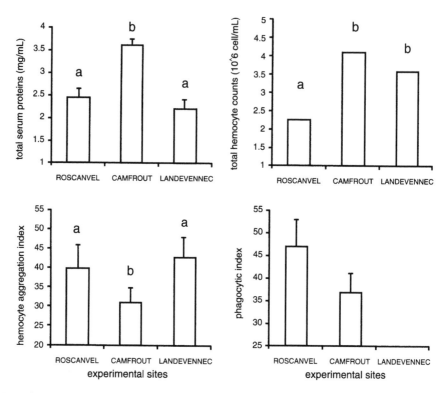

Figure 3.8   Several hemolymph parameters were measured in oysters, *Crassostrea gigas*, 1 year after being caged in three sites of the Rade de Brest, differing by their toxicological profile. Camfrout was highly and Landévennec moderately impacted. Roscanvel was considered as a clean area. Histogram values are the means and standard error established from experimental individuals (n = 10 to 15). Camfrout could be distinguished by conspicuous alterations of all parameters measured resulting from environmental, stressful conditions. Labels indicate homogenous statistical groups (analysis of variance). (From Auffret M. and R. Oubella, unpublished data.)

due to environmental exposure. Preliminary chemical analysis of water and sediments in this semi-enclosed bay had identified various levels of heavy metals and pesticides in estuarine sites receiving inputs from intensive farming and elevated concentrations of tributyltin in harbors. Consequently, considering the geographical location and the hydrodynamic conditions, the caging sites could be classified as low to high pollution. In oysters maintained for a year in the experimental sites, several parameters measured in the hemolymph showed a convergent "site" effect. Indeed, one of the most polluted areas could be distinguished from the others by stimulated hemolymph parameters as higher levels of total serum proteins and elevated total hemocyte counts, whereas functional parameters such as the spontaneous aggregation activity or the phagocytic capacity of hemocytes were depressed (Figure 3.8). Further observations included a reduction in growth and survival of the oysters maintained in this site, which suggested that hemolymph changes and immunotoxic responses accompanied an overall alteration of the oysters' health status.

As mentioned previously, experimental work in the laboratory has demonstrated that several functional parameters of immune cells are targeted by chemical contaminants. The transplantation of individuals in the field revealed several alterations of hemolymph parameters due to chronic exposure in polluted sites. However, evaluating the contribution of other environmental, biotic, or abiotic factors, remains a great challenge for immunotoxicologists. This gap underlines the need to run concurrently with caging studies complementary assays exploring other physiological functions in the same experimental populations.

3.5 BIVALVES AS SENTINEL SPECIES IN MARINE ECOSYSTEMS: BIOMARKER APPROACH

Increasing interest in aquatic ecosystem monitoring necessitates the development of operational, chemical, and biological tools suitable for medium- and long-term surveys. Marine benthic invertebrates, including several species of bivalve mollusks, can be suitable models in ecotoxicologic studies (Viarengo and Canesi, 1991; Burgeot et al., 1995). A bivalve biologic specificity is the ability to bioaccumulate and concentrate during their sedentary life many chemicals, especially those that are in particulate form or in suspension (Phillips, 1995). This attribute and their position in the trophic network of most coastal ecosystems, that is, being positioned between primary producers and secondary consumers, has made bivalves an attractive model for ecotoxicologists.

Recent progress in research on biological responses to chemical contaminants, especially the various forms of toxicity induced by exposure, has allowed the development of an increasing number of biomarkers. While mutagenicity and genotoxicity are among the most dramatic effects to be studied, there is an urgent need to develop biomarkers for the detection of early biological signals generated by pollution in animal populations. The increasing knowledge of immune alterations in lower animals has assisted researchers in developing new biomarkers for immunotoxicology (Zeeman, 1996). As mentioned above, the internal defense of bivalve mollusks appears to be simple and is based on innate mechanisms where cellular and humoral processes lead to cytotoxic and antimicrobial functions. However, increasing knowledge in this area indicates that this system works with a complex organization of hemolymph cells — that is, hemocytes. Therefore, examination of several complementary cellular functions will improve our assessment of the contamination level in areas of interest.

Based on this concept, Pipe et al. (1995) investigated immune responses in mussels from the Venice Lagoon, a highly contaminated site. Results obtained from this work are reported in Section 3.4.4.1. Changes in total cell counts and phagocytosis-related activities in bivalve mollusks appeared to be promising as biomarkers of pollution.

More recently we have been involved in a monitoring study whose objectives were to validate immunological alterations in the mussel *Mytilus galloprovincialis* and to use these as biomarkers of chemical contamination in the western Mediterranean Sea. Three polluted sites, each including several sampling stations with a wide range of pollution, were selected. These sites were sampled twice a year. Several immune system alterations were regularly found in mussel hemolymph. Furthermore, the estimated levels of contamination correlated with hemocyte counts in the hemolymph or cell membrane integrity (Figure 3.9). Functional parameters such as hemocyte phagocytic capacity and ROS production levels were altered in most of the contaminated stations. Overall, the observed combined responses strongly suggested that mussels from these sites suffered immunotoxic effects. However, the comprehensive synthesis of such a multiparametric survey can be influenced by intrinsic characteristics of the parameters measured, including the threshold of contamination necessary to trigger the responses and the range of assay variability. Furthermore, biological variables such as phagocytosis or ROS production were most often inhibited in impacted stations, whereas others such as hemocyte counts were stimulated. These limitations could be partly overcome by providing an overall assessment of the individual's immune status. In the study mentioned above, we proposed to calculate a "site immunotoxicologic index" from the entire set of immunological parameters measured (Auffret et al., submitted). Briefly, in each site (i.e., for each mussel sample), a log-transformed parameter was calculated from the mean value of each parameter measured in the sample. Normalization was achieved by using values obtained in the reference site. The transformed parameters were divided into two groups defined as parameters indicative of structural changes in the hemolymph compartment (immunopathology), or parameters indicative of functional alterations in hemocytes (immunocompetence). To increase the discriminative power of highly stress-responsive parameters such as hemocyte counts or phagocytic index,

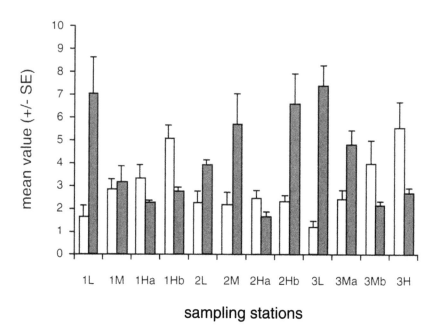

Figure 3.9   Variation of two immunological parameters measured in mussels, *Mytilus galloprovincialis,* sampled in three sites of the West Mediterranean in September 2001. White bars: total hemocyte counts ($\times 10^6$ cell/mL). Hatched bars: neutral red uptake by hemocytes (arbitrary units). Within each site (labeled 1, 2, 3), the location of the stations in relation to identified pollution sources allowed the following grading: the lowest polluted (labeled L) was distant from any sewage or release of contaminants, whereas the others ranged from medium impact to heavy pollution (M for medium or H for high). In Sites 1 and 3, an opposite gradient for the two parameters appeared, indicating severe environmental perturbation. (From Auffret et al., submitted.)

an "ecophysiological significance factor" was applied to transformed parameters, based on previous experimental results obtained through single-contaminant studies performed in the laboratory. Calculating this integrated immunotoxicologic index allowed us to identify, in two of three sites, a gradient of biological perturbation that correlated to the pollution gradient actually observed (Figure 3.10). This approach shows how data from a range of complementary assays obtained from individual bivalves could be processed in order to establish an appropriate immune diagnosis in the context of ecotoxicologic studies.

Correlations obtained in these studies, especially in mussels and oysters, between alterations in selected immune parameters and high pollutant levels indicate that several responses may now be considered as biomarkers of contamination in bivalves. Immunotoxic alterations have been classified as effective biomarkers to indicate that contaminants induced some physiological distur- bance when entering and accumulating in the individuals (Depledge, 1994). It has been demon- strated that there is a need to apply standardized and reproducible techniques in multiparametric studies. This of course implies there is a need for qualified personnel and properly equipped laboratories. The complexity of environmental factors including the toxicologic profile of the sites and the variability in bivalve biological functions complicate the findings and obscure cause–effect relationships. Seasonal cycles and every factor potentially generating natural changes in physio- logical functions have to be considered when the sampling schedule is programmed.

3.6 FUTURE PERSPECTIVES

Methodologies available to investigate most of the mechanisms supporting hemocyte functions are continuously improving. Most often these are based on protocols initially developed in vertebrates.

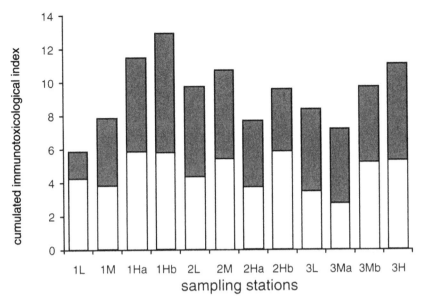

Figure 3.10 Set of integrated immunotoxicologic index calculated from standardized assays in mussels, *Mytilus galloprovincialis*, sampled in three sites of the West Mediterranean (see Figure 3.9). The bar represents cumulative relative index values, calculated as indicated in the text from two groups of parameters: immunopathology (in blank area) and hemocyte immunocompetence (hatched area). In Sites 1 and 3, a clear gradient of immunotoxicity was revealed. (From Auffret et al., submitted.)

Interindividual variability in all biologic and physiologic parameters has frequently been observed in these studies. The use of more homogenous animal groups, possibly obtained in hatcheries from genetically controlled genitors and standardized, reproducible techniques would, undoubtedly, minimize this variability.

Experimental protocols at different levels of complexity are currently used to study immunotoxic responses in bivalves. Exposures of isolated hemocytes or whole individuals were found to be complementary approaches. However, a gap remains between laboratory and field observations. The observation by Fournier et al. (2001) of a differential sensitivity in the manifestation of immuno-toxicity in clams toward different species of mercury suggests that the bioavailability of molecules is a major parameter in the environment. To address this issue, experiments in mesocosms could enhance the reliability of artificial exposure to chemicals. In particular, large basins, which contain sediment and associated animal and vegetation communities, allow complex processes to operate that induce the occurrence of biogeochemical cycles during long-term experiments.

In field studies, evidence for environmentally induced immunomodulation in bivalves has been demonstrated by applying multiassay protocols. As indicated by Coles et al. (1995), this approach is necessary due to the inherent variability in the biologic and physiologic effects produced by xenobiotics. Improving knowledge of the toxicity mechanisms in these organisms will help in understanding these apparently varying responses. Nevertheless, the use of a large battery of complementary assays to target a broad range of biological responses of hemocyte function will be necessary to detect immunosuppressive effects in polluted waters. This approach is especially needed for areas where low-level or chronic contamination is suspected. To increase the predictive capacity of assays for immunotoxicologic responses, it is important that these are performed at or near the threshold and before acute toxicity is induced. Since the function of the immune system is based on the recruitment of differentiated cells, all of which need to communicate to ensure an efficient response, the search for possible targets of xenobiotics in cell signaling appears to be a promising field of investigation. Recently, Canesi et al. (2003) explored the effects of organic contaminants in tyrosine kinase–mediated cell signaling in mussels. A variable immunotoxic activity of PCB congeners was demonstrated and several intracellular targets were identified. Other well-conserved mechanisms such as those based on ionic messengers would be interesting candidates. It has been demonstrated in bivalves that alterations in calcium homeostasis may induce cellular damage (Gnassia-Barelli et al., 1995). The demonstration that immune cells could undergo apoptosis (Burchiel et al., 1997) suggests that the effects of chemical contaminants on intracellular ionic fluxes should be further studied. Finally, more information is needed on specific interactions of chemicals and their cellular or subcellular targets in exposed bivalves to understand observed variations in the response to different stressors.

In addition to existing knowledge regarding the origin and mechanisms of xenobiotic toxicity in bivalve mollusks, cellular and molecular mechanisms that could interfere with the development of toxicity need to be further investigated so that the potential use of this zoological group in risk assessment is fully realized. Since the anatomical, biological, and physiological specificities of bivalve mollusks are different than those of higher animal groups, there is a need to elucidate molecular responses in bivalve hemocytes. Dyrynda et al. (1997) observed by using monoclonal antibodies that the expression of a 140-KDa epitope, whose function is yet unknown, was altered in hemocytes from mussels exposed to pollution. Several proteins called stress response proteins (SRPs) are known to play an important role in helping cells to withstand deleterious effects of physiochemical stressors. For example, heat shock proteins (HSPs) act as chaperones during protein synthesis and in cases of denaturation (Lindquist and Craig, 1988). Because they are overexpressed in mollusks, among other organisms, when these are exposed to chemical contaminants, their demonstration has been proposed in monitoring studies (Sanders, 1990). The presence of HSP has been first demonstrated in bivalve hemocytes by using immunological reagents directed against vertebrate proteins (Clegg et al., 1998). Corporeau and Auffret (2003) used oligonucleotidic probes synthesized from oysters, and demonstrated that an inducible response to sublethal, heavy metal concentrations corresponded to HSP gene expression. Other SRPs such as metallothioneins have been studied as detoxication mechanisms for heavy metals (Roesijadi, 1992). The genes coding for these proteins and factors controlling their expression have been recently investigated in bivalves (Tanguy and Moraga, 2001), and specific immunological reagents have been produced (Boutet et al., 2002). The development of such tools will allow us to learn more on tolerance mechanisms in bivalves exposed to low-intensity, chronic contamination.

Our knowledge related to responses of cellular components of bivalve hemolymph to chemical contamination is significant. Indeed, it seems that this hemocyte would be more responsive to stress than humoral factors. However, such a statement would probably be refuted when more research data on soluble hemolymph factors emerge. Recent developments in molecular genetics will assist in the identification of genes and in the study of their expression in response to environmental stressors.

3.7 SUMMARY

Data derived from *in vitro* and *in vivo* exposure experiments indicate that marine bivalve mollusks are susceptible to the immunotoxic effects of chemical contamination. Both organic and inorganic molecules were demonstrated to induce structural and functional alterations in hemolymph cells. For some of those pollutants, the deleterious effects of even environmentally realistic concentrations could be detected. However, dose–effect relationships often appeared to be complex when cell functions such as phagocytosis were assayed, since low doses could have effects opposite to high doses. Such findings demonstrate the need for complementary knowledge on cellular and molecular targets of chemicals. In addition, considering that the immune function relies on complex sequences of events, it is necessary to apply a multiparameter approach to detecting chemical-induced immunotoxicity.

The characterization of immunotoxic responses in the so-called sentinel species has allowed the development of biomarkers of effect for monitoring studies in polluted ecosystems. However, since species-to-species variations in sensitivity were regularly observed when individuals were exposed under similar experimental conditions to a given molecule, further investigations are needed in several species belonging to different families before general conclusions can be drawn on potential effects of pollution in the field.

As is the case in other domains of ecotoxicologic studies, applied immunotoxicology using bivalve species has to increase its ecological relevance. In this respect, even if immunosuppression is likely to lead to higher susceptibility to infectious diseases and other pathologies such as neoplasia, much work remains to be done to demonstrate how immunotoxicity may be related to the effective survival of individuals in field populations and to the possible disorganization of benthic communities.

ACKNOWLEDGMENTS

This contribution has been prepared from published data in journals and from personal communications with other immunotoxicologists working in the same field. The considerable volume of recent data contributed through the work of these immunotoxicologists is acknowledged. Other unpublished results obtained in our laboratory have also been included. Many thanks to Radouane Oubella for his contribution to the experimental work with oysters and mussels.

REFERENCES

Alvarez, M.R., and Friedl, F.E., Effects of a fungicide on *in vitro* hemocyte viability, phagocytosis and attachment in the American oyster, *Crassostrea virginica*, *Aquaculture*, 107, 135–140, 1992.

Alvarez, M.R., et al., Factors affecting *in vitro* phagocytosis by oyster hemocytes, *J. Invert. Pathol.*, 54, 233–241, 1989.

Anderson, R.S., Effects of anthropogenic agents on bivalve cellular and humoral defense mechanisms, *Am. Fish. Soc.*, 18(Suppl), 238–242, 1988.

Anderson, R.S., Modulation of nonspecific immunity by environmental stressors, in *Pathobiology of Marine and Estuarine Organisms*, Couch, J.A., and Fournie, J.W., Eds., CRC Press, Boca Raton, FL, 1993, 482–510.

Anderson, R.S., Oliver, L.M., and Jacobs, D., Immunotoxicity of cadmium for the eastern oyster (Crassostrea virginica, Gmelin 1791): effects on hemocyte chemiluminescence, *J. Shellfish Res.*, 11, 31–35, 1992.

Anderson, R.S., Hemocyte-derived reactive oxygen intermediate production in four bivalve mollusks, *Dev. Comp. Immunol*, 18, 89–96, 1994.

Anderson, R.S., Mora, L.M., and Brubacher, L.L., Luminol-dependent chemiluminescence in molluscs, in *Techniques in Fish Immunology*, vol. 4: *Immunology and Pathology of Aquatic Invertebrate*, Stolen, J.S., et al., Eds., SOS Publications, Fair Haven, NJ, 1995, chap. 13.

Anderson, R.S., et al., Effects of environmental pollutants on immunological competency of the clam Mercenaria mercenaria: impaired bacterial clearance, *Aquatic Toxicol.*, 1, 187–195, 1981.

Auffret, M., Bivalve hemocyte morphology, *Am. Fisher. Soc. Spec. Publ.*, 18, 169–177, 1988.

Auffret, M., Care and handling of marine bivalve mollusc species used in research, in *Techniques in Fish Immunology*, vol. 4, : *Immunology and Pathology of Aquatic Invertebrates*, Stolen, J.S., et al., Eds., SOS Publications, Fair Haven, NJ, 1995, pp. 67–70.

Auffret, M., and Oubella, R., Cytology and cytometric analysis of bivalve mollusc hemocytes, in *Techniques in Fish Immunology*, vol. 1, Stolen, J.S., et al., Eds., SOS Publications, Fair Haven, NJ, 1995, pp. 55–64.

Auffret, M., and Oubella, R., Cytometric parameters of bivalve molluscs: effect of environmental factors, in *Modulators of Fish Immune Responses*, Stolen, J.S., and Fletcher, T.C., Eds., SOS Publications, Fair Haven, NJ, 1994, pp. 23–32.

Auffret, M., and Oubella, R., Hemocyte aggregation in the oyster Crassostrea gigas: in vitro measurement and experimental modulation by xenobiotics, *Comp. Biochem. Physiol.*, 118A, 705–712, 1997.

Auffret, M., et al., Xenobiotic-induced immunomodulation in the European flat oyster, *Mar. Environ. Res.*, 54, 585–589, 2002.

Auffret, M., et al., Monitoring of immunotoxic responses in oysters reared in areas contaminated by the "Erika" oil spill, *Aquat. Living Resour.,* in press.

Bouchard, N., Fournier, M., and Pelletier, E., Effects of butyltin compounds on hemocytes phagocytosis activity of three marine bivalves, *J. Toxicol. Environ. Chem.,* 18, 519–522, 1999.

Boutet, I., et al., Immunochemical quantification of metallothioneins in marine molluscs: characterization of a metal exposure bioindicator, *Vet. Immunol. Immunopathol.,* 21, 1009–1014, 2002.

Brousseau, P. et al., Flow cytometry as a tool to monitor the disturbance of phagocytosis in the clam Mya arenaria hemocytes following in vitro exposure to heavy metals, *Toxicology,* 142, 145–156, 2000.

Burchiel, S.W., et al., Assessment of immunotoxicity by multiparameter flow cytometry, *Fund. Appl. Toxicol.,* 38, 38–54, 1997.

Burgeot, T., His, E., and Galgani, F., The micronucleus assays in Crassostrea gigas for the detection of seawater genotoxicity, *Mutations Res.,* 2094, 201–216, 1995.

Cajaraville, M.P., Olabarrieta, I., and Marigomez, I., In vitro activities in mussel hemocytes as biomarkers of environmental quality: a case study in the Abra estuary (Biscay Bay), *Ecotoxicol. Environ. Saf.,* 35, 253–260, 1996.

Canesi, L., et al., Effects of PCB congeners on the immune function of Mytilus hemocytes: alterations of tyrosine kinase-mediated cell signaling, *Aquatic Toxicol.,* 63, 293–306, 2003.

Chen, J.H., and Bayne, C.J., Bivalve mollusc hemocyte behaviors: Characterization of hemocyte aggregation and adhesion and their inhibition in the California mussel (Mytilus californianus), *Biol. Bull.,* 188, 255–266, 1995.

Chen, J.H., and Bayne, C.J., The roles of carbohydrates in aggregation and adhesion of hemocytes from the California mussel (Mytilus californianus), *Comp. Biochem. Physiol.,* 109A, 117–125, 1994.

Cheng, T.C., and Howland, K.H., Effects of colchicin and cytochalasin B on chemotaxis of oyster (*Crassostrea virginica*) hemocytes, *J. Invertebr. Pathol.,* 40, 150–152, 1982.

Cheng, T.C., Bivalves, in *Invertebrate Blood Cells,* vol. 1, Ratcliffe, N.A., and Rowley, A.F., Eds., Academic Press, London, 1981, pp. 233–300.

Chu, F.-L.E., and Hale, R.C., Relationship between pollution and susceptibility to infectious disease in the eastern oyster, Crassostrea virginica, *Mar. Environ. Res.,* 38, 243–256, 1994.

Cima, F., et al., Immunotoxic effects of organotin compounds in Tapes philippinarum, *Chemosphere,* 37, 3035–3045, 1998.

Clegg, J.S., et al., Induced thermotolerance and the heat-shock protein-70 family in the Pacific oyster Crassostrea gigas, *Mol. Mar. Biol. Biotechnol.,* 7, 21–30, 1998.

Clemons, E., Arkoosh, M.R., and Casillas, E., Enhanced superoxide anion production in activated peritoneal macrophages from English sole (Pleuronectes vetulus) exposed to polycyclic aromatic compounds, *Mar. Environ. Res.,* 47, 71–87, 1999.

Coles, J.A., Farley, S.R., and Pipe, R.K., Effects of fluoranthene on the immunocompetence of the common marine mussel, *Mytilus edulis, Aquatic Toxicol.,* 30, 367–379, 1994.

Coles, J.A., Farley, S.R., and Pipe, R.K., Alteration of the immune response of the common marine mussel *Mytilus edulis* resulting from exposure to cadmium, *Dis. Aquatic Organisms,* 22, 59–65, 1995.

Corporeau, C., and Auffret, M., In situ hybridization for flow cytometry: a molecular method for monitoring stress-gene expression in hemolymph cells of oysters, *Aquatic Toxicol.,* 64, 427–435, 2003.

Depledge, M., The rational basis for the use of biomarkers as ecotoxicological tools, in Fossi, M.C., and Leonzio, C., Eds., *Non-destructive Biomarkers in Vertebrates,* Lewis Publishers, Boca Raton, FL, 1994, pp. 271–295.

Dyrynda, E.A., Pipe, R.K., and Ratcliffe, N.A., Sub-populations of haemocytes in the adult and developing marine mussel, Mytilus edulis, identified by the use of monoclonal antibodies, *Cell Tissue Res.,* 289, 527–536, 1997.

Fagotti, A., et al., The effects of copper on actin and fibronectin organization in Mytilus galloprovincialis haemocytes, *Dev. Comp. Immunol.,* 20, 383–391, 1996.

Fent, K., and Bucheli, T.D., Inhibition of hepatic microsomal monooxyenase system by organotins in vitro in freshwater fish, *Aquatic Toxicol.,* 28,: 107–126, 1994.

Fisher, W.S., Structure and functions of oyster hemocytes, in *Immunity in Invertebrates,* Brehélin, M., Ed., Springer, Berlin/Heidelberg, 1986, pp. 25–35.

Fisher, W.S., and Tamplin, M., Environmental influence on activity and foreign-particle binding by hemocytes of American oysters, Crassostrea virginica, *Can. J. Fish. Aquatic Sci.,* 45, 1309–1315, 1988.

Fisher, W.S., Wishkovsky, A., and Chu, F.L.E., Effects of tributyltin on defense-related activities of oyster hemocytes, *Arch. Environ. Contam. Toxicol.*, 19, 354–360, 1990.

Ford, S.E., Ashton-Alcox, K.A., and Kanaley, S.A., Comparative cytometric and microscopic analyses of oyster hemocytes, *J. Invertebr. Pathol.*, 64, 114–122, 1994.

Fournier, M., et al., Effects of in vivo exposure of Mya arenaria to organic and inorganic mercury on phagocytic activity of hemocytes, *Toxicology*, 161, 201–211, 2001.

Fournier, M., et al., Effects of exposure of Mya arenaria and Mactromeris polynyma to contaminated marine sediments on phagocytic activity of hemocytes, *Aquatic Toxicol.*, 59, 83–92, 2002.

Fries, C.R., and Tripp, M.R., Depression of phagocytosis in *Mercenaria* following chemical stress, *Dev. Comp. Immunol.*, 4, 233–244, 1980.

George, S.G., Pirie, B.J.S., and Frazier, J.M., Effects of cadmium exposure on metal-containing ameobocytes of the oyster Ostrea edulis, *Mar. Biol.*, 76, 63–66, 1983.

Gnassia-Barelli, M., Romeo, M., and Puiseux-Dao, S., Effects of cadmium and copper contamination on calcium content of the bivalve *Ruditapes decussatus*, *Mar. Environ. Res.*, 39, 325–328, 1995.

Gomez-Mendikute, A., et al., Oxygen radical production and actin filament disruption in bivalve hemocyte treated with benzo(a)pyrene, *Mar. Environ. Res.*, 54, 431–436, 2002.

Hahn, M.E., The aryl hydrocarbon receptor: A comparative perspective, *Comp. Biochem. Physiol.*, C121, 23–53, 1998.

Hakem, R., et al., Differential requirement for caspase 9 in apoptotic pathways in vivo, *Cell*, 94, 339–352, 1998.

Harper, N., Connor, K., and Safe, S., Immunotoxic potencies of polychlorinated byphenyl (PCB), dibenzofuran (PCDF) and dibenzo-p-dioxin (PCDD) congeners in C57BL/6 and DBA/2 mice, *Toxicology*, 80, 217–227, 1993.

Hernroth, B., The influence of temperature and dose on antibacterial peptide response against lipopolysaccharide in the blue mussel, Mytilus edulis, *Fish Shellfish Immunol.*, 14, 25–37, 2003.

Larson, K.G., Roberson, B.S., and Hetrick, F.M., Effect of environmental pollutants on the chemiluminescence of hemocytes from the American oyster Crassostrea virginica, *Dis. Aquatic Organisms*, 6, 131–136, 1989.

Lindquist, S., and Craig, E.A., The heat shock proteins, *Ann. Rev. Genet.*, 22, 531–577, 1988.

McCormick-Ray, M. G., Hemocytes of Mytilus edulis affected by Prudhoe Bay crude oil emulsion, *Mar. Environ. Res.*, 22, 107–122, 1987.

Mitta, G., et al., Mytilin B and MGD2, two antimicrobial peptides of marine mussels: gene structure and expression analysis, *Dev. Comp. Immunol.*, 24, 381–393, 2000.

Moore, M.N., Cellular responses to pollutants, *Mar. Pollut. Bull.*, 16, 134–139, 1985.

Olabarrieta, I., et al., In vitro effects of cadmium on two different animal cell models, *Toxicology*, 15, 511–517, 2001.

Olafsen, J.A., Invertebrate lectins: biochemical heterogeneity as a possible key to their biological function, in *Immunity in Invertebrates*, Brehélin, M., Ed., Springer, Berlin, 1986, pp. 94–111.

Oliver, L.M., and Fisher, W.S., Comparative form and function of oyster Crassostrea virginica hemocytes from Chesapeake Bay (Virginia) and Apalachicola Bay (Florida), *Dis. Aquatic Organisms*, 22, 217–225, 1995.

Oubella, R., et al., Experimentally induced variations in hemocyte density in Ruditapes philippinarum and Ruditapes decussatus, *Dis. Aquatic Organisms*, 15, 193–197, 1993.

Pelletier, E., personal communication, 2004.

Petushok, N., et al., Comparative study of the xenobiotic metabolising system in the digestive gland of the bivalve molluscs in different aquatic ecosystems and in aquaria experiments, *Aquatic Toxicol.*, 61, 65–72, 2002.

Phillips, D.J.H., The chemistries and environmental fates of trace metals and organochlorines in aquatic ecosystems, *Mar. Pollut. Bull.*, 31, 193–200, 1995.

Pipe, R.K., et al., Evidence for environmentally derived immunomodulation in mussels from the Venice Lagoon, *Aquatic Toxicol.*, 32, 59–73, 1995.

Pipe, R.K., Differential binding of lectins to haemocytes of the mussel Mytilus edulis, *Cell Tissue Res.*, 261, 261–268, 1990.

Pipe, R.K., and Coles, J.A., Environmental contaminants influencing immune function in marine bivalve molluscs, *Fish Shellfish Immunol.*, 5, 581–595, 1995.

Pipe, R.K., et al., Copper induced immunomodulation in the marine mussel, Mytilus edulis, *Aquatic Toxicol.*, 46, 43–54, 1999.

Regoli, F., Lysosomal responses as a sensitive stress index in biomonitoring heavy metal pollution, *Mar. Ecol. Prog. Ser.*, 84, 63–69, 1992.

Renault, T., Appearance and spread of diseases among bivalve molluscs in the northern hemisphere in relation to international trade, *Revue Scientifique et Technique de l'Office International des Epizooties*, 15, 551–561, 1996.

Renwrantz, L., and Stahmer, A., Opsonizing properties of an isolated hemolymph agglutinin and demonstration of lectin-like recognition molecules at the surface of hemocytes from Mytilus edulis, *J. Comp. Physiol.*, 149, 535–546, 1983.

Renwrantz, L., Daniels, J., and Hansen, P.D., Lectin binding to hemocytes of Mytilus edulis, *Dev. Comp. Immunol.*, 9, 203–210, 1985.

Rice, C.D., Banes, M.M., and Ardelt, T.C., Immunotoxicity in channel catfish, Ictalurus punctatus, following acute exposure to tributyltin, *Arch. Environ. Contam.Toxicol.*, 28, 464–470, 1995.

Roch, P., Defense mechanisms and disease prevention in farmed marine invertebrates, *Aquaculture*, 172, 125–145, 1999.

Roesijadi, G., Metallothioneins in metal regulation and toxicity in aquatic animals, *Aquatic Toxicol.*, 22, 81–114, 1992.

Romeo, M., and Gnassia-Barelli, M., Effect of heavy metals on lipid peroxidation in the Mediterranean clam Ruditapes decussatus, *Comp. Biochem. Physiol.*, 118C, 33–37, 1997.

Roos, D., The metabolic response to phagocytosis, in *The Cell Biology of Inflammation*, L.E. Glyn., et al., Eds., Elsevier, Amsterdam, 1980, pp. 337–388.

Sami, S., Faisal, M., and Hugget, R.J., Alterations in cytometric characteristics of hemocytes from the American oyster Crassostrea virginica exposed to a polycyclic aromatic hydrocarbon (PAH) contaminated environment, *Mar. Biol.*, 113, 247–252, 1992.

Sanders, B.M., Stress proteins: potential as multitiered biomarkers, in *Biomarkers of Environmental Contamination*, McCarthy, J.F., and Shugart, L.R., Eds., CRC Press, Boca Raton, FL, 1990, pp. 165–191.

Sauvé, S., et al., Phagocytic activity of marine and freshwater bivalves: in vitro exposure of hemocytes to metals (Ag, Cd, Hg and Zn), *Aquatic Toxicol.*, 58, 189–200, 2002.

Shapiro, H.M., Natale, P.J., and Kamentsky, L.A, Estimation of membrane potentials of individual lymphocytes by flow cytometry, *Proc. Natl. Acad. Sci. U.S.A.*, 76, 5728–5730, 1979.

Sparks, A.K., Ed., *Synopsis of Invertebrate Pathology*, Elsevier, Amsterdam, 1985, 423 pp.

Tanguy, A., and Moraga, D., Cloning and characterization of a gene coding for a novel metallothionein in the Pacific oyster crassostrea gigas (CgMT2): a case of adaptive response to metal-induced stress? *Gene*, 273, 123–130, 2001.

Verhoef, J., and Visser, R., Neutrophil phagocytosis and killing: normal function and microbial evasion, in *The Neutrophil*, Abramson, J.S., and Wheeler, J.G., Eds., IRL Press, Oxford, 1993, pp. 109–137.

Viarengo, A., Heavy metals in marine invertebrates: mechanisms of regulation and toxicity at the cellular level, *Rev. Aquatic Sci.*, 1, 295–316, 1989.

Viarengo, A., and Canesi, L., Mussels as biological indicators of pollution, *Aquaculture*, 94, 225–243, 1991.

Wright, S.D., and Detmers, P.A., Adhesion-promoting receptors on phagocytes, *J. Cell Sci.*, 9 (Suppl.), 99–120, 1988.

Xue, Q., and Renault, T., Monoclonal antibodies to European flat oyster Ostrea edulis hemocytes: characterization and tissue distribution of granulocytes in adult and developing animals, *Dev. Comp. Immunol.*, 25, 187–194, 2001.

Zatta, P., et al., Evaluation of heavy metal pollution in the venetian lagoon by using Mytilus galloprovincialis as biological indicator, *Sci. Total Environ.*, 119, 29–41, 1992.

Zeeman, M., Comparative immunotoxicology and risk assessment, in *Modulators of Immune Responses*, Stolen, J.S., et al., Eds., SOS Publications, Fair Haven, NJ, 1996, pp. 317–329.

CHAPTER **4**

Approaches and Models for the Assessment of Chemical-Induced Immunotoxicity in Fish

Jessica E. Duffy and Judith T. Zelikoff

CONTENTS

4.1 INTRODUCTION

Fish represent a sensitive target for the toxic effects of aquatic pollutants such as metals, polycyclic aromatic hydrocarbons (PAHs), and halogenated aromatic hydrocarbons (HAHs). Given that they can be directly and chronically exposed to these types of contaminants in their natural environment, fish represent a relevant model for toxicological studies.

Over the past few decades, the immune system of fish has been increasingly defined and studied (Wester et al., 1994; Zelikoff, 1994; Carlson and Zelikoff, in press). Immunologically, fish share a number of aspects with their mammalian counterparts including morphologically and functionally identical immune cells; humoral, cell-mediated, and innate immune responses; and, similarly functioning primary and secondary lymphoid tissue (Wester et al., 1994; Zelikoff, 1994). Because of these similarities, immune assays, which were originally developed for mammals, have been successfully adapted and validated for use in a variety of fish species (Beaman et al., 1999; Carlson, Li and Zelikoff, 2002a, 2002b; Zelikoff et al., 2002).

The purpose of this chapter is to provide the reader with an overview of what is currently known regarding chemical-induced immunotoxicity in fish, specifically, metal-, PAH-, and HAH-induced immunotoxicity. For a more detailed description of the fish immune system, see other reviews focused specifically on this topic (Iwama and Nakanishi, 1996).

4.2 CHEMICAL-INDUCED IMMUNOTOXICITY IN FISH: AN OVERVIEW

4.2.1 Innate Immunity

4.2.1.1 In Vivo Exposure Studies

4.2.1.1.1 Metals

Heavy metals enter the environment, and particularly the aquatic environment, through a variety of sources such as mining, industrial wastes and discharges, ocean dumping, and atmospheric deposition (Carlson and Zelikoff, in press). The impact of certain metals on the biological function of exposed species has been found to include alterations in metabolism, reproduction, development, and immune function. Due to the abundance of heavy metals in contaminated aquatic environments, numerous laboratory, mesocosm and field studies have been conducted to examine the immunotoxic potential of these pollutants in fish. This chapter will review some of the more recent studies concerning the impact of metals on fish immunocompetence, particularly those examining cadmium (Cd), copper (Cu), mercury (Hg), tin (Sn) and zinc (Zn). For a more detailed discussion of metal-induced immunotoxicity in fish, see reviews by Zelikoff (1993) and Carlson and Zelikoff (in press).

Investigations into metal-induced alterations of the innate immune system of fish have been carried out for a variety of species. Phagocytosis, reactive oxygen intermediate (ROI) production, and natural cytotoxic cell (NCC) activity have all been shown to be altered in fish exposed *in vivo* to a variety of metals.

Phagocytic activity is frequently employed as a marker of metal-induced innate immune system dysfunction in fish. Cd, Cu, Zn, and Hg, have all been shown to reduce phagocytic activity in exposed fish. In one study, zebrafish exposed for 7 days to waterborne Cu at concentrations ranging from 0.05 to 0.15 ppm demonstrated a dose-dependent decrease in phagocytic activity (Rougier et al., 1996). In another study, head kidney macrophage from juvenile trout exposed to Cd for 30 days demonstrated depressed phagocytic activity (Voccia et al., 1996; Sanchez-Dardon et al., 1999). In the same study, exposure of juvenile trout to either Hg (0.1 or 0.5 μg/L) or Zn (10 or 50 μg/L) produced similar effects on phagocytosis (Sanchez-Dardon et al., 1999). However, in contrast to the latter study, macrophage-mediated phagocytosis was enhanced in rainbow trout exposed for 8 days to waterborne Cd; no effect on phagocytosis was observed in trout exposed to Cd for longer durations (i.e., 17 or 30 days) (Zelikoff et al., 1995). Differences between the two trout studies were most likely due to age of the fish at the time of Cd exposure.

Production of ROIs, including superoxide ($O_2^{\cdot-}$) and hydrogen peroxide (H_2O_2), is another sensitive innate immune endpoint for evaluating metal-induced alterations in exposed fish (Anderson and Zeeman, 1995; Zelikoff et al., 1996a, 1996b; Zelikoff, 1998). Rainbow trout injected with *Aeromonas salmonicidae* and then exposed to 2 μg Cd/L in the water for 30 days demonstrated significantly depressed phorbol myristate acetate (PMA)-stimulated $O_2^{\cdot-}$ and H_2O_2 production (Zelikoff et al., 1995). Similarly, macrophage-mediated PMA-stimulated H_2O_2 production was suppressed in rainbow trout following low-dose Cd (1 or 5 µg/L) or Hg (0.1 or 0.5 µg/L) exposure in the water (Sanchez-Dardon et al., 1999). Oxyradical production by phagocytes was also suppressed in sea bass following intraperitoneal (i.p.) injection with Cd (Bennani et al., 1996). In a comprehensive study by Zelikoff et al. (1996a), the effects of waterborne heavy metal exposure on ROI production were evaluated in Japanese medaka. Results from this study demonstrated effects of Cd on ROI production opposite to those reported in the aforementioned studies in trout and sea bass. Exposure of medaka for 5 days to Cd at concentrations ranging from 6 to 600 ppb significantly enhanced H_2O_2 and PMA-stimulated intra- and extra-cellular $O_2^{\cdot-}$ production by kidney phagocytes; effects persisted for up to 3 days after Cd exposure had ceased. Furthermore, ROI production by kidney phagocytes from Cd-exposed medaka fell significantly below those of controls when measured 10 days after placement in clean water. Mercury, nickel (Ni), and Zn were also evaluated in the same study for their effects on ROI production. While PMA-stimulated extracellular $O_2^{\cdot-}$ production by medaka kidney cells was increased following exposure to each of the three metals, intracellular $O_2^{\cdot-}$ production was increased only by exposure to Hg and Ni; H_2O_2 production was unaffected by exposure to Hg, Ni, or Zn (Zelikoff et al., 1996a).

Copper has also been shown to alter ROI production following exposure of fish *in vivo*. For example, $O_2^{\cdot-}$ production by kidney macrophages recovered from rainbow trout exposed to waterborne Cu at concentrations ranging from 6 to 27 µg/L was depressed compared to unexposed fish (Dethloff and Bailey, 1998); in addition, neutrophil numbers in Cu-exposed trout were significantly increased above control values. In another study, sea bass exposed to Cu via i.p. injection and examined 48 hours post-exposure demonstrated suppressed ROI production by bacterially stimulated kidney macrophage (Bennani et al., 1996). In contrast to the Cu-induced suppression of ROIs observed in some fish species, Cu exposure has also been shown to increase $O_2^{\cdot-}$ generation. For example, waterborne exposure of goldfish to 100 ppb of Cu for up to 11 days increased $O_2^{\cdot-}$ production in exposed fish; this effect persisted even after fish were maintained for 7 days in clean water (Jacobson and Reimschuessel, 1998).

Other nonspecific immune responses such as NCC activity have also been evaluated in fish in response to metal exposure. Like ROI production, NCC activity has also proven sensitive for demonstrating the immunotoxic effects of some metals. For example, zebrafish exposed to 0.05 to 0.15 ppm Cu for 7 days demonstrated a dose-dependent decrease in NCC activity as well as in kidney leukocyte cell numbers (Rougier et al., 1996).

4.2.1.1.2 *Halogenated Aromatic Hydrocarbons and Polycyclic Aromatic Hydrocarbons*

Halogenated (HAHs) and polycyclic aromatic hydrocarbons (PAHs) are well-known environmental contaminants, the health effects of which have been extensively studied in mammalian species. Halogenated aromatic hydrocarbons such as 2,3,7,8-tetrachlorodibenzo-p-dioxin (TCDD) and polychlorinated biphenyls (PCBs) are extremely persistent and have been detected in environmental compartments such as water, soil, and atmosphere long after discharges have occurred. Certain PAH family members, such as benzo(a)pyrene (BaP), methylcholanthrene (3-MC), and 7,12-dimethylbenzanthracene (DMBA), although not nearly as persistent as many HAHs, are constantly being released into the environment via incomplete combustion of organic matter. Thus, both classes of organic pollutants can pose a serious health threat for directly and indirectly exposed species.

The immunotoxic potential of HAHs such as TCDD and PCBs in exposed fish has not been adequately addressed despite the widespread nature of these aquatic pollutants. From studies performed to date, TCDD, a potent immunotoxicant in mammals, does not appear to have similar potency for fish. Of the few published studies, only a handful have reported alterations in immune function following TCDD exposure; some investigators have suggested that lymphoid tissues themselves may be a direct target for TCDD toxicity in certain fish species.

In contrast to TCDD, PCBs have been shown to induce immune alterations in both laboratory- and environment-exposed fish. However, the specific endpoint and the manner in which immune function is affected following PCB exposure is dependent on multiple factors, including fish species, duration of exposure, time of examination post-exposure, and particular PCB congener(s). For example, there are 209 possible PCB congeners, each differing in the number and positioning of chlorine molecules on the biphenyl rings. The toxic potential of a particular congener is generally thought to be dependent on which class it belongs to, with coplanar greater than monoortho coplanar, which in turn is greater than noncoplanar.

Reactive oxygen intermediate production has been shown to be a sensitive endpoint for assessing PCB-induced immunotoxicity in a variety of fish species. For example, i.p. exposure of channel catfish to doses of the coplanar PCB 126 congener ranging from 0.01 to 1.0 ppm suppressed the oxidative burst response, as well as reduced NCC activity in exposed fish (Rice and Schlenk, 1995; Regala et al., 2001). Given that NCC activity was only suppressed at the highest PCB dose while oxyradical production was affected by all PCB concentrations (at multiple post-exposure time points), ROIs appear more sensitive to the immunotoxic effects of PCB 126 than does NCC activity. Studies in this laboratory have demonstrated that exposure to PCB 126 (via i.p. injection) altered intracellular $O_2^{\cdot-}$ production by Japanese medaka (Duffy et al., 2003) and killifish (Duffy et al., in press). In these studies, PCB 126 modified both unstimulated and PMA-stimulated $O_2^{\cdot-}$ production in a time- and age-dependent manner (Duffy et al., 2003). Suppression of intracellular $O_2^{\cdot-}$ production by killifish exposed to PCB 126 was also time dependent and was only evident at that PCB dose and post-exposure timepoint that caused induction of hepatic cytochrome P450-1A (CYP1A) (Duffy et al., in press).

The immune response of feral fish environmentally exposed to PCBs has also been shown to be affected by this class of contaminants. For example, smallmouth bass collected from a PCB-contaminated site in the Great Lakes demonstrated suppressed ROI production, as well as reduced phagocytosis and superoxide dismutase (that enzyme responsible for conversion of $O_2^{\cdot-}$ to H_2O_2) activity compared to fish recovered from a relatively clean reference site (Anderson et al., 1997, 2003). Similar effects were also observed for brown trout recovered from a PCB-contaminated area near the smallmouth bass collection site (Stratus Consulting, personal communication, Colorado). In a study examining the impact of PCB-contaminated sediments (1500 ng total PCB/g dry weight [DW]) from Baie des Anglais on the St. Lawrence Estuary (Quebec), phagocytic activity by American plaice was reduced 1, 2, and 3 months following exposure (Lacroix et al., 2001).

PAHs, another widely distributed group of aquatic contaminants, have also been shown to be immunotoxic for fish. For example, BaP has been shown to reduce respiratory burst activity in tilapia (Holladay et al., 1998), as well as alter the phagocytic activity of rainbow trout (Walczak et al., 1987) and sea bass (Lemaire-Gony et al., 1995). In a comprehensive immunotoxicological study by Carlson et al. (2002b), ROI production was reduced in Japanese medaka following i.p. exposure to 200 μg BaP/g BW.

Studies examining the effects of other PAHs such as 3-MC and DMBA in fish are sparse, despite their toxic potential. However, of the existing studies it appears that the innate immune system is also sensitive to exposure to 3-MC. For example, carp exposed to 3-MC (40 mg/kg) via i.p. injection demonstrated significantly increased respiratory burst activity that appeared to coincide with hepatic and head kidney CYP1A activity (Reynaud et al., 2001). In a study with tilapia, exposure to DMBA (5 or 15 mg/kg) for 5 consecutive days altered lymphoid organ cellularity with a lesser effect on immune function (Hart et al., 1998).

Nonspecific immune functional endpoints such as phagocyte-mediated chemotaxis, phagocytosis, and respiratory burst, as well as NCC activity have all been shown to be altered in feral fish species collected from PAH-contaminated sites (Weeks and Warinner, 1984; Weeks et al., 1986, 1990; Weeks et al., 1988; Faisel et al., 1991a, 1991b). In one study, killifish collected from a PAH-contaminated river demonstrated reduced NCC activity (compared to fish collected from the reference site) against the mammalian tumor cell line K562 (Faisal et al., 1991a). Interestingly, the effect was reversed when fish were maintained for 28 weeks in the cleaner reference site (Faisal et al., 1991a).

Laboratory studies in which fish were exposed to PAH-contaminated environmental samples (i.e., contaminated sediments) further support the potent immunotoxic potential of these contaminants seen in fish exposed in the laboratory to a single chemical (Arkoosh et al., 2001). In a laboratory study in which female dab were exposed for 7 days to PAH-spiked sediments (nominally 10 mg total PAH/kg DW), H_2O_2 production by kidney phagocytes was suppressed, although no effects were observed on extracellular $O_2^{\cdot-}$ production (Hutchinson et al., 2003). In the same study, H_2O_2 production by dab kidney leukocytes was also suppressed following exposure to sediments spiked with both PAHs and PCBs. Interestingly, no effects on ROI production were observed when dab were exposed to only PCB-spiked sediments (nominally 200 μg total PCB/kg DW).

4.2.1.2 In Vitro Exposure Studies

4.2.1.2.1 Metals

Although not nearly as abundant as the *in vivo* studies, some investigations have examined the impact of metal exposure on innate immune functional endpoints *in vitro*. For certain metals, studies *in vitro* often contrast with results observed *in vivo*. For example, while bacterially stimulated ROI production was reduced in sea bass exposed to Cd by i.p. injection, *in vitro* Cd exposure of bass phagocytes produced a dose-dependent increase in oxyradical generation (Bennani et al., 1996). On the other hand, immunomodulating effects of Cu were similar for sea bass using both *in vivo* and *in vitro* exposure systems. In this study, the suppressed ROI response demonstrated by bacterially stimulated macrophage from sea bass exposed to Cu *in vivo* was also observed when bass macrophages were exposed to Cu *in vitro* (Bennani et al., 1996).

In vitro exposure of bluegill kidney phagocytes to selenium (Se), an essential trace element required by antioxidant enzymes for proper functioning, abrogated oxidant-induced effects on phagocytic activity and ROI production (Palchaudhuri et al., 2001). In this study, pretreatment of H_2O_2-exposed phagocytes with 0.5 to 10 μM SeO_3^{2-} for 48 hours altered both PMA-stimulated and unstimulated $O_2^{\cdot-}$ production, as well as kidney phagocytic activity. The authors of the study concluded that, depending on the metal concentration, Se can act as either a pro-oxidant or antioxidant in exposed fish. These results were similar to those previously reported for mammalian systems (Shamberger, 1983).

Another common metal found in contaminated waterways is Ni. A study by Bowser et al. (1994) demonstrated that *in vitro* exposure of rainbow trout peritoneal macrophage to either 100 or 250 μM nickel sulfate significantly increased unstimulated H_2O_2 production. The authors speculated that the observed changes in ROI production might have been due to Ni-induced alterations in cell membrane oxidases responsible for cellular respiration.

Natural cytotoxic cell activity is also altered by *in vitro* exposure to certain metals (Viola et al., 1996). For example, exposure to soluble Cd at 5 μM diminished catfish NCC activity (Viola et al., 1996). Results of this study are similar to what has been observed with mammalian NK cells also exposed to Cd *in vitro*. In addition, *in vitro* exposure of zebrafish NCC to either 10 or 20 μg soluble Cu/mL has also been shown to suppress cytotoxic activity (Rougier et al., 1996).

One metal that has been highly studied in fish for its *in vitro* effects on the immune response is Sn. Studies by Rice and Weeks (1989) have shown that *in vitro* exposure of oyster toadfish to tributyltin (TBT) can reduce phagocyte function in a time- and dose-dependent manner. In contrast, NCC from juvenile rainbow trout exposed to either TBT or dibutyltin (DBT) were unaffected by *in vitro* Sn exposure (O'Halloran et al., 1998).

4.2.2 Cell-Mediated Immunity

4.2.2.1 In Vivo Exposure Studies

4.2.2.1.1 Metals

Cell-mediated immunity has been reported in the literature to be altered in fish following exposure to a variety of different metals. For this chapter, lymphocyte proliferation was used as the primary immune parameter for which pollutant-induced effects on cell-mediated immunity were assessed. Altered lymphoproliferative responses of B-cells were also considered in this section since cell-mediated immunity can assist in this process via secretion of specific cytokines (Feldmann, 1998).

Rainbow trout exposed to waterborne Cd at a concentration of either 1 or 5 ppb demonstrated depressed head kidney and thymic B-lymphocyte proliferation at both doses; T-lymphocyte proliferation was depressed following exposure to Cd at 1 ppb, but increased at the highest Cd dose (Sanchez-Dardon et al., 1999). In the same study, exposure to Hg or Zn also suppressed T- and B-lymphocyte proliferation.

Similar effects on lymphoproliferation have also been reported for Cr. For example, freshwater catfish exposed to Cr for 28 days demonstrated decreased proliferative responses and eye-allograft rejection time (Khangarot and Tripathi, 1991). In another study using catfish, in vivo exposure to 20 µg Cd/L for 30 days inhibited T- and B-lymphocyte proliferation (Albergoni and Viola, 1995b). Taken together, effects on lymphocyte proliferation appear dependent on the metal type and concentration tested, as well as the particular fish species used.

4.2.2.1.2 Halogenated Aromatic Hydrocarbons and Polycyclic Aromatic Hydrocarbons

Very little data are available concerning the effects of HAHs on cell-mediated immunity in fish. Moreover, from the few investigations reported, data appear inconsistent between studies. While T- and B-lymphocyte proliferation were enhanced in rainbow trout following PCB exposure (Thuvander and Carlstein, 1991), studies from our laboratory have demonstrated suppression of the same response following i.p. exposure of bluegill sunfish to the noncoplanar PCB 153 (Duffy et al., 2004). Discrepancies between the two studies may have been due to differences in PCB congener(s), duration and/or method of PCB exposure, or fish species used.

As is the case for the HAHs, studies of the effects of PAHs on cell-mediated immunity are also extremely limited. However, in an investigation completed in our laboratory involving the impact of BaP on immune function of Japanese medaka, i.p. exposure to BaP inhibited T- and B-lymphocyte proliferation in exposed fish (Carlson et al., 2002b); inhibition of mitogen-induced lymphocyte proliferation was observed 48 hours post-injection at a dose as low as 2 µg BaP/g BW. In field studies, spot collected from a PAH-contaminated site (i.e., Lower Chesapeake Bay) also demonstrated suppressed T-lymphocyte proliferative responses (Faisal et al., 1991b); reduced lymphoproliferation was positively correlated with total PAH concentrations in the sediment (Faisal et al., 1991b).

4.2.2.2 In Vitro Exposure Studies

4.2.2.2.1 Metals

Studies examining the effects of in vitro exposure to Hg on lymphocyte proliferation have demonstrated that metal concentration is critical for determining its effect. While exposure of blue gourami head kidney lymphocytes to either 0.09 or 0.18 mg Hg/L suppressed mitogen-induced T-lymphocyte proliferation, exposure to 0.045 mg Hg/L increased the response (Low and Sin, 1998). In another study, lymphoproliferation by splenic and kidney B lymphocytes recovered from juvenile rainbow trout was reduced following in vitro exposure to 50 µg DBT/L (O'Halloran et al., 1998). The authors attributed the effect of DBT to a change in cell population profiles following Sn exposure. Finally, *in vitro* treatment of catfish peripheral blood lymphocytes with Cd (2 to 40 μM) inhibited lipopolysaccharide- and phytohemagglutinin-stimulated lymphocyte proliferation (Albergoni and Viola, 1995b).

4.2.3 Humoral-Mediated Immunity

4.2.3.1 *In Vivo Exposure Studies*

4.2.3.1.1 *Metals*

Catfish exposed to waterborne Cd at concentrations ranging from 10 to 30 μg/L demonstrated reduced nonspecific immunoglobulin (Ig) titers at 7 days but not 14 days postexposure (Albergoni and Viola, 1995a). In the same study, catfish exposed to 20 μg Cd/L and then immunized with sheep red blood cells (SRBCs) demonstrated peak IgM levels in a shorter timeframe than that of unexposed control fish; antibody levels increased when catfish were exposed to Cd for 2 weeks prior to SRBC immunization (Albergoni and Viola, 1995a). Studies by Thuvander (1989) and Robohm (1986) have demonstrated that exposure to Cd had a stimulatory effect on the humoral immune response of rainbow trout and striped bass, respectively.

In addition to Cd, exposure to Sn also appears to have a stimulatory effect on the humoral immune response of certain fish species. For example, channel catfish exposed to TBT (1.0 mg/kg) by a single i.p. injection demonstrated an increased plasma antibody response to *Vibrio anguillarum* at 21 days post-exposure; the same effect was also observed in catfish that received 6 i.p. injections of a lower TBT dose (1.7 μg/kg) over a 16-day period (Regala et al., 2001).

In contrast to Cd and TBT, Cr appears to suppress the humoral immune response of fish. For example, splenic- and kidney-antibody forming cell (AFC) numbers were reduced in freshwater catfish following waterborne Cr exposure for 28 days (Khangarot and Tripathi, 1991). Suppressive effects of waterborne Cr exposure on humoral immunity have also been observed for brown trout and carp (O'Neill, 1981). In contrast, waterborne exposure to 50 or 200 μg Cr/L for 1 month had no effect on the humoral immune response of rainbow trout (Viales and Calamari, 1984). It is difficult to determine whether differential effects of Cr in the aforementioned studies were due to differences in fish species, Cr dose, and/or antigen used to elicit the humoral immune response.

The effect of Hg on plasma-agglutinating antibody titers has been reported for blue gourami (Low and Sin, 1998). In this study, fish immunized with formalin-killed *Aeromonas hydrophilia* and then exposed to 0.01 or 0.09 mg Hg/L for up to 8 weeks demonstrated reduced antibody titers on subsequent bacterial challenge; the observed effect was seen for up to 7 and 8 weeks in fish treated with the high and low Hg concentrations, respectively (Low and Sin, 1998). A reduction in antibody titers in blue gourami was also observed when fish were treated with Hg (0.09 mg/L) prior to immunization (Low and Sin, 1998).

Of the existing data examining metal-induced effects on the humoral immune response of fish, outcome appears to be dependent on the particular metal being tested and fish species being examined.

4.2.3.1.2 *Halogenated Aromatic Hydrocarbons and Polycyclic Aromatic Hydrocarbons*

Suppression of humoral immunity is one of the most sensitive effects of TCDD on the mammalian immune response (Kerkvliet, 2002). However, the same outcome does not seem to hold true for fish. For example, yearling trout exposed via i.p. injection to 10 μg TCDD/kg failed to exhibit any alterations in the AFC response following immunization with SRBCs (Spitsbergen et al., 1986).

In comparison to the limited number of TCDD studies, investigations examining the effects of PCBs on humoral immune defense mechanisms of fish are more plentiful. Both juvenile (4 to

6 months) and aged (12 to 15 months) Japanese medaka exposed via i.p. injection to PCB 126 (1.0 µg/g BW) demonstrated significantly reduced AFC numbers as early as 3 days, and up to 14 days, post-injection (Duffy et al., 2002). A similar suppression of AFC numbers has been observed in juvenile chinook salmon following i.p. exposure to Aroclor 1254 (Arkoosh et al., 1994). In contrast to the effects observed in the latter study employing adult salmonids, AFC numbers were unaffected in juvenile rainbow trout exposed for 12 months to Aroclor 1254 (up to 300 ppm) in the diet (Cleland et al., 1988). Discrepancies between the studies may have been due to differences in PCB exposure routes (i.e., dietary vs. i.p. injection).

Specific antibody secreting cells (SASC) against *Edwardsiella ictaluri* were measured in channel catfish 14 days post-injection to PCB 126 (0.01 to 1.0 mg/kg) (Rice and Schlenk, 1995). While kidney leukocyte SASC numbers were enhanced in fish treated with the lowest PCB concentration (i.e., 0.01 mg/kg), no effects on SASC numbers were observed in catfish treated with either 0.1 or 1.0 mg PCB/kg. In a later study, channel catfish plasma antibody responses to *V. anguillarum* were suppressed following a single i.p. exposure to 1.0 mg PCB 126/kg (Regala et al., 2001). Interestingly, the same effect was not observed when catfish received an identical PCB 126 dose via 6 i.p. injections over a 16-day period.

PAHs are well-known suppressors of humoral immunity in a variety of mammalian species and similar effects have also been reported for fish; unfortunately, studies examining PAH-induced effects specifically on AFC numbers are sparse. The AFC response of Japanese medaka was significantly suppressed 48 hours following a single i.p. injection of BaP at a concentration of 20 or 200 µg/g BW (Carlson et al., 2002b); a similar effect of BaP was also observed in tilapia following i.p. exposure (Smith et al., 1999).

4.2.4   Overall Host Immunocompetence (Host Resistance Studies)

4.2.4.1   Metals

Despite the fact that studies examining host resistance against infection can serve as a meaningful indicator for determining overall immune status following chemical exposure, surprisingly few challenge studies have been performed in fish. Overall, metals appear to increase host susceptibility to challenge with an infectious agent. For example, freshwater catfish exposed to subtoxic levels of Cr for 28 days were more susceptible to infection with *A. hydrophilia* than their unexposed infected counterparts (Khangarot and Tripathi, 1991). Moreover, exposure of zebrafish to Cu for 7 days prior to, and 10 days after, infection with *Listeria monocytogenes* increased kidney bacterial burdens as well as altered resistance of zebrafish to listeriosis (Rougier et al., 1996).

4.2.4.2   Halogenated Aromatic Hydrocarbons and Polycyclic Aromatic Hydrocarbons

As it appears from the aforementioned studies that TCDD may not be as potent an immunomodulator in fish as it is for some mammalian species, it is not surprising that resistance of trout fingerlings against infection with hematopoetic necrosis virus was unaffected by dietary exposure to ≤1 µg of TCDD (Spitsbergen et al., 1988).

With regard to the effects of PCB exposure on fish resistance against infection, results appear ambiguous. Chinook salmon collected from a PCB/PAH-contaminated estuary had increased mortality following laboratory challenge with *V. anguillarum* compared to similarly challenged reference site fish (Arkoosh et al., 1998). However, juvenile chinook salmon fed a laboratory diet containing up to 10,000 µg Aroclor 1254/kg of food (wet weight) for 28 days and subsequently challenged with *Listonella anguillarum* demonstrated mortality rates similar to bacterially challenged fish fed a control diet (Powell et al., 2003). Moreover, the same juvenile salmon fed the aforementioned PCB-containing diet for 28 days, vaccinated subsequently with *L. anguillarum* and then challenged with *Listonella*, demonstrated a mortality incidence comparable to that exhibited by control fish. Possible explanations for these discrepancies include the route of PCB exposure (i.e., dietary vs. environmental), age of the fish at the time of exposure, and/or presence of other contaminants (i.e., PAH) in the field-tested salmon.

4.3 SUMMARY AND CONCLUSIONS

The immune system is exquisitely sensitive for assessing the toxic effects of chemicals and physical stresses of environmental concern. The sophistication and complexity of the immune system enables it to (potentially) be the most sensitive and, therefore, most prominent body function to detect harmful effects from environmental stressors. Sensitivity of this system seems to reside in the complex interactions that must occur for the mobilization of immune defenses. Disruption of any of these processes could offset the balance necessary for immunoregulation in the host and, thus, produce a cascade of detrimental secondary events, including compromised host resistance against infectious diseases and cancer.

Because of the sensitivity of the immune response to environmental toxicants, and its importance for maintaining host resistance against disease, chemical-induced immune dysfunction can be predictive of the toxicological hazards/risks associated with pollutant exposure. The previous establishment of highly sensitive immune assays to enumerate these alterations has culminated in a well-characterized battery of endpoints that can be used successfully to predict biological impact and adverse health outcomes in exposed populations. Such sensitivities also have major applications in efficacy-testing programs including those following remediation. Information on host immuno-competence generated from such programs could aid in management decisions regarding the effectiveness of any remedial activities, and the rates of recovery of affected sites. It could be assumed that changes in directions indicating decreased exposure/effects of affected sites precede an improvement in the ecological health of the environment. Thus, assays that measure immune dysfunction can serve as rapid indicators of the direction of change in toxic exposure and effects at a particular monitoring site.

It is important to recognize that although alterations in immune parameters are measured in a singe individual, these effects often result in decreased susceptibility to infectious agents and cancer and, thus, affect the population at large. Therefore, measures sensitive enough — and a monitoring system broad enough — to identify immunologic and environmental trouble in time to head off catastrophes are desperately needed.

4.4 FUTURE PERSPECTIVES

With rising social and political concern regarding the use of mammalian species for scientific studies, alternative animal models are actively being sought. Because of their versatility and immune system similarities with mammals (including humans), teleost species appear to be a logical choice as an alternate model for immunotoxicological investigations. Moreover, studies such as the ones described in this chapter that employ fish to assess chemical-induced immunotoxicity fit well into the newly emerging trends for immunotoxicity testing, which include more *in vitro* tests, greater use of computational methods, and the development and validation of nonmammalian alternative species (Karol, 1998).

The validity of cross-species extrapolation studies is improved by the ability to assess chemical-induced immunotoxicity in different animal models using the same assays. While more investigations are clearly needed before an accurate assessment of risk to inhabiting aquatic species can be made, the ability to evaluate fish health from an immunological standpoint may be invaluable, particularly to many countries whose populations have become increasingly more dependent on finfish produced by aquacultural methods.

A panel of assays, originally developed for use in rodents (Luster et al., 1988), has been adapted for use in a variety of wildlife species, including fish, to identify immune system changes brought about by chemical exposure. Some of the immune parameters used in this capacity include immuno-pathology (e.g., thymic atrophy); alterations in innate (i.e., macrophage activity/function) and acquired (e.g., antibody-forming cell numbers and lymphoproliferation) immunity; and, overall changes in immune-system-regulated functions (e.g., host resistance challenge models). For exam-ple, our laboratory has employed immune assays that measure: macrophage-mediated phagocytosis and oxyradical production; antibody-forming cell ability; lymphoproliferation in response to mito-gens and allogeneic cell stimulation; host resistance against bacterial infection; and, circulating leukocyte counts and lymphoid organ cellularity. Unfortunately, no one immune assay is going to provide the "magic indicator" for determining xenobiotic-induced immunotoxicity in fish. The greatest strength for predicting immunotoxicity in wildlife populations, as proved true for mammals (Luster et al., 1994), lies in the use of a battery of select immune assays with high predictive value. Taken together with data from those bioindicators measuring biochemical, physiologic, and/or histopathologic processes should prove most successful for predicting toxicological risks to species residing in contaminated aquatic environments.

In addition to the potential of fish to serve as alternate models for higher vertebrates, and the utility of the fish immune response to function as a biomarker of pollutant exposure, an important practical reason for investigating the effects of toxicant exposure on fish immunocompetence is that fish culture is a major industry. In the United States alone, state fish hatcheries spend an estimated $100 million annually to produce the half billion fish grown for anglers to catch. In such cases, the control of fish disease is a major concern. Furthermore, it is also beginning to be appreciated that fish can serve as vectors of human diseases. The role that fish immunotoxicology can play in the control of fish health has just begun to be appreciated.

ACKNOWLEDGMENTS

We would like to thank Yun Li for technical assistance. This work was supported, in part, by a Hudson River Graduate Fellowship and U.S. Army Contract no. DAMD 17-99-9011.

The views, opinions, and/or findings contained in this report are those of the author(s) and should not be construed as official Department of the Army position, policy, or decision, unless so designated by other official documentation. Research was conducted in compliance with the Animal Welfare Act, and other federal statues and regulations relating to animals and experiments involving animals and adheres to principles stated in the *Guide for the Care and Use of Laboratory Animals* (National Research Council, 1996) in facilities that are fully accredited by the Association for the Assessment and Accreditation of Laboratory Animal Care, International.

REFERENCES

Albergoni, V., and Viola, A., Effects of cadmium on catfish, *Ictalurus melas*, humoral immune response, *Fish Shellfish Immunol.,* 5, 89–95, 1995a.

Albergoni, V., and Viola, A., Effects of cadmium on lymphocyte proliferation and macrophage activation in catfish, *Ictalurus melas, Fish Shellfish Immunol.,* 5, 301–311, 1995b.

Anderson, D.P., and Zeeman, M.G., Immunotoxicology in fish, in *Fundamental Aquatic Toxicology,* 2nd ed., Rand, G.M., Ed., Taylor & Francis, London, 1995, pp. 371–402.

Anderson, M.J., et al., Biomarker selection for restoration monitoring of fishery resources, in *Environmental Toxicology and Risk Assessment: Modeling and Risk Assessment* (6th Volume), ASTM STP 1317, Dwyer, F. J., Doane, T.R., and Hinman, M.L., Eds., American Society for Testing and Materials, 1997, pp. 333–359.

Anderson, M.J., et al., Biochemical and toxicopathic biomarkers assessed in smallmouth bass recovered from a polychlorinated biphenyl-contaminated river, *Biomarkers*, 8, 371–393, 2003.

Arkoosh, M.R., Stein, J.E., and Casillas, E., Plaque-forming assays, in *Modulators of Fish Immune Responses*, vol. 1, Stolen, J.S., and Fletcher, T.C., Eds., SOS Publications, Fair Haven, NJ, 1994, pp. 33–48.

Arkoosh, M.R., et al., Effect of pollution on fish disease: potential impacts on salmonid populations, *J. Aquatic Anim. Health*, 10, 182–190, 1998.

Arkoosh, M.R., et al., Increased susceptibility of juvenile Chinook salmon to vibriosis after exposure to chlorinated and aromatic compounds found in contaminated urban estuaries, *J. Aquatic Anim. Health*, 13, 257–268, 2001.

Beaman, J.R., et al., Mammalian immunoassays for predicting the toxicity of malathion in a laboratory fish model, *J. Toxicol. Environ. Health A*, 56, 523–542, 1999.

Bennani, N., and Schimid-Alliana, A., and Lafaurie, M., Immunotoxic effects of copper and cadmium in the sea bass *Dicentrarchus Labrax*, *Immunopharmacol. Immunotoxicol*, 18, 129–144, 1996.

Bowser, D.H., and Frenkel, K., and Zelikoff, J.T., Effects of *in vitro* nickel exposure on the macrophage-mediated immune functions of rainbow trout (*Oncorhynchus mykiss*), *Bull. Environ. Contam. Toxicol.*, 52, 367–373, 1994.

Carlson, E.A., and Li, Y., and Zelikoff, J.T., The Japanese medaka (*Oryzias latipes*) model: applicability for investigating the immunosuppressive effects of the aquatic pollutant benzo[a]pyrene (BaP), *Mar. Environ. Res.*, 54, 1–4, 2002a.

Carlson, E.A., and Li, Y., and Zelikoff, J.T., Exposure of Japanese medaka (*Oryzias latipes*) to benzo(a)pyrene suppresses immune function and host resistance against bacterial challenge, *Aquatic Toxicol*, 56, 289–301, 2002b.

Carlson, E.A., and Li, Y., and Zelikoff, J.T., Benzo(a)pyrene-induced immunotoxicity in Japanese medaka (*Oryzias latipes*): relationship between lymphoid CYP1A activity and humoral immune suppression, *Toxicol. Appl. Pharmacol.*, in press.

Carlson, E.A., and Zelikoff, J.T., The immune system of fish: a target organ of toxicity, in *Toxicology of Fishes*, Di Giulio, R., and Hinton, D., Eds., CRC Press, Boca Raton, FL, in press.

Cleland, G.B., and McElroy, P.J., and Sonstegard, R.A., The effect of dietary exposure to Aroclor 1254 and/or mirex on humoral immune expression of rainbow trout (*Salmo gairdneri*), *Aquatic Toxicol.*, 2, 141–146, 1988.

Dethloff, G.M., and Bailey, H.C., The effects of copper on immune system parameters of rainbow trout (*Oncorhynchus mykiss*), *Vet. Immunol. Immunopathol.*, 17, 1807–1814, 1998.

Duffy, J., et al., Exposure to a coplanar PCB congener differentially alters the immune responsiveness of juvenile and aged fish, *Ecotoxicol.*, 12, 251–259, 2003.

Duffy, J.E., et al., Impact of polychlorinated biphenyls (PCBs) on the immune function of fish: age as a variable in determining adverse outcome, *Mar. Environ. Res.*, 54, 1–5, 2002.

Duffy, J.E., and Li, Y., and Zelikoff, J.T., Immunotoxicity of a coplanar and noncoplanar polychlorinated biphenyl (PCB) congener in a fish model, *Toxicol. Sci.*, 78, 1S, 272, 2004.

Duffy, J.E., and Li, Y., and Zelikoff, J.T., Hepatic CYP1A induction is associated with PCB-induced innate immune dysfunction in a feral teleost fish, in press.

Faisal, M., et al., Evidence of aberration in natural cytotoxic cell activity in *Fundulus heteroclitus* (Pisces: Cyprinodontidae) from the Elizabeth River, *Vet. Immunol. Immunopathol.*, 29, 339–351, 1991a.

Faisal, M., et al., Mitogen induced proliferative responses of lymphocytes from spot (*Leiostomus xanthurus*) exposed to polycyclic aromatic hydrocarbon contaminated environments, *Immunopharmacol. Immunotoxicol.*, 13, 311–327, 1991b.

Feldmann, M., Cell cooperation in the antibody response, in *Immunology*, 5th ed., Roitt, I., Brostoff, J., and Male, D., Eds., Mosby, London, 1998, pp. 139–153.

Hart, L.J., et al., Subacute immunotoxic effects of the polycyclic aromatic hydrocarbon 7, 12-dimethylbenzanthracene (DMBA) on spleen and pronephros leukocytic cell counts and phagocytic cell activity in tilapia (*Oreochromis niloticus*), *Aquatic Toxicol.*, 41, 17–29, 1998.

Holladay, S.D., et al., Benzo[a]pyrene-induced hypocellularity of the pronephros in tilapia (*Oreochromis niloticus*) is accompanied by alterations in stromal and parenchymal cells and by enhanced immune cell apoptosis, *Vet. Immunol. Immunopathol.*, 64, 69–82, 1998.

Hutchinson, T.H., and Field, M.D.R., and Manning, M.J., Evaluation of non-specific immune functions in dab, *Limanda limanda* L., following short-term exposure to sediments contaminated with polyaromatic hydrocarbons and/or polychlorinated biphenyls, *Mar. Environ. Res.*, 55, 193–202, 2003.

Iwama, G., and Nakanishi, T., Eds., *The Fish Immune System: Organism, Pathogen and Environment*, Academic Press, San Diego, 1996, pp. 1–380.

Jacobson, S.V., and Reimschuessel, R., Modulation of superoxide production in goldfish (*Carassius auratus*) exposed to and recovering from sublethal copper levels, *Fish Shellfish Immunol.*, 8, 245–259, 1998.

Karol, M.H., Target organs and systems: methodologies to assess immune system function, *Environ. Health Perspect.*, 106, 533–540, 1998.

Kerkvliet, N.I., Recent advances in understanding the mechanisms of TCDD immunotoxicity, *Int. Immunopharmacol.*, 2, 277–291, 2002.

Khangarot, B.S., and Tripathi, D.M., Changes in humoral and cell-mediated immune responses and in skin and respiratory surfaces of catfish, *Saccobranchus fossilis*, following copper exposure, *Ecotoxicol. Environ. Saf.*, 22, 291–308, 1991.

Lacroix, A., et al., Phagocytic response of macrophages from the pronephros of American plaice (*Hipoglossoides platessoides*) exposed to contaminated sediments from Baie des Anglais, Quebec, *Chemosphere*, 45, 599–607, 2001.

Lemaire-Gony, S., and Lemaire, P., and Pulsford, A., Effects of cadmium and benzo(a)pyrene on the immune system, gill ATPase and EROD activity of European sea bass *Dicentrarchus labrax*, *Aquatic Toxicol.*, 31, 297–313, 1995.

Low, K.W., and Sin, Y.M., Effects of mercuric chloride and sodium selenite on some immune responses of blue gourami, *Trichogaster trichopterus* (Passus), *Sci. Total Environ.*, 214, 153–164, 1998.

Luster, M.I., et al., Development of a testing battery to assess chemical-induced immunotoxicity: National Toxicology Program's guidelines for immunotoxicity evaluation in mice, *Fundam. Appl. Toxicol.*, 10, 2–19, 1988.

Luster, M.I., et al., Use of animal studies in risk assessment for immunotoxicology, *Toxicol.*, 92, 229–243, 1994.

National Research Council, *Guide for the Care and Use of Laboratory Animals*, National Academy Press, Washington, DC, 1996.

O'Halloran, K., and Ahokas, J.T., and Wright, P.F., Response of fish immune cells to *in vitro* organotin exposures, *Aquatic Toxicol.*, 40, 141–156, 1998.

O'Neill, J.G., Effect of intraperitoneal lead and cadmium on the humoral immune response of *Salmo trutta*, *Bull. Environ. Contam.*, 27, 42–48, 1981.

Palchaudhuri, S., et al., Cytotoxic and cytoprotective effects of selenium on bluegill sunfish (*Lepomis macrochirus*) phagocytic cells *in vitro*, *Bull. Environ. Contam. Toxicol.*, 67, 672–679, 2001.

Powell, D.B., et al., Immunocompetence of juvenile chinook salmon against *Listonella anguillarum* following dietary exposure to Aroclor 1254, *Vet. Immunol. Immunopathol.*, 22, 285–295, 2003.

Regala, R.P., et al., The effects of tribuyltin (TBT) and 3, 3, 4, 4, 5-pentachlorbiphenyl (PCB 12) mixtures on antibody responses and phagocyte oxidative burst activity in channel catfish, *Ictalurus punctatus*, *Arch. Eniviron. Contam. Toxicol.*, 4, 386–391, 2001.

Reynaud, S., and Duchiron, C., and Deschaux, P., 3-Methylcholanthrene increases phorbol 12-myristate 13-acetate-induced respiratory burst activity and intracellular calcium levels in common carp (*Cyprinus carpio* L.) macrophages, *Toxicol. Appl. Pharmacol.*, 175, 1–9, 2001.

Rice, C.D., and Weeks, B.A., Influence of tributyltin on *in vitro* activation of oyster toadfish macrophages, *J. Aquatic Anim. Health*, 1, 62–68, 1989.

Rice, C.D., and Schlenk, D., Immune function and cytochrome P4501A activity after acute exposure to 3, 3′, 4, 4′, 5-pentachlorbiphenyl (PCB 126) in channel catfish, *J. Aquatic Anim. Health*, 7, 195–204, 1995.

Robohm, R.A., Paradoxical effects of cadmium exposure on antibacterial antibody responses in two fish species: Inhibition in cunners (*Tautogolabrus adspersus*) and enhancement in striped bass (*Morone saxatilis*), *Vet. Immunol. Immunopathol.*, 12, 251–262, 1986.

Rougier, F., et al., Copper and zinc exposure of zebrafish, *Brachydanio rerio* (Hamilton-Buchaman): effects in experimental Listeria infection, *Ecotoxicol. Environ. Saf.*, 34, 134–140, 1996.

Sanchez-Dardon, J., et al., Immunomodulation by heavy metals tested individually or in mixtures in rainbow trout (*Oncorhynchus mykiss*) exposed *in vivo*, *Vet. Immunol. Immunopathol.*, 18, 1492–1497, 1999.

Shamberger, R.J., Toxicity of selenium, in *Biochemistry of Selenium*, Shamberger, R.J., Ed., Plenum, New York, 1983, pp. 185–206.

Smith, D.A., et al., The hemolytic plaque-forming cell assay in tilapia (*Oreochromis niloticus*) exposed to benzo(a)pyrene: Enhanced or depressed plaque formation depends on dosing schedule, *Toxicol. Methods*, 9, 57–70, 1999.

Spitsbergen, J.M., et al., Interactions of 2, 3, 7, 8-tetrachlordibenzo-p-dioxin (TCDD) with immune responses of rainbow trout, *Vet. Immunol. Immunopathol.*, 12, 263–280, 1986.

Spitsbergen, J.M., et al., Effects of 2, 3, 7, 8-tetrachloro-p-dioxin (TCDD) or Aroclor 1254 on the resistance of rainbow trout, *Salmo gairdneri* Richardson, to infectious haematopoietic necrosis virus, *J. Fish Dis.*, 11, 78–83, 1988.

Thuvander, A., Cadmium exposure of rainbow trout, *Salmo gairdneri* Richardson: Effects on immune functions, *J. Fish Biol.*, 35, 521–529, 1989.

Thuvander, A., and Carlstein, M., Sublethal exposure of rainbow trout (*Oncorhynchus mykiss*) to polychlorinated biphenyls: effects on the humoral immune response to *Vibrio anguillarum.*, *Fish Shellfish Immunol.*, 1, 77–86, 1991.

Viales, G., and Calamari, D., Immune response in rainbow trout *Salmo gairdneri* after long-term treatment with low levels of Cr, Cd, and Cu, *Environ. Pollut.*, 35, 247–257, 1984.

Viola, A., and Pregnolato, G., and Albergoni, V., Effects of *in vitro* cadmium exposure on natural killer (NK) cells of catfish, *Ictalurus melas*, *Fish Shellfish Immunol.*, 6, 167–172, 1996.

Voccia, I., et al., *In vivo* effects of cadmium chloride on the immune response and plasma cortisol of rainbow trout (*Oncorhynchus mykiss*), in *Modulators of Immune Responses: The Evolutionary Trail*, Stolen, J.S., Fletcher, T.C., Bayne, C.J., Secombes, C.J., Zelikoff, J.T., Twerdok, L.E., and Anderson, D.P., Eds., SOS Publications, Fair Haven, NJ, 1996, pp. 547–555.

Walczak, B., and Blunt, B., and Hodson, P., Phagocytic function of monocytes and hematological changes in rainbow trout injected intraperitoneally with benzo(a)pyrene and benzo(a)anthracene, *J. Fish Biol.*, 31 (Suppl. A), 251–253, 1987.

Weeks, B.A., and Warriner, J.E., Effects of toxic chemicals on macrophage phagocytosis in two estuarine fishes, *Mar. Environ. Res.*, 14, 327–334, 1984.

Weeks, B.A., et al., Functional evaluation of macrophages in fish from a polluted estuary, *Vet. Immunol. Immunopathol.*, 12, 313–320, 1986.

Weeks, B.A., and Warinner, J.E., and Mathews, E.S., Influence of toxicants on phagocytosis, pinocytosis and melanin accumulation by fish macrophages, *Aquatic Toxicol.*, 11, 395–438, 1988.

Weeks, B.A., et al., Immunological biomarkers to assess environmental stress, in *Biomarkers of Environmental Contamination*, McCarthy, J.F., and Shugart, L.R., Eds., Lewis Publishers, Boca Raton, 1990, pp. 193–230.

Wester, P.W., and Vethaak, A.D., and van Muiswinkel, W.B., Fish as biomarkers in immunotoxicology, *Toxicology*, 86, 213–232, 1994.

Zelikoff, J.T., Metal pollution-induced immunomodulation in fish, *Annu. Rev. Fish Dis.*, 2, 305–325, 1993.

Zelikoff, J.T., Fish immunotoxicology, in *Immunotoxicology and Immunopharmacology*, Dean, J.H., Luster, M.I., Munson, A.E., Kimber, I., Eds., Raven Press, New York, 1994, pp. 71–89.

Zelikoff, J.T., et al., Immunotoxicity of low level cadmium exposure in fish: alternative animal models for immunotoxicological studies, *J. Toxicol. Environ. Health*, 45, 235–248, 1995.

Zelikoff, J.T., et al., Assays of reactive oxygen intermediates and antioxidant enzymes: potential biomarkers for predicting the effects of environmental pollution, in *Techniques in Aquatic Toxicology*, Ostrander, G.K., Ed., Lewis Publishers, Boca Raton, FL, 1996a, pp. 287–306.

Zelikoff, J.T., et al., Heavy metal-induced changes in antioxidant enzymes and oxyradical production by fish phagocytes: application as biomarkers for predicting the immunotoxic effects of metal-polluted aquatic environments, in *Modulators of Immune Responses: A Phylogenetic Approach*, vol. 2, Stolen, J., Zelikoff, J.T., Twerdok, L.E., Anderson, D., Bayne, C., Secombes, C., and Fletcher, T., Eds., SOS Publications, Fair Haven, NJ, 1996b, pp. 135–148.

Zelikoff, J.T., Biomarkers of immunotoxicity in fish and other non-mammalian sentinel species: Predictive value for mammals? *Toxicology*, 129, 63–71, 1998.

Zelikoff, J.T., et al., Immunotoxicity biomarkers in fish: Development, validation and application for field studies and risk assessment, *Hum. Ecotoxicol. Risk Assessment*, 8, 253–263, 2002.

CHAPTER **5**

Immunotoxicology in Marine Mammals

Sylvain De Guise

CONTENTS

5.1 INTRODUCTION

Oceans have long been considered endless sinks where domestic and industrial wastes could be discarded. It is not surprising to now measure significant concentrations of such contaminants in the water, sediments, and inhabitants of the marine environment. There is special concern for stable compounds, such as organochlorines, which often persist long after their production and use have been banned. Marine mammals are usually at the top of complex food chains, integrating the pollutants in the ecosystem and accumulating large amounts of the stable contaminants. The persistence of the contaminants, as well as their lipophilic characteristics, is responsible for their biomagnification in the food chain. Organochlorines (PCBs, DDT, etc.), heavy metals, and polycyclic aromatic hydrocarbons (PAHs) have drawn attention because of their abundance and/or known toxicity. These pollutants have been found worldwide in marine mammal tissues, including Europe (Holden and Marsden, 1967; Reijnders, 1980; Baumann and Martinsen, 1983), North America (Gaskin et al., 1971; Addison et al., 1973; Muir et al., 1990), South America (Gaskin et al., 1974), Asia (Taruski et al., 1975), Arctic (Addison and Smith, 1974; Born et al., 1981), and the

Antarctic (Sladen et al., 1966). Not surprisingly, organochlorines have also been found in humans, and are most abundant in remote maritime populations, such as in the Canadian Arctic that rely heavily on seafood and marine mammals for their subsistence (Dewailly et al., 1989; Dewailly et al., 1993; Mulvad et al., 1996). It is now widely accepted that pollution has become a global problem, and that there are no more pristine environments. Among the most studied effects of organochlorines are their endocrine disruptor (including immunotoxic) effects in laboratory animals and wildlife (Brouwer et al., 1999; Vos et al., 2000). Nevertheless, the biological significance and potential health effects of these environmental contaminants are not fully understood.

Catastrophic viral epidemics have recently affected several populations of seals (Geraci et al., 1982; Osterhaus et al., 1988), porpoises (Kennedy et al., 1991), and dolphins (Domingo et al., 1990; Domingo et al., 1992), all severely contaminated by industrial pollutants. Influenza virus or morbilliviruses were undoubtedly the direct causes for the deaths of thousands of animals. However, it has been well demonstrated in experimental animals that chronic exposure to PCBs increases susceptibility to a wide range of viral infections (Friend and Trainer, 1970; Koller, 1977; Imanishi et al., 1980). Accordingly, a possible immunosuppressive role of organohalogens to explain the severity of the cetacean and pinniped epidemics has been suggested (Eis, 1989). Similarly, contaminant-induced immunosuppression has been suggested as an explanation for the high incidence, severity, and diversity of lesions often caused by opportunistic and mildly pathogenic bacteria that were found on postmortem examination of the endangered, small isolated population of beluga whales (*Delphinapterus leucas*) from the St. Lawrence Estuary (Martineau et al., 1988; De Guise et al., 1995b). It was noted that the blubber concentrations of organochlorines in harbor porpoises that had died of infectious diseases were higher than in those individuals that died of physical trauma, suggesting that organochlorines may affect immune functions and predispose to infectious diseases (Jepson et al., 1999).

Although many contaminants of the marine environment (such as organochlorines, heavy metals, and polycyclic aromatic hydrocarbons) are well characterized as immunotoxicants in laboratory rodents, the demonstration of immunotoxic effects in marine mammals exposed to these environmental pollutants represents a significant challenge. The main reasons include the relatively limited immunologic database presently existing, relatively limited assay development and reagent availability to evaluate immune function in marine mammals, genetic diversity naturally occurring in outbred populations such as marine mammals, and logistical and ethical considerations when working with marine mammals. In this chapter, a review of current knowledge in marine mammal immunology and immunotoxicology is presented.

5.2 MARINE MAMMAL IMMUNOLOGY

Marine mammal immunology is a relatively new, rapidly evolving discipline. Efforts have been invested toward a better understanding of marine mammal immunology as a part of the medical care necessary for animals kept under human care as well as for a better understanding of the health of free-ranging marine mammals. A better understanding of the effects of environmental pollutants on the health of marine mammals roaming in polluted waters has been a major impetus for the development of assays to quantitatively assess immune functions, a known target for several classes of pollutants. The recent deaths of thousands of marine mammals inhabiting polluted waters from viral epizootics (see above) brought some sense of urgency to the question. As a consequence, several laboratories have contributed to the fields of fundamental as well as applied immunology for the development of reagents and standardization of assays to evaluate various parts of the marine mammal immune system such as the humoral and cell-mediated arms of the innate and adaptive immunity. Nevertheless, it is important to understand the complexity of the immune system and its functions. No single assay can pretend to evaluate the immune system as a whole, but different assays can evaluate different components, mechanisms, or functions. The following section describes the various components of the mammalian immune system and briefly describes the assays and reagents available for marine mammals.

The mammalian immune system is composed of two branches: the innate or natural and the acquired or adaptive immune system. The main differences between the two arms consist of the specificity of the acquired immune system for very precise antigens and its "immunological memory" (Goldsby et al., 2003). Each lymphocyte is in fact able to recognize and respond to only one specific antigen, whereas the cells of the innate immune system recognize broad classes of antigens (such as lipopolysaccharide, which is present on all Gram-negative bacteria). Immunological memory represents the ability of the acquired immune system to respond more rapidly and more intensely to an antigen encountered for the second time, whereas the innate immune system always responds to an antigen as if encountered for the first time. Both the acquired and innate immune systems have been classically divided into two arms, the cell-mediated immunity and the humoral immunity, with intricately related functions. The fast-acting innate immune system that recognizes a broad specificity of antigens, and which is therefore ideally suited as a first line of defense, combines with the highly efficient acquired immune system, which needs time (7 to 10 days) to reach its peak activity, to make up the highly sophisticated mammalian immune system.

Immunoglobulins, or antibodies, are soluble proteins used by the immune system to recognize and help to eliminate antigens. Immunoglobulins were at least partially purified and characterized in several species of marine mammals (Boyden and Gemeroy, 1950; Travis and Sanders, 1972; Nash and Mack, 1977; Cavagnolo, 1979; Andresdottir et al., 1987; Suer et al., 1988). More recently, King et al., produced monoclonal antibodies to seal immunoglobulin subclasses (King, et al., 1993), allowing their quantification (King et al., 1994). Similarly, killer whale and sea otter immunoglobulin-specific polyclonal antibodies were generated and used to quantify serum immunoglobulin concentrations in those species (Taylor et al., 2002). Monoclonal antibodies to dolphin IgG were also generated and used for assessment of total immunoglobulins as well as for the detection of antigen-specific immunoglobulins in wild dolphins (Beck and Rice, 2003).

Other molecules important in signaling for the immune system have been studied in marine mammals. C-reactive protein (C-RP) and interleukin-6 (IL-6) were characterized in harbor seals and killer whale, respectively, and used for the generation of antibodies which were in turn used in the development of immunoassays to quantify the levels of those molecules in biological samples (Funke et al., 1997; Funke et al., 2003).

Lymphocytes are the central players for the acquired immune system. They consist of morphologically indistinguishable populations of T and B lymphocytes, which account for cell-mediated and humoral immunity, respectively. Since their functions differ so drastically, considerable efforts have been expended toward the generation of reagents to identify lymphocyte subsets based on their expression of surface differentiation antigens using monoclonal antibodies. The identification of the different cell subsets proved to be a useful diagnostic and prognostic tool for diseases such as human AIDS (Fahey et al., 1984). In view of the current efforts to classify morphologically indistinguishable lymphocyte subpopulations in several species, and of the relative success obtained with some cross-reactive monoclonal antibodies (Jacobsen, et al., 1993), attempts were first made to identify subclasses of lymphocytes in cetaceans using cross-reactive monoclonal antibodies. Cross-reactivity in marine mammals was demonstrated using antibodies to human major histocompatibility complex (MHC) class II in bottlenose dolphins (Romano et al., 1992), as well as when using antibodies to bovine CD2, T-cell receptor (TCR) gamma-delta, MHC class I and class II, human CD4, and mouse IgM in beluga whales (De Guise et al., 1997a). These studies used flow cytometry to quantify lymphocyte subpopulations from blood samples. Immunohistochemistry on paraffin-embedded tissues demonstrated the cross-reactivity of antibodies to human CD3, IgG, lysozyme, and MHC class II, as well as bovine S100 protein in striped dolphin lymph nodes and liver lesions (Jaber et al., 2003a), as well as (with the exception of lyzozyme) common dolphin liver lesions (Jaber et al., 2003b). Similarly, antibodies to human CD3, bovine and canine MHC class II, and an equine pan-leucocyte marker labeled paraffin-embedded lymphoid tissues from harbor porpoises (Beineke et al., 2001).

Further progress was made when species-specific monoclonal and polyclonal antibodies to marine mammal cell surface antigens were produced and characterized. Romano et al. (1992) produced an antiserum to dolphin immunoglobulins for further use in identification of B lymphocytes. Monoclonal antibodies were also generated against bottlenose dolphin peripheral blood leucocytes. This effort resulted in the characterization of antibodies to a homologue of CD2 (recognizing T lymphocytes), CD19 and CD21 (recognizing B lymphocytes), CD45R (a marker of T cell activation), and β2-integrin (a marker of activation of neutrophils, monocytes, and B cells) specific for cetacean cells (De Guise et al., 1998a; De Guise et al., 2002; De Guise et al., 2004). Those antibodies are currently used for a number of applications, including the recent identification of dendritic cells in the skin of bottlenose dolphins using immunohistochemistry (Zabka and Romano, 2003).

Several components of the immune system have recently been examined at the molecular level, including beluga whale CD4 (Romano et al., 1999), bottlenose dolphin IgM (Lundqvist et al., 2002), beluga whale and grey seal IL-2 (St-Laurent et al., 1999), as well as beluga whale IL-6 (St-Laurent and Archambault, 2000), IL-1β, and tumor necrosis factor alpha (Denis and Archambault, 2001). The major histocompatibility complex has been the target of cloning efforts in the beluga whale (Murray and White, 1998), California sea lion (Bowen et al., 2002), and southern elephant seal (Slade, 1992). Those efforts are likely to continue and the full extent of their benefits will probably become more apparent in view of the several genome projects currently ongoing.

Although it is important to recognize and differentiate cells and the molecules by which they function, there is a need to measure the ability of those cells to function adequately. Lymphocyte proliferation represents one of the events following activation. The *in vitro* mitogen-induced lymphoblastic transformation or blastogenesis measures, through the incorporation in the nucleus of dividing cells of labeled amino acids or synthetic analogues, the ability of lymphocytes to proliferate in response to a nonspecific polyclonal stimulation (usually using mitogens). This test is used to mimic the activation of the immune response after an antigenic stimulation *in vivo* (Kristensen et al., 1982). This assay has been performed previously with peripheral blood from a variety of species of pinnipeds and cetaceans (Mumford et al., 1975; Colgrove, 1978; Romano et al., 1992; de Swart et al., 1993; Ross et al., 1993; DiMolfetto-Landon et al., 1995; Erickson et al., 1995; Lahvis et al., 1995; De Guise et al., 1996a; Shaw, 1998). Over the years, there have been trends toward standardization and validation of methods and reagents. Methods have also been developed to evaluate an earlier and more subtle phenomenon following the activation of T lymphocytes, the expression of the receptor for interleukin-2, IL-2R, in both pinnipeds and cetaceans (DiMolfetto-Landon et al., 1995; Erickson et al., 1995).

Neutrophils represent the first line of defense of the innate immune system against invading agents, especially bacteria (van Oss, 1986). Ingestion of foreign material through the process of phagocytosis, and destruction of phagocytized particles through a series of biochemical events known as the respiratory burst, are the major functions of neutrophils (Tizard, 1992). Quantitative assays were developed to measure phagocytosis and respiratory burst in beluga whales using flow cytometry (De Guise et al., 1995a), and in bottlenose dolphins using microscopy and spectrophotometry, respectively (Noda et al., 2003). Respiratory burst was also evaluated in bottlenose dolphins using chemiluminescence, along with documentation of the enzymes involved (Itou et al., 2001). The flow cytometric assay has the advantage of examining large number of cells in short periods of time (compared to microscopy) and determining functions at the individual cell level (as opposed to spectrophotometry, which examines whole populations of cells).

NK cells are classically referred to as a heterogeneous population of large granular lymphocytes whose activity is mainly directed against tumor cells and virus-infected cells (O'Shea and Ortaldo, 1992). Unlike T cells, NK cells do not require an antigen to be presented through the MHC, do not need previous sensitization, and represent a first line of defense in early phases of virus infection (O'Shea and Ortaldo, 1992). Unlike most cells of the immune system, which are defined by their morphology or phenotype, NK cells have remained elusive because they are defined by their function. NK cell activity is typically measured as the killing of target (usually tumor cell lines or virally infected cells), and assays have been developed and standardized in harbor seals (Ross et al., 1996) and beluga whales (De Guise et al., 1997b).

More details have been uncovered about the maturation of the immune functions of developing marine mammals. The study of harbor seal pups showed that lymphocyte proliferation and total IgG levels in pups were low at birth and higher at the end of lactation, while the same functions were reduced in mothers at the end of gestation (Ross et al., 1993). A more detailed study suggested that colostrum intake represents an important source of transfer of antibodies in harbor seal pups, although newborn seal pups developed high specific antibody titers after immunization with rabies vaccine (Ross, et al., 1994). Not surprisingly, circulating IgG concentrations increased between birth and weaning (Hall, et al., 2003). In gray seals, lymphocyte proliferation was stable from week 2 to week 3 postweaning, and then increased significantly and reached a plateau 1 week later, while phagocytosis was shown to increase between week 2 and week 5 postweaning, at which time it reached a plateau (Lalancette et al., 2003).

Overall, a battery of reagents and assays has been developed to assess immune functions in marine mammals. Those assays will be important, as they represent the basis for any investigation of the effects of toxicants on immune functions in marine mammals.

5.3 MARINE MAMMAL IMMUNOTOXICOLOGY

Immunotoxicology is a relatively new scientific discipline with the aims to detect, quantify, and interpret direct or indirect alterations of the immune system that occur as a result of exposure to chemicals, pharmaceuticals, recombinant biologicals, or environmental and occupational pollutants (Burleson and Dean, 1995). The immune system is a complex network of cells with diverse functions, communicating through a wide array of messenger molecules, which has evolved to protect the host from potentially pathogenic agents including viruses, bacteria, parasites, fungi, neoplastic, and nonself cells (Roitt et al., 1998). Immunotoxicology studies deal with stimulatory or suppressive immune alterations, their mechanisms, and their resulting effects on susceptibility or duration of infectious, allergic, or autoimmune diseases (Burleson and Dean, 1995).

Considerable efforts have been made to standardize and validate methods and assays. A comprehensive testing panel composed of two tiers has been developed and validated to characterize immune alterations following *in vivo* chemical exposure in mice (Luster et al., 1988). Further studies demonstrated the wide diversity in the ability of assays to predict immunotoxicity when performed individually (Luster et al., 1992). Nevertheless, several combinations of two assays could predict immunotoxicity with a percentage concordance (sum of specificity and sensitivity) of 90% or more (Luster et al., 1992). Further studies comparing the relationship between the immune function and host resistance showed a good correlation between changes in the immune tests and altered host resistance, with no instances where host resistance was altered without affecting an immune test (Luster et al., 1993). These studies also showed that no single immune test could be identified that was fully predictive for altered host resistance, although several assays were relatively good individual indicators (Luster et al., 1993).

Although significant progress has been made in chemical-induced immunotoxicity in rodents, the predictive value of those studies for risk assessment in other species remains questionable. For this and other reasons, efforts have been made to develop models using *in vitro* exposure. Although such models do not allow evaluation of the pharmacokinetics of the compounds tested (including absorption, distribution, excretion, and long-term accumulation), they have several advantages. They allow the evaluation of the direct effects of a compound on the immune system, with a relatively rapid detection of immunotoxic effects (compared to *in vivo* systems), at a much lower cost, and with fewer animals used (Munson and LeVier, 1995). These reasons have prompted an attempt to validate and use *in vitro* studies for regulatory purposes (Prochazkova, 1993; Jackson, 1998). But most importantly, *in vitro* exposures allow studies in species such as marine mammals, where obvious logistical and ethical considerations would make experimental *in vivo* studies very difficult, in addition to allowing comparisons among species. Nevertheless, several of the assays for which predictive values were established *in vivo* could not be performed if the animals used have not been experimentally exposed or immunized previously.

In view of suspected contaminant-induced immune suppression underlying health problems and epizootics in marine mammals (see above), and of the abundant literature on the immunotoxicity of several chemicals in laboratory species, we review the evidence for immunotoxic effects of chemicals in marine mammals. The direct determination of the effects of environmental contaminants on the immune system of wild marine mammals is difficult because of logistical and ethical considerations. Nevertheless, various approaches have been adopted, including *in vitro* exposures and animal models, in addition to semi-field and field studies, yielding different levels of certainty.

5.3.1   Coincidental Evidence

Several studies have suggested that contaminants could have immunosuppressive effects on marine mammals. Field studies in wild populations of marine mammals are logistically difficult. As with humans, there is a plethora of confounding factors, such as age, sex, habitat, and all the other "lifestyle" factors (including favorite food, preferred habitat, etc.) that may vary from one individual to another. Nevertheless, they represent the best way to fully evaluate the effects of chronic exposure to environmental contaminants on a population. We examine several instances in which exposure to environmental contaminants was correlated to changes in immune functions. Note that given the difficulty in assessing the details of exposure and the temporal relationship of the onset of the immunological changes measured, the relationship between exposure and effects on the immune system upon natural exposure is coincidental at best.

In a study of a relatively low number of free-ranging bottlenose dolphins in Florida, lymphocyte proliferation was negatively correlated to blood concentrations of PCBs and DDT using regression analysis, that is, the animals with highest concentrations of orgnochlorines had the lowest lymphocyte proliferation (Lahvis et al., 1995). Similarly, Pacific harbor seal pups with higher concentrations of non- and mono-ortho coplanar PCBs and p,p'-DDE had lower lymphocytes proliferation (Shaw, 1998). Also, in a study of Northern fur seals and Steller sea lions in Alaska, the mitogen-induced proliferation of blood lymphocytes was negatively correlated with whole blood levels of OC contaminants in both species (Beckmen et al., 2002). Additionally, it was found that pups of primiparous dams, having the highest blood levels of OC contaminants, had significantly lower antibody production in response to vaccination with tetanus toxoid compared to pups of old dams with low concentrations of OCs in the blood (Beckmen et al., 2003). Live-captured harbor seal pups off the cost of British Columbia showed a positive correlation between B- and T-lymphocyte proliferation and blubber organochlorine concentrations, that is, the greater the organochlorines, the greater the lymphocyte proliferation (Levin et al., 2003a). Although interesting and suggestive of possible problems associated with long-term exposure to environmental pollutants in marine mammals, those correlations do not provide cause-and-effect relationships.

5.3.2   *In Vitro* Immunotoxicology

In vitro exposures have been used to assess the direct toxicity of chemicals on cells of the immune system in species, such as marine mammals, for which it is not practical to perform *in vivo* exposure studies for ethical, logistical, and economic reasons. *In vivo* exposures also allow for the comparison of susceptibility of different species to the immunotoxic effects of chemicals.

In vitro exposure of Arctic beluga whale immune cells to heavy metals showed that lymphocyte proliferation was significantly reduced upon exposure to 10^{-5} M $HgCl_2$ and $CdCl_2$, but not to lower concentrations (De Guise et al., 1996b). It is interesting to note that these concentrations are within the range of mercury and cadmium found in tissues of wild belugas (De Guise et al., 1996b). The immunotoxicity of mercury was also documented in juvenile grey seals, in which phagocytosis and lymphocyte proliferation were both decreased upon exposure to 10^{-5} M, with slight changes associated with the maturation of the different function (Lalancette et al., 2003). In similar *in vitro* experiments, exposure of Arctic beluga lymphocytes to 20 ppm or more PCB 138 and 50 ppm or more p,p′-DDT, significantly reduced their potential for proliferation, whereas PCB 153, 180, 169, and p,p′-DDE did not when similar concentrations were tested (De Guise et al., 1998b). Interestingly, PCB congeners at concentrations that had no effect on Arctic beluga lymphocyte proliferation when tested individually (5 ppm) were found to significantly reduce proliferation when three of them were mixed together, suggesting the possibility of synergistic interactions between individual chemicals (De Guise et al., 1998b) at concentrations that are well within the range of those measured in the wild in St. Lawrence beluga whale blubber (Muir et al., 1990). Similar studies demonstrated the relatively higher toxicity of butyltins, particularly dibutyltin and tributyltin, compared to the lack of toxicity of coplanar PCBs 77, 126, and 169 on Con-A–induced T-lymphocyte proliferation in several species of marine mammals including Dall's porpoise, bottlenose dolphin, California sea lion, and large seal (*Phoca larga*) upon *in vitro* exposure (Nakata et al., 2002). Comparison of the concentrations of chemical residues in the tissues of Dall's porpoises with those modulating lymphocyte proliferation *in vitro* resulted in concern for butyltin compounds, the tissue concentrations of which were higher than those inducing effects experimentally (Nakata et al., 2002). In the same study, exposure of bottlenose dolphin lymphocytes to a mixture of dibutyltin and PCB 77 or PCB 169, at concentrations at which they did not induce effects individually, resulted in significant reductions in lymphocyte proliferation, suggesting synergistic effects (Nakata et al., 2002). Using similar *in vitro* exposure studies in harbor seals and grey seals, an Arochlor mix was shown to modulate phagocytosis, respiratory burst, and cytotoxicity, the latter two being more severely affected in harbor seals than in grey seals (Hammond et al., 2003). Ongoing studies of the interactions of organochlorines in mixtures in different species of marine mammals using different immune functions suggest marked differences between species, which cannot be predicted using the mouse model, in addition to differences in toxicity for different assays and variable interactions in different species/assays (Levin et al., 2003b; Mori, et al., 2003). For example, it appears that phagocytosis in bottlenose dolphins and beluga whales was modulated specifically by noncoplanar PCBs, which are usually considered of low toxicity (Levin et al., 2004).

Overall, the results of these *in vitro* assays demonstrated the relative susceptibility of cells from different species of marine mammals to the toxicity of chemicals to which they are exposed. In addition, these experiments could demonstrate the interactions of chemicals when in mixtures, and the possibility of increased risk of effects on exposure to chemicals in mixtures. Although *in vitro* exposure only provides information on the direct toxicity of chemicals on immune cells, this information, especially in combination with exposure data (such as tissue concentrations of the chemicals of concern) in the species of interest may help focus further studies for more accurate risk assessment.

5.3.3 Animal Models of Marine Mammal Immunotoxicology

Again, *in vivo* exposures in marine mammals are difficult for ethical, logistical, and economic reasons. Whereas *in vitro* exposures allow the evaluation of the direct toxicity to immune cells, they do not allow for the evaluation of phenomena such as absorption, metabolism, and excretion. Animal models, when appropriate, can provide an intermediate step for the assessment of toxicity.

Animal models were developed to simulate exposure to complex, environmentally relevant mixtures of contaminants and evaluate health effects resulting from such exposure. In one of those models, rats were fed highly contaminated St. Lawrence beluga whale blubber versus much less contaminated Arctic beluga blubber (Lapierre et al., 1999). Rats fed a St. Lawrence beluga blubber diet or a mixture of Arctic and St. Lawrence beluga blubber diet were not different from control rats fed a diet containing Arctic beluga blubber (Lapierre et al., 1999). In a similar study using mice, peritoneal macrophage phagocytosis and spleen cell antibody production were suppressed by all diets containing beluga blubber, suggesting that the relatively low concentrations of pollutants present in Arctic beluga blubber were sufficient to account for effects on the immune system (Fournier et al., 2000). Taken together, those two studies highlight the importance of the choice of the model. Similarly, rats fed with oil extracted from highly contaminated Baltic Sea fish had impaired cellular immune response, including decreased mitogen-induced lymphocyte proliferation in spleen and thymus, lower CD4:CD8 ratios in thymus, and lower NK cell activity compared to those fed oil extracted from much less contaminated Atlantic Ocean fish (Ross et al., 1997).

While not perfect, those models can be useful in determining the *in vivo* effects of exposure to complex, environmentally relevant mixtures of toxicants similar to those marine mammals that would be exposed to. Nevertheless, one cannot exclude the possibility that differences exist in absorption, metabolism, or excretion of those pollutants that would account for differences between the animal model and the marine mammal species for which the model was established.

5.3.4   Marine Mammal Experimental Immunotoxicology

The obvious best strategy to assess the toxicity of chemicals on marine mammals is to experimentally expose marine mammals and measure the effects. Because of ethical, logistical, and economic concerns, this is not always possible.

In a hallmark study in marine mammal immunotoxicology, harbor seals were kept in semi-field conditions and experimentally fed a diet consisting of fish from either the contaminated Baltic Sea or the much less contaminated North Atlantic Ocean. Seals were sampled periodically over time for assessment of their immune functions. Exposure to fish from the polluted Baltic Sea resulted in impaired natural killer (NK) cell activity, T-lymphocyte function, and delayed-type hypersensitivity compared to seals fed the less contaminated North Atlantic fish (de Swart et al., 1994; Ross et al., 1995; Ross et al., 1996).

This probably represents the best experimental study design to investigate the effects of exposure to environmental contaminants in marine mammals through food consumption. Nevertheless, the perinatal transfer of contaminants (through the placenta and milk) that occurs in nature and that affects the highly susceptible developing immune system is not accounted for in such a semi-field experiment. Therefore, this study probably underestimated the "real-life" effects of pollutants on marine mammals.

5.4   RESEARCH NEEDS

5.4.1   Marine Mammal Immunology

Significant advances have been made in the field of marine mammal immunology in the last decade. Nevertheless, there is still a paucity of reagents and assays when compared to the traditional rodent models. There is clearly a need to generate more species-specific reagents for marine mammals, and to validate the use of cross-reacting reagents. The generation of new reagents and validation of new assays will certainly be useful to elucidate the intricacies of the immune system of marine mammals, including differences between species. Nevertheless, the elusiveness of the species of interest, combined with the difficulty of performing invasive procedures, will remain significant barriers to efforts aimed at unraveling the details of marine mammal immunology.

5.4.2   Models for Marine Mammal Immunotoxicology

Despite the intrinsic difficulties in using marine mammals as model species, with all the ethical, logistical, and economic constraints, these animals are nevertheless of great interest for their inherent social value (e.g., who would care if environmental pollutants endangered mosquitoes?). In addition, their relatively long life span, as well as their position at the top trophic level of complicated food chains, makes them a relatively good model for the assessment of chronic exposure to environmental pollutants in humans. Several lines of evidence point to the effects of pollutants in marine mammals with different degrees of certainty. In view of the multiple deaths associated with infectious diseases in the last few decades, there is reasonable evidence for concern regarding the deleterious health effects that these contaminants, which are often persistent in the environment, have on marine mammals in particular, and on whole ecosystems in general, including human health.

Efforts should be made to take advantage of improvements in the field of marine mammal immunology and immunotoxicology in general to continue to investigate and monitor the effects of pollutants in marine mammals living in different parts of the world through studies ranging from *in vitro* to *in vivo* to document effects and unravel the mechanisms through which they occur. This will allow the development of better tools for more accurate risk assessment and management.

5.5   SUMMARY

Taken together, the available data strongly suggest that environmental contaminants with known immunotoxic effects in laboratory animals and present in high concentrations in tissues of marine mammals may have immunotoxic effects in marine mammals in their natural habitats. Continued documentation and monitoring of those effects and investigations on the mechanisms through which they occur will provide data for more accurate assessment of the risk associated with exposure to pollutants, possibly resulting in improved management of resources to better protect not only marine mammals but entire ecosystems.

ACKNOWLEDGMENTS

I am grateful for the people in my laboratory and colleagues who inspire and challenge me in my studies on marine mammal immunotoxicology. Funding from the U.S. Environmental Protection Agency (EPA) STAR program (R-82836101-0) is instrumental to ongoing work in my laboratory, and allows a significant focus of my work on marine mammal immunotoxicology. Initial funding from the University of Connecticut Research Foundation is also acknowledged.

REFERENCES

Addison, R.F., et al., Variation of organochlorine residue levels with age in Gulf of St. Lawrence harp seals (Pagophilus groenlandicus), *J. Fish Res. Board Can.*, 30, 595–600, 1973.

Addison, R.F., and Smith, T.G., Organochlorine residue levels in Arctic ringed seals: variation with age and sex, *Oikos*, 25, 335–337, 1974.

Andresdottir, V., et al., Subclasses of IgG from whales, *Dev. Comp. Immunol.*, 11, 801–806, 1987.

Baumann, O.E., and Martinsen, K., Persistent organochlorine compounds in seals from Norwegian coastal waters, *Ambio*, 12, 262–264, 1983.

Beck, B.M., and Rice, C.D., Serum antibody levels against select bacterial pathogens in Atlantic bottlenose dolphins, Tursiops truncatus, from Beaufort NC USA and Charleston Harbor, Charleston, SC, USA, *Mar. Environ. Res.*, 55, 161–79, 2003.

Beckmen, K.B., et al., Organochlorine contaminant exposure and associations with hematological and humoral immune functional assays with dam age as a factor in free-ranging northern fur seal pups (Callorhinus ursinus), *Mar. Pollut. Bull.*, 46, 594–606, 2003.

Beckmen, K.B., et al., Effects of Environmental Contaminants on Immune Function and Health in Free-Ranging Pinnipeds in Alaska, in Proceedings, 23rd annual meeting of the Society for Environmental Toxicology and Chemistry, Salt Lake City, UT, 2002.

Beineke, A., et al., Immunohistochemical investigation of the cross-reactivity of selected cell markers from various species for characterization of lymphatic tissues in the harbour porpoise (Phocoena phocoena), *J. Comp. Pathol.*, 125, 311–7, 2001.

Born, E.W., Kraul, I., and Kristensen, T., Mercury, DDT and PCB in the Atlantic warlus (Odobenus rosmarus rosmarus) from the Thule district, North Greenland, *Arctic*, 34, 255–260, 1981.

Bowen, L., et al., Molecular characterization of expressed DQA and DQB genes in the California sea lion (Zalophus californianus), *Immunogenetics*, 54, 332–347, 2002.

Boyden, A., and Gemeroy, D., The relative position of the cetacea among the order of mammalia as indicated by precipitin tests, *Zoologica*, 35, 145, 1950.

Brouwer, A., et al., Characterization of potential endocrine-related health effects at low-dose levels of exposure to PCBs, *Environ. Health Perspect.*, 107 (Suppl. 4), 639–649, 1999.

Burleson, G.R., and Dean, J.H., Immunotoxicology: past, present, and future, in Munson, A.E., Ed., *Methods in Immunotoxicology*, Wiley-Liss, New York, 1995, pp. 3–10.

Cavagnolo, R.Z., The immunology of marine mammals, *Dev. Comp. Immunol.*, 3, 245–257, 1979.

Colgrove, G.S., Stimulation of lymphocytes from a dolphin (Tursiops truncatus) by phytomitogens, *Am. J. Vet. Res.*, 39, 141–144, 1978.

De Guise, S., Bernier, J., Dufresne, M.M., et al., Immune functions in beluga whales (Delphinapterus leucas): evaluation of mitogen-induced blastic transformation of lymphocytes from peripheral blood, spleen and thymus, *Vet. Immunol. Immunopathol.*, 50, 117–126, 1996a.

De Guise, S., et al., Phenotyping of beluga whale blood lymphocytes using monoclonal antibodies, *Dev. Comp. Immunol.*, 21, 425–433, 1997a.

De Guise, S., et al., In vitro exposure of beluga whale lymphocytes to selected heavy metals, *Vet. Immunol. Immunopathol.*, 15, 1357–1364, 1996b.

De Guise, S., et al., Characterization of a monoclonal antibody that recognizes a lymphocyte surface antigen for the cetacean homologue to CD45R, *Immunology*, 94, 207–212, 1998a.

De Guise, S., et al., Characterization of F21.A, a monoclonal antibody that recognize a leucocyte surface antigens for killer whale homologue to β-2 integrin, *Vet. Immunol. Immunopathol.*, 97, 195–206, 2004.

De Guise, S., et al., Monoclonal antibodies to lymphocyte surface antigens for cetacean homologues to CD2, CD19 and CD21, *Vet. Immunol. Immunopathol.*, 84, 209–221, 2002.

De Guise, S., et al., Immune functions in beluga whales (Delphinapterus leucas): evaluation of phagocytosis and respiratory burst with peripheral blood leukocytes using flow cytometry, *Vet. Immunol. Immunopathol.*, 47, 351–362, 1995a.

De Guise, S., et al., Possible mechanisms of action of environmental contaminants on St. Lawrence beluga whales (Delphinapterus leucas), *Environ. Health Perspect.*, 103 (Suppl. 4), 73–77, 1995b.

De Guise, S., et al., Effects of in vitro exposure of beluga whale leukocytes to selected organochlorines, *J. Toxicol. Environ. Health A*, 55, 479–493, 1998b.

De Guise, S., et al., Immune functions in beluga whales (Delphinapterus leucas): evaluation of natural killer cell activity, *Vet. Immunol. Immunopathol.*, 58, 345–354, 1997b.

de Swart, R.L., et al., UytdeHaag, F.G., and Osterhaus, A.D., Mitogen and antigen induced B and T cell responses of peripheral blood mononuclear cells from the harbour seal (Phoca vitulina), *Vet. Immunol. Immunopathol.*, 37, 217–230, 1993.

de Swart, R.L., et al., Impairment of immune function in harbour seals (Phoca vitulina) feeding of fish from polluted waters, *Ambio*, 23, 155–159, 1994.

Denis, F., and Archambault, D., Molecular cloning and characterization of beluga whale (Delphinapterus leucas) interleukin–1beta and tumor necrosis factor-alpha, *Can. J. Vet. Res.*, 65, 233–240, 2001.

Dewailly, E., et al., Inuit exposure to organochlorines through the aquatic food chain in arctic Quebec, *Environ. Health Perspect.*, 101, 618–620, 1993.

Dewailly, E., et al., High levels of PCBs in breast milk of Inuit women from arctic Quebec, *Bull. Environ. Contam. Toxicol.*, 43, 641–646, 1989.

DiMolfetto-Landon, L., et al., Blastogenesis and interleukin-2 receptor expression assays in the harbor seal (Phoca vitulina), *J. Wildl. Dis.*, 31, 150–158, 1995.

Domingo, M., et al., Morbillivirus in dolphins, *Nature*, 348, 21, 1990.

Domingo, M., et al., Pathologic and immunocytochemical studies of morbillivirus infection in striped dolphins (Stenella coeruleoalba), *Vet. Pathol.*, 29, 1–10, 1992.

Eis, D., Simplification in the etiology of recent seal deaths, *Ambio*, 18, 144, 1989.

Erickson, K.L., et al., Development of an interleukin-2 receptor expression assay and its use in evaluation of cellular immune responses in bottlenose dolphin (Tursiops truncatus), *J. Wildl. Dis.*, 31, 142–149, 1995.

Fahey, J.L., et al., Quantitative changes in T helper or T suppressor/cytotoxic lymphocyte subsets that distinguish acquired immune deficiency syndrome from other immune subset disorders, *Am. J. Med.*, 76, 95–100, 1984.

Fournier, M., et al., Immunosuppression in mice fed on diets containing beluga whale blubber from the St Lawrence estuary and the arctic populations, *Toxicol. Lett.*, 112–113, 311–317, 2000.

Friend, M., and Trainer, D.O., Polychlorinated biphenyl: interaction with duck hepatitis virus, *Science*, 170, 1314–1316, 1970.

Funke, C., et al., Harbor seal (Phoca vitulina) C-reactive protein (C-RP): purification, characterization of specific monoclonal antibodies and development of an immuno-assay to measure serum C-RP concentrations, *Vet. Immunol. Immunopathol.*, 59, 151–162, 1997.

Funke, C., et al., Expression and functional characterization of killer whale (Orcinus orca) interleukin-6 (IL-6) and development of a competitive immunoassay, *Vet. Immunol. Immunopathol.*, 93, 69–79, 2003.

Gaskin, D.E., Holdrinet, M., and Frank, R., Organochlorine pesticide residues in harbour porpoises from the Bay of Fundy region, *Nature*, 233, 499–500, 1971.

Gaskin, D.E., et al., Mercury, DDT, dieldrin, and PCB in two species of odontoceti (Cetacea) from St. Lucia, Lesser Antilles, *J. Fish Res. Board Can.*, 31, 1235–1239, 1974.

Geraci, J.R., et al., Mass mortality of harbor seals: pneumonia associated with influenza A virus, *Science*, 215, 1129–1131, 1982.

Goldsby, R.A., et al., *Immunology*, W.H. Freeman, San Francisco, 2003, chap. 1.

Hall, A.J., et al., The immunocompetence handicap hypothesis in two sexually dimorphic pinniped species: is there a sex difference in immunity during early development? *Dev. Comp. Immunol.*, 27, 629–637, 2003.

Hammond, J.A., Hall, A.J., and Dyrynda, E.A., In Vitro Exposure of Seal Leucocytes to PCB Contaminants and Their Effects on Innate Immunity, presented at 9th International Congress of the International Society for Developmental and Comparative Immunology, St. Andrews, Scotland, 2003.

Holden, A.V., and Marsden, K., Organochlorine pesticides in seals and porpoises, *Nature*, 216, 1274–1276, 1967.

Imanishi, J., et al., Effect of polychlorinated biphenyl on viral infections in mice, *Infect. Immunity*, 29, 275–277, 1980.

Itou, T., et al., Oxygen radical generation and expression of NADPH oxidase genes in bottlenose dolphin (Tursiops truncatus) neutrophils, *Dev. Comp. Immunol.*, 25, 47–53, 2001.

Jaber, J.R., et al., Cross-reactivity of human and bovine antibodies in striped dolphin paraffin wax-embedded tissues, *Vet. Immunol. Immunopathol.*, 96, 65–72, 2003a.

Jaber, J.R., et al., Immunophenotypic characterization of hepatic inflammatory cell infiltrates in common dolphins (Delphinus delphis), *J. Comp. Pathol.*, 129, 226–30, 2003b.

Jackson, M.R., Priorities in the development of alternative methodologies in the pharmaceutical industry, *Arch. Toxicol. Suppl.*, 20, 61–70, 1998.

Jacobsen, C.N., et al., Reactivities of 20 anti-human monoclonal antibodies with leucocytes from ten different animal species, *Vet. Immunol. Immunopathol.*, 39, 461–466, 1993.

Jepson, P.D., et al., Investigating potential associations between chronic exposure to polychlorinated biphenyls and infectious disease mortality in harbour porpoises from England and Wales, *Sci. Total Environ.*, 243–244, 339–348, 1999.

Kennedy, S., et al., Histopathologic and immunocytochemical studies of distemper in harbor porpoises, *Vet. Pathol.*, 28, 1–7, 1991.

King, D.P., et al., The use of monoclonal antibodies specific for seal immunoglobulins in an enzyme-linked immunosorbent assay to detect canine distemper virus-specific immunoglobulin in seal plasma samples, *J. Immunol. Methods*, 160, 163–171, 1993.

King, D.P., et al., Identification, characterisation, and measurement of immunoglobulin concentrations in grey (Haliocherus grypus) and common (Phoca vitulina) seals, *Dev. Comp. Immunol.*, 18, 433–442, 1994.

Koller, L.D., Enhanced polychlorinated biphenyl lesions in Moloney leukemia virus-infected mice, *Clin. Toxicol.*, 11, 107–116, 1977.

Kristensen, F., Kristensen, B., and Lazary, S., The lymphocyte stimulation test in veterinary immunology, *Vet. Immunol. Immunopathol.*, 3, 203–277, 1982.

Lahvis, G.P., et al., Decreased lymphocyte responses in free-ranging bottlenose dolphins (Tursiops truncatus) are associated with increased concentrations of PCBs and DDT in peripheral blood, *Environ. Health Perspect.*, 103 (Suppl. 4), 67–72, 1995.

Lalancette, A., et al., Contrasting changes of sensitivity by lymphocytes and neutrophils to mercury in developing grey seals, *Dev. Comp. Immunol.*, 27, 735–747, 2003.

Lapierre, P., et al., Immune functions in the Fisher rat fed beluga whale (Delphinapterus leucas) blubber from the contaminated St. Lawrence Estuary, *Environ. Res.*, 80, S104-S112, 1999.

Levin, M.J., De Guise, S., and Ross, P.S., Lymphocyte proliferation is associated with organochlorine contaminants in recently weaned free-ranging harbor seal (Phoca vitulina) pups, *Environ. Health. Perspect.*, 2003.

Levin, M.J., et al., Mechanisms of Phagocytosis Modulation Upon In Vitro Exposure to Organochlorines in Marine Mammals, in Proceedings of 24th annual meeting of the Society for Environmental Toxicology and Chemistry, Austin, TX, 2003b.

Levin, M.J., et al., Specific non-coplanar PCB-mediated modulation of bottlenose dolphin and beluga whale phagocytosis upon *in vitro* exposure, *J. Toxicol. Environ. Health* A, 67, 1–19, 2004.

Lundqvist, M.L., et al., Cloning of the IgM heavy chain of the bottlenose dolphin (Tursiops truncatus), and initial analysis of VH gene usage, *Dev. Comp. Immunol.*, 26, 551–562, 2002.

Luster, M.I., et al., Development of a testing battery to assess chemical-induced immunotoxicity: National Toxicology Program's guidelines for immunotoxicity evaluation in mice, *Fundam. Appl. Toxicol.*, 10, 2–19, 1988.

Luster, M.I., et al., Risk assessment in immunotoxicology. I. Sensitivity and predictability of immune tests, *Fundam. Appl. Toxicol.*, 18, 200–210, 1992.

Luster, M.I., et al., Risk assessment in immunotoxicology. II. Relationships between immune and host resistance tests, *Fundam. Appl. Toxicol.*, 21, 71–82, 1993.

Martineau, D., et al., Pathology of stranded beluga whales (Delphinapterus leucas) from the St. Lawrence Estuary, Quebec, Canada, *J. Comp. Pathol.*, 98, 287–311, 1988.

Mori, C., Levin, M., Morsey, B., and De Guise, S., Immunomodulatory effects of in vitro exposure to organochlorines in marine mammals and mice, in Proceedings of 24th annual meeting of Society for Environmental Toxicology and Chemistry, Austin, TX, 2003.

Muir, D.C., et al., Organochlorine contaminants in belugas, Delphinapterus leucas, from Canadian waters, *Can. Bull. Fish. Aquatic Sci.*, 224, 165–190, 1990.

Mulvad, G., et al., Exposure of Greenlandic Inuit to organochlorines and heavy metals through the marine food-chain: an international study, *Sci. Total Environ.*, 186, 137–139, 1996.

Mumford, D.M., et al., Lymphocyte transformation studies of sea mammal blood, *Experientia*, 31, 498–500, 1975.

Munson, A.E., and LeVier, D., Experimental design in immunotoxicology, in A.E. Munson, Eds., *Methods in Immunotoxicology*, Wiley-Liss, New York, 1995, pp. 11–24.

Murray, B.W., and White, B.N., Sequence variation at the major histocompatibility complex DRB loci in beluga (Delphinapterus leucas) and narwhal (Monodon monoceros), *Immunogenetics*, 48, 242–252, 1998.

Nakata, H., et al., Evaluation of mitogen-induced responses in marine mammal and human lymphocytes by in-vitro exposure of butyltins and non-ortho coplanar PCBs, *Environ. Pollut.*, 120, 245–253, 2002.

Nash, D.R., and Mack, J.-P., Immunoglobulin classes in aquatic mammals characterized by serologic cross-reactivity, molecular size and binding of human free secretory component, *J. Immunol.*, 107, 1424–1430, 1977.

Noda, K., et al., Evaluation of the polymorphonuclear cell functions of bottlenose dolphins, *J. Vet. Med. Sci.*, 65, 727–729, 2003.

O'Shea, J., and Ortaldo, J.R., The biology of natural killer cells: insights into the molecular basis of function, in McGee, J.O.D., Ed., *The Natural Immune System: The Natural Killer Cell*, IRL Press at Oxford University Press, New York, 1992, pp. 2–40.

Osterhaus, A.D., et al., Canine distemper virus in seals, *Nature*, 335, 403–404, 1988.

Prochazkova, J., Contribution of in vitro assays to preclinical and premarketing testing in immunotoxicology, *Central Eur. J. Public Health.*, 1, 101–105, 1993.

Reijnders, P.J.H., Organochlorine and heavy metal residues in harbour seals from the Wadden Sea and their possible effects on reproduction, *Netherlands J. Sea. Res.*, 14, 30–65, 1980.

Roitt, I.M., Brostoff, J., and Male, D.K., *Immunology*, 5th ed., London, Mosby, 1998, chap. 1.

Romano, T.A., et al., Molecular cloning and characterization of CD4 in an aquatic mammal, the white whale Delphinapterus leucas, *Immunogenetics*, 49, 376–383, 1999.

Romano, T.A., Ridgway, S.H., and Quaranta, V., MHC class II molecules and immunoglobulins on peripheral blood lymphocytes of the bottlenosed dolphin, Tursiops truncatus, *J. Exp. Zool.*, 263, 96–104, 1992.

Ross, P.S., et al., Contaminant-related suppression of delayed-type hypersensitivity and antibody responses in harbor seals fed herring from the Baltic Sea, *Environ. Health. Perspect.*, 103, 162–167, 1995.

Ross, P.S., et al., Suppression of natural killer cell activity in harbour seals (Phoca vitulina) fed Baltic Sea herring, *Aquatic Toxicol.*, 34, 71–84, 1996.

Ross, P.S., et al., Impaired cellular immune response in rats exposed perinatally to Baltic Sea herring oil or 2, 3, 7, 8-TCDD, *Arch. Toxicol.*, 71, 563–574, 1997.

Ross, P.S., et al., Relative immunocompetence of the newborn harbour seal, Phoca vitulina, *Vet. Immunol. Immunopathol.*, 42, 331–348, 1994.

Ross, P.S., et al., Immune function in free-ranging harbor seal (Phoca vitulina) mothers and their pups during lactation, *J. Wildl. Dis.*, 29, 21–29, 1993.

Shaw, S.D., Organochlorines and Biomarkers of Immune and Endocrine Effects in Pacific Harbor Seal and Northern Elephant Seal Pups, Thesis, Columbia University, New York, 1998.

Slade, R.W., Limited MHC polymorphism in the southern elephant seal: implications for MHC evolution and marine mammal population biology, *Proc. R. Soc. Lond. B Biol. Sci.*, 249, 163–171, 1992.

Sladen, W.J., Menzie, C.M., and Reichel, W.L., DDT residues in Adelie penguins and a crabeater seal from Antarctica, *Nature*, 210, 670–673, 1966.

St-Laurent, G., and Archambault, D., Molecular cloning, phylogenetic analysis and expression of beluga whale (Delphinapterus leucas) interleukin 6, *Vet. Immunol. Immunopathol.*, 73, 31–44, 2000.

St-Laurent, G., Beliveau, C., and Archambault, D., Molecular cloning and phylogenetic analysis of beluga whale (Delphinapterus leucas) and grey seal (Halichoerus grypus) interleukin 2, *Vet. Immunol. Immunopathol.*, 67, 385–394, 1999.

Suer, L.D., et al., Erysipelothrix rhusiopathiae. II. Enzyme immunoassay of sera from wild and captive marine mammals, *Dis. Aquatic Organisms*, 5, 7–13, 1988.

Taruski, A.G., Olney, C.E., and Winn, H.E., Chlorinated hydrocarbons in cetaceans, *J. Fish. Res. Board Can.*, 32, 2205–2209, 1975.

Taylor, B.C., et al., Measurement of serum immunoglobulin concentration in killer whales and sea otters by radial immunodiffusion, *Vet. Immunol. Immunopathol.*, 89, 187–195, 2002.

Tizard, I., *Veterinary Immunology: An Introduction*, W.B. Saunders, Philadelphia, 1992, chap. 3.

Travis, J.C., and Sanders, B.G., Whale immunoglobulins. II. Heavy chain structure, *Comp. Biochem. Physiol.*, 43B, 637, 1972.

van Oss, C.J., Phagocytosis: an overview, *Methods Enzymol.*, 132, 3–15, 1986.

Vos, J.G., et al., Health effects of endocrine-disrupting chemicals on wildlife, with special reference to the European situation, *Crit. Rev. Toxicol.*, 30, 71–133, 2000.

Zabka, T.S., and Romano, T.A., Distribution of MHC II (+) cells in skin of the Atlantic bottlenose dolphin (Tursiops truncatus): an initial investigation of dolphin dendritic cells, *Anat. Rec.*, 273A, 636–647, 2003.

CHAPTER **6**

Amphibian Models and Approaches to Immunotoxicology

Louise A. Rollins-Smith and Judit E.G. Smits

CONTENTS

6.1 INTRODUCTION

The first terrestrial vertebrates crawled out of their watery environments more than 200 million years ago (Duellman and Treub, 1986; Bolt 1991; Milner, 1993; Carroll, 2000). It is likely that the environment seemed hostile to these first primitive amphibians, but they survived and flourished to eventually inhabit aquatic and semi-aquatic environments on all continents except Antarctica. In rainforest habitats, where the amphibian fauna is richest, new species are still being discovered (Meegaskumbura et al., 2002). However, there is concern in recent years that many colorful species of amphibians are in danger of becoming extinct. The causes for global amphibian declines are complex, but most scientists would probably agree that one likely cause of amphibian declines is changed environments due, in part, to human activities that have resulted in overwhelming "stresses" to amphibian populations. Here, we review the unique and shared components of the amphibian immune system with respect to other vertebrate groups and examine how environmental contaminants may alter immune defenses in ways that make amphibians more susceptible to pathogenic organisms.

6.2 OVERVIEW OF AMPHIBIAN IMMUNE DEFENSES

6.2.1   Innate Immunity

The innate immune system provides rapid, nonspecific protection until the adaptive immune response develops. As observed in other vertebrate species, the first defenders against invading microorganisms are phagocytic cells, including macrophages and neutrophils. Macrophages have been characterized in a limited number of amphibian species, but the information available suggests that they share properties in common with this class of leukocytes in all other vertebrate groups (reviewed in Manning and Horton, 1982). Recent studies of Kupffer cells from the liver of representative amphibians show that these cells are very similar to mammalian Kupffer cells, based upon their phagocytic and pinocytotic abilities and nonspecific esterase activity (Corsaro et al., 2000; Sichel et al., 2002). MHC class II positive cells, some with dendritic morphology, have been reported in the skin of *Xenopus laevis* and *Rana pipiens* (Du Pasquier and Flajnik, 1990; Carillo et al., 1990; Castell-Rodriguez et al., 1999). Peritoneal macrophages from *X. laevis* produce an interleukin-1-like cytokine (Watkins et al., 1987). Neutrophilic leukocytes or heterophils have been described for several species (Campbell, 1970; Hadji-Azimi et al., 1987; Ottaviani et al., 1992), and it is generally assumed that they provide a critical first response to pathogens. Amphibian eosinophils are spectacular in their appearance in differentially stained blood smears. Their bright red-staining granules readily identify them. Eosinophils have been described for several amphibian species (Campbell, 1970; Hadji-Azimi et al., 1987). They are generally thought to play a role in control of parasitic helminths (roundworms and flatworms).

Natural cytotoxicity provided by natural killer (NK) cells is another important component of the innate cellular immune system of amphibians. NK cells may provide an immediate cytotoxic response against virus-infected or tumor targets (Horton et al., 2000, 2001). Splenocytes of adult *X. laevis* have NK-like activity, but NK cells are not yet evident in tadpole spleens (Horton et al., 1996; Horton et al., 1998; Horton et al., 2003). NK cells become a significant component of the splenocyte population at about 1 year of age (Horton et al., 2003). Monoclonal antibodies specific for this population identified candidate NK cells in the liver and gut, as well as spleen, of adult frogs; and the numbers were increased in animals that had been thymectomized as larvae (Horton et al., 2000; 2001).

Like other vertebrates, amphibians have a complement system that can kill bacteria directly by activation of the alternative pathway of complement and formation of the membrane attack complex (MAC) (Green and Cohen, 1977; Grossberger et al., 1989; Kato et al., 1994; Kato et al., 1995; Lambris et al., 1995; Shimazaki et al., 2001). Antibodies bound to pathogens can also activate complement via the classical pathway (Green and Cohen, 1977). Pathogens would be killed directly by the MAC or by macrophages that consume the opsonized pathogens. Very recently, evidence has been presented demonstrating that *X. laevis* has serum lectins called ficolins capable of binding to mannose-binding lectin-associated serine proteases (MASPs). Complement is activated by this unique lectin pathway (Kakinuma et al., 2003; Endo et al., 1998). It is thought that the lectins bind foreign sugars on invading microorganisms, thus activating complement.

In addition to these defenses that are shared with other vertebrate groups, amphibians possess unique innate defenses in the skin. Most anuran species possess granular glands (also called serous glands or poison glands) in the dermal layer of the epithelium that synthesize and release anti-microbial peptides. A growing number of antimicrobial peptides, with sizes ranging from 10 to 50 amino acid residues, have been described. These peptides have activity against Gram-positive and Gram-negative bacteria, fungi, protozoa, and viruses (reviewed in Nicolas and Mor, 1995; Simmaco et al., 1998; Rinaldi, 2002; Zasloff, 2002; Conlon et al., 2004).

6.2.2 Adaptive Immunity

The adaptive immune system of only two amphibian species has been studied in detail. The representative species are the South African clawed frog, *X. laevis*, and the Mexican axolotl, *Ambystoma mexicanum*. Collectively, the studies of these two species have revealed that the basic components of the adaptive immune system of anuran amphibians (frogs and toads) are the same as those of all other vertebrates except the jawless fishes. Although anurans lack organized lymph nodes, they do have T and B lymphocytes (Bleicher and Cohen, 1981) that express rearranging T-cell receptors (TCR) and immunoglobulin (Ig) receptors (Schwager et al., 1988; Fellah and Charlemagne, 1993; reviewed in Du Pasquier et al., 1989). They have a well-characterized major histocompatibility complex (MHC) (reviewed in Flajnik and Du Pasquier, 1990; Flajnik and Kasahara, 2001) encoding classical class Ia, class II, and class III molecules (C4, Bf, and HSP70), the proteosome components LMP2 and LMP7 (low-molecular-mass proteins), and transporters associated with antigen processing, TAP1 and TAP2. Amphibians express three Ig isotypes (Green and Steiner, 1976; Warr et al., 1982; Hsu and Du Pasquier, 1984; Hsu et al., 1985; Fellah and Charlemagne, 1988), and leukocyte-derived cytokines including IL-1, IL-2-like molecules, transforming growth factor-beta (TGF-β), and macrophage migration inhibitory factor (MIF) (Watkins and Cohen, 1987; Watkins et al., 1987; Haynes and Cohen, 1993; Koniski and Cohen, 1994; Zou et al., 2002; Suzuki et al., 2004). MHC-restricted cytotoxic and helper T-cell responses have been characterized (Blomberg et al., 1980; Flajnik et al., 1985; Harding et al., 1984; Robert et al., 2002). Adaptive immune responses in the skin have been investigated by examination of allograft rejection mechanisms. Skin allograft rejection mediated by thymus-derived (T) cells is a classical measure of cellular immunity in amphibians. It is dependent on recognition of MHC or minor histocompatibility antigens in adults (reviewed in Du Pasquier, 1973; Cohen et al., 1985), characterized by classical first- and second-set kinetics, and dependent on the environmental temperature of the hosts. Warmer temperatures accelerate graft rejection while cooler temperatures delay the response (reviewed in Cohen et al., 1985). Another aspect of adaptive immunity is the ability to resist virus infections. Viruses of the family Iridoviridae have been linked to global amphibian declines (reviewed in Carey et al., 1999; Daszak et al., 1999). Recent virus infection studies of *X. laevis* demonstrate that adults are relatively resistant to infection by frog virus 3 (an iridovirus), whereas tadpoles are very susceptible. Both cellular and humoral components of the adaptive immune system are necessary for clearance of infectious viruses (Gantress et al., 2003).

6.3 EFFECTS OF ENVIRONMENTAL CHEMICALS ON AMPHIBIAN IMMUNE DEFENSES

6.3.1 Altered Development of the Immune System in Embryos

The immune system, as it has evolved in all vertebrate species, protects the host from infection by environmental pathogens. The great diversity and flexibility of the adaptive immune response depends on a continuing process of self-renewal from a population of hematopoietic stem cells

(HSCs). Amphibians have long been the favorite choice for studies of embryonic development, and thus the location of embryonic stem cell compartments and the pattern of development of the hematopoietic system are very well understood (reviewed in Turpen, 1998). While there are only a limited number of studies that document the immunotoxic effects of xenobiotic chemicals on the mature immune system in amphibian species, there is even less available information about the effects of these chemicals on the immature or developing immune system. Most insults to the adult immune system cause temporary loss of function because many components of the immune system will renew themselves. If pluripotential HSC are destroyed, however (e.g., by gamma irradiation), the consequence is irreversible loss of renewal capability. Most immunotoxicology studies use rodent (mouse and rat) model systems. They are very good models of the human immune system. However, these models are less relevant to lower vertebrates such as fish or amphibians where exposure of embryos to hazardous chemicals is direct. Because amphibians develop free of maternal influences, they may be valuable models to study potentially adverse effects of environmental chemicals during early development. To determine whether acute embryonic exposure to a potentially immunotoxic agent can affect early hematopoiesis in *X. laevis*, we adopted a well-established amphibian diploid/triploid transfer model (Kau and Turpen, 1983; Maéno et al., 1985). Diploid embryos (16 to 20 hours of age) were exposed to test chemicals for 2 hours. At this time in development, HSC are moving to the locations where they will produce tadpole blood cells. The HSC are located in two embryonic regions, the dorsal lateral plate (DLP) and the ventral blood island (VBI) (reviewed in Turpen, 1998). After 2 hours of exposure to the test chemical, the VBI containing HSC was transplanted from a chemically treated or untreated control embryo to an untreated triploid host embryo. After 55 days, the contribution of the donor VBI to cell populations in the blood, thymus, and spleen was assessed by flow cytometry after staining of the cells with propidium iodide. Diazinon, but not lead acetate, apparently interfered with the ability of transferred stem cells to contribute to hematopoiesis (Rollins-Smith et al., 2004). At the concentration at which an effect on hematopoiesis was observed, diazinon was not directly toxic to the embryos (Rollins-Smith et al., 2004). Because of the small size of the embryos, this method employs very small amounts of each toxic agent delivered at a critical time in the development of the immune system. Amphibian embryos are very sensitive indicators of the toxic effects of chemicals. This VBI assay could be employed to test any toxic chemical that may affect development of the hematopoietic system (Rollins-Smith et al., 2004). There is some controversy in the literature as to whether common pesticides can penetrate the fertilization membrane and jelly coat of amphibian embryos. ^{14}C-labeled isoproturon was taken up over a 24-hour period by embryos of *Bombina bombina* with intact fertilization membranes and jelly coats (Greulich et al., 2002). However, Berrill et al. (1998) saw no evidence of uptake of endosulfan by *R. sylvatica* embryos with intact fertilization membranes and jelly coats, and they interpreted this result as evidence that the jelly coat is protective.

6.3.2   Altered Development of the Immune System in Tadpoles

Amphibian tadpoles are free-living and need a functional immune system to defend against naturally occurring pathogens in the water. In the model amphibian, *X. laevis*, B-and T-lymphocyte populations expand rapidly during the larval period (Du Pasquier and Weiss, 1973; Rollins-Smith et al., 1984; Cohen et al., 1985). Following the major expansion of lymphocytes in the larval period, a significant number of tadpole lymphocytes are eliminated by glucocorticoid hormone-driven apoptosis (Rollins-Smith et al., 1997; Barker et al., 1997), and then the number of lymphocytes expands greatly in the immediate postmetamorphic period. The postmetamorphic expansion of lymphocytes is dependent on thyroid hormones (Rollins-Smith and Blair, 1990a, 1990b). Agents that interfere with uptake of iodine from the blood, such as perchlorate, impair lymphocyte expansion (Rollins-Smith and Blair, 1990a, 1990b). Perchlorate is the primary ingredient in solid rocket propellant, and is present in some fertilizers (Susarla et al., 1999); thus, it is an important contaminant of some aquatic systems, especially in the western U.S. (Urbansky, 1998; Smith et al., 2001). Perchlorate was detectable in the tissues of bullfrog tadpoles (*R. catesbeiana*) from a contaminated site in Texas (Smith et al., 2001). Thus, the potential for inhibition of thyroid function and impaired development of the immune system of free-living amphibian tadpoles is evident. Likewise, agents that may interfere with the action of the glucocorticoids can inhibit the naturally occurring lymphocyte deletion at metamorphosis (Rollins-Smith et al., 1997). We have hypothesized that deletion of larval lymphocytes is necessary to avoid development of immunity to adult-specific self-antigens that emerge at the time of metamorphosis (Rollins-Smith et al., 1997). Because amphibian metamorphosis is driven by the combined actions of thyroid hormones and glucocorticoid hormones, it is likely that agents that might disturb the normal neuroendocrine control of metamorphosis could alter immune system development. Several groups have described delayed metamorphosis following treatment of tadpoles with malathion (*R. tigerina*, Mohanty-Hejmadi and Dutta, 1981; *R. catesbeiana*, Fordham et al., 2001) or atrazine (*X. laevis*, Sullivan and Spence, 2003). Although thyroid hormone levels were not measured, interference with thyroid hormone function by malathion is suggested as the possible cause for this developmental delay (Fordham et al., 2001). Whether this degree of inhibition would alter immune defenses is not yet determined. However, studies of the infection of wood frog (*R. sylvatica*) tadpoles by trematodes capable of inducing limb deformities (*Ribeiroia spp.* and *Telorchis spp.*) showed that increased malathion or atrazine resulted in a greater degree of parasitism that was associated with fewer circulating eosinophils (Kiesecker, 2002). In contrast to delayed metamorphosis induced by malathion and atrazine, the herbicide acetochlor was shown to accelerate metamorphic changes in *R. pipiens* (Cheek et al., 1999), *R. catesbeiana* (Veldhoen and Helbing, 2001), and *X. laevis* (Crump et al., 2002). We have previously shown that accelerating metamorphosis with thyroxine results in impaired allograft rejection capacity (Rollins-Smith et al., 1988).

6.3.3 Effects of Ultraviolet Irradiation on Tadpole Immunity

Human activities that have contributed to thinning of the protective ozone layer have increased the potential for exposure of amphibians to harmful solar ultraviolet (UV) radiation (Stolarski et al., 1986; Gleason et al., 1993; Madronich et al., 1998; Herman et al., 1999). Decreased dissolved organic mater or droughts related to climate change may cause increased UV exposure to animals in shallow upper latitude lakes (Schindler et al., 1996; Yan et al., 1996). The most harmful natural component of the UV radiation spectrum is UV-B (280 to 320 nm). Increased UV-B radiation has been associated with increased mortality of developing amphibians (Blaustein et al., 1995, Kiesecker and Blaustein, 1995; Blaustein et al., 1997) and amphibian population declines (Middleton et al., 2001). It is hypothesized that amphibians exposed to elevated levels of UV-B radiation may have impaired immune defenses in the skin that would result in greater susceptibility to invasive skin pathogens. To determine whether chronic low-level UV-B exposure alters immune defenses of developing frogs, we conducted experiments in which UV-B–sensitive *X. laevis* tadpoles were chronically exposed to defined sublethal UV-B treatment regimes, and their capacities to reject skin allografts were examined. The result was impaired allograft rejection capacity (Rollins-Smith, Pandelova, and Hays, unpublished). Field studies of amphibian populations in the western U.S. implicate increased UV-B in greater mortality of developing amphibians due to disease (Kiesecker et al., 2001).

6.3.4 Effects of Agricultural Chemicals on Immune Defenses of Juvenile and Adult Frogs and Salamanders

While the immune system in young postmetamorphic frogs is adult-like, it is still expanding rapidly (Du Pasquier and Weiss, 1973; Rollins-Smith et al., 1984) and may be more susceptible to toxic insult than the immune system of more mature adults. Juvenile leopard frogs (*R. pipiens*) exposed to environmentally relevant concentrations of six agricultural pesticides (atrazine, metribuzin, aldicarb, endosulfane, lindane, and dieldrin) showed evidence of impaired immune defenses. Lymphocytes from pesticide-treated frogs showed reduced proliferation *in vitro* in response to T-dependent mitogens concanavalin-A (Con-A) and phytohemagglutinin (PHA), and they were found to have increased parasitism when challenged with a parasitic nematode (*Rhabdias ranae*). Parasite- and pesticide-challenged frogs had reduced splenocyte numbers and reduced phagocytic activity in the spleen cell population (Christin et al., 2003). Further studies showed that the parasitic worms migrated more rapidly to the lungs and reproduced earlier in pesticide-exposed frogs (Gendron et al., 2003).

Although juvenile frogs may be especially susceptible to immunotoxic chemicals, other studies show that the immune defenses of mature adult frogs and toads are also affected by exposure to agricultural pesticides. Adult Woodhouse's toads (*Bufo woodhousii*) were more susceptible to development of hepatomegaly, development of clinical disease symptoms, and death due to experimental infection with the bacterium *Aeromonas hydrophila* if they were exposed to sublethal doses of malathion in comparison with toads that were not treated with the pesticide (Taylor et al., 1999b). Studies of the immune defenses of northern leopard frogs (*R. pipiens*) treated with malathion, DDT, or dieldrin showed that specific IgM antibody responses were suppressed in comparison with untreated controls. Oxidative burst products measured in whole blood were also decreased in malathion and dieldrin-treated frogs. Delayed-type hypersensitivity, measured by the nonspecific response to injected PHA, was enhanced in pesticide-treated frogs in comparison with control frogs. The pattern of altered immune responses observed in laboratory studies was also detected in wild frog populations collected in pesticide-exposed locations, but not collected from pesticide-free locations suggesting that the laboratory results reflect immunosuppression in nature (Gilbertson et al., 2003). Marine toads (*B. marinus*) and whistling frogs (*Eleutherodactylus johnstonei*) collected between 1995 and 1999 on the island of Bermuda had significant residues of p,p'-dichlorodiphenyldichloroethylene (DDE) in liver and fat bodies. Analyses of the toad livers also revealed significant concentrations of heavy metals (cadmium, chromium, copper, and zinc). Spleens from contaminated *B. marinus* showed a marked decrease in white pulp indicative of decreased lymphocyte numbers and the cells exhibited decreased proliferation in response to the B-cell mitogen lipopolysaccharide (LPS). The incidence of trematode infection in *B. marinus* during the years of collection ranged from 53.8% to 90%. The authors of this study suggested that this high level of parasitism may be due to the environmental pollutants that suppress natural immunity (Linzey et al., 2003).

A limited number of immunotoxicity studies have been done using urodele amphibians as subjects. In general, immune responses of urodeles have been more difficult to measure than those of anuran amphibians. Skin graft rejection responses are chronic (Cohen, 1968), and *in vitro* mitogen responses have been reported to be modest (Collins et al., 1975; Froese, 2002). While proliferation of newt (*Notophthalmus viridescens*) and Mexican axolotl (*A. mexicanum*) lymphocytes to PHA, Con-A, and allogeneic cells was poor, significant proliferation to relatively high concentrations of LPS was a consistent observation (Collins et al., 1975; Collins and Cohen, 1976). Froese (2002) also reported significant proliferation of Mexican axolotl lymphocytes in the presence of LPS, but not Con-A, whereas tiger salamander (*A. tigrinum*) lymphocytes were modestly stimulated by PHA and not by Con-A or LPS. Although mitogen-driven proliferation of urodele lymphocytes has sometimes been difficult to measure, Koniski and Cohen (1992) reported substantial proliferative responses of lymphocytes of the Mexican axolotl induced by PHA, Con-A, LPS, and phorbol 12-myristate, 13-acetate (PMA) when fetal bovine serum (FBS) was replaced by bovine serum albumin (BSA). Thus, *in vitro* lymphocyte proliferation could be a useful measure of immunotoxic effects if the lymphocyte culture conditions are optimized. In contrast to the results observed with leopard frogs (Gilbertson et al., 2003), intradermal injection of PHA into tiger salamanders did not stimulate a DTH-like response (Froese, 2002). Thus, this measure of *in vivo* T-cell–mediated responses may not be useful for studies of urodeles. For immunotoxicity studies in tiger salamanders, phagocytosis and oxidative burst assays have been modified from mammalian protocols (Froese, 2002; Froese et al., 2004). One useful finding was that large numbers of peritoneal neutrophils can be collected from salamanders injected with thioglycollate. These neutrophils readily engulf foreign material (phagocytic activity) and produce measurable amounts of hydrogen peroxide (oxidative burst activity). In this work, quantification of phagocytosis using flow cytometry proved to be well correlated with manual counts of the numbers of fluorescent latex beads that had been engulfed by the neutrophils and the total number of neutrophils phagocytosing beads. Response to the oxidative burst assay was generally less consistent than the phagocytosis assay, indicating that phagocytic activity of peritoneal neutrophils was the endpoint of choice in immunotoxicologic studies to evaluate the potential impact of environmental contaminants on innate defense mechanisms in urodele amphibians (Froese et al., 2004). Immune-related endpoints including hematology, phagocytosis, respiratory burst, and histopathology have been examined in tiger salamanders after experimental exposure to the military explosive and now soil contaminant, trinitrotoluene (Johnson et al., 2000), but only histopathology of the liver showed some evidence of contaminant-related change.

6.4 ASSOCIATION OF GLOBAL AMPHIBIAN DECLINES WITH DISEASE AND CONCERNS ABOUT SKIN EXPOSURE TO PESTICIDES

6.4.1 The Problem of Global Amphibian Declines

Recent declines in amphibian populations in many parts of the world have been linked to infectious skin diseases. The pathogen most frequently associated with recent die-offs is a chytrid fungus, *Batrachochytrium dendrobatidis*, named for the blue poison dart frog (*Dendrobates auratus*) from which it was isolated (Berger et al., 1998; Longcore et al., 1999; Pessier et al., 1999). A second fungus was isolated from clinically ill individuals from declining populations of Wyoming toads (*B. baxteri*) and was identified as *Basidiobolus ranarum* (Taylor et al., 1999; Taylor et al., 1999a). Mass die-offs of tiger salamanders (*A. tigrinum*) in Arizona and Saskatchewan and the common frog (*R. temporaria*) appear to be caused by iridoviruses (Cunningham et al., 1996; Jancovich et al., 1997; Bollinger et al., 1999). A fourth pathogen associated with amphibian declines is the bacterium *A. hydrophila*. This agent, often associated with the syndrome called "red leg," appears to be an opportunist found on the skin and digestive tracts of healthy frogs (Hird et al., 1981). It is capable of inducing disease in South African clawed frogs (*X. laevis*) (Hubbard, 1981) and American toads (*B. americanus*) (Dusi, 1949), especially when the animals are stressed (Carr et al., 1976; Taylor et al., 1999b). Although Koch's postulates have not been met for this disease, mortality in embryos and tadpoles has been attributed to the water mold *Saprolegnia* in the Pacific Northwest (Kiesecker and Blaustein, 1995, 1997, 1999; Kiesecker et al., 2001). This organism was associated with the death of developing salamanders in South Carolina (Lefcort et al., 1997). Thus, amphibian populations may suffer from disease epidemics caused by fungal, viral, or bacterial pathogens; however, most recent declines in the western U.S., Central America, and Australia are linked to *B. dendrobatidis* (Berger, et al., 1998; Lips, 1998, 1999; Lips et al., 2003; reviewed in Carey et al., 1999; Daszak et al., 1999).

6.4.2 Effects of Pesticides on Release of Antimicrobial Peptides Released from Amphibian Skin

Because many pathogens can infect amphibian skin, much attention has focused on the immune defenses of the skin. Early histological studies of amphibian skin identified so-called mucous glands and granular glands (also called poison or serous glands) in the dermal layer of amphibian skin

(Noble and Noble, 1944; Bovbjerg, 1963; Sjoberg and Flock, 1976; Mills and Prum, 1984). The mucous glands produce a watery material rich in mucopolysaccharides that serve to keep the skin moist (Duellman and Treub, 1986). Granular glands produce a large variety of bioactive peptides including antimicrobial peptides (reviewed in Erspamer, 1994; Daly, 1995; Nicolas and Mor, 1995; Simmaco et al., 1998; Zasloff, 2002; Conlon et al., 2004). Both mature mucous and granular glands are composed of a syncytial layer of glandular epithelium surrounding a secretory compartment. In granular glands, the center of the gland is filled with granules containing peptides (Giovannini et al., 1987). Granular glands are surrounded by a layer of myoepithelial cells or smooth muscle cells with sympathetic axons terminating in the regions between the contractile elements and in the gland epithelium (Sjoberg and Flock, 1976). The myoepithelial cells possess α-adrenoreceptors; and α-adrenoreceptor agonists, such as epinephrine or norepinephrine, induce contraction and release of secretory products (Benson and Hadley, 1969; Dockray and Hopkins, 1975; Holmes and Balls, 1978). An extensive body of literature characterizes the amino acid sequences and activity of a large number of antimicrobial peptides isolated from amphibian skin (reviewed in Nicolas and Mor, 1995; Simmaco et al., 1998; Rinaldi, 2002; Zasloff, 2002; Conlon et al., 2004). They have activity against Gram-positive and Gram-negative bacteria, fungi, protozoa, and viruses (reviewed in Nicolas and Mor, 1995; Simmaco et al., 1998; Rinaldi, 2002; Zasloff, 2002; Conlon et al., 2004). Each species appears to produce its own unique set of peptides that have activity against a variety of organisms (Amiche et al., 1999). The main families of antimicrobial skin peptides belong to a large group of linear amphipathic helical peptides. They are cationic, containing a variable number of positively charged residues. The positively charged amino acids are oriented on one face of the molecule while the hydrophobic residues are oriented on the opposite face. This structure provides them with an ability to disrupt biological membranes, and this seems to be the main mechanism of induction of death of their targets (reviewed in Nicolas and Mor, 1995; Simmaco et al., 1998; Rinaldi, 2002; Zasloff, 2002; Conlon et al., 2004). A limited number of amphibian antimicrobial peptides have been tested against amphibian pathogens. We have tested about 35 antimicrobial peptides derived from 22 amphibian species against the lethal chytrid fungus *B. dendrobatidis*, and six were tested against the fungal pathogen, *B. ranarum* (Rollins-Smith et al., 2002a, 2002b, 2002c; Rollins-Smith et al., 2003; D. Woodhams et al., unpublished). Many of the peptides are potent inhibitors of growth of *B. dendrobatidis*. Two of the peptides from *X. laevis* act synergistically to inhibit growth of both *B. dendrobatidis* and *Basidiobolus* (Rollins-Smith et al., 2002a), and natural mixtures of peptides, as they would be secreted on the skin surface, are highly effective and may be more effective than individual peptides (D. Woodhams et al., unpublished). The peptides are more effective against the zoospore transmission stage of the chytrid fungus than against mature stages (Rollins-Smith et al., 2002b, 2002c). Because the release of antimicrobial peptides from granular glands in the skin is dependent on stimulation of the sympathetic nerves that innervate these glands, we hypothesize that environmental chemicals capable of blocking neurotransmission would impair peptide secretion. The effects of the herbicide 2,4-D (2,4-dichlorephenoxyacetic acid) on neurotransmission at the sympathetic nerve and skin gland interface was tested in a Chilean amphibian, the water helmeted toad (*Caudiverbera caudiverbera*). Increasing concentrations of 2,4-D inhibited neurotransmission in a dose-dependent fashion. Concentrations tested were 0.01 to 1.0 mM (Suwalsky et al., 1999). Similar effects were observed when the nerve and skin preparations from *C. caudiverbera* were exposed to the agricultural pesticide dieldrin. Concentrations tested were 0.01 to 1.0 mM (Suwalsky et al., 2002). Thus, it is clear that when a lipid-soluble pesticide reaches high enough concentrations at the neurological synapse of frog skin, normal secretory processes are impaired. To what extent this may happen in nature is not known. Preliminary studies of the capacity of newly metamorphosed froglets of the foothill yellow-legged frog (*R. boylii*) to release peptides in response to norepinephrine stimulation showed that carbaryl applied at sublethal concentrations significantly inhibited release of skin peptides in comparison with untreated controls (C. Davidson and L. Rollins-Smith, 2004, unpublished).

6.5 RESEARCH NEEDS

Because amphibian populations are declining on a global scale, there is an urgent need to determine whether and to what extent introduced xenobiotic chemicals may be contributing to these declines. There is also a compelling need for additional laboratory studies and field studies of the effects of potentially immunotoxic compounds on development of the immune system and on the induction of immune responses at larval as well as at adult stages of life. Studies of the effects of acute and chronic exposures to each agent at environmentally relevant concentrations on innate and adaptive immunity would expose the most harmful agents. This research effort would ultimately lead to the development of safer chemicals that no longer compromise the health of amphibians and fish.

6.6 SUMMARY

The amphibian immune system is remarkably similar to that of mammalian species. The organs and cells that comprise the adaptive immune system serve similar functions as those of mammals. Amphibians have T and B lymphocytes that express rearranging T-cell receptors (TCR) and immunoglobulin (Ig) receptors. They have an MHC, express several Ig isotypes, and they can produce leukocyte-derived cytokines. Their innate immune system includes phagocytic cells, a complement system, and NK cells. In addition, amphibians possess granular glands in the dermis of the skin that release a variety of bioactive peptides including antimicrobial peptides. The immune system may be harmed during embryonic, larval, or adult stages by contamination from xenobiotic chemicals in the water. During the embryonic period, the developing hematopoietic system is at risk. During larval life, chemicals that alter the natural endocrine balance can alter immune system development. Chemicals taken up across the skin may affect adult amphibians by altering the function of lymphocytes and antigen-presenting cells in the skin or by inhibiting synthesis of antimicrobial peptides produced in granular glands of the skin. Environmental chemicals that contribute to ozone depletion may result in increased UV-B exposure to the skin. The loss of effective skin defenses may predispose amphibians to skin infection by pathogens such as the chytrid fungus, *B. dendrobatidis*, or the water mold, *Saprolegnia*. Understanding the impact of environmental chemicals and increased UV-B exposure on amphibian immune defenses is especially important because many amphibian species have experienced unprecedented declines that have been linked to disease.

ACKNOWLEDGMENTS

Research from the Rollins-Smith laboratory was supported by the National Science Foundation (grants DCB-8710234, DCB-904666, MCB-941349, IBN-9809876, and IBN-0131184).

REFERENCES

Amiche, M., et al., The dermaseptin precursors: a protein family with a common preproregion and a variable C-terminal antimicrobial domain, *FEBS Lett.*, 456, 352–356, 1999.

Barker, K.S., Davis, A.T., and Rollins-Smith, L.A., Spontaneous and corticosteroid-induced apoptosis of lymphocyte populations in metamorphosing frogs, *Brain Behav. Immun.*, 11, 119–131, 1997.

Benson, B.J., and Hadley, M.E., In vitro characterization of adrenergic receptors controlling skin gland secretion in two anurans *Rana pipiens* and *Xenopus laevis*. *Comp. Biochem. Physiol.*, 30, 857–864, 1969.

Berger, L., et al., Chytridiomycosis causes amphibian mortality associated with population declines in the rain forests of Australia and Central America, *Proc. Natl. Acad. Sci. U.S.A.*, 95, 9031–9036, 1998.

Berrill, M., et al., Toxicity of endosulfan to aquatic stages of anuran amphibians, *Vet. Immunol. Immunopathol.*, 17, 1738–1744, 1998.

Blaustein, A.R., et al., Ambient ultraviolet radiation causes mortality in salamander eggs, *Ecol. Appl.*, 5, 740–743, 1995.

Blaustein, A.R., et al., Ambient UV-B radiation causes deformities in amphibian embryos, *Proc. Natl. Acad. Sci. U.S.A.*, 94, 13735–13737, 1997.

Bleicher, P.A., and Cohen, N., Monoclonal anti-IgM can separate T-cell from B-cell proliferative responses in the frog *Xenopus laevis*, *J. Immunol.*, 127, 1549–1555, 1981.

Blomberg, B., Bernard, C.C.A., and Du Pasquier, L., In vitro evidence for T-B lymphocyte collaboration in the clawed toad, *Xenopus*, *Eur. J. Immunol.*, 10, 869–876, 1980.

Bollinger, T.K., et al., Pathology, isolation, and preliminary molecular characterization of a novel iridovirus from tiger salamanders in Saskatchewan, *J. Wildl. Dis.*, 35, 413–429, 1999.

Bolt, J.R., Lissamphibian origins, in Schultze, H.-P., and Trueb, L., Eds., *Origins of the Higher Groups of Tetrapods: Controversy and Consensus*, Cornell University Press, Ithaca, NY, 1991, pp. 194–222.

Bovbjerg, A.M., Development of the glands of the dermal plicae in *Rana pipiens*, *J. Morphol.*, 113, 321–243, 1963.

Campbell, F.R., Ultrastructure of the bone marrow of the frog, *Am. J. Anat.*, 129, 329–355, 1970.

Carey, C., Cohen, N., and Rollins-Smith, L., Amphibian declines: an immunological perspective, *Dev. Comp. Immunol.*, 23, 459–472, 1999.

Carillo, J., Castel, A., Perez, A., and Rondan, A., Langerhans-like cells in amphibian epidermis, *J. Anat.*, 172, 39–46, 1990.

Carr, A.A., et al., Aerobic bacteria in the intestinal tracts of bullfrogs (*Rana catesbeiana*) maintained at low temperatures, *Herpetologia*, 32, 239–244, 1976.

Carroll, R.L., The fossil record and large-scale patterns of amphibian evolution, in Heatwole, H., and Carroll, R.L., Eds., *Amphibian Biology*, vol. 4, Surrey Beatty and Sons Ltd., Chipping Norton, Oxfordshire, England, 2000, pp. 973–978.

Castell-Rodriguez, A.E., Hernandez, P.A., Sampedro-Carrillo, W.A., Herrera-Enriquez, M.A., Alvarez-Perez, S.J.U., and Rondan-Zarate, A., ATPase and MHC class II molecules co-expression in *Rana pipiens* dendritic cells, *Dev. Comp. Immunol.*, 23, 473–485, 1999.

Cheek, A.O., et al., Alteration of leopard frog (*Rana pipiens*) metamorphosis by the herbicide acetochlor, *Arch. Environ. Contam. Toxicol.*, 37, 70–77, 1999.

Christin, M-S., et al., Effects of agricultural pesticides on the immune system of *Rana pipiens* and on its resistance to parasitic infection, *Vet. Immunol. Immunopathol.*, 22, 1127–1133, 2003.

Cohen, N., Chronic skin graft rejection in the Urodela I. A comparative study of first-and-second-set allograft reactions, *J. Exp. Zool.*, 167, 36–48, 1968.

Cohen, N., et al., The ontogeny of allo-tolerance and self-tolerance in larval *Xenopus laevis*, in Balls, M., and Bownes, M., Eds., *Metamorphosis*, Oxford University Press, Oxford, UK, 1985, pp. 388–419.

Collins, N.H., and Cohen, N., Phylogeny of immunocompetent cells: II In vitro behavior of lymphocytes from the spleen, blood and thymus of the Urodele Ambystoma mexicanum, in *Phylogeny of Thymus and Bone Marrow-Bursa Cells*, Wright, R.K., and Cooper, E.L., Eds., North-Holland, Amsterdam, 1976, pp. 169–182.

Collins, N.K., Manickavel, V., and Cohen, N., In vitro responses of urodele lymphoid cells: Mitogenic and mixed lymphocyte culture reactivities, in *Immunologic Phylogeny*, Hildemann, W.H., and Benedict, A.A., Eds., Plenum, New York, 1975, pp. 305–314.

Conlon, J.M., Kolodziejek, J., Nowotny, N., Antimicrobial peptides from ranid frogs: taxonomic and phylogenetic markers and a potential source of new therapeutic agents, *Biochim. Biophys. Acta*, 1696, 1–14, 2004.

Corsaro, C., et al., Characterisation of Kupffer cells in some amphibian, *J. Anat.*, 196, 249–261, 2000.

Crump, D., et al., Exposure to the herbicide acetochlor alters thyroid hormone-dependent gene expression and metamorphosis in *Xenopus laevis*, *Environ. Health Perspect.*, 110, 1199–1205, 2002.

Cunningham, A.A., et al., Pathological and microbiological findings from incidents of unusual mortality of the common frog (*Rana temporaria*), *Phil. Trans. R. Soc. London B*, 351, 1539–1557, 1996.

Daly, J.W., The chemistry of poisons in amphibian skin, *Proc. Natl. Acad. Sci.*, 92, 9–13, 1995.

Daszak, P., et al., Emerging infectious diseases and amphibian population declines, *Emerg. Infect. Dis.*, 5, 735–748, 1999.

Dockray, G.J., and Hopkins, C.R., Caerulein secretion by dermal glands in *Xenopus laevis, J. Cell Biol.*, 64, 724–733, 1975.

Du Pasquier, L., and Weiss, N., The thymus during the ontogeny of the toad *Xenopus laevis:* growth, membrane-bound immunoglobulins and mixed lymphocyte reaction, *Eur. J. Immunol.*, 3, 773–777, 1973.

Du Pasquier, L., and Flajnik, M., Expression of MHC class II antigens during *Xenopus* development, *Dev. Immunol.*, 1, 85–95, 1990.

Du Pasquier, L., Ontogeny of the immune response in cold-blooded vertebrates, *Curr. Top. Micro. Immunol.*, 61, 37–88, 1973.

Du Pasquier, L., Schwager, J., Flajnik, M.F., The immune system of *Xenopus, Annu. Rev. Immunol.*, 7, 251–275, 1989.

Duellman, W.E., and Treub, L., *Biology of Amphibians*, McGraw-Hill, New York, 1986.

Dusi, J.L., The natural occurrence of "redleg," *Pseudomonas hydrophila*, in a natural population of American toads, *Bufo americanus, Ohio J. Sci.*, 49, 70–71, 1949.

Endo, Y., et al., Two lineages of mannose-binding lectin-associated serine protease (MASP) in vertebrates, *J. Immunol.*, 161, 4924–4930, 1998.

Erspamer, V., Bioactive secretions of the amphibian integument, in *Amphibian Biology*, vol. 1, Heatwole H., and Barthalmus, B.T., Eds., Surrey Beatty and Sons, Ltd., Chipping Norton, Oxfordshire, England, 1994, pp. 178–350.

Fellah, J.S., and Charlemagne, J., Characterization of an IgY-like low molecular weight immunoglobulin class in the Mexican axolotl, *Mol. Immunol.*, 25, 1377–1386, 1988.

Flajnik, M.F., and Kasahara, M., Comparative genomics of the MHC: glimpses into the evolution of the adaptive immune system, *Immunity*, 15, 351–362, 2001.

Flajnik, M.F., and Du Pasquier, L., The major histocompatibility complex of frogs, *Immunol. Rev.*, 150, 47–63, 1990.

Flajnik, M.F., Du Pasquier, L., and Cohen, N., Immune responses of thymus/lymphocyte embryonic chimeras: studies on tolerance and MHC restriction in *Xenopus, Eur. J. Immunol.*, 15, 540–547, 1985.

Fordham, C.L., et al., Effects of malathion on survival, growth, development, and equilibrium posture of bullfrog tadpoles (*Rana catesbeiana*), *Vet. Immunol. Immunopathol.*, 20, 179–184, 2001.

Froese, J.M., Effects of Dietary Deltamethrin Exposure on the Immune System of Adult Tiger Salamanders *Ambystoma tigrinum*, Master's thesis, University of Saskatechewan, Saskatoon, SK, Canada, 2002.

Froese, J.M., Smits, J.E., and Wickstrom M.L., Evaluation of two methods for assessing mechanisms of non-specific immunity in tiger salamanders (*Ambystoma tigrinum*), *J. Wildl. Dis.*, in press, 2004.

Gantress, J., et al., Development and characterization of a model system to study amphibian immune responses to iridoviruses, *Virology*, 311, 254–262, 2003.

Gendron, A.D., et al., Exposure of leopard frogs to a pesticide mixture affects life history characteristics of the lungworm *Rhabdias ranae, Oecologia*, 135, 469–476, 2003.

Gilbertson, M.K., et al., Immunosuppression in the northern leopard frog (*Rana pipiens*) induced by pesticide exposure, *Vet. Immunol. Immunopathol.*, 22, 101–110, 2003.

Giovannini, M.G., et al., Biosynthesis and degradation of peptides derived from *Xenopus laevis* prohormones, *Biochem. J.*, 243, 113–120, 1987.

Gleason, J.F., et al., Record low global ozone in 1992, *Science*, 260, 523–526, 1993.

Green, C., and Steiner, L.A., Isolation and preliminary characterization of two varieties of low molecular weight immunoglobulin in the bullfrog, *Rana catesbeiana, J. Immunol.*, 117, 364–374, 1976.

Green, N., and Cohen, N., Effect of temperature on serum complement levels in the leopard frog *Rana pipiens, Dev. Comp. Immunol.*, 1, 59–64, 1977.

Greulich K., Hoque E., and Pflugmacher, S., Uptake, metabolism, and effects on detoxication enzymes of isoproturon in spawn and tadpoles of amphibians, *Ecotoxicol. Environ. Saf.*, 52, 256–66, 2002.

Grossberger, D., et al., Conservation of structural and functional domains in complement component C3 of *Xenopus* and mammals, *Proc. Natl. Acad. Sci. U S A.*, 86, 1323–1327, 1989.

Hadji-Azimi, I., Coosemans, V., and Canicatti, C., Atlas of adult *Xenopus laevis laevis* hematology, *Dev. Comp. Immunol.*, 11, 807–874, 1987.

Harding, F.A., Flajnik, M.F., and Cohen, N., MHC restriction of T cell proliferative responses in *Xenopus, Dev. Comp. Immunol.*, 17, 425–437, 1984.

Haynes, L., and Cohen, N., Further characterization of an interleukin–2-like cytokine produced by *Xenopus laevis* T lymphocytes, *Dev. Immunol.*, 23, 1–23, 1993.

Herman, J.R., et al., Distribution of UV radiation at the Earth's surface from TOMS-measured UV-backscattered radiances, *J. Geophysical Res.*, 104, 12059–12076, 1999.

Hird, D.W., et al., *Aeromonas hydrophila* in wild-caught frogs and tadpoles (*Rana pipiens*) in Minnesota, *Lab. Anim. Sci.*, 31, 166–169, 1981.

Holmes, C., and Balls, M., In vitro studies on the control of myoepithelial cell contractions in the granular glands of *Xenopus laevis* skin, *Gen. Comp. Endocrinol.*, 36, 255–263, 1978.

Horton, T., et al., Ontogeny and phylogeny of NK and NK/T cells, *Scand. J. Immunol.*, 54 suppl.1, 78, 2001.

Horton, T.L., et al., *Xenopus* NK cells identified by novel monoclonal antibodies, *Eur. J. Immunol.*, 30, 604–613, 2000.

Horton, T.L., et al., NK-like activity against allogeneic tumor cells demonstrated in the spleen of control and thymectomized *Xenopus, Immunol. Cell Biol.*, 74, 365–373, 1996.

Horton, T.L., et al., Natural cytotoxicity towards allogeneic tumor targets in *Xenopus* mediated by diverse splenocyte populations, *Dev. Comp. Immunol.*, 22, 217–230, 1998.

Horton, T.L., et al., Ontogeny of *Xenopus* NK cells in the absence of MHC class I antigens, *Dev. Comp. Immunol.*, 27, 715–26, 2003.

Hsu, E., and Du Pasquier, L., Studies on *Xenopus* immunoglobulins using monoclonal antibodies, *Mol. Immunol.*, 21, 257–270, 1984.

Hsu, E., Flajnik, M.F., and Du Pasquier, L., A third immunoglobulin class in amphibians *J. Immunol.*, 135, 1998–2004, 1985.

Hubbard, G.B., *Aeromonas hydrophila* infection in *Xenopus laevis*, *Lab. Anim. Sci.*, 31, 297–300, 1981.

Jancovich, J.K., et al., Isolation of a lethal virus from the endangered tiger salamander *Ambystoma tigrinum stebbinsi, Dis. Aquatic Organisms*, 31, 161–167, 1997.

Johnson, M.S., et al., Effects of 2, 4, 6-trinitrotoluene in a holistic environmental exposure regime on a terrestrial salamander, *Ambystoma tigrinum, Toxicol. Pathol.*, 28, 334–341, 2000.

Kakinuma, Y., et al., Molecular cloning and characterization of novel ficolins from *Xenopus laevis, Immunogenetics*, 55, 29–37, 2003.

Kato, Y., et al., Isolation of the *Xenopus* complement factor B complementary DNA and linkage of the gene to the frog MHC, *J. Immunol.*, 153, 4546–54, 1994.

Kato, Y., et al., Duplication of the MHC-linked *Xenopus* complement factor B gene, *Immunogenetics*, 42, 196–203, 1995.

Kau, C.L., and Turpen, J.B., Dual contribution of embryonic ventral blood island and dorsal lateral plate mesoderm during ontogeny of hemopoietic cells in *Xenopus laevis, J. Immunol.*, 131, 2262–2266, 1983.

Kiesecker, J.M., and Blaustein, A.R., Influences of egg laying behavior on pathogenic infection of amphibian eggs, *Conserv. Biol.*, 12, 214–220, 1997.

Kiesecker, J.M., and Blaustein, A.R., Synergism between UV-B radiation and a pathogen magnifies amphibian embryo mortality in nature, *Proc. Natl. Acad. Sci. U.S.A.*, 92, 11049–11052, 1995.

Kiesecker, J.M., and Blaustein, A.R., Pathogen reverses competition between larval amphibians, *Ecology*, 80, 2442–2448, 1999.

Kiesecker, J.M., Blaustein, A.R., and Belden, L.K., Complex causes of amphibian population declines, *Nature*, 410, 681–684, 2001.

Kiesecker, J.M., Synergism between trematode infection and pesticide exposure: a link to amphibian limb deformities in nature? *Proc. Natl. Acad. Sci. U.S.A.*, 99, 9900–9904, 2002.

Koniski, A., and Cohen, N., Mitogen-activated axolotl (*Ambystoma mexicanum*) splenocytes produce a cytokine that promotes growth of homologous lymphoblasts, *Dev. Comp. Immunol.*, 18, 239–50, 1994.

Koniski, A.D., and Cohen, N., Reproducible proliferative responses of salamander (*Ambystoma mexicanum*) lymphocytes cultured with mitogens in serum-free medium, *Dev. Comp. Immunol.*, 16, 441–451, 1992.

Lambris, J.D., et al., The third component of *Xenopus* complement: cDNA cloning, structural and functional analysis, and evidence for an alternate C3 transcript, *Eur. J. Immunol.*, 25, 572–578, 1995.

Lefcort, H., et al., The effects of used motor oil, silt, and the water mold *Saprolegnia parasitica* on the growth and survival of mole salamanders (Genus Ambystoma), *Arch. Environ. Contam. Toxicol.*, 32, 383–388, 1997.

Linzey, D., et al., Role of environmental pollutants on immune functions, parasitic infections and limb malformations in marine toads and whistling frogs from Bermuda, *Int. J. Environ. Health Res.*, 13, 125–148, 2003.

Lips, K.R., Decline of tropical montane amphibian fauna, *Conserv. Biol.*, 12, 106–117, 1998.

Lips, K.R., Green, D.E., and Papendick, R., Chytridiomycosis in wild frogs from southern Costa Rica, *J. Herpetol.*, 37, 215–218, 2003.

Lips, K.R., Mass mortality and population declines of anurans at an upland site in western Panama, *Conserv. Biol.*, 13, 117–125, 1999.

Longcore, J.E., Pessier, A.P., and Nichols, D.K. *Batrachochytrium dendrobatidis* gen. et sp. nov., a chytrid pathogenic to amphibians, *Mycologia*, 91, 219–227, 1999.

Madronich, S., et al., Changes in biologically active ultraviolet radiation reaching the Earth's surface, *J. Photochem. Photobiol. B*, 46, 5–19, 1998.

Maéno, M., Tochinai, S., and Katagiri, C., Differential participation of ventral and dorsolateral mesoderms in the hemopoiesis of *Xenopus*, as revealed in diploid-triploid or interspecific chimeras, *Dev. Biol.*, 110, 503–508, 1985.

Manning, M.J., and Horton, J.D., RES structure and function of the amphibia, in Cohen, N., and Sigel, M.M., Eds., *The Reticuloendothelial System*, Plenum Press, New York, 1982, pp. 423–459.

Meegaskumbura, M., et al., Sri Lanka: an amphibian hot spot, *Science*, 298, 379, 2002.

Middleton, E.M., et al., Evaluating ultraviolet radiation exposures determined from satellite data at sites of amphibian declines in Central and South America, *Conserv. Biol.* 15, 914–929, 2001.

Mills, J.W., and Prum, B.E., Morphology of the exocrine glands of the frog skin, *Am. J. Anat.*, 171, 91–106, 1984.

Milner, A.R., Amphibian-grade tetrapoda, in M.J. Benton, Ed., *The Fossil Record 2*, Chapman and Hall, London, 1993, pp. 665–679.

Mohanty-Hejmadi, P., and Dutta, S.K., Effects of some pesticides on the development of the Indian bullfrog *Rana tigerina*, *Environ. Pollut. Ser. A Ecol. Biol.*, 24, 145–161, 1981.

Nicolas, P., and Mor, A., Peptides as weapons against microorganisms in the chemical defense system of vertebrates, *Annu. Rev. Microbiol.*, 49, 277–304, 1995.

Noble, G.A., and Noble, E.R., On the histology of frog skin glands, *Trans. Am. Microscopy Soc.*, 63, 254–263, 1944.

Ottaviani, E., Trevisan, P., and Pederzoli, A., Immunocytochemical evidence for ACTH- and beta-endorphin-like molecules in phagocytic blood cells of urodelan amphibians, *Peptides*, 13, 227–231, 1992.

Pessier, A.P., et al., Cutaneous chytridiomycosis in poison dart frogs (*Dendrobates* spp.) and White's tree frogs (*Litoria caerulea*), *J. Vet. Diagn. Invest.*, 11, 194–199, 1999.

Rinaldi, A.C., Antimicrobial peptides form amphibian skin: an expanding scenario, *Curr. Opin. Chem. Biol.*, 6, 799–804, 2002.

Robert, J., et al., Minor histocompatibility antigen-restricted CD8 T-cell responses elicited by heat shock proteins, *J. Immunol.*, 168, 1697–1703, 2002.

Rollins-Smith L. A., et al., Antimicrobial peptide defenses of the Tarahumara frog, *Rana tarahumarae*, *Biochem. Biophys. Res. Comm.*, 297, 361–367, 2002c.

Rollins-Smith L.A., Parsons, S.C., and Cohen, N., During frog ontogeny, PHA and Con A responsiveness of splenocytes precedes that of thymocytes, *Immunology*, 52, 491–500, 1984.

Rollins-Smith, L.A., et al., Activities of temporin family peptides against the chytrid fungus (*Batrachochytrium dendrobatidis*) associated with global amphibian declines, *Antimicrob. Agents Chemother.*, 47, 1157–1160, 2003.

Rollins-Smith, L.A., et al., Activity of antimicrobial skin peptides from ranid frogs against *Batrachochytrium dendrobatidis*, the chytrid fungus associated with global amphibian declines, *Dev. Comp. Immunol.*, 26, 471–479, 2002b.

Rollins-Smith, L.A., et al., Antimicrobial peptide defenses against pathogens associated with global amphibian declines, *Dev. Comp. Immunol.*, 26, 63–72, 2002a.

Rollins-Smith, L.A., and Blair, P., Contribution of ventral blood island mesoderm to hematopoiesis in post-metamorphic and metamorphosis-inhibited *Xenopus laevis*, *Dev. Biol.*, 142, 178–83, 1990a.

Rollins-Smith, L.A., and Blair, P., Expression of class II major histocompatibility complex antigens on adult T cells in *Xenopus* is metamorphosis-dependent, *Dev Immunol.*, 1, 97–104, 1990b.

Rollins-Smith, L.A., Hopkins, B.D., and Reinert, L.K., An amphibian model to test the effects of xenobiotic chemicals on development of the hematopoietic system, *Environ. Toxicol. Chem.*, in press, 2004.

Rollins-Smith, L.A., Barker, K.S., and Davis, A.T., Involvement of glucocorticoids in the reorganization of the amphibian immune system at metamorphosis, *Dev. Immunol.*, 5, 145–152, 1997.

Rollins-Smith, L.A., Parsons, S.C.V., and Cohen, N., Effects of thyroxine-driven precocious metamorphosis on maturation of adult-type allograft rejection responses in early thyroidectomized frogs, *Differentiation*, 37, 180–185, 1988.

Schindler, D.W., et al., Consequences of climate warming and lake acidification for UV-B penetration in North American boreal lakes, *Nature*, 379, 705–708, 1996.

Schwager, J., Grossberger, D., and Du Pasquier, L., Organization and rearrangement of immunoglobulin M genes in the amphibian *Xenopus, EMBO J.*, 7, 2409–2415, 1988.

Shimazaki, Y., Maeyama, K., and Fujii, T., Isolation of the third component of complement and its derivative with anaphylatoxin-like activity from the plasma of the newt *Cynops pyrrhogaster, Dev. Comp. Immunol.*, 25, 467–474, 2001.

Sichel, G., Scalia, M., and Corsaro, C., Amphibia Kupffer cells, *Micros. Res. Tech.*, 57, 477–490, 2002.

Simmaco, M., Mignogna, G., and Barra, D., Antimicrobial peptides from amphibian skin: what do they tell us? *Biopolymers (Peptide Science)*, 47, 435–450 1998.

Sjoberg, E., and Flock, A., Innervation of skin glands in the frog, *Cell Tiss. Res.*, 172, 81–91, 1976.

Smith, et al., Preliminary assessment of perchlorate in ecological receptors at the Longhorn Army Ammunition Plant (LHAAP), Karnack, Texas, *Ecotoxicology*, 10, 305–313, 2001.

Stolarski, R.S., et al., Nimbus 7 satellite measurements of the springtime Antarctic ozone decreases, *Nature*, 322, 808–810, 1986.

Sullivan, K.B., and Spence, K.M., Effects of sublethal concentrations of atrazine and nitrate on metamorphosis of the African clawed frog, *Vet. Immunol. Immunopathol.*, 22, 627–635, 2003.

Susarla, S., et al., Perchlorate identification in fertilizers, *Environ. Sci. Technol.*, 33, 3469–3472, 1999.

Suwalsky, L., et al., Toxic action of the herbicide 2, 4-D on the neuroepithelial synapse and on the nonstimulated skin of the frog *Caudiverbera caudiverbera, Bull. Environ. Contam. Toxicol.*, 62, 570–577, 1999.

Suwalsky, M., Norris, B., and Benites, M., The toxicity of exposure to the organochlorine, dieldrin, at a sympathetic junction and on the skin of the frog, *Caudiverbera caudiverbera, Hum. Exp. Toxicol.*, 21, 587–591, 2002.

Suzuki, M., et al., *Xenopus laevis* macrophage migration inhibitory factor is essential for axis formation and neural development, *J. Biol. Chem.*, 279, 21406–21414, 2004.

Taylor, S.K., et al., Causes of mortality of the Wyoming toad, *J. Wildl. Dis.*, 35, 49–57, 1999.

Taylor, S.K., Williams, E.S., and Mills., K.W., Experimental exposure of Canadian toads to *Basidiobolus ranarum, J. Wildl. Dis.*, 35, 58–63, 1999a.

Taylor, S.K., Williams, E.S., and Mills, K.S., Effects of malathion on disease susceptibility in Woodhouse's toads, *J. Wildl. Dis.*, 35, 536–541, 1999b.

Turpen, J.B., Induction and early development of the hematopoietic and immune systems in *Xenopus, Dev. Comp. Immunol.*, 22, 265–278, 1998.

Urbansky, E.T., Perchlorate chemistry: implications for analysis and remediation, *Biorem. J.*, 2, 81–95, 1998.

Veldhoen, N., and Helbing, C.C., Detection of environmental endocrine-disruptor effects on gene expression in live *Rana catesbeiana* tadpoles using a tail fin biopsy technique, *Vet. Immunol. Immunopathol.*, 20, 2704–2708, 2001.

Warr, G.W., Ruben, L.N., and Edwards, G.J., Evidence for low molecular weight antibodies in the serum of a urodele amphibian, *Ambystoma mexicanum, Immunol. Lett.*, 4, 99–102, 1982.

Watkins, D., and Cohen, N., Mitogen-activated *Xenopus laevis* lymphocytes produce a T-cell growth factor, *Immunology*, 62, 119–125, 1987.

Watkins, D., Parsons, S.C., and Cohen, N., A factor with interleukin-1-like activity is produced by peritoneal cells from the frog, *Xenopus laevis, Immunology*, 62, 669–673, 1987.

Yan, N.D., et al., Increased UV-B penetration in a lake owing to drought-induced acidification, *Nature*, 381, 141–143, 1996.

Zasloff, M., Antimicrobial peptides of multicellular organisms, *Nature*, 415, 389–395, 2002.

Zou, J., et al., Molecular cloning of the gene for interleukin-1β from *Xenopus laevis* and analysis of expression in vivo and in vitro, *Immunogenetics*, 51, 332–338, 2002.

CHAPTER 7

Earthworms as Ecosentinels for Chemical-Induced Immunotoxicity

Arthur J. Goven, Barney J. Venables, and Lloyd C. Fitzpatrick

CONTENTS

7.1 INTRODUCTION

In this chapter we explore the use of earthworms for modeling chemical-induced immunotoxic potential of xenobiotics in terrestrial ecosystems. We begin by listing the credentials of earthworms for use in ecoimmunotoxicology. Next, we briefly describe the earthworm immune system and highlight the immune functions that we believe qualify them as attractive candidates for use as immune markers (i.e., biomarkers of immunotoxicity) in ecological risk assessment. In the remainder of the chapter, we offer our thoughts on what should guide the selection of and research on earthworm immune markers for use in ecotoxicology by focusing on four essential elements: (1) normalization of exposure metrics across taxon-specific toxicokinetics; (2) consideration of and control for natural variation in immune marker performance, and influence of ambient conditions on the performance; (3) use of pathogen-challenge assays to document significance of immune marker effect; and (4) affirmative linkage of immune markers to endpoints having ecological significance.

7.2 EARTHWORM CREDENTIALS AS ECOSENTINELS OF IMMUNOTOXICITY

Ecotoxicologists should agree that prediction of chemical-induced immunotoxic risks across key wildlife taxa and to ecosystems requires use of sentinel and surrogate species. But, by what criteria should we judge the appropriateness of candidates for these roles? For ecological relevance the sentinel should be among the first to experience the chemical challenge and exhibit an easily diagnosed response of a meaningful biological performance. Early detection is as important for ecological health as it is for public health. Species at or near the base of the food chain or web should satisfy the first-responder criterion. Selection among these depends on which species has an immune marker that comes closest to satisfying a set of rigorous criteria (Table 7.1). The value of an immune marker may be judged according to its intended use or role in immunotoxicology. The proximate role of an immune marker includes screening chemicals before use (e.g., agrochemicals), testing existing environmental contaminants to determine their toxic profile and mode of action, and as an *in situ* red flag alerting the regulatory community of a possible environmental threat. The ultimate role is to predict ecologically relevant immunotoxic effects of chemical xenobiotics. That role includes affirmatively linking the effects on the surrogate's immune marker to an immune marker in higher wildlife and then to an important ecology-level performance measure.

For terrestrial ecosystems, including hazardous waste sites (HWSs), earthworms are ideal candidates for both use as ecological sentinels and surrogates. As sentinels, earthworms are excellent indicators of the bioavailability potential of chemical contaminants in the soil. Their high body surface/body mass ratio and feeding behavior provide two effective exposure routes: dermal and gastrointestinal. Close to the base of the food chain or web, earthworms can be useful in modeling transfer of chemicals from the soil to specific higher wildlife. As important soil organisms, earthworms are themselves potential ecological endpoints of chemical threats to terrestrial ecosystems. Earthworms' potential for predicting immunotoxic risks to higher wildlife is based on their well-studied immune system, known to have functions sufficiently analogous and homologous to those in vertebrates for use as surrogate immune markers (see Sections 7.3 and 7.4). Earthworms' wide use in standardized toxicity protocols (Organisation for Economic Cooperation and Development, 1984; European Economic Community, 1985; Greene et al., 1989) and established toxic endpoints (e.g., mortality, growth, and reproduction) enhance their status as sentinels and surrogates for immunotoxic studies. Consequently, earthworms offer an eclectic array of ecoindicators of chemical threats throughout terrestrial ecosystems. Earthworm attributes relevant to their use in ecotoxicology are summarized in Table 7.2.

Table 7.1 Criteria for Ecologically Relevant Immune Markers

Broadly applicable or predictive of chemical-induced immunotoxicity across key wildlife taxa
Readily measured and standardized
Sensitive to chemical exposures/doses significantly below lethal concentrations
Low natural (nonchemical linked) variability
Diagnostic of relevant chemical xenobiotics
Capable of exposure/dose–response profiles
Capable of *in vivo* and *in vitro* use
Affirmatively linked to whole organism-level performance(s) having ecological relevance
Cost-effective

Table 7.2 Earthworm Attributes for Ecoimmunotoxicology

Intimate contact with the soil and its contaminants
High dermal surface area to volume or mass ratio
Both dermal and gastrointestinal exposure routes for chemicals
Widespread in terrestrial ecosystems
Existing toxicity protocols
History of use in ecotoxicology
Inexpensive
Well-known biology
Well-studied immune system
Immune functions that are analogous or homologous to those in higher wildlife
Easily exposed to chemicals and chemical mixtures
Socially noncontroversial

7.3 EARTHWORM IMMUNE SYSTEM

Seminal studies by Cooper (Hostetter and Cooper, 1974; Cooper, 1976) have defined the earthworm immune system, providing the foundation for its use in ecoimmunotoxicology. These studies have shown that earthworm immunity involves coelomocytes, which are housed in the coelomic cavity together with coelomic fluid. The coelomocyte population is composed mainly of phagocytic cells that have functions similar to those of macrophages and granulocytes in higher vertebrates. Coelomocytes are active in (1) the innate immune functions such as phagocytosis, and the inflammatory response, (2) the more complex cell-mediated responses mediating reactions such as recognition of foreign or altered cells, and (3) humoral immune responses involving synthesis and secretion of agglutinins and lytic factors. The earthworm's immune system is compared with other taxa in Table 7.3.

7.3.1 Innate Immune Functions in Earthworms

The nonspecific, phagocyte-mediated immune response that forms the earthworm's frontline defense against foreign material is homologous to that in other animals, including vertebrates. Coelomocytes seek out phagocytose, and then destroy nonself material through oxygen-dependent mechanisms including hydrogen peroxide and superoxide production (Chen et al., 1991) and oxygen-independent mechanisms such as lysozyme production (Goven et al., 1994). Coelomocytes also exhibit a well-defined inflammatory response, sequentially infiltrating injured and infected tissues as do leukocytes in vertebrates (Valembois, 1974).

7.3.2 Cell-Mediated Immune Functions in Earthworms

The principal role of cell-mediated immunity is detection and elimination of altered self and foreign components, and cells harboring intracellular parasites. Transplantation studies using auto-, allo-, and xeno-geneic tissue demonstrate the presence of cell-mediated immunity in earthworms. Earthworms are capable of recognizing and rejecting foreign tissue grafts while accepting autografts (Cooper, 1971). Xenografts are rejected more vigorously than allografts. Second-set graft rejection is accelerated, suggesting a memory component (Cooper and Roch, 1986). Graft rejection kinetics imply that the earthworm has a cell-mediated defense system that possesses memory and nonself recognition, or at minimum a capacity to respond to the absence of self.

7.3.3 Humoral Immunity in Earthworms

The vertebrate humoral immune response, mediated by soluble antibody, protects against bacteria and foreign macromolecules, mainly toxins. Derived antibody has not been reported for invertebrates.

Table 7.3 Taxonomic Comparison of Immune Response

Taxonomic Group	Nonspecific Innate Immunity	Specific Adaptive Immunity	Protective Enzymes and Enzyme Cascades	Phagocytosis	Antimicrobial Factors	Rejection of Foreign/ Altered Tissue	T and B Lymphocytes	Immunoglobulin
Invertebrates								
Porifera	+	–	?	+	?	+	–	–
Annelids (earthworms)	+	–	?	+	+	+	–	–
Arthropods	+	–	+	+	+	?	–	–
Vertebrates[a]	+	+	+	+	+	+	+	+

[a] Elasmobranchs, teleosts, amphibians, reptiles, birds, and mammals.

Key: + = definitive demonstration; – = failure to demonstrate thus far; ? = presence or absence remains to be established.

Source: Goldsby, R.A., et al., Eds., *Immunology*, 5th ed., W.H. Freeman and Co., New York, 2000.

However, earthworm coelomocytes, as well as cells from other invertebrates, synthesize and secrete an array of partially characterized opsinins, agglutinins, and lysins that exhibit properties functionally analogous to vertebrate immunoglobulin. Agglutinins, shown to aggregate and opsonize foreign material, thereby facilitating phagocytosis, are specifically induced by and react with antigen (Wojdani et al., 1982). Lytic factors, which inhibit the growth of bacteria, are important to earthworm defense against pathogens (Lasseques et al., 1989). Bacteriostatic and bactericidal activity can be induced by inoculating earthworms with sublethal numbers of bacteria, resulting in immunization and resistance to challenge (Stein and Cooper, 1982).

7.4 EARTHWORM IMMUNEMARKERS

7.4.1 Potential Earthworm Immunomarker Candidates

Since the late 1980s, when the earthworm's immune system was proposed for use in immunotoxicology (Goven et al., 1988; Fitzpatrick et al., 1990; Venables et al., 1992), a number of its immune components have been proposed and investigated for their potential as biomarkers of chemical-induced threats (Table 7.4). Although several of these show promise as immunemarkers of chemical-induced effects, we believe that only a few have sufficient potential for clear affirmative linkage to effects of ecological relevance to merit intensive investigation at this time. Consequently, we suggest that progress in developing earthworm immunemarkers for ecotoxicology would be best served by focusing in depth on only the most promising candidates, not merely increasing the number of red flags. Thus, we heartily agree with De Coen (2000a) that "[r]ather than continuously focusing on development of new biomarkers [immunemarkers], more effort should be conducted towards the ecological evaluation of the existing set of endpoints." As such, we disagree somewhat with Recommendation 34 from the Sheffield Workshop on earthworm ecotoxicology (in which we were participants): "[E]ncouragement should be given to ... develop new sensitive endpoints..." (Greig-Smith, 1992).

Table 7.4 Earthworm Immune Markers Having Potential to Assess Chemical Immunotoxicity

Class	Response Monitored	Earthworm Immunoassay
Immunopathology	Cytology	Complete and differential coelomocyte counts
	Body mass	Total body weight
	Histology	Coelomatopoietic tissue (somatopleure)
	Inflammatory	Wound healing
	Immune cell quantitation	Flow cytometric analysis of coelomocytes
Nonspecific immunity	Enzyme function	Lysozyme activity (coelomic fluid, coelomocyte extracts)
	Natural killer cell activity	Allogeneic and xenogeneic target cells
	Phagocytosis	Erythrocytes and fluorescent beads
		Coelomocyte bacterial ingestion
	Bacterial activity	Coelomocyte bacterial killing
	Enzyme function	Nitroblue-tetrazolium dye reduction by coelomocytes
Humoral-mediated immunity	Lytic factors	Erythrocyte lysis
	Agglutinins	Secretory rosette formation
	Challenge response	Erythrocyte agglutination
	Antimicrobial action	Bacterial agglutination; lytic factor erythrocyte plaque-forming cells; bacteriostatic effect of coelomic fluid
Cell-mediated immunity	Alloantigen recognition	Allogeneic tissue recognition and rejection
	Blast transformation	Mitogen/antigen blastogenesis
	Mixed leukocyte response	Allogeneic/xenogeneic stimulation
Host-resistance challenge models	Bacterial models	Resistance to bacterial challenge

7.4.2   Recommended Earthworm Immune Marker Candidates

The most promising components of the earthworm's immune system for predicting risks to other wildlife (i.e., surrogate immune markers) are the ones homologous across animal taxa (i.e., phylogenetically conserved). Of these, we believe that phagocytosis by coelomocytes and the associated biochemical responses involved in destroying foreign material, and perhaps the inflammatory response involving coelomocytes, plus host-resistance challenge assays (Table 7.5) should be given the greatest immediate investigative attention.

We include several cytological parameters — that is, total/complete cell count (TCC), differential cell count (DCC), cell viability (CV), and coelomocyte repopulation rates (CRR) — in our list of candidate immune markers (Table 7.5), because they can provide an overview of and effects of chemicals on the general health of the structural parts of the earthworm's immune system (Eyambe et al., 1991). Collectively, they are a logical set of easily measured, sensitive, and stable markers for assessing the overall health of an earthworm, potential problems in its coelomatopoietic tissue and/or coelomocyte immune function, much as a complete blood count is used in vertebrates.

Phylogenetically conserved nonspecific immune responses of phagocytosis, and subsequent destruction of foreign material through formation of oxygen radicals and release of lysozyme, along with inflammation, provide fundamental cellular level information about an earthworm's immune status, which individually and collectively can be effective in forcasting chemical effects across animal taxa. Phygocytosis of various items, including yeast cells, red blood cells, bacteria, and fluorescent beads, by earthworm coelomocytes has been shown to be perhaps the most reliable and cost-effective of our recommended immunemarker candidates (Eyambe et al., 1991; Fitzpatrick et al., 1992; Goven et al., 1993; Giggleman et al., 1998; Burch et al., 1999). However, phagocytosis is part of a sequence that ultimately results in the destruction of nonself material. Stimulation and activation of coeolomocytes are prerequisite to the actual ingestion of foreign material. As such, chemical effects on coelomocyte spreading, which indicates membrane activation, could provide an easier and quicker assessment of the chemical's potential effect on the phagocytic ability of coelomocytes than actually measuring phagocytosis. We suggest exploring the use of spreading for its potential as an immune-marker adjunct to actual phagocytic assays.

Antimicrobial actions of coelomocytes and other phagocytic cells across taxa are principally carried out by oxygen-dependent respiratory burst mechanisms involving superoxide anion (O_2^-) and hydrogen peroxide (H_2O_2) production, and by oxygen-independent mechanisms such as lysozyme production. The nitroblue-tetrazolium (NBT) dye reduction assay is commonly used to

---

Table 7.5   Recommended Earthworm Immune Marker Candidates

Cytological (coelomocyte) parameters
Total/complete cell count (TCC)
Differential cell count (DCC)
Cell viability (CV)
Coelomocyte repopulation rates (CRR)

Nonspecific Cellular Immunity
Phagocytosis
Coelomocyte spreading
Ingestion of bacteria, yeast, fluorescent beads
Anti-microbial action (colelomocyte mediated)
Lysozyme activity of cell extracts
Neutral red dye retention
Nitroblue tetrazolium dye reduction by coelomocytes (H_2O_2, O_2^-)
Inflammation
Wound healing
Host resistance
Whole organism challenge
Intracellular bacterial killing

---

evaluate the ability of phagocytes to catabolize and kill phagocytosed bacteria through an oxidative process (Braunde, 1981). This assay, which indirectly measures superoxide and hydrogen peroxide production, was developed to detect metabolic defects in neutrophils associated with chronic granulomatous disease in humans. Earthworm coelomocytes have been shown to reduce NBT dye in a near linear manner, similar to that of mammalian leukocytes (Chen et al., 1991). The NBT assay has potential to identify chemicals and concentrations that do not affect cell stimulation, activation, and/or actual ingestion, but interfere with the general intracellular oxidative bactericidal activity common to all phagocytes.

Lysozyme is an enzyme capable of bactericidal activity through action on peptidoglycan of Gram-positive bacterial cell walls, and functions as a component of an organism's nonspecific antibacterial defense system (Salton, 1975). Lysozyme activity, present in coelomic fluid and coelomocyte extracts, appears to be homologous to that found in mammalian serum and leukocyte extracts (Goven et al., 1994). Phylogenetic conservation indicates that lysozyme is a constituent of the primitive immune defense mechanism associated with the granulocyte, monocyte–macrophage system in vertebrates, and with coelomocytes in earthworms. Reduced lysozome activity in earthworms from chemical exposure should be predictive of chemical immunotoxicity in higher organisms, resulting in a suppressed ability to kill Gram-positive bacteria. The neutral, red dye retention assay indirectly measures lysosomal function by assessing membrane stability (Booth and O'Halloran, 2001). Healthy, unstressed cells retain neutral red dye, in contrast to chemically stressed cells that leak the dye from lysosomes into the cytoplasm. As such, lowered retention time could be an effective indicator of reduced lysosomal function.

The inflammation response, as we have measured through wound healing in earthworms (Cikutovic et al., 1999), has potential as a sensitive marker of chemical toxicity across taxa. This immune marker, unlike the others mentioned, directly assesses a chemical's threat potential to the entire earthworm, and has the ability to heal a lesion in the integument. The other immune markers, regardless of their sensitivity, reliability, or cross-taxa comparability, are only suggestive of a chemical's effect on the earthworm's immunocompetence. Simply reducing phagocytosis or production of oxygen radicals and/or lysozyme does not mean the earthworm is immunocompromised. As such, the ultimate assessment of chemical effects on one or more of these immune markers is through challenge studies using earthworms or coelomocytes exposed to chemicals demonstrated to significantly alter an immune marker. After chemical exposure, earthworms or cells are exposed to or dosed with appropriate bacterial, viral, or parasitic agents. Challenge agents may be different across species, but should be selected to provide comparable interspecies information. Assays measure survivorship or cell viability and ability to elicit a protective immune response.

7.5 INTERACTION OF EXOGENOUS CHEMICALS WITH EARTHWORM IMMUNE MARKER CANDIDATES

Studies have shown that our recommended earthworm immune marker candidates are sensitive to exposure to several classes of chemicals. Unfortunately, virtually nothing is known about the molecular basis of earthworm responses to exposure to these chemicals. For that matter, relatively little detail on the molecular mechanisms of immunotoxicity is known for even the well-studied rodent models. In the following, we attempt to place some of what is known about mechanisms of immunotoxicity in traditional animal models into the context of what is known about earthworm biology.

7.5.1 Heavy Metals

Metal toxicity often involves multiple sites of action affecting specific enzymes and membrane integrity at the cellular and organelle levels. Lead, arsenic, mercury, and cadmium have been best studied for immunotoxic effects (Burns-Naas et al., 2001). Specific effects are influenced by varying toxicokinetic patterns affecting disposition. Earthworms sequester heavy metals with metallothionein-like metal binding proteins or MBPs (Scott-Fordsmann and Weeks, 1998), and are known to accumulate high concentrations of heavy metals in specialized organelles found in chloragocytes lining the gut wall; chloragosomes are constitutive, whereas cadmosomes are apparently inducible organelles. Several studies have documented the importance of these organelles in heavy metal disposition in earthworms as well as their value as biomarkers of exposure (Morgan et al., 2002). Earthworm studies have demonstrated immunotoxic effects for a variety of heavy metals, including mercury, cadmium, copper, nickel, and zinc (Fitzpatrick et al., 1996; Fugere et al., 1996; Burch et al., 1999; Nusetti et al., 1999; Reinecke et al., 2002; Sauve et al., 2002).

7.5.2 Polycyclic Aromatic Hydrocarbons

The immunotoxicity of polycyclic aromatic hydrocarbons (PAHs) observed in mammals is primarily associated with the formation of toxic oxygenated metabolites. The process has been studied in detail for some PAHs, most notably benzo(a)pyrene (BaP). Formation of BaP toxic metabolites results from cytochrome P450 metabolism initiated via the aryl hydroxylase (Ah) receptor. The 7,8,9,10-diol epoxide of BaP that results forms protein and nucleic acid adducts and results in a range of mutagenic, carcinogenic, and immunotoxic effects. Earthworms have a high potential for exposure to PAHs and possess cytochrome P450 activity (Achazi et al., 1998; Eason et al., 1998; Lee, 1998). Increased DNA adducts have been reported for earthworms from PAH-contaminated habitats (Van Shooten et al., 1995). To our knowledge, the relationship between the formation of cytochrome P450 metabolites of PAHs and immune function in earthworms has not been reported.

7.5.3 Halogenated Aromatic Hydrocarbons

This class of potential immunotoxicants includes a variety of persistent chlorinated and brominated aromatics with high potential for bioaccumulation. Chlorinated biphenyls (PCBs), chlorinated dioxins, and chlorinated furans have received particular attention in immunotoxicology. Tetrachlorodibenzo-p-dioxin (TCDD), extremely immunotoxic in mice (Holsapple et al., 1991), like PAHs has a high affinity for the Ah receptor. The toxicity of these compounds has been described as falling into two categories: (1) congeners that share structural similarities with TCDD (PCB congeners with coplanar structures); and (2) nonplanar congeners with activities as hormone mimics and/or calcium ionophoric or membrane-perturbing properties (Rice, 2001). Earthworms readily bioaccumulate these compounds (Matscheko et al., 2002). The commercial PCB mix Aroclor 1254 has been shown to have immunotoxic effects on earthworm coelomocytes (Rodriguez-Grau et al., 1989; Fitzpatrick et al., 1992; Goven et al., 1993; Suzuki et al., 1995; Burch et al., 1999).

7.5.4 Pesticides

This category includes a wide variety of organic compounds, most of which are neurotoxic. Many of these compounds are also lipophilic, bioaccumulate, and produce a variety of immunotoxic effects (Voccia et al., 1999). Many of the pathways involved in the metabolism of pesticides in vertebrates also occur in earthworms (Stenerson, 1984, 1992). Suppression of earthworm immunological response has been demonstrated for pentachlorophenol (Giggleman et al., 1998), chlordane (Giggleman, 1997), and diazinon and chlorpyrifos (Booth and O'Halloran, 2001). Purified pentachlorophenol exhibits broad toxicity characteristic of phenolic compounds that uncouple oxidative metabolism. The immunosuppressive characteristics of technical-grade pentachlorophenol are probably due to nonpentachlorophenol contaminants acting through other mechanisms (Kerkvliet et al., 1982). The primary target for immunosuppressive effects of chlordane in mammals is the macrophage

(Burns-Naas, 2001). Mechanisms responsible for effects on this mammalian phagocytic cell may also be responsible for chlordane's suppression of earthworm coelomocyte phagocytic activity (Giggleman, 1997).

7.6 RESEARCH DIRECTIONS

Metazoan defense includes a wide variety of interacting biochemical events that provide both constitutive protection as well as pathogen-induced responses that can be ramped up in the face of attack. Constitutive defenses vary widely among taxa and depend heavily on the basic biology, habitat, and ecology of a given species. Pathogen-induced responses also may vary widely in nature of their effector molecules and effector cells, but share a common denominator in the similarity of pathways associated with pathogen perception and up-regulation of genes responsible for the coordinated pathogen-induced response. A great deal of evidence from molecular immunology supports the view that most metazoans (and perhaps plants) coordinate these basic defense responses through membrane surface receptors (pattern recognition receptors) sensitive to generalized patterns of pathogen biochemical signatures or pathogen-associated molecular patterns (PAMPs). These receptors appear to be linked to similar transcription factors (nuclear factor kappa B or NFkB-like transcription factors) through a signal cascade shown to be remarkably consistent in a wide variety of taxa (the Toll pathway). These pathways can be activated in a variety of ways, but the common outcome is the activation of genes resulting in production of reactive oxygen and nitrogen species, including hydrogen peroxide, superoxide, and nitric oxide (Nappi and Ottaviani, 2000). This common thread of defense-related biochemical responses is among the most conserved features in the biology of multicellular organisms (Janeway et al., 2001; Janeway and Medzhitov, 2002).

If the core of innate immune function is phylogenetically conserved through processes of recognition and effector molecules coordinated through similar receptors and related transcription factors, these processes should be useful as biomarkers of chemical effects on immune function across taxa. This extrapolation should extend to a wide variety of invertebrates and vertebrate innate immunity.

In suggesting the form that such extrapolations might take for successful translation from a laboratory model to field conditions, we propose the following guidelines for further development of earthworm immune markers to improve their value in ecological risk assessment.

7.6.1 Exposure Metrics: Normalization of Immunotoxicokinetics Across Taxa

Valid extrapolation of immunotoxic effects of chemicals in earthworms to other species requires leveling the playing field for expressing exposure (or dose/)response among taxa having different biologic attributes. We can find guidance for doing this in the considerable body of information on cross-taxa comparisons of chemical lethality or LC_{50s}. Two primary requirements for interspecific comparisons have emerged as paradigms over the many years of using survivorship data for assessment of chemical risks. First, data must be normalized to reflect true chemical exposure by using whole body or tissue concentrations — the critical body residue approach (CBR) proposed by McCarty and MacKay (1993) — rather than relying on environmental concentrations to estimate exposure. Second, data must be normalized for the steady-state endpoint response (Lanno and McCarty, 1997). In a practical sense, we must be able to accurately relate a measure of true chemical dosage to endpoint response as a function of exposure time such that (1) a transient response is reliably captured, or (2) the point at which the relationship no longer varies with time can be estimated. The second condition represents the sublethal effect–concentration equivalent of the incipient lethal level (ILL) (Sprague, 1969), and avoids uncertainties associated with transient responses. Data sets that allow normalization in this fashion have demonstrated that organisms with drastically different LC_{50} estimates may have similar chemical sensitivities when compared on the basis of lethal body burdens or CBRs (Lanno and McCarty, 1997). Valid comparisons of sublethal immune impairment concentrations across taxa require this approach.

7.6.2   Control of Natural Sources of Variability in Immune Marker Response

Very little information is available regarding the performance of immune markers in natural earthworm populations (see Svendsen et al., 1998, for a promising exception). Obviously, mean-ingful interpretation of potential chemical effects requires sufficient knowledge of background variability attributable to adaptation, acclimatization, and other confounding factors (Kammenga et al., 2000) to permit reasonable linkage of a chemical cause and an immune marker effect. Such studies will require significant investment in field studies that encompass at least annual cycles of data collection from control and contaminated sites of comparable habitats. Estimates of field effects can be made in shorter-term laboratory studies using field-collected soils, but at the expense of ecological relevance. Laboratory studies using natural soils or artificial substrates amended with neat toxicants allow much greater control of experimental variables (e.g., temperature, moisture) and are appropriate for some kinds of testing and immune marker development, but are of dubious relevance to risk assessment in the field.

7.6.3   Linkage of Immune Marker to Ecologically Significant Endpoints

Although progress has been made in the development of immune markers in earthworms, consid-erable experimental work remains before even the most promising immune markers can be used effectively as early warning ecosentinels. To take these immune markers beyond that proximate role to the surrogate status where they can be used to assess risk to higher wildlife is a daunting task. Morgan et al. (1999), upon summarizing others, stated that "from the point of view of environmental protection it is ecological-level changes that are paramount, but that these effects are evoked through stress-induced disturbances of cellular and molecular functions in individual organisms. Thus it should theoretically be possible to use biomarkers not only as early warning stress indicators, but also to anticipate or predict population and community responses." They further state that "the [biomarker] approach would gain wider acceptance ... if biomarker responses could be linked, preferably mechanistically, to higher organizational level performance indicators." However, as De Coen (2000a) rightly concludes, there has been "little progress in solving [this] 'linear ecotoxicology paradigm'...." This is expected from the principle of functional integration: new properties emerge (emergent properties) at each level of organization that cannot be predicted from lower levels. This is the essence of Fry's paradigm: "You take the properties of a level of organization and use,those observations to analyze the next level of organization below it. If you take the properties too many steps down, you're being stupid; and you cannot go the other direction" (cited in McCarty, 1996 from Kerr, 1976). So, how do we take earthworm immune markers from the sentinel and screening status to higher levels? Or, is it even possible to extrapolate upward from chemical-induced alterations in the performance of immune markers?

Because risk assessment relies heavily on use of biomarkers and organism-level bioassays, it is important to establish affirmative linkage between immune markers and whole earthworm performance measures that in turn have potential linkage to at least their own populations, which are important unto themselves to terrestrial ecosystems. We caution investigators that just because an observed immune marker effect is statistically significant, immunologic significance must be confirmed through pathogen challenge. Also, the observed effect may not be immune specific, but part of a general toxicity. And, no observable effect level does not assure us that there is no real effect. As discussed above, we advocate taking the CBR approach for establishing the affirmative linkage to higher wildlife (i.e., the cross-taxa dose–response correspondence between homologous immune markers) and linkage to relevant ecological endpoints. Clearly, demonstration of risk to important wildlife has ecological relevance, but mere suppression of an immune marker does not translate linearly to an ecological effect. Using successive levels of organization, we look first at the whole organism (earthworm) to see if there is linkage to the organism's level of performance that has potential relevance to the next level, that is, the population. Effects on populations have ecological relevance, portending ramifications throughout the biotic community. Individual endpoints that could be affirmatively linked to demographic performance measures are survival, growth, and reproduction. Population measures include primary characteristics (mortality, immigration, and emigration) and derived (emergent) secondary characteristics (density, dispersion, age structure, and secondary production).

The principle of allocation (Levins, 1968) provides a conceptual framework for linking immunological stress, as measured by an immune marker, in earthworms to the key organism-level performance of growth and reproduction. Essentially, all organisms exist on a finite resource budget that can be evaluated in terms of energy apportionment between or among competing activities during the organism's life history. Energy input through feeding minus losses through ingestion and excretion is net metabolizable energy (NME), which is available for production ($P = P_{growth}$, $P_{reproduction}$, $P_{secretion}$, $P_{replacement}$, ...) and performance of work. Metabolic costs (R) associated with all processes from the cell to the whole organism are integrated as the organism's metabolic heat production, usually measured using O_2 respirometry. Thus, because $NME - R = P$, any increase in metabolic costs should be reflected in reduced production. If a chemical-induced immunological stress increases the metabolic costs to an earthworm, it may be possible to link that to a compromise in the earthworm's growth or production or both. De Coen (2000a), referring to the metabolic cost hypothesis (see Forbes et al., 1996), states that "changes in energy metabolism, in general, will ultimately influence the future life characteristics of an organism. Responses to contaminants are considered to be a metabolic cost for an organism." The common currency in which costs of chemical damage are ultimately expressed is energy availability for activities, which improve fitness. The ultimate price paid by an organism in response to toxic insult is decreased energy availability for activities contributing to reproductive success. This is true regardless of taxon or toxicant. Thus the most proximate and universally applicable manifestation of a sublethal biomarker effect (immune marker in this case) that has true ecological significance, is disruption of patterns of energy flow at the individual organism level that decrease its fitness. The first task is to measure the cost of a sublethal immunological stress to an earthworm's growth or reproduction. De Coen (2000b) outlines a fruitful approach to this problem using a cellular energy allocation (CEA) biomarker in *Daphnia magna*. The CEA involves measuring changes in energy reserves in carbohydrates, lipids, and proteins, and energy consumption by electron transport activity. De Coen and Jansson (2003) provide an excellent example of linking biomarkers to population effects in *D. magna*. We believe that the CEA approach is conceptually sound and provides a model for use in developing the affirmative linkage between immune markers in earthworms and key demographically relevant endpoints. But, at present we are a long way from doing so for earthworms such as *L. terrestris*. It may be easier using *Eisenia fetida*, which is easy to culture, but has less relevance to terrestrial communities than other earthworms. Reinecke et al. (2002) have made an exemplary effort in that direction by linking the neutral retention red dye marker to life-cycle traits in *E. fetida*. Regardless of the encouraging steps to affirmatively link a chemical-induced effect on an immune marker to a whole earthworm performance measure, we still have a difficult issue to resolve: "Is the chemical's effect actually immunotoxic or is the immune marker response just a manifestation of a more systemic sublethal effect (i.e., an artifact of the choice of the biomarker)?" We believe that resolution to this issue requires using a pathogen challenge assay to test for reduced immunocompetence. To date, such an assay has not been well developed for earthworms. There is a critical need for development of a rapid, easily conducted assay of this type that can be validated across a variety of combinations of immunotoxicants and immune markers.

7.7 SUMMARY

Herein we argue, as have others elsewhere, that earthworms have great potential for use as sentinels and surrogates in ecotoxicology. Specifically, we believe that earthworms and several of their immune functions show a promising proximate or sentinel role in screening chemicals (e.g., agrochemicals) prior to application or use, and evaluating existing chemical contaminants in terrestrial ecosystems. Bringing these immune markers on line as reliable screening and/or early warning endpoints requires rigorous experimental evaluation. First, the inherent variability of each immune marker's performance must be assessed. This is not an easy task when using earthworms collected from native populations. For a single species (e.g., *L. terrestris*), we would expect baseline performance (e.g., phagocytosis) to vary within an individual over time (e.g., age/developmental influence, seasonal influence), between individuals within the same population (intrademic), and between different populations (interdemic or ecotypic). We would also expect variations in performance related to laboratory conditions, maintenance protocol, and experimental procedures. Although researchers establish controls for each experiment, it is essential to know the chemical/contaminant-independent performance of the immune marker and how other factors influence that performance. Additionally, when assessing the influence of the chemical or contaminant, it is important to know something about the chemicals in the soil from where the earthworms originate to avoid using earthworms that may be acclimatized or adapted to the chemicals of interest. In evaluating the immune marker's sensitivity to chemicals and its response variability, it is essential to use reference toxicants. These can be used as positive controls during experiments on different groups of earthworms. For both reference toxicants and chemicals to be tested, exposure-dose (i.e., uptake) profiles and subsequent dose–immune response or performance relationships must be determined. The importance of doing this is clearly developed and instructively outlined in McCarty and MacKay (1993), McCarty (1996), and Lanno and McCarty (1997). These papers address the critical body residue (CBR) and incipient lethal level (ILL) approaches, respectively. We are suggesting taking a CBR and an incipient immunotoxic level (IIL) approach in developing and using earthworm immune markers, affirmatively linking the immune markers to whole earthworm performance measures, and validating the chemical-induced changes in the markers through a pathogen challenge assay.

The process might be initiated by a detailed examination of relatively well-studied endpoints (e.g., phagocytosis) in combination with selected reference immunotoxicants (e.g., a heavy metal and an organic toxicant) and a small suite of model organisms (e.g., earthworm, bivalve, fish, *Xenopus*, rodent). With sufficient data collection focused on a single such biomarker, driving factors important in making cross-taxa predictions could be revealed. Such studies would go a long way toward establishment of a user's manual (sensu De Coen, 2000b) necessary to usefully employ the immune marker as a risk assessment tool.

REFERENCES

Achazi, R.K., et al., Cytochrome P450 and dependent activities in unexposed and PAH-exposed terrestrial annelids, *Comp. Biochem. Physiol. C Pharmacol. Toxicol. Endocrinol.*, 121, 339–50, 1998.

Booth, L.H., and O'Halloran, K., A comparison of biomarkers responses in the earthworm Aporrectodea caligenosa to the oreganophosphorus insecticides diazinon and chlorpyrifos, *Environ. Toxicol. Chem.*, 11, 2494–2502, 2001.

Braunde, A.I., Mechanisms of natural resistance to infection, in *Medical Microbiology and Infectious Disease*, Braunde, A.I., Davis, C.E., and Fierer, J., Eds., Saunders, Philadelphia, 1981, pp. 739–756.

Burch, S.W., et al., In vitro earthworms Lumbricus terrestris coelomocyte assay for use in terrestrial toxicity identification evaluation, *Environ. Contam., Toxicol.*, 62, 547–54, 1999.

Burns-Naas, L.A., Meade, B.J., and Munson, A.E., Toxic responses of the immune system, in *Casarett and Doull's Toxicology the Basic Science of Poison*, Klassen, C.D., Ed., McGraw-Hill, New York, 2001, pp. 419–470.

Chen, S.C., et al., Nitroblue tetrazolium dye reduction by earthworm Lumbricus terrestris coelomocytes: an enzyme assay for non-specific immunotoxicity of xenobiotics, *Environ. Toxicol. Chem.*, 10, 1037–1043, 1991.

Cikutovic, M.A., et al., Wound healing in earthworms Lumbricus terrestris: a cellular-based immunity biomarker for assessing sublethal chemical toxicity, *Bull. Environ. Contam. Toxicol.*, 62, 508–514, 1999.

Cooper, E.L., et al., Phylogeny of transplantation immunity: graft rejection in earthworms, *Transpl. Proc.*, 3, 214–16, 1971.

Cooper, E.L. The earthworm coelomocyte: a mediator of cellular immunity, in *Phylogeny of Thymus and Bone Marrow Cells*, Wright, R.K., and Cooper, E.L., Eds., Elsevier/North-Holland, Amsterdam, 1976, pp. 9–18.

Cooper, E.L., and Roch, P., Second-set allograft responses in the earthworm Lumbricus terrestris: kinetics and characteristics, *Transplantation*, 4, 514–20, 1986.

De Coen, W., Have biomarkers lost their charm? "Old" concepts put into a "new" perspective, *SETAC Globe*, 1, 35–37, 2000a.

De Coen, W., Multivariate analysis of biomarker endpoints: a promising tool to assess their ecological relevance, *SETAC Globe*, 1, 37–39, 2000b.

De Coen, W., and Jansson, C., The missing biomarker link: relationships between effects on the cellular energy allocation biomarkers of toxicant-stressed *Daphnia magna* and corresponding population characteristics, *Vet. Immunol. Immunopathol.*, 22, 1632–1641, 2003.

Eason, C.T., Booth, L.H., Brennan, S., and Ataria, J., Cytochrome P450 activity in 3 earthworm species, in *Advances in Earthworm Toxicology*, Sheppard, S., et al., Eds., SETAC Press, Pensacola, FL, 1998, pp. 191–98.

European Economic Community, EEC Directive 79/831, Annex V, Part C: Methods for the determination of ecotoxicity, Level 1, Toxicity for Earthworms. Commission of the European Communities, Doc. EUR 9360 EN, 1985.

Eyambe, G.S., et al., Extrusion protocol for use in chronic immunotoxicity studies with earthworm Lumbricus terrestris coelomic leukocytes, *Lab. Anim.*, 25, 61–67, 1991.

Fitzpatrick, L.C., et al., Earthworm immunoassay for evaluating biological effects of exposure to hazardous materials, in *In Situ Evaluation of Biological Hazards of Environmental Pollutants*, Sandu, S.S., Ed., Plenum Press, New York, 1990, pp. 119–29.

Fitzpatrick, L.C., et al., Comparative toxicity in earthworms Eisenia fetida and Lumbricus terrestris exposed to cadmium nitrate using artificial soil and filter paper protocols, *Environ. Contam. Toxicol.*, 57, 63–68, 1996.

Fitzpatrick, L.C., et al., Comparative toxicity of polychlorinated biphenyls to the earthworms Eisenia foetida and Lumbricus terrestris, *Environ. Pollut.*, 77, 65–79, 1992.

Forbes, V.E., et al., Costs of living with contaminants: implications for assessing low-level exposures, *Belle News Lett.*, 4, 1–8, 1996.

Fugere, N., Heavy metal-specific inhibition of phagocytosis and different in vitro sensitivity of heterogeneous coelomocytes from Lumbricus terrestris (Oligochaeta), *Toxicol.*, 109, 157–66, 1996.

Giggleman, M., Phagocytosis by Earthworm Coelomocytes: A Biomarker for Immunotoxicity of Hazardous Waste Sites Soils, Ph.D. thesis, University of North Texas, Denton, Texas, 1997.

Giggleman, M.A., et al., Effects of pentachlorophenol on survival of earthworms (Lumbricus terrestris) and phagocytosis by their immunoactive coelomocytes, *Environ. Toxicol. Chem.*, 17, 2391–94, 1998.

Goldsby, R.A., Kindth, T.J., Osborne, B.A., and Kerby, J., Eds., *Immunology*, 5th ed., W.H. Freeman, New York, 2000 pp.

Goven, A.J., et al., An invertebrate model for analyzing effects of environmental xenobiotics on immunity, *Clin. Ecol.*, 4, 150–54, 1988.

Goven, A.J., et al., Cellular biomarkers for measuring toxicity of xenobiotics effects of polychlorinated biphenyls on earthworm Lumbricus terrestris coelomocytes, *Environ. Toxicol. Chem.*, 12, 863–70, 1993.

Goven, A.J., et al., Suppression of lysozyme activity in Lumbricus terrestris coelomocytes and coelomic fluid collected from earthworms exposed to copper sulfate, *Environ. Toxicol. Chem.*, 13, 607–13, 1994.

Greene, J.C., et al., Protocols for Short-Term Toxicity Screening of Hazardous Waste Sites, U.S. Environmental Protection Agency, Corvalis, OR, 1989, EPA/600/3–88/029.

Greig-Smith, P.W., Recommendation of an international workshop on ecotoxicology of earthworms, in *Ecotoxicology of Earthworms*, Greig-Smith, P.W., Becker, H., Edwards, P.J., and Heimbach, F., Eds., Intercept Ltd., Andover, England, 1992, pp. 247–262.

Holsapple, M.P., et al., A review of 2, 3, 7, 8- tetrachloro-p-dioxin-induced changes in immunocompetence: 1991 update, *Toxicology*, 69, 219–55, 1991.

Hostetter, R.K., and Cooper, E.L., Earthworm coelomocyte immunity, in *Contemporary Topics in Immunobiology*, E.L. Cooper, Ed., Plenum Press, New York 1974, pp. 91–107.

Janeway, C.A., and Medzhitov, R., Innate immune recognition, *Annu. Rev. Immunol.*, 20, 197–216, 2002.

Janeway, C.A., et al., *Immunobiology: The Immune System in Health and Disease*, Garland Publishing, New York, 2001.

Kammenga, J.E., et al., Biomarkers in terrestrial invertebrates for ecological soil risk assessment, *Rev. Environ. Contam. Toxicol.*, 164, 93–147, 2000.

Kerkvliet, N.I., Baecher-Steppan, L., and Schmitz, J.A., Immunotoxicity of pentachlorophenol (PCP): increased susceptibility to tumor growth in adult mice fed technical PCP-contaminated diets, *Toxicol. Appl. Pharm.*, 62, 55–64, 1982.

Kerr, S.R., Ecological analysis and the Fry paradigm, *J. Fish. Res. Board Canada*, 33, 29–39, 1976.

Lanno, R.P., and McCarty, L.S., Earthworm bioassays: Adopting techniques from aquatic toxicity testing, *Soil Biol. Biochem.*, 29, 693–697, 1997.

Lasseques, M., Roch P., and Valembois, P., Antibacterial activity of Eisenia fetida andrei coelomic fluid: evidence, induction and animal protection, *J. Invertebr. Pathol.*, 53, 1–6, 1989.

Lee, R. F., Annelid cytochrome P–450, *Comp. Biochem. Physiol. C Pharmacol. Toxicol. Endocrinol.*, 121, 173–179, 1998.

Levins, R., *Evolution in Changing Environments*, Princeton University Press, Princeton, NJ, 1968.

Matscheko, N., et al., Application of sewage sludge to arable land-soil concentrations of polybrominated diphenyl ethers and polychlorinated dibenzo-p-dioxins, dibenzofurans, and biphenyls, and their accumulation in earthworms, *Vet. Immunol. Immunopathol.*, 21, 2515–2525, 2002.

McCarty, L.S., and MacKay, D., Enhancing ecotoxicological modeling and assessment, *Environ. Sci. Technol.*, 27, 1719–1728, 1993.

McCarty, L.S., Comments on the significance and use of tissue residues in sediment toxicology and risk assessment, *SETAC News*, 12–14, November 1996.

Morgan, A.J., Stürzenbaum, S.R., and Kille, P., A short overview of molecular biomarker strategies with particular regard to recent developments in earthworms, *Pedob.*, 43, 574–84, 1999.

Morgan, A.J., Turner, M.P., and Morgan, J.E., Morphological plasticity in metal- sequestering earthworm chloragocytes: morphometric electron microscopy provides a biomarker of exposure in field populations, *Vet. Immunol. Immunopathol.*, 21, 610–18, 2002.

Nappi, A.J., and Ottaviani, E., Cytotoxicity and cytotoxic molecules in invertebrates, *BioEssays*, 22, 469–80, 2000.

Nusetti, O., et al., Acute-sublethal copper effects on phagocytosis and lysozyme activity in earthworm *Amynthas hawayanus*, *Bull. Environ. Contam. Toxicol.*, 63, 350–56, 1999.

Organisation for Economic Cooperation and Development, Guideline for Testing Chemicals, No. 207, Earthworm Acute Toxicity Tests, adopted 4 April 1984, OECD, Paris.

Reinecke, S.A., Helling, B., and Reinecke, A.J., Lysosomal response of earthworm (Eisenia fetida) coelomocytes to the fungicide copper oxychloride and relation to life-cycle parameters, *Vet. Immunol. Immunopathol.*, 21, 1026–31, 2002.

Rice, C.R., Fish immunotoxicology, in *Target Organ Toxicity in Marine and Freshwater Teleosts*, Schlenk, D., and Benson, W.H., Eds., Taylor & Francis, London, 2001, pp. 96–138.

Rodriguez-Grau, J.B., et al., Suppression of secretory rosette formation by PCBs in Lumbricus terrestris: an earthworm immunoassay for humoral immunotoxicity of xenobiotics, *Environ. Toxicol. Chem.*, 8, 1201–1207, 1989.

Salton, M.R.J., The properties of lysozyme and its action on microorganisms, *Bacteriol. Rev.*, 21, 82–99, 1975.

Sauve, S., et al., Phagocytic response of terrestrial and aquatic invertebrates following in vitro exposure to trace elements, *Ecotoxicol. Environ. Saf.*, 52, 21–29, 2002.

Scott-Fordsmann, J., and Weeks, J.M., Review of selected biomarkers in earthworms, in *Advances in Earthworm Toxicology*, Sheppard, J., et al., Eds., SETAC Press, Pensacola, FL, 1998, pp. 173–189.

Sprague, J.B., Measurement of pollutant toxicity to fish I. Bioassay methods for acute toxicity, *Water Res.*, 3, 793–821, 1969.

Stein, E.A., and Cooper, E.L., Agglutinins as receptor molecules: a phylogenetic approach, in *Developmental Immunology: Clinical Problems and Aging*, Cooper, E.L., and Brazier, M.A.B., Eds., Academic Press, New York, 1982, pp. 85–98.

Stenerson, J., Detoxication of xenobiotics by earthworms, *Comp. Biochem. Phys.*, 78C, 249–252, 1984.

Stenerson, J., Uptake and metabolism of xenobiotics by earthworms, in *Ecotoxicology of Earthworms*, Greig-Smith, P.W., et al., Eds., Intercept Ltd., Andover, England, 1992, pp. 129–138.

Suzuki, M.M., et al., Polychlorinated biphenyls (PCBs) depress allogeneic natural cytotoxicity by earthworm coelomocytes, *Environ. Toxicol. Chem.*, 14, 1697–1700, 1995.

Svendsen, C., et al., Lysosomal membrane permeability and earthworm immune-system activity: field testing on contaminated land, in *Advances in Earthworm Toxicology*, Sheppard, S., et al., Eds., SETAC Press, Pensacola, FL, 1998, pp. 225–232.

Valembois, P., Cellular aspects of graft rejection in earthworms and some other metazoa, in *Contemporary Topics in Immunology*, Cooper, E.L., Ed., Plenum Press, New York, 1974, pp. 75–90.

Van Schooten, F.J., et al., DNA dosimetry in biological indicator species living on PAH-contaminated soils and sediments, *Ecotoxicol. Environ. Saf.*, 30, 171–179, 1995.

Venables, B.J., Fitzpatrick, L.C., and Goven, A.J., Earthworms as indicators of ecotoxicity, in *Ecotoxicology of Earthworms*, Greig-Smith, P.W., Becker, H., Edwards, P.J., and F. Heimback, Eds., Intercept Ltd., Andover, England, 1992, pp. 197–206.

Voccia, I., et al., Immunotoxicity of pesticides: a review, *Toxicol. Ind. Health*, 15, 119–132, 1999.

Wojdani, A., et al., Agglutinins and proteins in the earthworm, Lumbricus terrestris, before and after injection of erythrocytes, carbohydrates and other materials, *Dev. Comp. Immunol.*, 6, 407, 1982.

CHAPTER 8

Reptiles: The Research Potential of an Overlooked Taxon in Immunotoxicology*

Andrew A. Rooney

CONTENTS

---

* This report has been reviewed by the Environmental Protection Agency's Office of Research and Development, and approved for publication. Approval does not signify that the contents necessarily reflect the views and policies of the agency, nor does mention of trade names or commercial products constitute endorsement or recommendation for use.

8.1 INTRODUCTION

Cross-species comparative studies that include reptiles can contribute significantly to the field of immunotoxicology due to their similarities and diversity from the more commonly studied birds and mammals. In spite of this, reptilian taxa are generally overlooked in the field of toxicology. Even within the most comprehensive ecotoxicology studies, reptiles are conspicuously absent (Campbell and Campbell, 2002). Such studies have commonly included invertebrates, fish, amphibians, birds, and mammals — taxa for which good background data are readily available (Chapman, 2002). One area of toxicology where reptiles have increased prominence is in endocrine or reproductive toxicology and the study of endocrine-disrupting chemicals (EDCs). In fact, turtles and crocodilians have been proposed as sentinels of EDC exposure because of their sensitivity to EDCs during the process of sex determination (Bergeron et al., 1994). This chapter outlines the phylogenetic/evolutionary relationships of reptiles to other animal taxa and reviews the existing background information on reptilian taxa. Selected functional immune assays are discussed and research gaps are identified for future action. Finally, the potential advantages and disadvantages in using reptiles as experimental models in immunotoxicology are presented and discussed.

8.2 REPTILE PHYLOGENY/EVOLUTIONARY RELATIONSHIPS

Reptiles represent a pivotal phylogenetic group that is the evolutionary progenitor of both mammals and birds. As a result, modern reptiles display characteristics that may be ancestral to each group, although extant reptiles are not the direct ancestors of mammals or birds. They also possess two key evolutionary changes relevant to toxicology that divide amniotes (reptiles, birds, and mammals) from anamniotes (fish and amphibians): (1) the cleidoic or amniotic egg, which allowed a new relative freedom from water and drastically altered excretion; and (2) a skin that is relatively impervious to the environment and therefore changed absorption. The class Reptilia is a diverse evolutionary group, and extant species are divided into four orders (Pough, 1998). Today, Rhynchocephalia is only present as a single species (or possibly two subspecies) of lizard-like tuatara, and therefore is a minor group for research purposes. Major extant orders include Testudinata, Crocodylia, and Squamata. Turtles and tortoises comprise Testudinata, the most disparate and ancient group among the reptilian orders, which is currently comprised of about 300 species. There are only 23 modern representatives of Crocodylia, the alligators and crocodiles, an order with a close evolutionary relationship to birds. Squamata is the most recently evolved, and by far the largest, reptile order with over 7700 recognized species of lizards and snakes.

The relative independence of reptilian reproduction from water, when compared with amphibians and fish, results in a fundamentally different level of interaction with water during development. Although reptile and bird eggs may accumulate xenobiotics from contaminated nest materials (Canas and Anderson, 2002), they do not have the open exchange to waterborne xenobiotics commonly experienced by fish (Gonzalez-Doncel et al., 2003) and amphibian eggs (Bridges, 2000). Whether early development takes place within a reptile egg (in egg-laying or oviparous species) or within the mother (in non-egg laying or viviparous species), waste cannot be excreted as dilute urine characteristic of fish and amphibians, and the kidneys must concentrate wastes while retaining water. The allantois membrane unique to amniotes (reptiles, birds, and mammals) collects and stores waste during early development, exchanges gasses, and retains water, allowing freedom from water; the allantois is a key advance in reptile and bird eggs and is critical to the formation of the placenta in mammals (Guillette, 1993). The importance of this membrane to gas and water management suggests an important distinction between amniote and nonamniote excretion and may be relevant to toxicology studies.

Reptilian respiration also contributes to a fundamentally different level of environmental interaction than their more aquatic ancestors. Fish and many amphibians respire through gills, and therefore require intimate contact between the surrounding water and highly vascular tissue. This creates enhanced vulnerability for fish and amphibians for aqueous toxins that are not faced by more terrestrial vertebrates. Even in amphibians that do not rely on gills for part or all of their life stages, their highly permeable skin is often sited as a special concern and perhaps contributing factor in recent amphibian declines (Lambert, 1997). Reptiles have a skin that is characteristically impermeable to water loss, consisting of a highly keratinous epidermal layer, and sometimes also consisting of a thick, hydrophobic cuticle (Frye, 1991a). In this aspect, reptiles are more similar to mammals and birds; in fact, the advanced cuticle in reptiles may make them more resistant to dermal toxin exposure than even mammal or birds. Interestingly, the necessity for reptiles to periodically shed their relatively impermeable skin to allow for growth, may also allow for toxic metal excretion in the discarded skin of some species (Burger, 1992).

8.3 REPTILIAN TOXICOLOGY AND IMMUNOLOGY: BACKGROUND DATA NEEDED FOR IMMUNOTOXICOLOGY STUDIES

The immunotoxicity testing guidelines published by the U.S. Environmental Protection Agency in 1998 suggest the use of three major assays: antibody response to antigen administration analyzed using (1) either a plaque-forming cell (PFC) assay or immunoglobulin quantification by ELISA; (2) natural killer (NK) cell activity; and (3) enumeration of splenic or peripheral blood B and T cells and T-cell subpopulations (U.S. EPA, 1998). A more extensive list of tests to assess immunotoxicity was put forth in the consensus workshop on methods to evaluate developmental toxicity held in Washington, D.C. in 2001 (Luster et al., 2003). The suggested methods include the EPA guidelines plus four additional tests: (1) thymus, spleen, and lymph node weights, (2) mitogenic response of lymphocytes, (3) delayed-type hypersensitivity (DTH) response, and (4) macrophage function. These analyses include both structural and functional measures of the immune system and cover biomarkers of toxicant exposure within the immune system. Historically, these measures have been separated into a two-tier system with the first tier consisting of screening parameters, and the second tier of comprehensive assays. The goal of immunotoxicity assays is for tests to have a high degree of predictability for host resistance, which is often considered the third tier of immunotoxicity testing (Luster et al., 1992).

In this chapter, a three-tiered system is used to sort the following brief review of published research on reptilian immunity for the purpose of detailing and evaluating the background information necessary to apply immunotoxicology protocols to reptiles. From the eight assays listed above, two survey assays will be included in Tier 1 (organ weight/histology and phenotypic analysis of lymphoid subpopulations). Lymphocyte blastogenesis and four assays measuring immune function (antibody response, NK cell activity, macrophage function, and DTH response) will make up the comprehensive tests of Tier 2 (Vos et al., 1989). Host resistance challenge assays will be the focus of Tier 3.

Although much of the immune system of reptiles remains largely uninvestigated, a sound foundation in histology and cytology of reptilian immune tissues is present in publications from the 1960s to the 1980s. Research on reptilian immunity has not kept pace with similar research on fish or even amphibians since the mid-1980s, with the exception of major work in the area of seasonal fluctuations in reptilian immunity (El Masri et al., 1995; Nelson and Demas, 1996). Significant research gaps in the current knowledge of reptile immunology may make it difficult to effectively perform immunotoxicology assays on reptiles. Therefore, a considerable portion of the following review will be devoted to highlighting the lack of existing research and to suggesting future studies to fill gaps and provide necessary background for each of the following eight immunotoxicology assay areas.

When evaluating data from any test of immunotoxicity, it is important to consider the compounding influence of biotic (especially age and gender) and abiotic (principally season and temperature) variables on the immune system, and this may be particularly true for ectothermal groups such as the reptiles. Physiological properties internal to the animal such as age and gender are important considerations for immunotoxicity studies because of their influence on immune function. For example, in humans and rodents, many pesticides have a more pronounced or persistent immunosuppressive effect when exposure occurs in developing animals rather than in adults (Holladay and Smialowicz, 2000). The importance of age in the relationship between immunotoxicity and persistence has not been studied in reptiles, but similarity between the maturation of immune function in mammals and reptiles is suggested by ontogenetic studies such as El Deeb and Saad's (1987) study of T-cell function in the ocellated lizard, *Chalcides ocellatus*. Therefore, it may be important to address age or control for ontogenetic periods of increased susceptibility in reptile immunotoxicology studies.

Gender specificity is also present in the immunotoxic effects of some pesticides. For example, the herbicide atrazine suppresses cell-mediated and humoral immunity in male rats exposed during development, but does not affect females (Rooney et al., 2003b). Although gender differences in the reptilian response to immunotoxins have not been investigated, there are examples of gender disparity in immune function in reptiles. For example, the proliferative response of lymphocytes to the mitogen concanavalin A (Con-A) and the antibody response to sheep red blood cell (SRBC) challenge were consistently greater in female olive sand snakes, *Psammophis sibilans* (Saad and Shoukrey, 1988). In addition, female *P. sibilans* consistently had a greater mitogenic response than males to Con-A, phytohemagglutinin (PHA), and lipopolysaccharide (LPS), except during pregnancy (Saad, 1989a). To consider the role of gender in reptilian immune function, data are required on both sexes, but as is the case in much of mammalian immunotoxicology, only a single sex has been examined in most studies related to reptilian immunity.

The importance of considering gender when studying immunotoxicology in reptiles is also suggested by the ability of endogenous and exogenous hormones to act as immune suppressers. Corticosteroids are the major immunosuppressive hormones produced in all vertebrates in response to stressors (Guillette et al., 1995a). In Selye's characterization of the stress response, he described a collection of physiological and anatomical changes characteristic of animals exposed to severe external forces or stressors (Selye, 1936, 1949). Selye's original description of stress included immune suppression characterized by atrophy of the thymus and lymph nodes, and endocrine changes such as hypertrophy of the adrenal glands that were related to corticosteroid secretion. In his early immunotoxicology research, Selye (1936) included toxic xenobiotic compounds among the first stressors he examined. The immunosuppressive effects of corticosteroids are well-studied in reptiles (Saad, 1988), demonstrating that basal concentrations of corticosteroids are sexually dimorphic, which suggests that corticosteroids may contribute to a sexually dimorphic immune response in some reptiles. As in mammals (Olsen and Kovacs, 1996), the sex hormone testosterone is also immunosuppressive in reptiles (Saad et al., 1991; Varas et al., 1992). The concentration of plasma testosterone is sexually dimorphic in reptiles, and even juvenile males of some reptilian species such as alligators have more testosterone than females (Rooney et al., 2004). Studies of the immunosuppressive effects of exogenous testosterone or corticosteroids will be highlighted in the following discussions of traditional immunotoxicity assays and the existence of data from reptiles. However, the use of hormones as positive controls or standards for immune suppression is complicated by the presence of endogenous hormones, compensation in endogenous hormone secretion by regulatory feedback loops, and nonimmune roles for hormones used to produce immunosuppression. Validation of the efficacy of traditional immunosuppressive agents such as dexamethasone, cyclophosphamide, and cyclosporine A (Dean et al., 1998) in reptiles must be undertaken to produce better positive controls for all of the immunotoxicity assays below.

Endogenous hormones associated with seasonal hormonal cycles have also been linked to pronounced seasonality and sex-associated differences in reptilian immunity. For example, the mitogenic response to Con-A and PHA is seasonal in the Caspian turtle, *Mauremys caspica*, with maximal response in spring (Munoz and De la Fuente, 2001a) when concentrations of both corticosterone and testosterone are low in *M. caspica* (Leceta and Zapata, 1985). On the other hand, the mitogenic response to LPS, presumably that of B cells, displayed no seasonal variation in *M. caspica*, suggesting that seasonality is a property of specific aspects of immune function, not a universal property of the immune system. In general, however, immune function is affected by the season of the year, and this is particularly apparent in reptiles (Nelson and Demas, 1996). Reptiles exhibit well-documented seasonal changes in function (e.g., NK cell activity in the Caspian turtle, *M. caspica*) and morphology (e.g., the blue tailed skink, *Mabuya quinquetaeniata* and the Egyptian spiny tailed lizard, *Uromastyx aegyptia*) of the immune system.

The seasonal changes in the immune system of reptiles have also been linked to temperature. Generally, winter is characterized by involution of the spleen and thymus, and the spring is a period of seasonal recrudescence of immune tissue and immune function for reptiles living in temperate climates — a fact that long ago led to the discovery of optimal temperatures for reptilian immune function (Avtalion et al., 1976; Wright et al., 1978). Optimal temperature must be a primary concern for both animal housing and cellular culture conditions. Clearly, a seasonally depressed immune response could create an annual period of vulnerability that must be addressed when measuring immunotoxicity. Therefore, standards must be developed for the appropriate period in a seasonal or reproductive cycle that would allow measurements to determine potential immunotoxic effects in reptiles.

8.3.1   Tier 1: Present Status and Research Needs

8.3.1.1   *Structural Measurements/Major Organs of Reptile Immune System*

The reptilian immune system is grossly similar to that of other vertebrates — the central lymphoid organs of all reptiles are the thymus and bone marrow. The reptilian immune system is detailed elsewhere (Cooper et al., 1985; El Ridi et al., 1992; Frye, 1991a). Bone marrow is the major site of hemopoiesis in vertebrates and B-cell maturation in mammals; it is also the likely site of B-cell maturation in reptiles (with a possible contribution from the bursa in turtles; see below). T-cell maturation takes place in the thymus of all vertebrates examined (Matsunaga and Rahman, 2001). The thymus of reptiles is separated into many smaller lobules that are incompletely divided due to connections between the central or medulary regions of adjacent lobules. Within each lobe there are lymphoid cells within an epithelial cell framework separated into cortical and medullary areas characteristic of thymic organization in mammals (Saad and Zapata, 1992).

In addition to the thymus and bone marrow, some turtles also have a lymphoid aggregate or bursa at the base of the tail, a structure not found in other reptiles or in any mammals (Shields, 1976). The bursa in turtles is structurally similar to the bursa of Fabricius in birds (Cooper et al., 1985). In birds, the bursa of Fabricius serves as the principal site of B-cell differentiation (Sharma, 1991), but the function of the bursa in turtles is unknown. The bursa of Fabricius played an important role in the history of immunology research, as it facilitated the identification of B and T lymphocytes as functionally and physiologically distinct cell populations. Comparative research, in this case the study of birds, took advantage of the convenient compartmentalization of B-cell maturation in birds within the bursa of Fabricius rather than bone marrow-based maturation of mammals to study the dichotomy between B and T cells (Bona and Bonilla, 1990).

Reptiles only have a single well-organized peripheral lymphoid organ, the spleen, whereas mammals and birds also have discrete lymph nodes. The reptilian spleen is similar to that of other vertebrates; it is a site of hemopoiesis, antigen trapping, and initiation of the cellular immune response (Kroese et al., 1985; Pitchappan, 1980; Tanaka and Elsey, 1997). The fact that reptiles lack additional peripheral lymph organs suggests that the reptile spleen is more central to immune function than it is in mammals. Although reptiles lack discrete lymph nodes, they do have less structured lymphoid cell aggregates within intestinal and pharyngeal mucosa (El Ridi et al., 1981b), as do fish (Matsunaga and Rahman, 2001), amphibians (Ardavin et al., 1982), mammals (Wang et al., 1997), and birds (Weill et al., 2002). Little is known about the function of this mucosa-associated lymphoid tissue in reptiles; however, it seems to represent a phylogenetic precursor to the more organized mammalian Peyers patches and avian bursa (Cooper et al., 1985).

Published examples of immunotoxicology studies exploring morphology or histology of reptilian immune organs are severely limited, as are all studies of reptile immunotoxicology. A recent exception is a paper documenting histological changes in the spleen and thymus of alligators living in Lake Apopka, Florida, in a study intended to determine if previously detected endocrine abnormalities associated with contaminant exposure might also be reflected in immune tissue (Rooney et al., 2003a). A relatively large body of literature exists on the xenobiotics such as organochlorines and metals (Burger et al., 2000; Heinz et al., 1991; Woodward et al., 1993) and the associated endocrine and reproductive abnormalities (Crain et al., 1997; Guillette et al., 1995b; Guillette et al., 1996; Gunderson et al., 2001) of the alligator population living in Lake Apopka. Alligators living in Lake Apopka had smaller thymic ratios (medulla/cortex), smaller B-cell–associated Malpighian body areas within the spleen, and females had smaller T-cell-associated lymphocyte sheaths (Rooney et al., 2003a). In contrast to that observed for wild-caught animals, no difference was found in the thymic medulla/cortex ratio or Malpighian body areas of captive-raised female alligators hatched from eggs collected from the shores of Lake Apopka and the reference lake. The results point to population variation in the immune structures associated with the thymus and spleen that is associated with living in contaminated water (Rooney et al., 2003a). However, there was also evidence of an organizational effect of the eggs laid on the shores of Lake Apopka. Captive-raised female alligators from Lake Apopka did have the reduced lymphocyte sheath characteristic of wild alligators living at Lake Apopka, which suggested that the reduced lymphocyte sheath was altered developmentally and may be related to contaminant load deposited in the eggs (Rooney et al., 2003a). The measurement of immune tissue weights and histology may be among the tools most readily adapted to reptiles; however, these measures are not definitive for functional immunotoxicity and are known to have poor predictive value for functional immunosuppression in rodents (Dean, 1997). Studies of reptilian immune tissue morphology or histology after exogenous administration of chemicals are restricted to experimental administration of testosterone in lizards (Saad et al., 1990) and turtles (Varas et al., 1992), and corticosteroids in lizards (El Deeb et al., 1993) and alligators (Morici et al., 1997), not xenobiotic compounds. The established effects of immunosuppressive steroids on immune tissues may serve as preliminary positive controls for immunotoxicity research, but the functional studies remain to be done in reptiles.

8.3.1.2 General Parameters of Immune System

8.3.1.2.1 Major Immune Organs

Descriptive research on the histology of reptile immune organs under normal environmental conditions is adequate, but as yet no experimental studies have taken advantage of that background research and tested the effects of environmental chemicals on the immune system. There is also a need to create standards for collection of bone marrow in species with extensive skeletal and dermal bone such as turtles (Garner et al., 1996). Inclusion of gross organ weights and histology — the most basic of immunotoxicologic measures — in basic toxicology studies of reptiles would, undoubtedly, enrich the literature base for this taxon.

8.3.1.2.2 Characterization of Leukocyte Populations

One of the most commonly reported endpoints relating to reptilian immunity is complete blood counts (CBCs) or evaluation of peripheral blood smears. The histological appearance of reptile immune cells resembles those described in mammals, and studies of comparative hematology in reptiles date back to the mid-1800s (Cooper et al., 1985; Frye, 1991a, 1991b). This simple characterization of leukocytes is readily performed on reptile blood, but further quantitation of reptile leukocyte populations is limited by the availability of tools to discriminate among populations. In fact, no conclusive evidence of heterogeneity of lymphocytes into T- and B-cell populations exists in reptiles, although evidence of antigen trapping and plasma cell development suggest that reptiles share this organization (Borysenko, 1976; Muthukkaruppan et al., 1983). Lymphocyte heterogeneity has been demonstrated in the Egyptian ratsnake (*Spalerosophis diadema*) where anti-snake thymocyte antibody bound to 98% of thymocytes and only 72% of splenocytes (Mansour et al., 1980). Surface-bound immunoglobulin was also detected on 29% of splenocytes, while thymocytes were negative for surface immunoglobulin in *S. diadema*, as is the case in mammalian and avian lymphocytes (Mansour et al., 1980). Other studies identified an epitope shared by brain and thymus cells (Thy-1+) in the ocellated lizard, *C. ocellatus* (El Deeb et al., 1988), and partially characterized lymphocyte populations in the snapping turtle (*Chelydra serpentina*) using polyclonal antibodies (Mead and Borysenko, 1984). In addition, peanut agglutinin (PNA)-binding glycoproteins distinguish a subpopulation of lymphocytes in the thymus, peripheral blood, bone marrow, and a small population within the spleen of the ocellated lizard (Mansour et al., 1995).

8.3.1.2.3 Leukocyte Epitopes

Although background research has been performed on the immunosuppressive effects of endogenous steroids (Saad and El Ridi, 1988), immunotoxicology studies have not been performed to examine chemical-induced changes in leukocyte subpopulations. Several studies have demonstrated that hormones which are immunosuppressive in mammals cause changes in specific lymphocyte populations in reptiles. For example, exogenous testosterone propionate decreased PNA- Thy-1+ lymphocytes and hydrocortisol reduced the PNA(+)-Thy-1- subpopulation in *C. ocellatus* (El Masri et al., 1995). The ability to discriminate between reptilian leukocyte subpopulations is necessary before detailed characterization of leukocyte subpopulations. The maturational sequence of B and T cells is not explicitly documented in reptiles, and much of the basic work of identifying epitopes that mark immune cell differentiation (i.e., clusters of differentiation [CD]) has not been completed in reptiles. The T cells of mammals and birds express epitopes characteristic of distinct T-cell populations obtained within the thymic cortex, expressing the T-cell receptor and population defining CD expression (Bona and Bonilla, 1990; Sharma, 1997). These epitopes identify specific subpopulations of leukocytes, and enable population-level changes in leukocytes to be tracked using antibodies to the epitopes. While the enumeration of leukocyte populations is a more in-depth endpoint than CBC, the technique is largely unavailable in reptiles because of the lack of basic data on population-specific epitopes in reptilian leukocytes. The development of specific antibodies to reptilian immune cells would be very helpful to immunotoxicology studies in reptiles. The most important of these specific antibodies would include separate, specific antibodies binding to epitopes on T and B cells, and subsets of T cells, monocyte/macrophages, NK cells, mast cells, neutrophils, eosinophils, and basophils. As early as 1989, antibodies were available for these cells in mice, but were not available to the latter four cell types for rats (Vos et al., 1989). Reptiles may present a difficult case for generating antibodies due to the diversity of reptilian groups and the presumed species specificity of epitopes.

8.3.2 Tier 2: Present Status and Research Needs

8.3.2.1 Cell-Mediated Immune Assays

8.3.2.1.1 Blastogenic Response

Lymphocyte blastogenesis includes the mitogenic response of T and B cells and the mixed leukocyte reaction (MLR). The mitogenic response of lymphocytes has been assayed in all three major reptile orders and a number of different species. For example, the mitogenic response of alligator lymphocytes was first tested in 1979 (Cuchens and Clem, 1979). The proliferative response to T- and B-cell mitogens has also been examined in the Caspian turtle, *M. caspica* (Munoz and De la Fuente, 2001b), the ocellated lizard, *C. ocellatus* (Saad and El Deeb, 1990) and the olive sand snake, *P. sibilans* (Farag and El Ridi, 1986). The response to alloantigens has also been measured in reptiles using a MLR assay in *P. sibilans* (Farag and El Ridi, 1985) and *C. ocellatus* (El Deeb and Saad, 1987).

Few studies have investigated the effects of exogenous hormones on the mitogenic response in reptiles, and no studies have been published on mitogenic responses after administration of xenobiotic compounds to a reptile. In Horsfield's tortoise, *Testudo horsfieldi*, a single injection of hydrocortisone (100 mg/kg) increased the mitogenic response of lymphocytes to Con-A and PHA (Alimkhodzhaeva et al., 2002). Only limited conclusions can be drawn from this study, however, due to the high dose of hydrocortisone used and the absence of data relating to the timing of the hydrocortisone dose relative to sampling of the lymphocytes. A study of the mitogenic response was also completed on groups of alligators raised under identical housing conditions but derived from eggs collected at either a contaminated site (Lake Apopka, Florida) or a "clean," reference site (Rooney, 1998). The mitogenic response of Lake Apopka alligators to Con-A (20 µg/ml) was greater than that of the alligators from the reference lake, although the response to PHA did not differ between the groups of alligators from the two sites (Rooney, 1998). These data suggest that developmental exposure to chemicals through the egg (either from the mother or from the substrate prior to collection) may have caused the observed increase in the mitogenic response to Con-A.

The successful culture of reptilian lymphocytes often requires the addition of autologous reptile serum to the culture media, whereas mammalian lymphocytes are routinely cultured in fetal bovine serum (FBS). Cuchens and Clem (1979) could not successfully culture alligator lymphocytes without adding alligator serum to the media. Ulsh et al. (2000) were successful in culturing turtle lymphocytes from the red-eared slider (*Trachemys scripta*) using FBS, but obtained only a 1.5- to 2.75-fold increase in the mitogenic response by utilizing autologous serum. The requirement for autologous serum is problematic because it will increase the variability of mitogenic assays between laboratories, and even between studies, as many separate serum sources must be relied on. Therefore, the establishment of commercially available serum would be desirable. Albumax-I (Gibco, Gaithersburg, MD) has been successfully used in the culture of the green sea turtle (*Chelonia mydas*) lymphocytes; however, comparisons were not made between FBS, autologous serum, and Albumax-I for *C. mydas* (Work et al., 2000). In addition to the standardization of culture conditions for reptilian lymphocytes, studies are needed on lymphocyte responses to additional mitogens such as the pokeweed mitogen (PWM).

8.3.2.1.2 Antibody Response

The antibody response is a measure of specific immune function that is relatively well-studied in reptiles; comparatively simple tests provide evidence of antibody production. Hemagglutination assays involve immunization with washed red blood cells (RBCs) from a different species, often sheep RBCs (SRBCs) (El Ridi et al., 1981a). Serum from immunized animals is collected at postimmunization intervals, serially diluted *in vitro* and mixed with a fixed quantity of RBCs from the same animal used for immunization. Antibodies to the RBCs cross-link RBCs, resulting in agglutination. The standard techniques for measuring specific antibodies in mammals are either the plaque-forming cell (PFC) assay or immunoglobulin quantification by ELISA. The PFC assay enumerates the antibody secreting cells within cultured splenocytes rather than measuring the presence of antibodies in serum as in the haemagglutination assays. Complement is added to PFC cultures, and the presence of antibody-secreting cells is determined by detecting lysis of co-cultured RBCs. Although PFC assays have been applied to measure antibody production in the garden lizard, *Calotes versicolor* (Pillai and Muthukkaruppan, 1977) and the Caspian turtle, *M. caspica* (Leceta and Zapata, 1986), the assays require a source of complement (generally guinea pig), and there are questions as to the efficacy of mammalian complement in a reptile.

Antibody production in reptiles is slower (first detected after 1 to 2 weeks) than antibody production in mammals (first detected after 1 to 2 days) (Ambrosius et al., 1969), as might be predicted when comparing physiological reactions in an ectotherm and an endotherm (Evans, 1963). Several studies have attempted to link total antibody concentrations (not antigen-specific antibodies) to disease states or stressful conditions in reptiles. Increasing severity of fibropapillomatosis in Hawaiian green turtles was associated with a decrease in total antibody concentrations (Work et al., 2001). Turton (1997), however, found no relationship between low, presumably stressful temperatures, and circulating corticosterone or total antibody concentrations in young saltwater crocodiles (*Crocodylus porosus*).

A seasonal pattern, similar to the fluctuations of the mitogenic response, and an inverse relationship between corticosteroids and antigen-specific antibody production has also been observed in reptiles. In fact, periods of reduced antibody production in the Caspian turtle (*M. caspica*) (Leceta and Zapata, 1986), the Schneider's skink (*Eumeces schneideri*) (El Deeb et al., 1980), and the olive sand snake (*P. sibilans*) (Saad and Shoukrey, 1988) roughly correspond to periods with increased corticosterone or testosterone. Exogenous hydrocortisone acetate suppressed the antibody response of *C. ocellatus* during a period of otherwise enhanced antibody response (Saad and El Ridi, 1988). However, data on the antibody production in reptiles following xenobiotic exposure are scarce. A comparative study of alligators from Lake Apopka (Gross, 1997, 1998) suggested that hatchling alligators from Lake Apopka may be immunosuppressed — that is, displaying reduced antibody production relative to alligators from a reference lake.

The antibody response of rats and mice can be measured readily by ELISA. The major isotypes of immunoglobulin (Ig) released are IgM in the primary response and IgG during the secondary response in mammals. The Ig isotypes of reptiles have only been partially characterized and antibodies to identify individual isotypes, required for ELISA assays, are not widely available. Reptiles appear to have one or two classes of low-molecular-weight Ig's, designated as IgY for their similarity to avian and amphibian Ig's, and one high-molecular-weight Ig designated as IgM for its similarity to mammalian IgM (Ambrosius and Hädge, 1983). Recent advances in sea turtle research have led to the generation of monoclonal antibodies to two IgY isotypes and IgM in *C. mydas* (Herbst and Klein, 1995). These monoclonal antibodies have some reactivity to additional sea turtle species, but the antibodies have not been tested for reactivity to fresh water turtles, or other more distantly related reptiles. Widely available antibodies to reptilian Ig's would allow antibody response to be measured by ELISA, and increase the availability of these assays for use in immunotoxicology studies. Recently the mRNA of the joining chain of the red-eared slider, *T. scripta*, IgM has been sequenced and cloned (Iwata et al., 2002). Although advances at the molecular level will allow a more detailed analysis of reptilian Ig's, research at the protein level and the generation of antibodies to reptilian Ig's would be more useful in measuring the antibody response in reptiles. Until antibodies to reptilian Ig's are available, hemagglutination and PFC assays can be used to measure antibody production and do not require the development of new techniques. As mentioned above, the suitability of mammalian complement for a reptilian PFC assay is questionable, and must be firmly established or an alternative source of complement must be determined and made readily or commercially available. Finally, the choice of antigen for testing antibody response is an issue. Although SRBC is the generally accepted antigen for mammalian immunotoxicology assays, there is variability in SRBCs from different sources and different bleeds. The use of keyhole limpet hemocyanin (KLH) has been suggested as an alternative antigen for use in rodents (Exon et al., 1986) because more consistency may be obtained among laboratories. KLH is one of the antigens that have been successfully used to test antibody production in reptiles (Cooper et al., 1985).

8.3.2.1.3 Delayed-Type Hypersensitivity Response

The DTH response, a measurement of specific cell-mediated immunity by T cells, is evaluated in rodents by recording footpad swelling 1 day following antigen challenge subsequent to a period of antigen sensitization (Exon et al., 1986). Footpad swelling may be difficult to measure due to the relatively tough (and possibly rigid) dermal layers in reptiles. Cope et al. (2001) was able to quantify a DTH response to horseshoe crab hemocyanin in the green anole (*Anolis carolinensis*), using measurements of swelling in a skin flap beneath the chin, the throat fan. The DTH response in *A. carolinensis* was unaffected by ultraviolet (UV) irradiation as high as 192 kJ/m^2 UV-B-10 to 100 times the immunosuppressive dose in humans and mice, respectively. The authors suggest that the frequent basking necessary for temperature regulation in *A. carolinensis* causes this species to invariably be exposed to high UV irradiation, which may induce the development of a protective mechanism against immunotoxicity associated with UV radiation in mammals (Cope et al., 2001). However, a known immunosuppressive agent for DTH has not been tested in reptiles, nor was one used on *A. carolinensis* by Cope et al. (2001). The only other published DTH response of a reptile was the cutaneous response to tuberculin in the white spotted gecko, *Tarentola annularis* (Badir et al., 1981). The DTH responses in both *A. carolinensis* and *T. annularis* were characterized by edema, capillary dilation, and mononuclear cell infiltration at the site of antigen challenge (Badir et al., 1981; Cope et al., 2001).

Among all of the traditional assays of functional immunity used in toxicology studies, the DTH response is probably the most underexplored assessment of reptilian physiology that would be useful in immunotoxicology studies. Basic research is required to develop standard methods for DTH assays in turtles and alligators, and to expand on the initial groundwork in lizards. The DTH response of snakes and legless lizards would obviously require a different means of standardization than measuring swelling of the footpad as in tetrapods. Could the swelling of sides of the tail at a fixed distance from the cloaca be measured in place of footpad swelling? In measuring DTH, the contralateral footpad is generally challenged with saline and used as a control for the swelling caused by injection rather than a true inflammation-induced hypersensitivity. In snakes, would the contralateral side of the body be a suitable control? Another issue is the choice of antigens. In mammals, KLH and bovine serum albumin (BSA) are the most common antigens for DTH; the question of what are the best antigens for reptiles requires further investigation.

8.3.2.2 Nonspecific Immune Assays

8.3.2.2.1 NK Cell Activity

Tier 2 assays listed above measure specific immune function while the remaining assays (NK and macrophage function) measure innate immunity (i.e., those that are not directed against a specific antigen), and do not result in the generation of immunologic memory. Specific immune function is more thoroughly investigated in reptiles than innate immunity. The homology between mammalian NK cells and reptilian leukocytes that perform a similar function has not been established. Therefore, some authors (e.g., Munoz and De la Fuente, 2001a) refer to the reptilian cells as NK-like cells. From this point forward in this chapter, reptilian cytotoxic cells will be referred to as having NK activity. NK activity was described in the olive sand snake (*P. sibilans*) using a ^{51}Cr-labeled human erythroleukemia cell line (K-562) as target cells (Sherif and El Ridi, 1992). In an attempt to verify that the cells with NK activity in *P. sibilans* were a separate population, macrophages and B cells were removed and NK activity was then evaluated. Removing macrophages did not alter NK activity, but removing B cells with an anti-*P. sibilans* Ig antibody decreased the NK activity by approximately 50%, suggesting the involvement of B cells but also cytotoxicity independent of macrophages and B cells (Sherif and El Ridi, 1992). In separate studies, the ^{51}Cr-labeled murine lymphoma cell line (YAC-1) was utilized as target cells for NK cytotoxicity assays in the Caspian turtle (*M. caspica*) at an effector:target cell ratio of 25:1 (Munoz et al., 2000). The NK activity of *M. caspica* fluctuated seasonally with decreased NK cytotoxicity in spring and fall (Munoz and De la Fuente, 2001a) when other aspects of the immune system in *M. caspica* are also suppressed. To the author's knowledge, NK cell activity has not yet been used as a measure of immunotoxicity in reptiles.

With only two species examined for NK cell activity, further studies are required to establish the usefulness of NK assays for measuring immunotoxicity in reptiles. Although no new research tools or techniques need to be developed, optimal culture conditions for reptilian NK cells must be established, including incubation time, temperature, and media requirements. There are also questions as to the appropriate target cells (YAC-1 and/or K-562). Each target cell line has been used for one reptile species. The usefulness of the cell lines for other species remains to be determined. Additionally, the use of alternatives to the classical ^{51}Cr measurement of cell lysis may be considered for use in reptile immunotoxicology. Nonradioactive methods have been established that use green fluorescent protein (EGFP) (Kantakamalakul et al., 2003) or 3,3'-dioctadecyloxa-carbocyanine perchlorate (DiO) (Brousseau et al., 1999) in place of [51]Cr.

8.3.2.2.2 Macrophage Function

Although simple techniques are available to test phagocytosis, rarely have they been applied to reptiles. Standard blood smears were made after incubating heat-killed yeast with 20 µl of whole blood from the Chinese soft-shelled turtle (*Trionyx sinensis*) for 30 minutes at 37°C (Zhou et al., 2002). By determining the percentage of leukocytes that had phagocytized yeast, a measure of phagocytosis was generated. It is unclear whether attempts were made to discriminate the phagocytic cell population in *T. sinensis* as macrophages or neutrophils. Macrophages were removed from pooled spleen cell suspensions of the yellow-bellied house gecko (*Hemidactylus flaviviridis*) before evaluating phagocytosis in those animals (Mondal and Rai, 2001). The targets for the phagocytosis assay in *H. flaviviridis* were also heat-killed yeast cells; however, the *in vitro* incubation period was longer (90 minutes at 25°C) (Mondal and Rai, 2001, 2002). Addition of glucocorticoids to the culture medium suppressed the phagocytic activity of *H. flaviviridis* macrophages by over 30% (Mondal and Rai, 2002). Although the *in vitro* administration of chemicals may suggest what will happen in the living animal, *ex vivo* research (administration of chemicals *in vivo* and then testing of response or toxicity *in vitro*) is a much better predictor of immunotoxicity. *Ex vivo* investigation was done for the immunotoxic effects of UV radiation on phagocytosis of *Escherichia coli* by splenocytes of the green anole, *A. carolinensis* (Cope et al., 2001). High doses (up to 192 KJ/m[2]) of UV were not immunosuppressive in *A. carolinensis*. Although macrophage phagocytosis has not been examined in alligators, the phagocytic response of neutrophils has been measured in alligators from Lake Apopka, Florida. The uptake of fluorescent latex beads by peripheral blood neutrophils *in vitro* was compared in samples from captive-raised and wild-caught juvenile alligators from Lake Apopka or a reference lake. No difference in the phagocytic ability of cells was detected between animals from Lake Apopka and the reference lake (Rooney, 1998).

The phagocytic response of macrophages is another area of reptile immunotoxicology that does not necessarily require the development of a new assay, but rather the adaptation of existing techniques. Methods for leukocyte separation using whole blood have been developed for turtles (Harms et al., 2000). In addition, the ability of macrophages to avidly adhere to glass and plastic surfaces in a manner similar to mammalian macrophages has also been used in reptiles (Sherif and El Ridi, 1992). Presently, there is a need to optimize culture conditions for macrophage function and to select the most suitable assay targets.

8.3.3 Tier 3: Present Status and Research Needs

8.3.3.1 Host Resistance (Infectivity) Models

Reptiles represent an interesting group of animals in which to examine host resistance because of their behaviorally regulated body temperature. In mammals and birds, elevation of body temperature is part of the response to bacterial infection. In reptiles, body temperature is adjusted behaviorally to also create fever subsequent to challenge infections. In response to infection with the bacterium *Aeromonas hydrophila*, the lizard *Dipsosaurus dorsalis* adjusted its basking behavior and elevated its preferred body temperature by 2°C (Kluger et al., 1975). The febrile response to bacterial pyrogens are characteristic of all lizards examined thus far (Muchlinski et al., 1999). Reptiles produce a substance similar to mammalian interferon in response to viral challenge (Mathews and Vorndam, 1982), and are susceptible to a range of virus types. Antibodies have been detected to disease-associated viruses such as the paramyxovirus and reovirus (Marschang et al., 2002), as well as the herpes virus (Coberley et al., 2001).

Although reptiles clearly are infected with a range of viruses and bacteria, experimental infections are rarely used to measure their resistance to such infections. In one of the few published examples, repeated challenge of broad-headed skink (*Eumeces laticeps*) with larvae of the black-legged tick (*Ixodes scapularis*) failed to result in any measurable host resistance; however, house mice and cotton mice also failed to develop resistance while guinea pigs successfully developed protective antibodies (Galbe and Oliver, 1992). Additional host resistance studies are required in order to select and standardize useful challenge models for use in reptiles.

8.4 LIFE HISTORY (DOES IT HELP OR HINDER IMMUNOTOXICITY STUDIES?)

Several aspects of reptilian biology make reptiles an interesting group of animals for toxicology studies. Recently, the state of the science of reptilian toxicology was reviewed for the two most populous groups of reptiles/lizards (Campbell and Campbell, 2000) and snakes (Campbell and Campbell, 2001). The advantages of reptiles for ecotoxicology studies highlighted in these and other papers include the typically long life span of some reptiles and the preponderance of relatively site-faithful species, which makes local and long-term studies of contamination possible (Hopkins, 2000). This discussion highlights advantages and challenges presented to immunotoxicology research by the seasonal nature of reptilian immunity, characteristics of reptilian sex determination, and use of closely related species that differ in the degree of maternal/fetal interactions (oviparous vs. viviparous) to study maternal influence on developmental immunotoxicology.

8.4.1 Seasonality

The pronounced seasonality of the reptilian immune response is the most striking aspect of reptilian immunity when comparing immune function in mammals and reptiles. Although seasonal changes in mammalian and avian immune function have been described (Nelson and Demas, 1996), they are surpassed in scope and degree by the changes characteristic of reptiles. The pronounced variation in reptilian immunity may prove a hindrance in comparing immunotoxicity data from one study to the next. However, the seasonal fluctuations in reptilian immune function will encourage researchers to consider animals in the context of their natural environment and may be more relevant to real-world immunotoxic effects of xenobiotics such as insecticides that are applied seasonally. It will be of biological significance to know whether the lowest effective dose of an immunotoxicant observed during the summer season when the immune response is robust, will still be the lowest effective dose of that immunotoxicant during the winter months when immune response is relatively lower.

8.4.2 Gender and Immunotoxicity

Sexually dimorphic responses are common to the immune system and especially to diseases of the immune system and autoimmunity (Grossman et al., 1991; Olsen and Kovacs, 1996). There are a growing number of examples in which perinatal toxicant exposure influences the sexes differently. For example, an *in ovo* exposure to lead stimulated increased antibody production in male but not in female chickens (Bunn et al., 2000). In Sprague-Dawley rats, developmental exposure to atrazine depressed antibody production and the DTH response in males, but did not adversely affect immune responses in females (Rooney et al., 2003b). Species where sex can be readily controlled may prove beneficial to the study of the role of gender in immunotoxicology. In all crocodilians and in certain turtles and lizards, the sex of developing embryos is not predetermined genetically; instead, sex is determined by the incubation temperature experienced during a critical developmental window or temperature-sensitive period (Pieau et al., 1994). Under laboratory conditions, this temperature-dependent sex determination (TSD) allows researchers to control the sex of developing embryos and thus produce animals of the desired sex. For example, in the leopard gecko (*Eublepharis macularius*), a lizard with TSD, lizards of each sex were produced and then raised under varied temperature regimes to dissect the different roles of gonadal sex and incubation temperature on later physiological characteristics (Tousignant and Crews, 1995). Reptilian species where the mechanisms of sex determination are well studied and readily controlled may aid in discovering which specific aspects of gender are linked to regulating gender-specific immunotoxicity.

8.4.3 Viviparity, Oviparity, and Developmental Immunotoxicology

Developmental immunotoxicology studies have generally shown that exposure during early developmental stages causes more severe and persistent effects than exposure at the adult stage (Dietert et al., 2002; Holladay and Luster, 1996; Holladay and Smialowicz, 2000). The importance of the mother in metabolizing xenobiotics into either a more toxic or less toxic compound can be more readily studied in species where the mother and offspring can be separated early in development. The availability of closely related viviparous and oviparous species within several lizard genera (e.g., *Sceloporus*) allows the role of maternal exposure to be examined in comparative studies of developmental immunotoxicity. Comparative studies can also address the consequences of evolutionary changes in the immune system associated with the development of viviparity when exposed to immunotoxic chemicals (oviparity vs. viviparity). The evolutionary changes that lead to internal development are known to include suppression of the maternal immune system (Guillette, 1993; Ribbing et al., 1988). Does this maternal immunosuppression make viviparous species more vulnerable to immunotoxic compounds during gestation than are oviparous species during incubation? As a model, viviparous lizards and snakes represent a similar, but less invasive maternal/fetal interaction than mammals where maternal toxicant exposure directly impacts the developing offspring. In both viviparous and oviparous species, lipid-soluble xenobiotics are "dumped" into the yolk, and are probably higher in a female's first clutch as the initial body burden is cleared (Matter et al., 1998). The oviparous species (including all crocodilians, turtles, and tortoises, some lizards, and some snakes) are more similar to avian reproduction in that the developing embryo is sequestered from maternal contact early in development. The external egg itself has been used to considerable advantage in many developmental toxicology studies. External eggs can be exposed directly to a toxicant without concern for biotransformation or degradation of the toxicant within the mother. In turtles and alligators, this technique has been developed so that compounds can be either directly injected into the egg (Matter et al., 1998) or painted on the egg shell in alcohol to draw the toxicant inside (Wibbels et al., 1991). Egg-treatment studies can test immunotoxic effects of parent compounds and their derivatives while controlling the exact doses given to each egg. The ability of adult females to biotransform the parent compound can be examined separately from the biotransformation that takes place *in ovo*.

8.5 CRITICAL NEEDS FOR STANDARD REPTILIAN MODELS

8.5.1 The Reptilian Lab Rat

There are a vast number of species that could be used for ecotoxicology studies and each species would have its own advantages and disadvantages. For laboratory studies, a case can also be made for a variety of species, but the significance of a short generation time and the existence of baseline data can be used to draw some preliminary conclusions. Taxa or species are evaluated in broad terms, and suggestions are made keeping in mind that the following list is neither exhaustive nor unbiased. However, an attempt was made to select the reptile species for which there is the most published data on immune system morphology, immune system function, and toxicology.

8.5.1.1 *Egg Treatment and Developmental Studies*

Crocodilian species are too large, too slow to mature sexually, and too difficult to maintain breeding populations to make a good reptilian model for laboratory studies on adults or those requiring successive generations. However, the relative abundance of data on reproductive and developmental toxicology in alligators (Crain et al., 2000) and the commercial production of crocodilians worldwide make them an attractive group of species for developmental toxicology studies (Rooney and Guillette, 2001). Several hundred alligator eggs can be obtained by cooperation with state and federal agencies in the southeastern United States. Even larger numbers of turtle eggs can be from commercial sources. Turtles possess similar advantages to crocodilians with the exception that turtles are generally not the apex predator in their food web, and therefore turtles should have slightly less bioaccumulation of lipid-soluble xenobiotics. Additionally, there are both herbivorous and carnivorous turtle species so the influence of diet on exposure to or bioaccumulation of xenobiotics can be addressed using turtle species. Turtles are also much more widely distributed, and field studies on turtles can therefore span a larger geographic area than the more tropically restricted crocodilians. Between the two orders, Crocodylia and Testudinata, there are many good models for developmental immunotoxicology studies. With each species, however, most of the background research has yet to be done. It is also important to note that neither group makes a good model if adults are required after developmental exposure due to size at maturity and slow sexual maturation. This is certainly true for crocodilians, although there are some turtle species (e.g., the stinkpot, *Sternotherus odoratus*) that are both small and can reach sexual maturity within 3 years under ideal conditions. Snakes and lizards could also be used for developmental studies, but to the author's knowledge there is no commercial production of snake or lizard eggs, and therefore availability may not be as widespread.

8.5.1.2 *Generational Studies*

A good laboratory model must survive and reproduce well in captivity, but it must also reproduce rapidly. Certain species of both snakes and lizards reach sexual maturity in 1 year; it is these species that may hold the greatest promise as a practical reptilian laboratory model or models. It may also be possible to shorten generation time as reptiles have very plastic growth rates linked to temperature and food availability. Consideration and testing for reproductive or immunological effects of artificially reducing time to reproductive maturity must be included to standardize husbandry practices. Lizards have been more widely studied than snakes for baseline endocrine and immune parameters, and so will be discussed here. Small lizard populations from two different species of the genus *Sceloporus*, the western fence lizard (*S. occidentalis*), and eastern fence lizard (*S. undulatus*), were recently evaluated for their suitability as a reptile laboratory toxicology model (Talent et al., 2002). In a single study involving capture of three populations of *S. occidentalis* and four populations of *S. undulatus*, one population of *S. occidentalis* was found to be superior in survival and fecundity in captivity to the F1 generation (Talent et al., 2002). To my knowledge, a recent paper on the suppressing effects of exogenous 17α-ethinylestradiol or corticosterone in *S. occidentalis* is the only example of an immunotoxicology study in *Sceloporus* (Burnham et al., 2003). Antibody response, MLR, and differential leukocyte counts were all measured in *S. occidentalis* (Burnham et al., 2003). The Talent et al. (2002) study included a microinjection endocrine toxicology component to demonstrate susceptibility to endocrine disrupting chemicals within *Sceloporus*. Published background data in *Sceloporus* includes basic physiology, i.e., endocrine control of reproduction (Guillette et al., 1991), innervation (Rooney et al., 1997), and stress (Dunlap, 1995), but this leaves *Sceloporus* as a potential model for which no baseline immune data or immunotoxicology data are yet available. The green anole, *A. carolinensis*, is the second lizard species with rapid maturity, relative ease of maintenance in captivity, and some published data on immune or endocrine physiology. Functional immune research in *Anolis* includes measures of DTH and phagocytic response (Cope et al., 2001). A number of important variables have been measured in *Anolis* that will serve as background for immunotoxicology studies. These include, reproductive hormones (Tokarz et al., 1998), stress (Greenberg et al., 1984), neurotransmitters (Summers et al., 2000), and the response to organochlorines and UV light (Cope et al., 2001; Hall and Clark, 1982). The third lizard species, *C. ocellatus* (the ocellated lizard), is perhaps the most encouraging candidate for laboratory immunotoxicology studies, and is currently being cultivated for immune studies by several laboratories. Published *C. ocellatus* research includes extensive histological studies of the spleen, thymus, bone marrow, and gut-associated lymphoid tissue (El Deeb and Saad, 1985; Hussein et al., 1978a; Saad, 1989b) and several initial studies of antigenic determinants on lymphocytes (El Deeb et al., 1986; El Deeb et al., 1988). Published data on functional immune tests in *C. ocellatus* are also extensive. These include studies on the lymphocyte blastogenic response to various mitogens (Saad and El Deeb, 1990), MLR (El Deeb and Saad, 1987), rat RBC-specific antibody production (Saad and El Ridi, 1988), and skin allograft rejection (Saad and El Ridi, 1984). Although immune suppression has been demonstrated in *C. ocellatus* associated with exogenous and endogenous hormone (corticosteroids, testosterone) administration (El Masri et al., 1995; Saad et al., 1984), the immunotoxic effects of xenobiotics have not been examined in *C. ocellatus*. Chalcides species have been used to study the effects of a pesticide spill, but the study did not include immune measures (Lambert, 1997).

8.6 CONCLUSIONS

Published data on reptile immunotoxicology are conspicuously scarce. Therefore, there is a need for studies and the development/adaptation of techniques and models that can be used to determine immunotoxicity in this phylum. In 2000, William Hopkins of the Savannah River Ecology Laboratory wrote a letter to the editor of *Environmental Toxicology and Chemistry* detailing the challenges and potential rewards of developing reptile models in ecotoxicology (Hopkins, 2000). Between 1990 and 1999 only 1% of the vertebrate studies published in *Environmental Toxicology and Chemistry* covered reptiles. Thus, basic immune and toxicology research is required. This will enable immunotoxicologists to better understand the potential advantages of developing a model in immunotoxicology based on these long-lived, geographically faithful, profoundly seasonal amniotes.

ACKNOWLEDGMENTS

The author gratefully acknowledges the generous review, comments, mentorship, and discussion provided by Dr. Bob Luebke. The author also thanks D. Bermudez, Dr. D.K. Burnham, and Dr. L.S. Birnbaum for providing helpful suggestions on the manuscript. Critical support was provided by A.S. Galle. Support was also provided in part by the NCSU/EPA Cooperative Training Program in Environmental Sciences Research, Training Agreement CT826512010 with North Carolina State University under the helpful administration of Dr. K. Adler.

REFERENCES

Alimkhodzhaeva, P.R., Usmanova, A.S., and Gil'dieva, B.S., Effect of hydrocortisone on spontaneous and mitogen-dependent activity of peripheral blood lymphocytes in some vertebrates, *Bull. Exp. Biol. Med.*, 133, 471–474, 2002.

Ambrosius, H., and Hädge, D., Phylogeny of low molecular weight immunoglobulins, *Dev. Comp. Immunol.*, 7, 721–724, 1983.

Ambrosius, H., et al., Immunoglobulins and the dynamics of antibody formation in poikilothermic vertebrates, in *Developmental Aspects of Antibody Formation and Structure*, Sterzl, J., and Riha, I., Eds., Academia, Prague, 1969, pp. 727–774.

Ardavin, C.F., et al., Gut-associated lymphoid tissue (GALT) in the amphibian urodele Pleurodeles waltl, *J. Morphol.*, 173, 35–41, 1982.

Avtalion, R.R., Weiss, E., and Moalem, T., Regulatory effects of temperature upon immunity in ectothermic vertebrates, in *Comparative Immunology*, Marchalonis, J.J., Ed., Blackwell Scientific, Oxford, 1976, pp. 227–239.

Badir, N., Afifi, A., and El Ridi, R., Cell-mediated immunity in the gecko, Tarentola annularis, *Folia Biol.*, 27, 28–36, 1981.

Bergeron, J.M., Crews, D., and McLachlan, J.A., PCBs as environmental estrogens: turtle sex determination as a biomarker of environmental contamination, *Environ. Health Perspect.*, 102, 780–781, 1994.

Bona, C.A., and Bonilla, F.A., *Textbook of Immunology*, Hardwood Academic Publishers, New York, 1990.

Borysenko, M., Ultrastructural analysis of normal and immunized spleen of the snapping turtle, Chelydra serpentina, *J. Morphol.*, 149, 243–263, 1976.

Bridges, C.M., Long-term effects of pesticide exposure at various life stages of the southern leopard frog (Rana sphenocephala), *Arch. Environ. Contam. Toxicol.*, 39, 91–96, 2000.

Brousseau, P., et al., *Manual of Immunological Methods*, CRC Press, Boca Raton, FL, 1999.

Bunn, T.L., Marsh, J.A., and Dietert, R.R., Gender differences in developmental immunotoxicity to lead in the chicken: analysis following a single early low-level exposure in ovo, *J. Toxicol. Environ. Health A*, 61, 677–693, 2000.

Burger, J., Trace element levels in pine snake hatchlings: tissue and temporal differences, *Arch. Environ. Contam. Toxicol.*, 22, 209–213, 1992.

Burger, J., et al., Metals and metalloids in tissues of American alligators in three Florida lakes, *Arch. Environ. Contam. Toxicol.*, 38, 501–508, 2000.

Burnham, D.K., et al., Effects of 17alpha-ethinylestradiol on immune parameters in the lizard Sceloporus occidentalis, *Environ. Toxicol.*, 18, 211–218, 2003.

Campbell, K.R., and Campbell, T.S., Lizard contaminant data for ecological risk assessment, *Rev. Environ. Contam. Toxicol.*, 165, 39–116, 2000.

Campbell, K.R., and Campbell, T.S., The accumulation and effects of environmental contaminants on snakes: a review, *Environ. Monitoring Assessment*, 70, 253–301, 2001.

Campbell, K.R., and Campbell, T.S., A logical starting point for developing priorities for lizard and snake ecotoxicology: a review of available data, *Vet. Immunol. Immunopathol.*, 21, 894–898, 2002.

Canas, J.E., and Anderson, T.A., Organochlorine contaminants in eggs: the influence of contaminated nest material, *Chemosphere*, 47, 585–589, 2002.

Chapman, P.M., Integrating toxicology and ecology: putting the "eco" into ecotoxicology, *Mar. Pollut. Bull.*, 44, 7–15, 2002.

Coberley, S.S., et al., Detection of antibodies to a disease-associated herpesvirus of the green turtle, Chelonia mydas, *J. Clin. Microbiol.*, 39, 3572–3577, 2001.

Cooper, E.L., Klempau, A.E., and Zapata, A.G., Reptilian immunity, in *Biology of the Reptilia*, Vol. 14, Development A, Gans, C., Billet, F., and Maderson, P., Eds., John Wiley & Sons, New York, 1985, pp. 599–678.

Cope, R.B., et al., Resistance of a lizard (the green anole, Anolis carolinensis; Polychridae) to ultraviolet radiation-induced immunosuppression, *Photochem. Photobiol.*, 74, 46–54, 2001.

Crain, D.A., et al., Alterations in steroidogenesis in alligators (Alligator mississippiensis) exposed naturally and experimentally to environmental contaminants, *Environ. Health Perspect.*, 105, 528–533, 1997.

Crain, D.A., et al., Endocrine-disrupting contaminants and hormone dynamics: lessons from wildlife, in *Endocrine Disrupting Contaminants: Evolutionary and Comparative Approaches*, Guillette, Jr., L.J. and Crain, D.A., Eds., Taylor & Francis, New York, 2000, pp. 1–21.

Cuchens, M.A., and Clem, L.W., Phylogeny of lymphocyte heterogeneity III. Mitogenic responses of reptilian lymphocytes, *Dev. Comp. Immunol.*, 3, 287–297, 1979.

Dean, J.H., Issues with introducing new immunotoxicology methods into the safety assessment of pharmaceuticals, *Toxicology*, 119, 95–101, 1997.

Dean, J.H., Hincks, J.R., and Remandet, B., Immunotoxicology assessment in the pharmaceutical industry, *Toxicol. Lett.*, 102–103, 247–255, 1998.

Dietert, R.R., Lee, J.E., and Bunn, T.L., Developmental immunotoxicology: emerging issues, *Hum. Exp. Toxicol.*, 21, 479–485, 2002.

Dunlap, K.D., External and internal influences on indices of physiological stress: II. Seasonal and size-related variations in blood composition in free-living lizards, Sceloporus occidentalis, *J. Exp. Zool.*, 272, 85–94, 1995.

El Deeb, S., El Ridi, R., and Badir, N., Effect of seasonal and temperature changes on humoral response of Eumeces schneideri (Reptilia, Sauria, Scincidae), *Dev. Comp. Immunol.*, 4, 753–758, 1980.

El Deeb, S., El Ridi, R., and Zada, S., The development of lymphocytes with T- or B- membrane derterminants in the lizard embryo, *Dev. Comp. Immunol.*, 10, 353–364, 1986.

El Deeb, S., Saad, A.H., and Zapata, A.G., Detection of Thy−1+ cells in the developing thymus of the lizard, Chalcides ocellatus, *Thymus*, 12, 3–9, 1988.

El Deeb, S., and Saad, A.H.M., Ontogeny of hemopoietic and lymphopoietic tissues in the lizard, Chalcides ocellatus, *J. Morphol.*, 185, 241, 1985.

El Deeb, S., and Saad, A.H.M., Ontogeny of con A responsiveness and mixed leucocyte reactivity in the lizard, Chalcides ocellatus, *Dev. Comp. Immunol.*, 11, 595–604, 1987.

El Deeb, S., et al., Neuroimmunomodulation in reptiles III. Morphological and immunological changes in the spleen of juvenile lizards following steroid hormones treatment, *J. Egypt. Ger. Soc. Zool.*, 12, 489–518, 1993.

El Masri, M., et al., Seasonal distribution and hormonal modulation of reptilian T cells, *Immunobiology*, 193, 15–41, 1995.

El Ridi, R., Badir, N., and El Rouby, S., Effect of seasonal variations on the immune system of the snake, Psammophis schokari, *J. Exp. Zool.*, 216, 357–365, 1981a.

El Ridi, R., El Deeb, S., and Zada, S., The gut-associated lympho-epithelial tissue (GALT) of lizards and snakes, in *Aspects of Developmental and Comparative Immunology*, Solomon, J.B., Ed., Pergamon, Oxford, 1981b, pp. 233–239.

El Ridi, R., El Masry, M., and Badir, N., Reptilian immune system, in *Encyclopedia of Immunology*, Roitt, I.M., and Delves, P.J., Eds., Academic Press, New York, 1992, pp. 1313–1316.

Evans, E.E., Comparative immunology: antibody response in Dipsosauus dorsalis at different temperatures, in *Proc. Soc. Exp. Biol. Med.*, 112, 531–533, 1963.

Exon, J.H., et al., Immunotoxicity testing: an economical multiple-assay approach, *Fundam. Appl. Toxicol.*, 7, 387–397, 1986.

Farag, M.A., and El Ridi, R., Mixed leucocyte reaction (MLR) in the snake Psammophis sibilans, *Immunology*, 55, 173–181, 1985.

Farag, M.A., and El Ridi, R., Proliferative response of snake lymphocytes to concanavalin A., *Dev. Comp. Immunol.*, 10, 561–569, 1986.

Frye, F.L., Comparative histology, in *Reptile Care: An Atlas of Disease and Treatments*, TFH Publications, Neptune City, NJ, 1991a, pp. 470–512.

Frye, F.L., Hematology as applied to clinical reptile medicine, in *Reptile Care: An Atlas of Disease and Treatments*, TFH Publications, Neptune City, NJ, 1991b, pp. 209–279.

Galbe, J., and Oliver, J.H., Jr., Immune response of lizards and rodents to larval Ixodes scapularis (Acari: Ixodidae), *J. Med. Entomol.*, 29, 774–783, 1992.

Garner, M.M., et al., Staining and morphologic features of bone marrow hematopoietic cells in desert tortoises (Gopherus agassizii), *Am. J. Vet. Res.*, 57, 1608–1615, 1996.

Gonzalez-Doncel, M., et al., Stage sensitivity of medaka (Oryzias latipes) eggs and embryos to permethrin, *Aquatic Toxicol.*, 62, 255–268, 2003.

Greenberg, N., Chen, T., and Crews, D., Social status, gonadal state and the adrenal stress response in the lizard, Anolis carolinensis, *Horm. Behav.*, 18, 1–11, 1984.

Gross, D., et al., Potential Contaminant-Induced Immuno-Suppression in Neonatal Alligators from Contaminated and Control Lakes in Central Florida, presented at Society of Environmental Toxicology and Chemistry 18th annual meeting, San Francisco, 1997.

Gross, D.A., Thymus, spleen, and bone marrow hypoplasia and decreased antibody responses in hatchling Lake Apopka alligators, *Veterinary Medicine*, University of Florida, Gainesville, FL, 1998, pp. 1–47.

Grossman, C.J., Roselle, G.A., and Mendenhall, C.L., Sex steroid regulation of autoimmunity, *J. Steroid Biochem. Mol. Biol.*, 40, 649–659, 1991.

Guillette, L.J., Jr., The evolution of viviparity in lizards, *BioScience*, 43, 742–751, 1993.

Guillette, L.J., Jr., Cree, A.C., and Rooney, A.A., Biology of stress: interactions with reproduction, immunology, and intermediary metabolism, in *Health and Welfare of Captive Reptiles*, Warwick, C., Frye, F.L., and Murphy, J.B., Eds., Chapman & Hall, New York, 1995a, pp. 32–81.

Guillette, L.J., Jr., DeMarco, V., and Palmer, B.D., Exogenous progesterone or indomethacin delays parturition in the viviparous lizard Sceloporus jarrovi, *Gen. Comp. Endocrinol.*, 81, 105–112, 1991.

Guillette, L.J., Jr., et al., Gonadal steroidogenesis in vitro from juvenile alligators obtained from contaminated or control lakes, *Environ. Health Perspect.*, 103 (Suppl. 4), 31–36, 1995b.

Guillette, L.J., Jr., et al., Reduction in penis size and plasma testosterone concentrations in juvenile alligators living in a contaminated environment, *Gen. Comp. Endocrinol.*, 101, 32–42, 1996.

Gunderson, M.P., LeBlanc, G.A., and Guillette, L.J., Jr., Alterations in sexually dimorphic biotransformation of testosterone in juvenile American alligators (Alligator mississippiensis) from contaminated lakes, *Environ. Health Perspect.*, 109, 1257–1264, 2001.

Hall, R.J., and Clark, D.R.J., Responses of the iguanid lizard anolis carolinensis to four organophosphorus pesticides, *Environ. Pollut. Ser. A*, 28, 45–52, 1982.

Harms, C.A., Keller, J.M., and Kennedy-Stoskopf, S., Use of a two-step Percoll gradient for separation of loggerhead sea turtle peripheral blood mononuclear cells, *J. Wildl. Dis.*, 36, 535–540, 2000.

Heinz, G.H., Percival, H.F., and Jennings, M.L., Contaminants in American alligator eggs from Lake Apopka, Lake Griffin, Lake Okeechobee, Florida (USA), *Environ. Monitoring Assessment*, 16, 277–286, 1991.

Herbst, L.H., and Klein, P.A., Monoclonal antibodies for the measurement of class-specific antibody responses in the green turtle, Chelonia mydas, *Vet. Immunol. Immunopathol.*, 46, 317–335, 1995.

Holladay, S.D., and Luster, M.I., Alterations in fetal thymic and liver hematopoietic cells as indicators of exposure to developmental immunotoxicants, *Environ. Health Perspect.*, 104 (Suppl. 4), 809–813, 1996.

Holladay, S.D., and Smialowicz, R.J., Development of the murine and human immune system: differential effects of immunotoxicants depend on time of exposure, *Environ. Health Perspect.*, 108 (Suppl. 3), 463–473, 2000.

Hopkins, W.A., Reptile toxicology: challenges and opportunities on the last frontier in vertebrate ecotoxicology, *Vet. Immunol. Immunopathol.*, 19, 2391–2393, 2000.

Hussein, M.F., et al., Differential effect of seasonal variation on lymphoid tissue of the lizard, Chalcides ocellatus, *Dev. Comp. Immunol.*, 2, 297–310, 1978a.

Hussein, M.F., et al., Effect of seasonal variation on lymphoid tissues of the lizards, Mabuya quinquetaeniata Licht., and Uromastyx aegyptia Forsk, *Dev. Comp. Immunol.*, 2, 469–478, 1978b.

Iwata, A., et al., Cloning and expression of the turtle (Trachemys scripta) immunoglobulin joining (J)-chain cDNA, *Immunogenetics*, 54, 513–519, 2002.

Kantakamalakul, W., Jaroenpool, J., and Pattanapanyasat, K., A novel enhanced green fluorescent protein (EGFP)-K562 flow cytometric method for measuring natural killer (NK) cell cytotoxic activity, *J. Immunol. Methods*, 272, 189–197, 2003.

Kluger, M.J., Ringler, D.H., and Anver, M.R., Fever and survival, *Science*, 188, 166–168, 1975.

Kroese, F.G.M., et al., Dendritic immune complex trapping cells in the spleen of the snake, Python reticulatus, *Dev. Comp. Immunol.*, 9, 641–652, 1985.

Lambert, M.R., Environmental effects of heavy spillage from a destroyed pesticide store near Hargeisa (Somaliland) assessed during the dry season, using reptiles and amphibians as bioindicators, *Arch. Environ. Contam. Toxicol.*, 32, 80–93, 1997.

Leceta, J., and Zapata, A.G., Seasonal changes in the thymus and spleen of the turtle, Mauremys caspica: a morphometrical, light microscopical study, *Dev. Comp. Immunol.*, 9, 653–668, 1985.

Leceta, J., and Zapata, A.G., Seasonal variations in the immune response of the tortoise Mauremys caspica, *Immunology*, 57, 483–487, 1986.

Luster, M.I., Dean, J.H., and Germolec, D.R., Consensus workshop on methods to evaluate developmental immunotoxicity, *Environ. Health Perspect.*, 111, 579–583, 2003.

Luster, M.I., et al., Risk assessment in immunotoxicology. I. Sensitivity and predictability of immune tests, *Fundam. Appl. Toxicol.*, 18, 200–210, 1992.

Mansour, M.H., El Ridi, R., and Badir, N., Surface markers of lymphocytes in the snake, Spalerosophis diadema. I. Investigation of lymphocyte surface markers, *Immunology*, 40, 605–611, 1980.

Mansour, M.H., et al., Identification of peanut agglutinin-binding glycoproteins on lizard lymphocytes, *Zool. Sci.*, 12, 79–85, 1995.

Marschang, R.E., et al., Paramyxovirus and reovirus infections in wild-caught Mexican lizards (Xenosaurus and Abronia spp.), *J. Zool. Wildl. Med.*, 33, 317–321, 2002.

Mathews, J.H., and Vorndam, A.V., Interferon-mediated persistent infection of Saint Louis encephalitis virus in a reptilian cell line, *J. Gen. Virol.*, 61, 177–186, 1982.

Matsunaga, T., and Rahman, A., In search of the origin of the thymus: the thymus and GALT may be evolutionarily related, *Scand. J. Immunol.*, 53, 1–6, 2001.

Matter, J.M., et al., Development and implementation of endocrine biomarkers of exposure and effects in American alligator (Alligator mississippiensis), *Chemosphere*, 37, 1905–1914, 1998.

Mead, K.F., and Borysenko, M., Turtle lymphocyte surface antigens in, Chelydra Serpentina as characterized by rabbit anti-turtle thymocyte sera, *Dev. Comp. Immunol.*, 8, 351–358, 1984.

Mondal, S., and Rai, U., In vitro effect of temperature on phagocytic and cytotoxic activities of splenic phagocytes of the wall lizard, Hemidactylus flaviviridis, *Comp. Biochem. Physiol. Part A Mol. Integrated Physiol.*, 129, 391–398, 2001.

Mondal, S., and Rai, U., Dose and time-related in vitro effects of glucocorticoid on phagocytosis and nitrite release by splenic macrophages of wall lizard Hemidactylus flaviviridis, *Comp. Biochem. Physiol. C Pharmacol. Toxicol.*, 132, 461–470, 2002.

Morici, L.A., Elsey, R.M., and Lance, V.A., Effects of long-term corticosterone implants on growth and immune function in juvenile alligators, Alligator mississippiensis, *J. Exp. Zool.*, 279, 156–162, 1997.

Muchlinski, A.E., Gramajo, R., and Garcia, C., Pre-existing bacterial infections, not stress fever, influenced previous studies which labeled Gerrhosaurus major an afebrile lizard species, *Comp. Biochem. Physiol. Part A Mol. Integrated Physiol.*, 124, 353–357, 1999.

Munoz, F.J., and De la Fuente, M., The effect of the seasonal cycle on the splenic leukocyte functions in the turtle Mauremys caspica, *Physiol. Biochem. Zool.*, 74, 660–667, 2001a.

Munoz, F.J., and De la Fuente, M., The immune response of thymic cells from the turtle Mauremys caspica, *J. Comp. Physiol. B*, 171, 195–200, 2001b.

Munoz, F.J., et al., Seasonal changes in peripheral blood leukocyte functions of the turtle Mauremys caspica and their relationship with corticosterone, 17-beta-estradiol and testosterone serum levels, *Vet. Immunol. Immunopathol.*, 77, 27–42, 2000.

Muthukkaruppan, V., Borysenko, M., and El Ridi, R., RES structure and function of the reptilia, in *Phylogeny and Ontogeny of the Reticuloendothelial System*, Cohen, N., and Sigel, M.M., Eds., Plenum Press, New York, 1983, pp. 461–492.

Nelson, R.J., and Demas, G.E., Seasonal changes in immune function, *Q. Rev. Biol.*, 71, 511–548, 1996.

Olsen, N.J., and Kovacs, W.J., Gonadal steroids and immunity, *Endocrinol. Rev.*, 17, 369–384, 1996.

Pieau, C., et al., Environmental control of gonadal differentiation, in The Differences Between the Sexes, Short, R.V., and Balaban, E., Eds., Cambridge University Press, New York, 1994, pp. 433–450.

Pillai, P.S., and Muthukkaruppan, V., The kinetics of rosette-forming cell response against sheep erythrocytes in the lizard, *J. Exp. Zool.*, 199, 97–104, 1977.

Pitchappan, R., Review on the phylogeny of splenic structure and function, *Dev. Comp. Immunol.*, 4, 395–416, 1980.

Pough, F.H., Ed., *Herpetology*, Prentice-Hall, New York, 1998.

Ribbing, S.L., Hoversland, R.C., and Beaman, K.D., T-cell suppressor factors play an integral role in preventing fetal rejection, *J. Reprod. Immunol.*, 14, 83–95, 1988.

Rooney, A.A., Variation in the endocrine and immune system of juvenile alligators: environmental influence on physiology, *Zoology*, University of Florida, Gainesville, FL, 1998, 1–143.

Rooney, A.A., Bermudez, D.S., and Guillette, L.J., Jr., Altered histology of the thymus and spleen in contaminant-exposed juvenile American alligators, *J. Morphol.*, 256, 349–359, 2003a.

Rooney, A.A., et al., Seasonal variation in plasma sex steroid concentrations in juvenile American alligators, *Gen. Comp. Endocrinol.*, 135, 25–34, 2004.

Rooney, A.A., Donald, J.A., and Guillette, L.J., Jr., Adrenergic and peptidergic innervation of the oviduct of Sceloporus jarrovi during the reproductive cycle, *J. Exp. Zool.*, 278, 45–52, 1997.

Rooney, A.A., and Guillette, L.J., Jr., Biotic and abiotic factors in crocodilian stress: the challenge of a modern environment, in *Crocodilian Biology and Evolution*, Grigg, G.C., Seebacher, F., and Franklin, C.E., Eds., Surrey Beatty and Sons PTY Limited, New South Wales, Australia, 2001, pp. 214–228.

Rooney, A.A., Matulka, R.A., and Luebke, R.W., Developmental atrazine exposure suppresses immune function in male, but not female Sprague–Dawley rats, *Toxicol. Sci.*, 76, 366–375, 2003b.

Saad, A.H., Sex-associated differences in the mitogenic responsiveness of snake blood lymphocytes, *Dev. Comp. Immunol.*, 13, 225–229, 1989a.

Saad, A.H.M., Corticosteroids and immune systems of non-mammalian vertebrates: a review, *Dev. Comp. Immunol.*, 12, 481–494, 1988.

Saad, A.H.M., Pregnancy-related involution of the thymus in the viviparous lizard, Chalcides ocellatus, *Thymus*, 14, 223–232, 1989b.

Saad, A.H.M., and El Deeb, S., Immunological changes during pregnancy in the viviparous lizard, Chalcides ocellatus, *Vet. Immunol. Immunopathol.*, 25, 279–286, 1990.

Saad, A.H.M., and El Ridi, R., Mixed leukocyte reaction, graft-versus-host reaction, and skin allograft rejection in the lizard, Chalcides ocellatus, *Immunobiology*, 166, 484–493, 1984.

Saad, A.H.M., and El Ridi, R., Endogenous corticosteroids mediate seasonal cyclic changes in immunity of lizards, *Immunobiology*, 177, 390–403, 1988.

Saad, A.H.M., et al., Effect of hydrocortisone on immune system of the lizard, Chalcides ocellatus. I. Response of lymphoid tissues and cells to in vivo and in vitro hydrocrotisone, *Dev. Comp. Immunol.*, 8, 121–130, 1984.

Saad, A.H.M., Khalek, N.A., and El Ridi, R., Blood testosterone level: a season-dependent factor regulating immune reactivity in lizards, *Immunobiol.*, 180, 184–194, 1990.

Saad, A.H.M., and Shoukrey, N., Sexual dimorphism on the immune responses of the snake, Psammophis sibilans, *Immunobiology*, 177, 404–419, 1988.

Saad, A.H.M., et al., Testosterone induces lymphopenia in turtles, *Vet. Immunol. Immunopathol.*, 28, 171–180, 1991.

Saad, A.H.M., and Zapata, A.G., Reptilian thymus gland: an ultrastructural overview, *Thymus*, 20, 135–152, 1992.

Selye, H., A syndrome produced by diverse nocuous agents, *Nature*, 138, 32, 1936.

Selye, H., The general-adaptation-syndrome and the diseases of adaptation, in *Texbook of Clinical Endocrinology*, Selye, H., Ed., Acta Endocrinologica, Montreal, 1949, pp. 837–866.

Sharma, J.M., Overview of the avian immune system, *Vet. Immunol. Immunopathol.*, 30, 13–17, 1991.

Sharma, J.M., The structure and function of the avian immune system, *Acta Vet. Hung.*, 45, 229–238, 1997.

Sherif, M., and El Ridi, R., Natural cytotoxic cell activity in the snake, Psammophis sibilans, *Immunobiology*, 184, 348–358, 1992.

Shields, J.W., Bursal dissections and gill pouch hormones, *Nature*, 259, 373–376, 1976.

Summers, C.H., et al., Serotonergic responses to corticosterone and testosterone in the limbic system, *Gen. Comp. Endocrinol.*, 117, 151–159, 2000.

Talent, L.G., et al., Evaluation of western fence lizards (Sceloporus occidentalis) and eastern fence lizards (Sceloporus undulatus) as laboratory reptile models for toxicological investigations, *Vet. Immunol. Immunopathol.*, 21, 899–905, 2002.

Tanaka, Y., and Elsey, R.M., Light microscopic study of the alligator (Alligator mississippiensis) spleen with special reference to vascular architecture, *J. Morphol.*, 233, 43–52, 1997.

Tokarz, R.R., et al., Plasma corticosterone and testosterone levels during the annual reproductive cycle of male brown anoles (Anolis sagrei), *Physiol. Zool.*, 71, 139–146, 1998.

Tousignant, A., and Crews, D., Incubation temperature and gonadal sex affect growth and physiology in the leopard gecko (Eublepharis macularius), a lizard with temperature-dependent sex determination, *J. Morphol.*, 224, 159–170, 1995.

Turton, J.A., et al., Relationship of blood corticosterone, immunoglobulin and haematological values in young crocodiles (Crocodylus porosus) to water temperature, clutch of origin and body weight, *Aust. Vet. J.*, 75, 114–119, 1997.

Ulsh, B.A., et al., Culture methods for turtle lymphocytes, *Methods Cell Sci.*, 22, 285–297, 2000.

U.S. Environmental Protection Agency, Health Effects Test Guidelines: OPPTS 870.7800 Immunotoxicity, Office of Prevention, Pesticides and Toxic Substances, U.S. Environmental Protection Agency, Washington, DC, 1998.

Varas, A., Torroba, M., and Zapata, A.G., Changes in the thymus and spleen of the turtle Mauremys caspica after testosterone injection: a morphometric study, *Dev. Comp. Immunol.*, 16, 165–174, 1992.

Vos, J., et al., Toxic effects of environmental chemicals on the immune system, *Trends Pharmacol. Sci.*, 10, 289–292, 1989.

Wang, J., Whetsell, M., and Klein, J.R., Local hormone networks and intestinal T cell homeostasis, *Science*, 275, 1937–1939, 1997.

Weill, J.C., Cocea, L., and Reynaud, C.A., Allelic exclusion: lesson from GALT species, *Semin. Immunol.*, 14, 213–215; discussion 227–228, 2002.

Wibbels, T., Bull, J.J., and Crews, D., Synergism between temperature and estradiol: a common pathway in turtle sex determination? *J. Exp. Zool.*, 260, 130–134, 1991.

Woodward, A.R., et al., Low clutch viability of American alligators on Lake Apopka, *Florida Scientist*, 56, 52–63, 1993.

Work, T.M., et al., Assessing humoral and cell-mediated immune response in Hawaiian green turtles, Chelonia mydas, *Vet. Immunol. Immunopathol.*, 74, 179–194, 2000.

Work, T.M., et al., Immune status of free-ranging green turtles with fibropapillomatosis from Hawaii, *J. Wildl. Dis.*, 37, 574–581, 2001.

Wright, R.K., Eipert, E.F., and Cooper, E.L., Regulating role of temperature on the development of ectothermic vertebrate lymphocyte populations, in *Animal Models of Comparative and Development Immunity and Disease*, Gershwin, M.B., and Cooper, E.L., Eds., Pergamon Press, New York, 1978, pp. 80–92.

Zhou, X., et al., The effect of vitamin C on the non-specific immune response of the juvenile soft-shelled turtle (Trionyx sinensis), *Comp. Biochem. Physiol. Part A Mol. Integrated Physiol.*, 131, 917–922, 2002.

CHAPTER 9

Immunotoxicology in Terrestrial Wildlife

Judit E.G. Smits and David Janz

CONTENTS

9.1 INTRODUCTION

9.1.1 Background

This overview of immunotoxicology in terrestrial wildlife will include birds, rodents, mink, deer, and related ungulates. One of the major challenges when dealing with immunotoxicology in wildlife is the lack of standardized or validated methodologies, although Fournier et al. (2000) have produced a valuable handbook toward meeting this deficit. Brousseau et al. (1998) have presented various protocols in immunotoxicology with attention to species differences. The authors included adjustments in protocols being applied to a wide range of nondomestic species from marine mammals to herpetiles and birds. For example, protocols for measuring phagocytosis by peripheral blood neutrophils (heterophils in birds and herpetiles) and monocytes in mink, seals, and waterfowl are detailed. The oxidative burst assay provides information about the functional capacity of cells in the innate immune system. The assays allow the detection of reactive oxygen species such as hydrogen peroxide and superoxide anions that are produced during inflammation and destruction of foreign antigen. In general, specific reagents and validated protocols in immunotoxicology are limited for wildlife species, but the standard endpoints include immunopathology, B-cell (antibody) and T-cell mediated immunity. Natural killer (NK) cell activity, and macrophage and granulocyte phagocytosis indicate the functional capacity of the innate immune system. Regardless of the focus of individual studies, interactions between immunological responses and hormones must be considered. The final, most meaningful endpoint in immunotoxicology is actual host resistance in the face of bacterial, viral, parasitic, or neoplastic challenges.

9.1.2   Molecular Techniques in Wildlife Species

By determining the expression and activation of immunoregulatory cytokines and their receptors at the mRNA level, potential immunotoxins can be studied *in vitro* (Meredith and Miller, 1994). A large number of cytokines, interferons, and growth factors have been cloned at the cDNA and genomic levels enabling molecular analysis of gene expression at the RNA level. Linking *in vivo* and *in vitro* assays may become possible through analysis of cytokine mRNA levels in specific cells following *in vivo* exposure to potentially immunotoxic compounds.

The MHC complex, the genetic locus controlling T-lymphocyte recognition, is present in higher amphibians, fishes, birds, and mammals and is largely devoted to the processing and presentation of antigens. Thus, it has an essential role in the immune response against pathogens. The MHC is highly polymorphic in many species (Jarvi et al., 1995), and is thought to contribute to resistance or susceptibility to infectious disease (Camarena et al., 2001). Results from Pruett et al. (2003) suggest that MHC class II expression is the best predictor of stress (corticosteroid)-related immunosuppression. Differential immunosuppression from endogenous versus exogenous corticosteroids has been described in birds (Fairbrother et al., 2004) and appears to be related to overactivation of steroid receptors on specific immune cells by exogenous steroids (Sitteri et al., 1982).

Chickens and ring-necked pheasants (*Phasianus colchicus*), both galliforms, have small and compact MHC compared to that of mammals (Westerdahl et al., 2000). Recent work with wild birds, Savannah sparrow (*Passerculus sandwichensis*) (Freeman-Gallant et al., 2002), starlings (*Sturnus vulgaris*), great reed warblers (*Acrocephalus arundinaceus*), willow warblers (*Phylloscopus trochilus*) (Westerdahl et al., 2000), and great snipes (*Gallinago media*) (Ekblom et al., 2003), indicates that MHC genes differ markedly across species. In part because of correlations between MHC haplotypes and disease resistance, Jarvi et al. (1995) studied MHC in Florida sandhill cranes (*Grus canadensis pratensis*), which may be used with other species of cranes threatened with extinction. MHC-related immune expression has yet to be studied in the context of exposure to xenobiotics, although with the rapid advances in this area and expanding work with wildlife, it is a good candidate to add to the repertoire of immunotoxicology screens in the near future.

9.2   *IN VIVO* TESTS IN CURRENT USE

9.2.1   Birds

Blood sampling for basic hematology includes manual white blood cell (WBC) counts using one of several methods, none of which is as consistent or reliable as cell counts conducted on automated equipment, which is routine for mammalian blood. The nucleated red blood cells (RBCs) of avian and herpetile species preclude the use of automated techniques. Therefore, counts are conducted manually on a hemacytometer using blood diluted with Natt and Herricks diluent (Gross, 1984) or using the eosinophil Unopette method (Dein, 1986). Differential WBC counts are readily made from blood smears on glass slides. In the case of wild birds where there are no published normal values, the reference birds in the study provide the normal range. Total plasma or serum protein levels measured using a refractometer can be conducted under field conditions, while more detailed information including distinguishing between albumin, α, β, and γ globulins can provide useful information about the health and physiological status of individuals (Duncan et al., 1994).

The phytohemagglutinin (PHA) skin test is the most commonly applied *in vivo* test of immune function in wild birds. The lectin, PHA, derived from the bean *Phaseolus vulgaris*, is injected subcutaneously in a part of the bird that is easily accessible for measuring (patagium, wattle, dewlap, or interdigital web). The injection site is measured before, and 12 to 24 hours after the injection to determine the proliferative response of circulating T lymphocytes that have accumulated at the injection site. Twenty-four hours has been the standard for most PHA tests, but 12 hours is sufficient for some species. With wild-caught, adult passerines, holding the birds overnight and using a 12-hour response time has proven sufficient for a PHA response (Tella and Smits, unpublished). The stimulation index may be determined using two methods. In one method, the increase in skin thickness in the PHA stimulated site is subtracted from the opposite site which has been injected with the carrier alone. Smits et al. (1999, 2001) presented a simplified method based on analysis of seven studies and over 600 birds from species including passerines, raptors, waterfowl, and upland game birds. The technique entails one injection site only, the control being the "before injection" measurement. Apart from reducing the stress associated with handling time of wild birds, another obvious advantage is that the other wing (wattle, toe web) is spared for other uses such as the delayed-type hypersensitivity (DTH) test.

The antigen-specific DTH test measures an integrated response, dependent on T lymphocytes, antigen-presenting cells (macrophages, B lymphocytes, or dendritic cells) and their cytokine-based intercellular communication (Abbas et al., 2000). Because the DTH test depends on recall and memory T cells, it requires sensitization of the individuals with the test antigen (e.g., mycobacterium, bovine serum albumin, dinitophenol-keyhole limpet hemocyanin [DNP-KLH]) days to weeks before they are challenged with the DTH skin test. The optimal antigen and sensitization protocol would need to be determined for the species of interest. These constraints are likely responsible for the dearth of DTH tests used in studies in wild birds, although some have been carried out in captive wild birds (Rose and Bradley, 1977; Fairbrother and Fowles, 1990).

Although several studies with wild birds claim to be evaluating DTH, there is ongoing confusion about distinguishing DTH from PHA responses. To illustrate this point, in a thoughtful and valuable paper on the relationships and interpretation of various immunotoxicity tests and life history strategies in birds, Norris and Evans (1999) have quoted several studies as having investigated the DTH response, while in fact only a skin response to the mitogen PHA was tested (Saino et al., 1997; Christe et al., 1998; Zuk and Johnsen, 1998). Collette and co-workers (2000) showed a similar misconception about the nature of the DTH test. The skin test is applied in a similar manner in both tests, but the cellular interactions and the implications of the two tests are very different. The DTH response is generally more subtle, and measures a more complex immune interaction dependent on immunological memory. In young animals such as nestling birds, the species-specific age at which they can mount a DTH response would need to be determined.

The humoral, or antibody-mediated B-cell response, commonly tested in wild birds measures the other major arm of the adaptive immune system. There are T-cell dependent (e.g., SRBC, bovine serum albumin, DNP-KLH) and T-cell independent (e.g., bacterial cell wall lipopolysaccharide) antibody responses. When a naive animal is first immunized with a foreign antigen, the primary response will result in maximal circulating IgM antibodies in 5 to 7 days (species-specific responses should be determined with a pilot study whenever possible), and then quickly fall back to undetectable levels. After a booster vaccination, the secondary (memory) IgG-dominated response will peak in 5 to 10 days and sustain high circulating levels for months to years depending on the antigen and the species being tested.

RBC from foreign species (commonly SRBC, occasionally chukar RBC) have been widely used in wild birds (Apanius, 1998). This test can be carried out with relatively small volumes of peripheral blood (500 to 800 μl), using a hemagglutination assay well described by Fairbrother et al. (2003), which is relatively simple to perform. It has proven useful in many immunological studies with birds (Table 9.1) but it suffers from some disadvantages that have not been well recognized (Smits and Baos, 2004). Responses are inconsistent to foreign blood cell sensitization. Grasman and Scanlon (1995) determined that Japanese quail responded poorly to SRBC but well to chukar partridge RBC. Others working with adult zebra finch have found that half the birds produced no detectable response (Deerenberg et al., 1997). Pigeons younger than 7 days (Koppenheffer et al., 1980) and zebra finch nestlings at 11 days (Smits and Williams, 1999) do not respond to SRBC. In other research on wild birds in which SRBC has been used to evaluate immunocompetence in adults and young, nestlings fail to produce detectable antibodies (Apanius, 1998; Grasman, 2002; Moller et al., 2001). Smits and Bortolotti (2001) found nestling American kestrels (*Falco sparverius*) also produced a markedly weaker antibody response to DNP-KLH than did the adult birds.

Other antigens used to promote a humoral response in wild birds are dinitrophenol (DNP) conjugated to keyhole limpet hemocyanin (KLH), a respiratory pigment of the keyhole limpet (Smits and Bortolotti, 2001). Other antigens successfully used in birds are KLH alone (Hasselquist et al., 1999), diphtheria-tetanus vaccine (Svensson et al., 1998; Ilmonen et al., 2000), and Newcastle disease virus (Fair and Ricklefs, 2002). Interpretation of the antibody response to RBC is complicated by the fact that blood cells are antigenically complex, with hundreds of possible antigenic determinants against which antibodies may be made (Tizard, 1996). Unvaccinated birds also can show hemagglutination with SRBC because of naturally occurring, preexisting, cross-reacting antigen. This is in contrast to DNP-KLH, the DNP component having a single antigen, and the KLH being a series of simple glucose moieties (Coligen et al., 1994). One recent study with domestic chickens has proven that birds do have a differential antibody response depending on the antigen used. El-Lethy et al. (2003) showed that while antibody production against tetanus toxoid and SRBC were suppressed in stressed chickens, the response to human serum albumin was not affected. This discovery further emphasizes the importance of solid background work for studies of immunotoxicology in wildlife.

Enzyme-linked immunosorbent assays (ELISA) are commonly used to measure antibody response in many species. The ELISA is considerably more sensitive than the hemagglutination (HA) assay used for SRBCs. With the ELISA, serum diluted to 1:800 or higher can readily demonstrate differences among individuals, whereas the HA assay usually picks up differences at dilutions of 1:16 to 1:64. It clearly would be more difficult to detect subtle differences in the humoral response using an HA assay than it would be using an ELISA. In work on small birds, another benefit of the ELISA is that it requires much less serum or plasma than does the HA (5 μl vs. 50 μl). On the down side, ELISAs often require species-specific antibodies against immunoglobulins that are not commercially available for wildlife. Either a cross-reacting antibody has to be found, or specific antibodies must be raised for the species being studied. By lucky coincidence, rabbit antikestrel antibodies raised against kestrel immunoglobulin (Smits and Bortolotti, 2001) cross-react with IgG from white storks, *Ciconia ciconia* (Smits, unpublished). This finding emphasizes the unpredictability of cross-reactivity of blood group antigens. Martinez et al. (2003) found commercially available antichicken antibodies to be useful for detecting serum antibodies in several species of wild birds.

Table 9.1 Immunotoxicants Studied in Wildlife Species

Xenobiotic	Species	Endpoint	Effect	Citation
Lead shot	Bluebird	PHA skin test	Decreased	Fair and Myers (2002)
		Anti-SRBC antibodies	No effect	
		Anti-Newcastle disease virus antibodies	No effect	
Lead shot	Quail	WBC differential (heterophils)	Increased	Fair and Ricklefs (2002)
		PHA skin test	No effect	
		Antichukar RBC antibodies	No effect	
		Anti-Newcastle disease virus antibodies	No effect	
Lead shot	Mallard duck	Anti-SRBC antibodies	Decreased	Trust et al. (1990)
		PHA skin test	No effect	
		Lymphocyte blastogenesis — LPS	No effect	
Lead shot	Mallard duck	WBC count and differential ♂ only	Decreased	Rocke and Samuel (1991)
		Spleen weight ♂ only	Decreased	
		Splenocyte PFC assay ♂ and ♀	Decreased	
Lead acetate	Quail	WBC differential H:L ratio	Increased	Grasman and Scanlon (1995)
		PHA skin test	Decreased	
		Antichukar RBC antibodies	Decreased	
Lead acetate	Red-tailed hawk	Hematology	No effect	Redig et al. (1991)
		Anti-SRBC antibodies	No effect	
		Lymphocyte blastogenesis	No effect	
Mercury	Western grebe	WBC differential — heterophils	Increased	Elbert and Anderson (1998)
		Eosinophils	Decreased	
	Clark's grebe	WBC differential — heterophils	Increased	Elbert and Anderson (1998)
		Eosinophils	Decreased	
	Great egret	WBC total and differential	No effect, no effect	Spalding et al. (2000)
		Antibody titer to EEE, antibody titer to BSA	No effect	
		PHA wing web test	Atrophied	
		Histology — bursa and thymus	Lymphoid depletion	
		Bone marrow		
	Bald eagle nestling	Lymphocyte stimulation	No effect (highly variable)	Audet et al. (2000)
Selenium	Mallard	WBC total and differential	No effect	Fairbrother and Fowles (1990)
		PHA wing web test	No effect	
		DTH test	Decreased	
		Macrophage function (in vivo)	No effect	
		PFC and SRBC antibody titer	No effect	
	Mallard adult	Macrophage phagocytosis	Increased	Whitely and Yuill (1989)
		PFC and SRBC antibody titer	No effect	
	Mallard duckling	Mortality duck viral hepatitis	Increased	

Table 9.1 (continued) Immunotoxicants Studied in Wildlife Species

Xenobiotic	Species	Endpoint	Effect	Citation
Selenium arsenic	Avocet chicks	WBC differential — H:L	Increased	Fairbrother et al. (1994)
		Macrophage phagocytosis	Decreased	
		T lymphocyte stimulation	Increased	
Zinc	Mallard duck	WBC differential H:L	Increased	Levengood et al. (2000)
PCBs	Mallard duck	Mortality duck viral hepatitis	Increased	Friend and Trainer (1970)
	American kestrel	DNP-KLH antibody titer	Increased (♀), decreased (♂)	Smits and Bortolotti (2001)
		PHA wing web test	Increased	Smits et al. (2002)
	Double-crested cormorant	Bacterial eye infections	Increased	Ludwig et al. (1996)
	Mallard	Lymphocyte stimulation	No effect	Fowles et al. (1997)
		PHAS	No effect	
		RBC antibody titer	No effect	
		NK cell activity	No effect	
Organochlorines	Herring gull	WBC differential	No effect	Grasman et al. (1996)
		PHA wing web test	Decreased	
		Hematology	No effect	Grasman et al. (2000)
		WBC total and differential	Increased lymphocytes	
	Tern young	WBC differential	No effect	Grasman and Fox (2001)
		PHA wing web test	Decreased	
		SRBC antibody titer	Increased	
	Glaucous gull	Parasite load	No effect	Sagerup et al. (2000)
Azinphos methyl	Vole	Lymphocyte count	Decreased	Galloway and Handy (2003)
		Neutrophil count	Equivocal	
Chlorfenvinphos	Quail	Lymphocyte count	Decreased	Galloway and Handy (2003)
		Phagocytosis	Increased	
Malathion	Pheasant	Lymphoid organ mass	Decreased	Galloway and Handy (2003)
		Immunopathology	Increased	
Malathion	Ring-necked pheasant	Hematology	No effect	Day et al. (1995)
		WBC total and differential	Increased variability	
		Histopathology — bursa and thymus	Increased	
		Organ mass — bursa and thymus	Decreased	
Fungicides + insecticides (mixed orchard spray)	Tree swallows	Hematology	No effect	Bishop et al. (1998)
		Lymphocyte proliferation PWM Con-A, PHA, LPS	Decreased	
			No effect	
		Macrophage respiratory burst, macrophage phagocytosis	No effect	
		Bursal mass	Lower in sprayed group	
		Thymic involution	Delayed	
Bunker C	Mallard	Organ mass, - spleen only	Decreased	Rocke et al. (1984)
		PFC assay	No effect	
		Mortality — bacterial challenge	Increased	

Contaminant	Species	Assay/parameter	Effect	Reference
South Louisiana Crude	Mallard	Organ weight	No effect	Rocke et al. (1984)
	Mallard	PFC assay	No effect	Goldberg et al. (1990)
		Mortality — bacterial challenge	Increased	
		Duck plaque virus challenge	No effect	
Prudhoe Bay Crude	Herring gull	WBC total — lymphocytes and heterophils	Decreased	Leighton (1993)
		Histopathology — lymphoid organs	Reduced cellularity	
Crude oil + rehabilitation	American coot	Hematology	No effect	Newman et al. (1999, 2000)
		WBC total	Increased	
Bitumen	Tree swallow	PHA wing web test	Decreased	Smits et al. (2000)
		Spleen mass	Decreased	
Petrochemical waste "landfarms"	Cotton rat	WBC total	Increased	McMurry et al. (1999); Wilson et al. (2003)
		Splenocyte proliferation	Increased	
7,12-DMBA	European starling	Plasma protein	No effect	Trust et al. (1994)
		WBC total and differential	Decreased lymphocytes; Increased H:L ratio	
		Splenocyte viability	Decreased	
		Lymphocyte proliferation	Decreased	
		Macrophage phagocytosis	Decreased	
		Nestlings	Decreased	
		SRBC antibody titer −, adults	No effect	
		Nestlings	No effect	
Bleached pulp effluent	Mink	Lymphocyte blastogenesis	Increased	Smits et al. (1996a, 1996b)
		Antibody response	Decreased	
		Delayed-type hypersensitivity		
Radiation	Barn swallow	WBC total and differential	H:L ratio increased	Camplani et al. (1999)
		Splenic mass	Decreased	
		Ectoparasite abundance	No effect	
Explosives (TNT)	Bobwhite quail	WBC total and differential	No effect	Gogal et al. (2002)
		Plasma proteins	Decreased	
		Lymphocyte stimulation		
		Con-A	No effect	
		Phorbol	No effect	
		Spleen weight	No effect	
		Organ histopathology	No effect	
Aflatoxins	Wild turkey poult	WBC total and differential	Slight increase in total WBC	Quist et al. (2000)
		Organ weight	No effect	
		Lymphocyte stimulation	No effect	
		CD8+ lymphocytes	Increased	
		CD4+:CD8+ ratio	decreased	

Note: BSA, bovine serum albumin; DNP-KLH, 2,4-dinitrophenol-keyhole limpet hemocyanin; LPS, lipopolysaccharide; PFC, plaque-forming cells; PHA, phytohaemagglutinin-P; PWM, pokeweed mitogen; RBC, red blood cells; SRBC, sheep red blood cells; WBC, white blood cells.

9.2.2 Mammals

9.2.2.1 Rodents

Effects of toxicants on immune responses have been studied in wild deer mice (*Peromyscus maniculatus*) and cotton rats (*Sigmodon hispidus*) in the laboratory. However, the use of wild rodent species in field studies to evaluate immunotoxicity risks associated with contaminated terrestrial ecosystems has been slow to develop. This is surprising given the logistical advantages of using wild rodents as *in situ* bioindicator species to investigate wildlife immunotoxicity. Advantages of using resident rodent species include local abundance; ease of collection, handling, and transport; sufficient size for collection of fluids and tissues; small home range; and ecological importance in terrestrial communities. This section focuses on recent immunotoxicological field research using cotton rats.

In earlier laboratory studies, plant growth regulators used in agriculture were reported to cause dose-dependent immunomodulatory effects in deer mice, including alterations in lymphocyte viability, WBC counts, PFC responses to SRBC challenge, thymus weights, and hemolysin titers (Olson and Hinsdill, 1984). Oral exposure to plant growth regulators was also reported to increase mortality in deer mice challenged with arboviruses such as Venezuelan equine virus (Fairbrother et al., 1984, 1986). Age and nutritional status were shown to greatly influence the ability of the deer mouse immune system to respond to toxicants (Porter et al., 1984; Fairbrother et al., 1986). Similar approaches to developing *in situ* rodent bioindicators of immunotoxicity have been evaluated using cotton rats. Exposure of adult cotton rats to graded doses of benzene or cyclophosphamide resulted in increased spleen weights, splenocyte yields, WBC counts, and PFC responses to SRBC, but no cell-mediated immune suppression (McMurry et al., 1991, 1994). Further laboratory studies in cotton rats found elevated dietary Pb to cause reduced spleen mass (McMurry et al., 1995). Low-level arsenic exposure via drinking water decreased hypersensitivity reactions by 30% in cotton rats (Savabieasfahani et al., 1998). Collectively, these studies demonstrated the feasibility of using wild rodents as immunotoxicological bioindicator species in the field.

Petrochemical waste contains both organic (e.g., aromatic hydrocarbons) and inorganic (e.g., metals) contaminants that can pollute soil and may pose significant immunotoxicological risks to terrestrial wildlife. Cotton rats inhabiting petrochemical waste treatment units ("landfarms") in Oklahoma have been used to investigate immunological, hematological, and pathological effects of such contamination (McMurry et al., 1999; Rafferty et al., 2001; Schroder et al., 2003; Carlson et al., 2003; Wilson et al., 2003). In one study, cotton rats were sampled seasonally (summer and winter) from five landfarms and from five ecologically matched reference sites for 2 years (in 1998–2000), with follow-up immunological and hematological assays (Wilson et al., 2003). Analysis indicated that rats inhabiting landfarms exhibited decreased relative spleen size compared to rats collected from reference sites, with one landfarm showing the greatest reduction. Cotton rats collected from landfarms also had increased hemoglobin, hematocrit and platelet levels, and decreased blood leukocytes during summer. During winter, an increase in the number of popliteal node white blood cells was observed from rats collected on landfarms. No difference was detected in lymphocyte proliferation in response to concanavalin A (Con-A), pokeweed mitogen, or interleukin-2. Lymphokine-activated killer cell lytic ability showed a seasonal pattern, but no treatment differences. No differences between landfarm and reference sites were detected in the hypersensitivity reaction of rats given an intradermal injection of phytohemagglutinin (PHA-P). These results suggested that residual petrochemical waste affects the immune system and blood parameters of cotton rats living on landfarms, particularly during summer, and is complicated by variation in the contaminants found on individual petroleum sites (Wilson et al., 2003). Although an earlier study attempted to control such variation by using cotton rats caged in mesocosms on contaminated sites (Propst et al., 1999), it was concluded that the use of wild-caught rodents was a more relevant approach. To investigate potential mechanisms of immunotoxicity in cotton rats, the extent of apoptosis in thymus, spleen, and bone marrow was determined. In comparison to a reference site, thymic cell apoptosis was elevated at a petrochemical landfarm (Savabieasfahani et al., 1999). Although spleen and bone marrow cell apoptosis was elevated approximately twofold at the landfarm site, the large variability in apoptosis in these tissues precluded statistical significance. Overall, the considerable effort dedicated to the development of cotton rats for wildlife immuno-toxicology research clearly illustrates the validity of rodents as in situ bioindicator species for such studies.

9.2.2.2 Mink

Pseudomonas pneumonia, which causes high morbidity and mortality in mink (*Mustela vison*), has stimulated research into mink immunology (Elsheikh et al., 1988; Rivera et al., 1994). Evaluation of the humoral immune response in mink using protein A as a capture antigen with strong affinity for mink immunoglobulin has been described (Rivera et al., 1994; Smits and Godson, 1996b). Very little of the mink research is related to contaminant exposure. One study describes immunological consequences in mink chronically exposed to bleached pulp mill effluent. In mink vaccinated with live mycobacterial, bacillus Camette-Guerin (BCG), antibody production was enhanced and DTH was suppressed in effluent-exposed mink (Smits et al., 1996a, 1996b). The DTH response was evaluated using skin thickness measurements, histopathological assessments, and image analyzer technology.

9.2.2.3 Cervids

Immunological testing in cervids has been driven by the need to monitor disease status in captive animals in zoological collections, on game farms, and in free-ranging wild deer that are being translocated for wildlife management purposes. There are limited, if any, immunotoxicology studies involving wild or captive cervids. However, efforts to develop reliable tests for detecting infection by *Mycobacterium bovis* (tuberculosis) have resulted in the refining of numerous tests that could have immunotoxicological applications in the future. Elk (*Cervus elaphus canadensis*) vaccinated with live (Waters et al., 2003) and deer vaccinated with killed (Griffin et al., 1993) BCG produce strong DTH skin responses as well as specific antibody responses.

Immunoglobulin G against white-tailed deer (*Odocoileus virginianus*), elk, and moose (*Alces alces*) have been produced to facilitate ELISA as well as other diagnostic serological techniques (see Ogunremi et al., 1999). Peripheral blood mononuclear cells and lymphocyte subpopulations have been identified by phenotype, and lymphocyte subpopulations have been differentiated in elk (Waters et al., 2002). These protocols could be used in relation to exposure to environmental contaminants.

9.3 *IN VITRO* TESTS IN CURRENT USE

9.3.1 Birds

From a large initial battery of *in vitro* tests that were recommended for immunotoxicology studies, current protocols have been pared down to fewer assays (U.S. Environmental Protection Agency, 1998), with the NK cell assay and mitogen-induced lymphocyte proliferation tests favored. This recommendation is odd in view of the fact that Luster et al. (1993) showed that *in vitro* proliferative responses to mitogens were relatively poor predictors of immunomodulation or increased suscep-tibility to disease. The advantage of *in vitro* tests is that they normally require only one-time capture and handling of the animals, but a disadvantage is that any assay based on splenocytes, thymocytes, or bursal lymphocytes requires cells after the animal is sacrificed. The smaller the bird, the more difficult it is to collect sufficient blood from the live animal to conduct *in vitro* assays. In rats, the function of cytotoxic natural killer (NK) cells using small volumes of peripheral blood can be determined using flow cytometry (Marcusson-Stahl and Cederbrant, 2003). This could have valuable applications in small wildlife such as passerines and small mammals, because most current NK cell assays require up to 20 ml of blood. Assays may be based on whole blood or purified lymphocytes and monocytes from avian blood. Whole blood cultures require less blood and avoid the requirement to isolate mononuclear cells, which is more difficult than in mammalian blood. In many avian species, when using density gradient techniques, thrombocytes, which have no immunological role, contaminate the mononuclear cell layer. Methods to deal with these problems are presented in the review by Fairbrother et al. (2004).

The comet assay is used to detect contaminant-induced damage of nucleic acids in peripheral blood cells including leucocytes (Pastor et al., 2001; Petras et al., 1995). If leucocytes sustain genetic damage, this could be expected to interfere with immune responsiveness. A practical consideration for *in vitro* assays in wild birds is how best to deal with samples collected during fieldwork at locations remote from the laboratory. Shipping of preserved samples is risky at best, leaving cryopreservation as one viable alternative for future lymphoproliferation, phagocytosis (Finkelstein et al., 2003), and comet assays (Smits and Papp, unpublished).

Cytotoxic T cells defend the body by destroying cells that are foreign, or that have been altered by virus infection or neoplastic changes. The cell-mediated cytotoxicity assay requires immunization of the test animals *in vivo*. Days to weeks later, using peripheral blood or splenocytes from immunized animals, *in vitro* culture of the cytotoxic T cells with target cells will result in lysing of the target cell and release of detectable molecules (Tizard, 1996). The cytoxicity test is the specific cell-mediated immune response equivalent to the NK assay that measures the innate response of NK cells to lyse target cells without prior sensitization. NK cell toxicity has been detected in mallards (Fowles et al., 1997) and Japanese quail (Yamada et al., 1980).

Phagocytic activity by cells of the innate immune system is detected with flow cytometry. The number of phagocytes ingesting fluorescent beads or labeled yeast, the number of particles ingested per cell, or the release of reactive oxygen species by activated phagocytes (Laudert et al., 1993; Trust et al., 1994; Finkelstein et al., 2003) is determined. Although *in vitro* tests of immune responsiveness in wild birds are limited, with the rapid developments in molecular techniques and the desire for more mechanistic answers, this area of immunotoxicology is likely to expand in the future.

9.3.2 Rodents

Comparison of a comprehensive set of hematological and immunological parameters in cotton rats found that the most consistent and reliable *in vitro* measure of immunotoxicity in cotton rats exposed to petrochemical wastes was spontaneous or Con-A–induced proliferation of splenocytes (McMurry et al., 1999; Wilson et al., 2003). Potential toxicological interactions of petrochemical wastes with bone marrow were also investigated by determining colony formation of granulocyte-macrophage progenitor cells (CFU-GM) in cotton rats (Kim et al., 2001). Comparison of this response in rats collected from nine petrochemical landfarm sites and nine ecologically matched reference sites showed a consistent 21 to 39% decrease in the number of colony-forming units of CFU-GM. Since these effects on CFU-GM occurred with no significant changes in hematological or histopathological variables, it was concluded that bone-marrow progenitor cell culture may represent a sensitive technique to assess hematopoietic toxicity in wildlife (Kim et al., 2001).

The effectiveness of tumor necrosis factor-α (TNF-α) produced by splenocytes harvested from cotton rats collected from petrochemical landfarms and reference sites has also been evaluated (Wilson and Janz, unpublished). Addition of actinomycin d to test samples increased lysis of test cells (WEHI 164 murine-fibrosarcoma cells) by 100 times, making the assay unusable. A range of lipopolysaccharide (LPS) concentrations (0.0, 0.05, 0.5, 5.0, and 50.0 μg/ml) was tested to determine the appropriate concentration to stimulate TNF-α production in cotton rat splenocytes. Fifty μg/ml of LPS was found to be the most effective concentration for stimulating TNF-α production. Cotton rats collected from petrochemical landfarms had a 12.6% increase in lytic activity associated with TNF-α production over rats from reference sites. By stimulating splenocytes with an LPS concentration > 50.0 μg/ml, TNF-α can be used effectively to detect changes in the immune system of mammals inhabiting contaminated sites.

9.3.3 Mink

In mink peripheral blood lymphocytes, exposure to pulp effluent did not affect the proliferative response to (Con-A), pokeweed mitogen and PHA, although other immunotoxicological endpoints were compromised from contaminant exposure (Smits et al., 1996a). Mink lymphocytes do not respond to bacterial LPS. The insensitivity of the *in vitro* proliferative response of lymphocytes to immunotoxins has once again been demonstrated. Mouse monoclonal antibodies can distinguish sub-populations of T and B lymphocytes in mink, with flow cytometry or immunohistochemistry (Miyazawa et al., 1994). Cytokine profiles for mink leucocytes (Jensen et al., 2003) and commercially available monoclonal antibodies (ovine IL-8 and TNF-α, bovine IL-4 and IFN-γ) which cross-react with mink cytokines (Pedersen et al., 2002) will facilitate immunotoxicology research in mink.

9.3.4 Cervids

Zoonotic diseases (tuberculosis and brucellosis) and parasitic infections that threaten the national herd have stimulated progress in immunological techniques applicable to cervids. Subpopulations of lymphocytes have been identified, and nitric oxide and IFN-γ production by macrophages has been measured in deer using flow cytometry (Waters et al., 2002, 2003). ELISA tests in elk can distinguish animals vaccinated with *Brucella* strain 19, from those challenged with a pathogenic strain of *B. abortus* (Van Houten et al., 2003).

9.4 LIMITATION AND RELEVANCE OF IMMUNOTOXICITY TESTS

9.4.1 Limitations in Application or Interpretation

In many passerines and other small terrestrial wildlife, a major constraint to immunotoxicology or other diagnostic and survey work is the limited volume of blood samples that are either compatible with life (in research involving *in vivo*, nonlethal studies) or ultimately available even if the anesthetized animal is euthanized via exsanguination. In an 18-g song bird, maximal sample volume is approximately 1.1 ml. In the available protocols for tests such as phagocytosis, the blood volume that yields enough phagocytic cells to conduct a phagocytosis assay is not available. Blood protein levels, hematocrit, WBC differentials, and ELISA-based antibody responses can be measured. The HA test commonly used with SRBC assays requires relatively high volumes of serum or plasma, limiting what remains for testing other related endpoints such as hormone levels. Thyroid hormones and retinol, for example, are known to be closely related with immunological function, and therefore would be valuable to analyze, so plasma-sparing techniques are very desirable.

The NK assay has been difficult to achieve because of the need for large volumes of blood, but based on flow cytometric techniques described in Marcusson-Stahl and Cederbrant (2003), it may be possible to determine NK cell activity in small-bodied wildlife. The clinical relevance of WBC counts must be considered with care. There is a wide variation in the normal leucograms among birds of the same species, so only values that differ greatly from normal are of diagnostic importance (Campbell, 1994). Because the nucleated erythrocytes and thrombocytes interfere with leucocyte detection using electronic cell counters, WBC counts in birds are obtained from blood smears using manual techniques. Subjectiveness is a potential problem, as is consistency and accuracy of the count. Inconsistent antibody responses are seen against SRBC in contaminant-exposed birds (Grasman and Scanlon, 1995; Fox, 2001). Smits and Baos (2004) suggest developmental immaturity as an explanation for the unpredictable response of nestling birds to SRBC.

9.4.2 Sex-Specific Considerations

Sex-specific effects are frequently seen in studies involving immunotoxicity. In chickens exposed to lead acetate as embryos and vaccinated as juveniles, males and females had opposite antibody responses, as well as different DTH responses (Bunn et al., 2000). Sex differences have also been seen in polychlorinated biphenyl (PCB)-exposed American kestrels with antibody production increasing in females and decreasing in males (Smits and Bortolotti, 2001). In PCB-exposed ring doves, *Streptopelia risori*, retinol concentrations, which are integrated with immune function (Ross et al., 1992; Simms and Ross, 2000) were more consistently and dramatically decreased in male birds (Spear et al., 1989). Animals likely perceive stress generically, whether it is from physiological or social stress or subclinical toxicological exposure. Differences in immunology between stressed males and females are observed. Total leucocyte counts in male but not female chickens were depressed after lead exposure (Bunn et al., 2000), whereas in PCB-exposed kestrels, leucocyte counts driven by lymphocytosis were increased in males only (Smits et al., 2002).

9.5 SUMMARY

Phocine morbillivirus was diagnosed in seals, in conjunction with high tissue body burdens of PCBs and other organochlorine contaminants providing compelling evidence of the role of xenobiotics in the massive outbreak of distemper in harbor seals (*Phoca vitulina*) in the late 1980s. *In vivo* and *in vitro* immunosuppression were related to tissue contaminant levels in the seals (De Swart et al., 1994, 1995). The now notorious neoplasia problems in the heavily polluted beluga whale (*Delphinapterus leucas*) population in the St. Lawrence River (Martineau et al., 1994, 1999) provide circumstantial evidence of contaminant-induced suppression involving at least cell-mediated immunity such as NK cell antitumor surveillance. In a cormorant colony with high prevalence of severe ocular infections in the chicks, high PCB contamination supports the hypothesis of contaminant-related immunosuppression (Ludwig et al., 1996).

The immune system has evolved to protect individuals against foreign invaders that have the potential to cause disease. The ultimate concern underlying the entire discipline of immunotoxicology is whether exposure to xenobiotics increases the risk of disease in animals. Studies that have combined xenobiotic exposure with challenge from viruses, bacteria, or parasites provide practical information about the potential impact of contaminants on free-ranging populations. A substantial body of work has been conducted on the immunotoxicity of various xenobiotics. Since a review of avian immunotoxicology has recently been published (Fairbrother et al., 2004), many of the avian references provided in this table have been gleaned from their review. In the studies of organophosphates (OPs), almost all the *in vivo* and *in vitro* studies have been conducted on laboratory rodents (Galloway and Handy, 2003). The few exceptions are listed in Table 9.1. The most pressing needs in wildlife immunotoxicology are validated techniques for commonly used wildlife species to promote more consistent application of tests, which hopefully would spawn the development of commercially available reagents. Priority should also be given to developing and using testing techniques that are based upon nonlethal sampling.

REFERENCES

Abbas, A.K., Lichtman, A.H., and Pober, J.S., *Cellular and Molecular Immunology*, 4th ed., W.B. Saunders, Philadelphia, 2000.

Apanius, V., Ontogeny of immune function, in *Avian Growth and Development*, Stark, J.M., and Ricklefs, R.E., Eds., Oxford University Press, Oxford, 1998, pp. 203–222.

Audet, M., et al., Immune status of wild nestling bald eagle from Great Lakes, *Conf. Great Lakes. Res.*, 43, A–7, 2000.

Bishop, C.A., et al., Health of tree swallows (Tachycineta bicolor) nesting in pesticide-sprayed apple orchards in Ontario, Canada., I. Immunological parameters, *J. Toxicol. Environ. Health A*, 55, 531–559, 1988.

Brousseau, P., et al., *Manual of Immunological Methods*, CRC Press, Boca Raton, FL, 1998.

Bunn, T.L., Marsh, J.A., and Dietert, R.R., Gender differences in developmental immunotoxicity to lead in the chicken: Analysis following a single early low-level exposure in ovo, *J. Toxicol. Environ. Health A*, 61, 677–693, 2000.

Camarena, A., et al., Major histocompatibility complex and tumor necrosis factor-alpha polymorphism in pigeon breeder's disease, *Am. J. Respir. Crit. Care Med.*, 163 , 1513–1514, 2001.

Campbell, T.W., Hematology, in *Avian Medicine: Principles and Applications*, Ritchie B.W., Harrison G.J., and Harrison, L.R., Eds., Winger Publishing, Lake Worth, FL, 1994, pp. 176–198.

Camplani, A., Saino, N., and Moller, A.P., Carotenoids, sexual signals and immune function in barn swallows from Chernobyl, *Proc. R. Soc. London B*, 266, 1111–1116, 1999.

Carlson, R.I., et al., Ecotoxicological risks associated with land treatment of petrochemical wastes: II. Effects on hepatic Phase I and Phase II detoxification enzymes in cotton rats, *J. Toxicol. Environ. Health*, 66, 327–343, 2003.

Christe, P., Moller, A.P., and de Lope, F., Immunocompetence and nestling survival in the house martin: the tasty chick hypothesis, *Oikos*, 83, 175–179, 1998.

Cohen, N., Amphibian transplantation reactions: a review, *Am. Zool.*, 11, 193–205, 1971.

Coligen, J.E., Kruisbeek, A.M. Margulies, D.H. Shevach, E.M. Strober, W., *Current Protocols in Immunology*, vol. I, National Institutes of Health, John Wiley & Sons, New York, 1994.

Collette, J.C., et al., Neonatal handling of Amazon parrots alters the stress response and immune function, *Appl. Anim. Behav. Sci.*, 66, 335–349, 2000.

Day, B.L., et al., Immunopathology of 8-week-old ring-necked pheasants (Phasianus colchicus) exposed to malathion, *Vet. Immunol. Immunopathol.*, 14, 1719–1726, 1995.

De Swart, R.L., et al., Impairment of immune function in harbor seals (Phoca vitulina) feeding on fish from polluted waters, *Ambio*, 23, 155–159, 1994.

De Swart, R.L., et al., Impaired cellular immune response in harbour seals (Phoca vitulina) feeding on environmentally contaminated herring, *Clin. Exp. Immunol.*, 101, 480–486, 1995.

Deerenberg, C., et al., Reproductive effort decreases antibody responsiveness, *Proc. R. Soc. London B*, 264, 1021–1029, 1997.

Dein F.J., Hematology, in *Clinical Avian Medicine and Surgery*, Harrison, G.J., and Harrison, L.R., Eds., W.B. Saunders, Philadelphia, 1986, pp. 174–191.

Duncan, J.R., Prasse, K.W., and Mahaffey, E.A., *Veterinary Laboratory Medicine*, 3rd ed., Iowa State University Press, Ames, 1994.

Ekblom, R., Grahn, M., and Hoglund, J., Patterns of polymorphism in the MHC class II of a non-passerine bird, the great snipe (Gallinago media), *Immunogenetics*, 54, 734–741, 2003.

Elbert, R.A., and Anderson, D.W., Mercury levels, reproduction, and hematology in western grebes from three California lakes, USA, *Vet. Immunol. Immunopathol.*, 17, 210–213, 1998.

El-Lethy, H., Huber-Eicher, B., and Jungi, T.W., Exploration of stress-induced immunosuppression in chickens reveals both stress-resistant and stress-susceptible antigen responses, *Vet. Immunol. Immunopathol.*, 95, 91–101, 2003.

Elsheikh, E.L., et al., Induction of protective immune response by vaccination against Pseudomonas pneumonia of mink, *J. Vet. Med. B*, 35, 256–263, 1988.

Fair, J., and Myers, L.P., The ecological and physiological costs of lead shot and immunological challenge to developing western bluebirds, *Ecotoxicology*, 11, 199–208, 2002.

Fair, J., and Ricklefs, R., Physiological, growth, and immune responses of Japanese Quail chicks to the multiple stressors of immunological challenge and lead shot, *Arch. Environ. Contam. Toxicol.*, 42, 77–87, 2002.

Fairbrother, A., et al., Impairment of growth and immune function of avocet chicks from sites with elevated selenium, arsenic and boron, *J. Wildl. Dis.*, 30, 222–233, 1994.

Fairbrother, A., and Fowles, J., Subchronic effects of sodium selenite and selenomethionine on several immune functions in mallards, *Arch. Environ. Contam. Toxicol.*, 19, 836–844, 1990.

Fairbrother, A., Smits, J.E., and Grasman, K., Avian immunotoxicology, *J. Toxicol. Environ. Health B Crit. Rev.*, 7, 105–137, 2004.

Fairbrother, A., Yuill, T.M., and Olson, L.J., Effects of ingestion of chlorocholine chloride and cyclophosphamide on Venezuelan equine encephalitis virus infections in deer mice (Peromyscus maniculatus), *Toxicology*, 31, 67–71, 1984.

Fairbrother, A., Yuill, T.M., and Olson, L.J., Effects of three plant growth regulators on the immune response of young and aged deer mice Peromyscus maniculatus, *Arch. Environ. Contam. Toxicol.*, 15, 265–281, 1986.

Finkelstein, M., et al., Immune function of cryopreserved avian peripheral white blood cells: potential biomarkers of contaminant effects in wild birds, *Arch. Environ. Contam. Toxicol.*, 44, 502–509, 2003.

Fournier, M., et al., Phagocytosis as a biomarker of immunotoxicity in wildlife species exposed to environmental xenobiotics, *Am. Zool.*, 40, 412–420, 2000.

Fowles, J.R., et al., Effects of Aroclor 1254 on the thyroid gland, immune function, and hepatic cytochrome P450 activity in mallards, *Environ. Res.*, 75, 119–129, 1997.

Fox, G., Wildlife as sentinels of human health effects in the Great Lakes-St. Lawrence basin, *Environ. Health Perspect.*, 109 (Suppl. 6), 853–861, 2001.

Freeman-Gallant, C.R., et al., Variation at the major histocompatibility complex in Savannah sparrows, *Mol. Ecol.*, 11, 1125–1130, 2002.

Friend, M., and Trainer, D.O., Polychlorinated biphenyl: interaction with duck hepatitis virus, *Science*, 170, 1314–1316, 1970.

Galloway, T., and Handy, R., Immunotoxicity of organophosphorous pesticides, *Ecotoxicology*, 12, 345–363, 2003.

Gogal, R.M. Jr., et al., Influence of dietary 2, 4, 6-trinitrotoluene exposure in the northern bobwhite (Colinus virginianus), *Vet. Immunol. Immunopathol.*, 21, 81–86, 2002.

Goldberg, D.R., and Yuill, T.M., Effects of sewage sludge on the immune defences of mallards, *Environ. Res.*, 51, 209–217, 1990.

Goldberg, D.R., Yuill, T.M., and Burgess, E.C., Mortality from duck plague virus in immunosuppressed adult mallard ducks, *J. Wildl. Dis.* 26, 299–306, 1990.

Grasman, K.A., Assessing immunological function in toxicological studies of avian wildlife, *Integrative Comp. Biol.*, 42, 34–42, 2002.

Grasman, K.A., and Fox, G.A., Associations between altered immune function and organochlorine contamination in young Caspian terns (Sterna caspia) from Lake Huron, 1997–1999, *Ecotoxicology*, 10, 101–114, 2001.

Grasman, K.A., et al., Geographic variation in blood plasma protein concentrations of young herring gulls (Larus argentatus) and Caspian terns (Sterna caspia) from the Great Lakes and Lake Winnipeg, *Comp. Biochem. Physiol. C*, 125, 365–375, 2000.

Grasman, K.A., et al., Organochlorine-associated immunosuppression in prefledgling Caspian terns and herring gulls from the Great Lakes: An ecoepidemiological study, *Environ. Health Perspect.*, 104 (Suppl. 4), 829–841, 1996.

Grasman, K.A., and Scanlon, P.F., Effects of acute lead ingestion and diet on antibody and T-cell-mediated immunity in Japanese quail, *Arch. Environ. Contam. Toxicol.*, 28, 161–167, 1995.

Griffin, J.F., et al., BCG vaccination in deer: distinctions between delayed type hypersensitivity and laboratory parameters of immunity, *Immunol. Cell Biol.*, 71, 559–570, 1993.

Gross, W.B., Differential and total avian blood cell counts by the hemacytometer method, *Avian/Exotic Pract.*, 1, 31–36, 1984.

Hasselquist, D., et al., Is avian immunocompetence suppressed by testosterone? *Behav. Ecol. Sociobiol.*, 45, 167–175, 1999.

Ilmonen, P., Taarna, T., and Hasselquist, D., Experimentally activated immune defense in female pied flycatchers results in reduced breeding success, *Proc. R.. Soc. London B*, 267, 665–670, 2000.

Jarvi, S.I., et al., A complex alloantigen system in Florida sandhill cranes, Grus canadensis pratensis: evidence for the major histocompatibility (B) system, *J. Hered.*, 86, 348–353, 1995.

Jensen, P.V., Castelruiz, Y., and Aasted, B. Cytokine profiles in adult mink infected with Aleutian mink disease parvovirus, *J. Virol.*, 77, 7444–7451, 2003.

Kim, S., et al., Evaluation of myelotoxicity in cotton rats (Sigmodon hispidus) exposed to environmental contaminants. II. Myelotoxicity associated with petroleum industry wastes, *J. Toxicol. Environ. Health*, 62, 97–105, 2001.

Koppenheffer, T.L., Ford, J.W., and Robertson P.B., The ontogeny of immune responsiveness to sheep erythrocytes in young pigeons, in *Aspects of Developmental and Comparative Immunology*, Solomon, J.B., Ed., Permagon, Oxford, 1981, pp. 535–536.

Laudert, E., Sivanandan, V., and Halvorson, D., Effect of an H5N1 avian influenza virus infection on the immune system of mallard ducks, *Avian Dis.*, 37, 845–853, 1993.

Levengood, J., et al., Influence of diet on the hematology and serum biochemistry of zinc-intoxicated mallards, *J. Wildl. Dis.*, 36, 111–123, 2000.

Ludwig, J.P., et al., Deformities, PCBs, and TCDD-equivalents in double-crested cormorants (Phalacrocorax auritus) and Caspian terns (Sterna caspia) of the upper Great Lakes, 1986–1991: testing cause-effect hypothesis, *J. Great Lakes Res.*, 22, 172–197, 1996.

Luster, M.I., et al., Risk assessment in immunotoxicology. II. Relationships between immune and host resistance tests, *Fundam. Appl. Toxicol.*, 21, 71–82, 1993.

Marcusson-Stahl, M., and Cederbrant, K., A flow-cytometric NK-cytotoxicity assay adapted for use in rat repeated dose toxicity studies, *Toxicology*, 193 , 269–279, 2003.

Martineau, D., et al., Pathology and toxicology of beluga whales from the St. Lawrence Estuary, Quebec, Canada. Past, present and future, *Sci. Total Environ.*, 154, 201–215, 1994.

Martineau, D., et al., Cancer in beluga whales from the St. Lawrence Estuary, Quebec, Canada: a potential biomarker of environmental contamination, *J. Cetacean Res. Manage.*, 1 (special publication), 249–265, 1999.

Martinez, J., et al., Detection of serum immunoglobulins in wild birds by direct ELISA: a methodological study to validate the technique in different species using antichicken antibodies, *Functional Ecol.*, 17, 700–706, 2003.

McMurry, S.T., et al., Sensitivity of selected immunological, hematological, and reproductive parameters in the cotton rat (Sigmodon hispidus) to subchronic lead exposure, *J. Wildl. Dis.*, 31, 193–204, 1995.

McMurry, S.T., et al., Indicators of immunotoxicity in populations of cotton rats (Sigmodon hispidus) inhabiting an abandoned oil refinery, *Ecotoxicol. Environ. Saf.*, 42, 223–235, 1999.

McMurry, S.T., et al., Immunological responses of weanling cotton rats (Sigmodon hispidus) to acute benzene and cyclophosphamide exposure, *Bull Environ. Contam. Toxicol.*, 52, 155–162, 1994.

McMurry, S.T., et al., Acute effects of benzene and cyclophosphamide exposure on cellular and humoral immunity of cotton rats, Sigmodon hispidus, *Bull Environ. Contam. Toxicol.*, 46, 937–945, 1991.

Meredith, C., and Miller, K., Molecular immunotoxicology testing in vitro, *Toxicology*, 8, 1001–1005, 1994.

Miyazawa, M., et al., Production and characterization of new monoclonal antibodies that distinguish subsets of mink lymphoid cells, *Hybridoma* 13, 107–113, 1994.

Moller, A.P., et al., Immune defense and host sociality: a comparative study of swallows and martins, *Am. Naturalist*, 158, 136–14, 2001.

Newman, S.H., et al., Experimental release of oil-spill rehabilitated American coots (Fulica Americana): effects on health and blood parameters, *Pac. Seabirds*, 26, 41–42, 1999.

Newman, S.H., et al., An experimental soft-release of oil-spill rehabilitated American coots (Fulica Americana): II. Effects on health and blood parameters, *Environ. Pollut.*, 107, 295–304, 2000.

Norris, K., and Evans, M., Ecological immunology: life history trade-offs and immune defense in birds, *Behav. Ecol.*, 11, 19–26, 1999.

Ogunremi, O., et al., Evaluation of excretory-secretory products and somatic worm antigens for the serodiagnosis of experimental Parelaphonstrongylus tenuis infection in white-tailed deer, *J. Vet. Diagn. Invest.*, 11, 515–521, 1999.

Olson, L.J., and Hinsdill, R.D., Influence of feeding chlorocholine chloride and glyphosine on selected immune parameters in deer mice, Peromyscus maniculatus, *Toxicology*, 30, 103–114, 1984.

Pastor, N., et al., Assessment of genotoxic damage by the comet assay in white storks (Ciconia ciconia) after the Doñana ecological disaster, *Mutagenesis*, 16 , 219–223, 2001.

Pedersen, L.G., et al., Identification of monoclonal antibodies that cross-react with cytokines from different animal species, *Vet. Immunol. Immunopathol.*, 25, 111–122, 2002.

Petras, M., et al., Biological monitoring of environmental genotoxicity in southwestern Ontario, in *Biomonitors and Biomarkers as Indicators of Environmental Change*, Butterworth, F., Corkum, L., and Guzman-Rincon, J., Eds., Plenum Press, New York, 1995, pp. 115–137.

Porter, W.P., et al., Toxicant-disease-environment interactions associated with suppression of immune system, growth and reproduction, *Science*, 224, 1014–1017, 1984.

Propst, T.L., et al., In situ (mesocosm) assessment of immunotoxicity risks to small mammals inhabiting petrochemical waste sites, *Chemosphere*, 38, 1049–1067, 1999.

Pruett, S.B., et al., Modeling and predicting immunological effects of chemical stressors: characterization of a quantitative biomarker for immunological changes caused by atrazine and ethanol, *Toxicol. Sci.*, 75, 343–354, 2003.

Quist, C.F., et al., The effect of dietary aflatoxin on wild turkey poults, *J. Wildl. Dis.*, 36, 436–444, 2000.

Rafferty, D. P., et al., Immunotoxicity risks associated with land-treatment of petrochemical wastes revealed using an in situ rodent model, *Environ. Pollut.*, 112, 73–87, 2001.

Redig, P.T., et al., Effects of chronic exposure to sublethal concentrations of lead acetate on heme synthesis and immune function in Red-tailed hawks, *Arch Environ. Contam. Toxicol.*, 21, 72–77, 1991.

Rivera, E., et al., Evaluation of protein A and protein G as indicator system in an ELISA for detecting antibodies in mink to Pseudomonas aeruginosa, *Vet. Microbiol.*, 42, 265–271, 1994.

Rocke, T.E., and Samuel, M.D., Effects of lead shot ingestion on selected cells of the mallard immune system, *J. Wildl. Dis.*, 27, 1–9, 1991.

Rocke, T.E., Yuill, T.M., and Hinsdill, R.D., Oil and related toxicant effects on mallard immune defenses, *Environ. Res.*, 33, 343–352, 1984.

Rose, M., and Bradley, J., Delayed hypersensitivity in the fowl, turkey and quail, *Avian Pathol.*, 6, 313–326, 1977.

Ross, P.S., et al., Antibodies to phocine distemper virus in Canadian seals, *Vet. Rec.*, 130, 514–516, 1992.

Sagerup, K., et al., Intensity of parasitic nematodes increases with organochlorine levels in the glaucous gull, *J. Appl. Ecol.*, 37, 532–539, 2000.

Saino, N., Calza, S., and Moller, A.P., Immunocompetence of nestling barn swallows in relation to brood size and parental effort, *J. Anim. Ecol.*, 66, 827–836, 1997.

Savabieasfahani, M., et al., Sensitivity of wild cotton rats (Sigmodon hispidus) to the immunotoxic effects of low-level arsenic exposure, *Arch. Environ. Contam. Toxicol.*, 34, 289–296, 1998.

Savabieasfahani, M., Lochmiller, L.R., and Janz, D.M., Elevated ovarian and thymic cell apoptosis in wild cotton rats inhabiting petrochemical-contaminated terrestrial ecosystems, *J. Toxicol. Environ. Health*, 57, 101–107, 1999.

Schroder, J.L., et al., Ecotoxicological risks associated with land treatment of petrochemical wastes: I. Residual soil contamination and bioaccumulation by cotton rats (Sigmodon hispidus), *J. Toxicol. Environ. Health*, 66, 305–325, 2003.

Simms, W., and Ross, P., Vitamin A physiology and its application as a biomarker of contaminant-related toxicity in marine mammals: a review, *Toxicol. Ind. Health*, 16, 291–302, 2000.

Sitteri, P.K., et al., The serum transport of steroid hormones, *Recent Prog. Horm. Res.*, 38, 457–510, 1982.

Smits, J.E., unpublished.

Smits, J.E., Blakley, B.R., and Wobeser, G.A., Immunotoxicity studies in mink (Mustela vison) chronically exposed to dietary bleached kraft pulp mill effluent, *J. Wildl. Dis.*, 32, 190–208, 1996a.

Smits, J.E., and Godson, D.L., Assessment of humoral immune response in mink (Mustela vison): antibody production and detection, *J. Wildl. Dis.*, 32, 358–361, 1996b.

Smits, J.E., Bortolotti, G.R., and Tella, J.L., Simplifying the phytohemagglutinin skin-testing technique in studies of avian immunocompetence, *Functional Ecol.*, 13, 567–572, 1999.

Smits, J.E., et al., Reproductive, immune, and physiological end points in tree swallows on reclaimed oil sands mine sites, *Environ. Toxicol. Chem.*, 19, 2951–2960, 2000.

Smits, J.E., and Williams, T.D., Validation of immunotoxicological techniques in passerine chicks exposed to Oil Sands tailings water, *Ecotoxicol. Environ. Saf.*, 44, 105–112, 1999.

Smits, J.E., and Bortolotti, G.R., Antibody mediated immunotoxicity in American Kestrels (Falco sparverius) exposed to polychlorinated biphenyls, *J. Toxicol. Environ. Health A*, 62, 101–110, 2001.

Smits, J.E., et al., Thyroid hormone suppression and cell mediated immunomodulation in American Kestrels (Falco sparverius) exposed to PCBs, *Arch. Environ. Contam. Toxicol.*, 43, 338–344, 2002.

Smits, J.E., and Baos, R., Evaluation of antibody mediated immunity in nestling American kestrels (Falco sparverius), *Dev. Comp. Immunol.*, 29, 161–170, 2004.

Spalding, M.G., et al., Histologic, neurologic, and immunologic effects of methylmercury in captive great egrets, *J. Wildl. Dis.*, 36, 423–435, 2000.

Spear, P.A., et al., Dove reproduction and retinoid (vitamin A) dynamics in adult females and their eggs following exposure to 3, 3, 4, 4'-tetrachlorobiphenyl, *Can. J. Zool.*, 67, 908–913, 1989.

Svensson, E., et al., Energetic stress, immunosuppression and the costs of an antibody response, *Functional Ecol.*, 12, 912–919, 1998.

Tella, J.L. and Smits, J.E., unpublished.

Tizard, I.R., Drugs and other agents that affect the immune system, in *Veterinary Immunology: An Introduction*, 5th ed., Tizard, I.R., Ed., W.B. Saunders, Philadelphia, 1996, pp. 470–478.

Trust, K.A., Effects of ingested lead on antibody production in mallards (Anas platyrhynchos), *J. Wildl. Dis.*, 26, 316–322, 1990.

Trust, K.A., Fairbrother, A., and Hooper, M.J., Effects of 7, 12-Dimethylbenz{a}Anthracene on immune function and mixed-function oxygenase activity in the european starling, *Environ. Toxicol. Chem.*, 13, 821–830, 1994.

U.S. Environmental Protection Agency, Health Effects Test Guidelines, OPPTS 870.7800 Immunotoxicity, EPA, Washington, DC, 1998.

Van Houten, C.K., Jr., et al., Validation of a Brucella abortus competitive enzyme-linked immunosorbent assay for use in Rocky Mountain elk (Cervus elaphus nelsoni), *J. Wildl. Dis.*, 39 , 316–322, 2003.

Waters, W.R., et al., Analysis of mitogen-stimulated lymphocyte subset proliferation and nitric oxide production by peripheral blood mononuclear cells of captive elk (Cervus elaphus), *J. Wildl. Dis.*, 38 , 344–351, 2002.

Waters, W.R., et al., Experimental Infection of Reindeer (Rangifer tarandus) with Mycobacterium bovis: Diagnostic Implications, Proceedings abstract 76, 52nd Annual Wildlife Disease Association Conference, August 11–14, 2003, Saskatoon, Saskatchewan, Canada.

Westerdahl, H., Wittzell, H., and von Schantz, T., Mhc diversity in two passerine birds: no evidence for a minimal essential Mhc, *Immunogenetics*, 52, 92–100, 2000.

Whitely, P.L., and Yuill, T.M., Effects of Selenium on Mallard Duck Reproduction and Immune Function, U.S. Environmental Protection Agency, Environmental Research Laboratory, Corvallis, OR, ERL-COR–574 (EPA/600/3–89/078), 1989.

Wilson, J.A. and Janz, D., unpublished.

Wilson, J.A., et al., Ecotoxicological risks associated with land treatment of petrochemical wastes: III. Immune function and hematology of cotton rats, J. *Toxicol. Environ. Health*, 66, 345–363, 2003.

Yamada, A., et al., Detection of natural killer cells in Japanese quail, *Int. J. Cancer*, 26, 381–385, 1980.

Zuk, M., and Johnsen, T.S., Seasonal changes in the relationship between ornamentation and immune response in red jungle fowl, *Proc. R. Soc. London B*, 265, 1631–1635, 1998.

CHAPTER **10**

Immune Markers in Ecotoxicology: A Comparison across Species

Harri Salo, Claire Dautremepuits, Judit E.G. Smits, Pauline Brousseau, and Michel Fournier

CONTENTS

10.1 INTRODUCTION

One of the main problems of the past and present century has been to preserve environmental quality. Continually expanding industrial and agricultural production has resulted in ever-increasing impacts from contaminants on ecosystems. Contaminant release into the environment can be quite varied and complex. Dispersal, whether deliberate or not, of manufactured pesticides including polycyclic aromatic hydrocarbons (PAHs), polychlorinated biphenyls (PCBs), and heavy metals, even at relatively low concentrations, cause contamination of the biosphere locally and at distant sites. Ecosystems are being negatively affected by human activity worldwide (Bols et al., 2001).

Ecotoxicology is the study of the impact of environmental contaminants on populations and ecosystems. In the discipline of ecotoxicology, environmental quality is monitored through the detection and quantification of environmental contaminants, and eventually, the evaluation of their impact on living organisms at different levels of biological organization within the affected ecosystem.

The structure of ecosystems can be defined as successive layers of biological organization with each compartment controlling the composition of the subsequent level. For example community structure is dictated by input from constituent populations of species, the expression of variation in the populations, and the response of these populations to environmental stress. The community is the most popular level of investigation for environmental assessment (Warwick et al., 1993), and has been suggested as the most important level for impact studies (Clements and Kiffney, 1994; Martin and Richardson, 1995). It is therefore important to explore mechanisms linking the various levels of biological organization to gain an understanding of how individual toxicological responses in certain species may be expressed at the community level. In risk assessment, it is essential to determine which level of organization provides the most sensitive and robust method of evaluating overall environmental health (Attrill and Depledge, 1997; Bols et al., 2001; Clements and Kiffney, 1994; Martin and Richardson, 1995; Warwick et al., 1993).

Ecologists frequently note the importance of modeling entire ecosystems rather than individual species. For practical reasons and due to the immense complexity of modeling an ecosystem with all its intra- and inter-connections, the current literature focuses on particular species, their preferred habitats, and local climatic conditions. This approach has required the amassing of knowledge of many and varied interactions among species, their local habitat, and the greater environment. Interspecies relationships also provide an indication of the relative sensitivity of different species to toxicants or classes of toxicants, and consequently provide valuable information for selecting species best able to represent the wider range of biodiversity present in the environment of interest. In ecotoxicology, the goal of such a process is to select an organism, to study its responses and to look for possible adaptations of local populations when they are exposed to non-point source pollution.

The potential utility of biomarkers for monitoring both environmental quality and animal health in ecosystems has received increasing attention during the last few years (Lagadic and Caquet, 1998). Biomarkers or bioindicators have been identified at suborganism levels of organization (molecular, biochemical, cellular, and tissue), or at the level of the individual (physiological, behavioral). Bioindicators are measurable biological variables that change in response to toxicants, as well as other environmental or physiological stressors, such as pathogen pressure, change in temperature, and ageing. The ecological validation of biomarkers includes identifying a mechanical or causal relation between variation in the parameter of interest in individuals and the impact seen on the corresponding population, and more broadly, on the community from which the population is extracted. Ecotoxicological studies have focused on immunological variables because of their high sensitivity to changes in environmental conditions. Recent advances in identifying new and better biomarkers have depended on understanding the inter- and intra-individual variability of these markers, how their responses relate to ecologic disturbances, and whether increasing incidence of disease is associated with exposure to pollutants (Zelikoff, 1997).

The aim of this chapter is to provide an overview of the use of immunological biomarkers to predict ecotoxicologic risks in disrupted ecosystems. Considering the immune system as an important element in biomarker research, we will examine selected vertebrate and invertebrate species from aquatic and terrestrial ecosystems and compare their sensitivity to toxicants using their immune systems as biomarkers.

10.2 IMMUNE MARKERS IN ECOTOXICOLOGY

Assessments of chemical toxicity have regularly been based on laboratory animals, and the potential of chemicals to alter immune function has traditionally been performed in laboratory rodents (see Section III of this book). While the potential immunotoxic effects of some xenobiotics are well documented in experimental animals and humans, relatively little is known in terrestrial mammals and birds, or in freshwater and marine organisms. In addition, few attempts have been made to compare the effects of mixtures of xenobiotics on the immune system *in vivo*, or on immunologically active cells *in vitro* in these species. Comparative studies would provide valuable information on the potential adverse effects of chemicals on the immune function of domestic and wildlife species. The prospect of using a more diverse range of species in toxicity testing of chemicals is of value, because despite fundamental similarities in the immune system of mammals and birds, interspecies (Haley, 2003) as well as age-related differences exist. For these reasons, it is important to study the immunotoxicity of environmental contaminants in domestic animals and wildlife to predict the potential hazard these may pose to the health of these species, and by extrapolation to human health.

The basic function of the immune system is to defend the host against infection. For practical purposes, the immune system is divided into two types: innate (or nonspecific) and adaptive (or specific) immunity. Recent reviews on the evolution of the immune system have reached the consensus that many components of innate immunity appear to be evolutionarily conserved (Hoffmann et al., 1999; Roitt et al., 2001; Turner, 1994; Ulevitch, 2000). This implies that the sensitivity of innate immune responses to a particular toxicant may be similar among different species and could therefore, be used as a general biomarker.

10.2.1 Innate Immunity

Innate immunity serves as a first line of defense, and is found in all animal phyla. Invertebrates, which make up more than 95% of known animal species, rely solely on innate immune mechanisms. In a broad sense, the first defense against foreign antigens is the outer surface of the host, which forms a barrier against invading pathogens. Internal defense mechanisms are available to control and destroy the pathogen once it has passed through the barrier defenses. The main effector cells of this protective activity are the phagocytic cells that patrol the tissues of all the species discussed in this book. Phagocytic cells are able to (1) ingest foreign particles via phagocytosis, (2) mediate inflammatory responses, (3) enhance humoral immune responses that entail the synthesis and secretion of microbicial peptides, or, in vertebrates (4) stimulate production of specific antibodies. All animals have a set of both constitutive and inducible antimicrobial peptides and other humoral factors. For example, lysozyme is a nonspecific antimicrobial enzyme able to digest mucopeptides in the bacterial cell wall. It is present in all vertebrates and in many invertebrates as well. Natural killer (NK) cells are part of the innate immunity in vertebrates. They recognise and kill tumour cells and some virus-infected cells, without requiring previous exposure to that virus. Natural cytotoxic cells (NCCs) are NK-like cells in fish, which are also involved in defense against parasites.

10.2.2 Evolution of Adaptive Immunity

Adaptive immunity, which entails memory of exposure to specific antigens, has evolved only in vertebrates, except the earliest vertebrates, the jawless fish. Adaptive immunity is mediated by B and T lymphocytes that circulate throughout the body, and are "educated" to respond to specific antigens in the primary lymphoid tissues. Primary and secondary lymphoid organs and tissues with immune function have emerged during the evolution of vertebrates. Even jawless fish have gut-associated lymphoid tissues (GALT). Jawed vertebrates possess a greater variety of lymphoid tissues, and have B and T lymphocytes that display adaptive immune responses. In addition to GALT, jawed fish develop immunologically active cells within the thymus, spleen, and kidney. In amphibians, bone marrow appears. Birds can form lymphoid tissue with the capacity to develop responsive germinal centers, and they have a unique lymphoid organ, the bursa of Fabricius. In mammals, the primary lymphoid organs are thymus and bone marrow, and secondary organs are spleen, lymph nodes, and mucosa-associated lymphoid tissue.

10.2.3 Immunological Markers

The assessment of immunological responses can serve as a valuable tool for ecotoxicologic studies. Immunological methods used in ecotoxicology are listed in Table 10.1. The innate immune system is a sensitive target for many contaminants in both vertebrates and invertebrates. For example, phagocytosis and lysozyme activity are well conserved throughout the animal kingdom, are relatively easy to measure, and are sensitive to the effects of environmental contaminants. They thus offer good insight into the immunotoxic and ecotoxic potential of a given substance or pollutant. Also, wound healing, as a measure of the inflammatory response within the integument, has potential as a sensitive marker of the animal's ability to deal with chemical toxicity across taxa. In vertebrates, weights of lymphoid organs, as well as NK cell activity, are tools to assess immunological competence. Measurement of specific antibody response to foreign antigens and proliferation of lymphocytes in response to mitogen stimulation can be used in all jawed vertebrates. Some immune functions such as the antibody mediated immune response, lymphocyte blastogenesis, and delayed-type hypersensitivity (DTH) are dependent on the interplay of several types of immunologically competent cells, and so are regarded as sensitive markers of immunotoxicity. Cells that deliver the immune response communicate with each other, as well as with other cells of the body to maintain homeostasis, especially in relation to components of the neuroendocrine system. It is important to recognize that the response of the immune system is never simple because of the multiple, complex, and interacting mechanisms comprising the immune response. The ultimate assessment of immunotoxicity of a chemical is with pathogen challenge studies in animals that have been exposed to immunotoxicants. Resistance to infection provides a meaningful measure of immunological defense in the context of the whole animal since it is the result of an interplay of the host's immune system with hormonal, neuronal, and other physiological systems.

10.3 IMMUNOTOXICITY OF SELECTED SUBSTANCES ACROSS SPECIES

10.3.1 Heavy Metals

Heavy metals are widely distributed in the environment and have had major ecological consequences in many ecosystems. In aquatic and terrestrial ecosystems, sediments are the most frequently used matrix to monitor pollution by heavy metals. The impact of metal contamination of any site is related to the bioavailability of the metal species, which determines possible effects on the health of exposed organisms, which in turn, can induce population perturbation. The uptake of excess metals leads to either direct toxicity to tissues or cells with associated inflammation, or may cause more subtle alterations in homeostatic mechanisms. Because immunological endpoints are sensitive to environmental changes, they may constitute suitable biomarkers for detecting heavy metal contamination at an early stage and at low concentrations.

To determine the potential toxicity of exposure to chronic or sublethal concentrations of heavy metals, environmental studies have focused on lower trophic levels of organisms (Macdonald and Bewers, 1996). As lower organisms are easily acclimatized to laboratory conditions and are relatively sedentary, the toxicity and otherwise harmful potential of metal-contaminated soils have been assessed using earthworms as sentinels. In many ecosystems, earthworm populations are key species in decomposer communities and therefore are usually affected by accumulation of heavy metals that are chelated to soil particles. Effects of metals on the health of earthworms has been evaluated using outcomes such as the lysosomal activity of coelomocytes (Lukkari et al., 2004; Reinecke et al., 2002; Spurgeon et al., 1994). The toxicity of contaminated soils can be assessed through studies of the nonspecific immunological responses in earthworms. Metals such as copper, cadmium, and zinc induce immunosuppression in earthworms. In many cases, the earliest detectable immunological changes are associated with subcellular organelles such as lysosomes. For example,

Table 10.1 Successfully Employed Immunological Markers across Various Animal Kingdom Taxonomic Groups

Taxonomic Group	Soluble Substances			Cellular Responses				Tissue Responses			Host Responses
	Antimicrobial Agents (Numerous)	Lysozyme Activity	Antibody Response	Phago-cytosis	Natural Killer Cell Activity	Blastogenesis of Lymphocytes	Delayed-Type Hyper-sensitivity	Differential Counts of Immune Cells	Inflammatory Response (e.g., Wound Healing)	Weight of Lymphoid Organs	Host Resistance Model
Invertebrates:											
Porifera (sponges)	+			+							+
Annelids (earthworms)	+	+		+					+		+
Arthropods (insects, crustaceans)	+	+		+				+	+		+
Vertebrates:											
Elasmobranch (sharks, rays)	+	+		+	+			+	+		+
Teleosts (salmon)	+	+	+	+	+	+		+	+	+	+
Amphibians	+	+	+	+	+	+		+	+	+	+
Reptiles	+	+	+	+	+	+	+	+	+	+	+
Birds	+	+	+	+	+	+	+	+	+	+	+
Mammals	+	+	+	+	+	+	+	+	+	+	+

the ability of earthworm coelomocytes to incorporate neutral red dye intracellularly into their lysosomes decreases significantly after chronic exposure to heavy metals, but this is not seen with acute exposure to metals. As the innate immune system of earthworms has sufficient homology with that of vertebrate animals, this response may be expected to occur at higher phylogenetic levels. However, determining the toxicity of metal contamination in sediments using earthworms as a sentinel is not yet completely understood, because the behavior of metals is highly dependent on their chemical form, or speciation, which dictates bioavailability for organisms.

The persistence of heavy metals in terrestrial ecosystems is related to the nature of the soil, water table, and relative proximity of bodies of water. Runoff water and ground infiltration tend to concentrate metals in local aquatic environments (Steinberg et al., 1994). To evaluate the status of ecosystem health in marine or fresh water systems, prolific populations of molluscs, especially mussels and oysters, make practical bioindicators for monitoring metal contaminations in both controlled laboratory exposures and field collection experiments (Andersen et al., 1996; Cossa, 1988; Rojas de Astudillo et al., 2002; Scanes, 1996; Szefer et al., 2002). Because of their morphological and metabolic characteristics (i.e., they are sessile and filter feeders with low metabolic and excretion rates), these species are highly sensitive to environmental changes (Rodriguez-Ariza et al., 1992). Due to their ubiquitous distribution in the aquatic environment, and their benthic and sedentary lifestyle, which continuously exposes them to fluctuations in environmental factors (temperature, salinity, particulates, etc.) and contaminants, molluscs have long been known to accumulate both essential and nonessential trace elements in marine and estuarine ecosystems (Dallinger and Rainbow, 1993). Physiological features of bivalves make them excellent sentinels of marine and estuarine ecosystem health. Ecotoxicological studies have focused on the response of the innate immune system of bivalves exposed to heavy metal contamination. Immunosuppression in bivalves may be induced by infectious (viral, bacterial) as well as pollutant (pesticides, heavy metals) stressors. One of the major immunological mechanisms responsible for critical physiological responses of bivalves is dependent on hemocytes (blood cells). Hemocytes have numerous physiological functions, including phagocytosis and binding and storage of metals within intracellular granules; their competent performance is important to bivalve survival (George et al., 1978). This immunological activity is the predominant internal defense mechanism available to mollusks (Pipe et al., 1999). During laboratory experimentation as well as in field studies, sublethal metal contamination has proven to cause dysfunction of phagocytosis and lysosomal activity, thus suppressing immunocompetence. At higher concentrations, metal accumulation in tissues can induce an inflammatory response progressing to cellular or widespread tissue damage leading to death. Therefore, alterations of innate immune responses should be considered as biomarkers of a variety of stressors in mollusks. While the bioaccumulation of metals and other contaminants in the tissues of these filter feeders provides a good reflection of local contaminant loads, their immunological status gives an indication of potential biological effects associated with that environment.

A research program initiated by the Immunology Theme Team at the Canadian Network of Toxicology Centre assessed dose–response effects of exposure to heavy metals on immunological endpoints using multiple species. This comparative immunotoxicological study was performed on a broad range of invertebrate and vertebrate species. The effects of CdC_2, CH_3HgCl, and $HgCl_2$ over a wide dose range (10^{-9} to 10^{-4} M) were determined on phagocytic activity by cells of the innate immune system collected from earthworms, bivalves, starfish, mummichog, trout, plaice, frog, chicken, kestrel, rat, mouse, sheep, rabbit, horse, cat, dog, llama, musk ox, elk, seal, monkey, and human. Phagocytic function was evaluated by *in vitro* uptake of fluorescent microspheres using flow cytometry. For species with vascular systems and circulating lymphocytes, effects of exposure to heavy metals was also evaluated through the lymphoproliferative response to concanavalin A (Con-A) mitogen. For both assays, results were expressed as a percentage of activity measured in cells from unexposed control animals. Differences in sensitivity were estimated by logit plot graphical determination of the concentration of $CdCl_2$, $HgCl2_2$, and CH_3HgCl that induced 50% suppression of phagocytosis or blastogenesis in each species (IC_{50}). In this chapter, we present

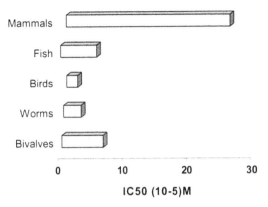

Figure 10.1   Dispersion of IC_{50} in various groups of organisms. The estimated inhibitory concentrations of CH_3HgCl that result in decreased phagocytosis in 50% of the samples tested were obtained by logit plot analysis. These results show that mammals are much less sensitive to the immunotoxic effect of methyl mercury.

examples of the results obtained and summarize them in a dispersion graph (Figure 10.1), since many of the results have been detailed in published scientific papers or those in preparation (Brousseau et al., 1999; Sauve et al., 2002a; Sauve and Fournier, 2004; Sauve et al., 2002b). We present conclusions drawn from this work. It is not the objective of this chapter to discuss mechanisms of immunotoxicity of metals, which has been thoroughly covered in published research (Bernier et al., 1995; Borella et al., 1990; Cifone et al., 1989; Zelikoff and Thomas, 1998).

Immunotoxicology of heavy metals on the blastogenic response of Con-A-stimulated lymphocytes from domestic and wild mammals and avian species (percent normal Con-A mitogenic response) are presented in Tables 10.2, 10.3, and 10.4. At lower concentrations (10^{-9} to 10^{-8} M), $HgCl_2$ tended to increase blastogenesis in musk ox, horse, elk, and sheep leukocytes to Con-A (Table 10.2), but not in other species tested. This work presented the first indication of major differences among species in the sensitivity to metals. In general, $HgCl_2$ at $\geq 10^{-5}$ M significantly suppressed Con-A-induced proliferation of musk ox, horse, sheep, and llama leukocytes. In cats and dogs, significant suppression of Con-A–induced proliferation only occurred at 10^{-4} M $HgCl_2$. In contrast, suppression of blastogenesis to Con-A was observed in elk and chicken lymphocytes at all concentrations of $HgCl_2$ tested. Exposure to lower concentrations of CH_3HgCl tended to enhance proliferation of musk ox, chicken, elk, and llama lymphocytes to Con-A (Table 10.3).

Table 10.2   Effect of $HgCl_2$ on Con-A–Induced Blastogenesis of Lymphocytes from Domestic and Wild Animal Species *In Vitro*

Species (*n*)	Concentration of $HgCl_2$ (M)					
	10^{-9}	10^{-8}	10^{-7}	10^{-6}	10^{-5}	10^{-4}
Musk ox (4)	176 ± 88	129 ± 49	83 ± 16	109 ± 24	2 ± 2	2 ± 2
Dog (4)	96 ± 27	116 ± 12	126 ± 9	309 ± 76	110 ± 18	5 ± 4
Cat (5)	86 ± 15	89 ± 15	116 ± 18	150 ± 29	249 ± 56	0.9 ± 0.5
Chicken (8)	97 ± 16	86 ± 18	96 ± 15	128 ± 10	109 ± 11	107 ± 25
Horse (9)	145 ± 50	133 ± 64	146 ± 38	73 ± 34	5 ± 8	2 ± 0.8
Elk (3)	200 ± 83	127 ± 14	123 ± 50	123 ± 23	74 ± 43	100 ± 46
Sheep (5)	292 ± 331	266 ± 296	137 ± 50	34 ± 10	20 ± 21	2 ± 2
Llama (3)	88 ± 46	172 ± 158	102 ± 23	113 ± 88	2 ± 2	2 ± 1

Notes: Optimal concentrations of Con-A were predetermined for each species tested. *n* = sample size. Cells (5×10^5/well) were incubated with a range of concentrations of $HgCl_2$ in the presence of predetermined and species specific (optimal) concentrations of Con-A for 72 h, pulsed with [^{3}H]thymidine during the last 18 h. Amount of radioactive thymidine incorporated into the DNA of dividing cells was determined using a scintillation counter. Data are expressed as percentages of control.

Table 10.3   Effect of CH₃HgCl on Con-A-Induced Blastogenesis of Lymphocytes from Domestic and Wild Animal Species *In Vitro*

Species (n)	Concentration of CH_3HgCl (M)					
	10^{-9}	10^{-8}	10^{-7}	10^{-6}	10^{-5}	10^{-4}
Musk ox (4)	153 ± 64	86 ± 27	19 ± 36	17 ± 20	0.4 ± 0.2	1 ± 0.3
Dog (4)	117 ± 23	113 ± 26	69 ± 38	57 ± 14	4 ± 3	4 ± 2
Mink (4)	104 ± 71	125 ± 78	138 ± 100	85 ± 51	86 ± 35	93 ± 87
Cat (5)	104 ± 23	102 ± 26	94 ± 37	98 ± 30	1 ± 0.3	3 ± 2
Chicken (8)	196 ± 98	228 ± 128	148 ± 79	3 ± 3	2 ± 1	2 ± 0.8
Horse (9)	92 ± 14	85 ± 14	65 ± 15	15 ± 9	0.1 ± 0.1	0.2 ± 0.1
Elk (3)	130 ± 45	152 ± 27	125 ± 83	40 ± 17	44 ± 26	22 ± 8
Sheep (5)	82 ± 22	58 ± 26	116 ± 76	3 ± 4	7 ± 11	2 ± 1.4
Llama (3)	114 ± 53	175 ± 146	12 ± 10	18 ± 18	2 ± 1	1 ± 1

Notes: Optimal concentrations of Con-A were predetermined for each species tested. Cells (5 × 10⁵/well) were incubated with various concentrations of CH₃HgCl in the presence of optimal concentrations of Con-A for 72 h, pulsed with [³H]thymidine during the last 18 h. Amount of radioactive thymidine incorporated into the DNA of dividing cells was determined using a scintillation counter. Data are expressed as percentages of control. n = sample size

Table 10.4   Effect of CdCl₂ on Con-A-Induced Blastogenesis of Lymphocytes from Domestic and Wild Animal Species *In Vitro*

Species (n)	Concentration of $CdCl_2$ (M)					
	10^{-9}	10^{-8}	10^{-7}	10^{-6}	10^{-5}	10^{-4}
Musk ox (4)	146 ± 57	128 ± 73	113 ± 120	149 ± 37	71 ± 28	3 ± 6
Dog (4)	96 ± 6	97 ± 13	103 ± 15	100 ± 9	114 ± 29	3 ± 1
Cat (5)	89 ± 17	92 ± 13	104 ± 21	102 ± 12	151 ± 24	3 ± 3
Chicken (9)	82 ± 24	121 ± 29	112 ± 32	117 ± 31	66 ± 28	1.2 ± 1
Horse (8)	65 ± 42	81 ± 42	76 ± 38	69 ± 43	33 ± 47	0.2 ± 0
Elk (4)	190 ± 67	123 ± 85	136 ± 118	98 ± 55	100 ± 21	57 ± 53
Sheep (5)	58 ± 27	58 ± 16	30 ± 31	69 ± 12	24 ± 26	2 ± 1
Llama (3)	370 ± 35	244 ± 64	75 ± 61	41 ± 23	29 ± 4	2 ± 2

Notes: Optimal concentrations of Con-A were predetermined for each species tested. Cells (5 × 10⁵/well) were incubated with various concentrations of CdCl₂ in the presence of optimal concentrations of Con-A for 72 h, pulsed with [³H]thymidine during the last 18 h. Amount of radioactive thymidine incorporated into the DNA of dividing cells was determined using a scintillation counter. Data are expressed as percentages of control. n = sample size.

CH₃HgCl significantly suppressed Con-A–induced proliferative responses of lymphocytes from llama and musk ox at 1^{-7} M, and those of chicken, horse, elk, and sheep leukocytes at 10^{-6} M (Table 10.3). At 10^{-5} M, CH₃HgCl₂ inhibited the Con-A–stimulated proliferative responses of leukocytes of all species tested except mink.

Generally, in all the species tested, significant inhibition of lymphocyte proliferation to the T-cell mitogen Con-A occurred at 10^{-4} M CdCl₂, with the exception of llama and sheep where a significant inhibition was evident at 10^{-5} M (Table 10.4). Suppression of blastogenesis was not observed in elk leukocytes at any concentration of CdCl₂ tested, whereas lower concentrations of CdCl₂ tended to increase blastogenic responses of lymphocytes from musk ox, elk, and llama suggesting a hormetic response (Table 10.4).

We demonstrated interesting species differences in their sensitivity to low-dose exposure to metals with some species responding with enhanced lymphocyte proliferation. The same diverse pattern occurred for the phagocytic activity of peripheral blood neutrophils. These results emphasise differences in sensitivity among species, especially in response to higher doses of metals. To more closely examine these species-specific responses, we estimated the IC₅₀ of each metal by logit

Table 10.5 IC_{50} for Blastogenesis and Phagocytosis Following Exposure to CH_3HgCl Blastogenesis

Blastogenesis			
Species	IC_{50} (M)	Standard Deviation	n
Musk ox	2.090×10^{-7}	1.580×10^{-7}	4
Dog	1.077×10^{-6}	3.950×10^{-7}	4
Llama	4.608×10^{-6}	5.310×10^{-6}	3
Sheep	2.958×10^{-6}	3.772×10^{-6}	5
Horse	1.770×10^{-7}	1.160×10^{-7}	9
Cat	2.608×10^{-6}	2.654×10^{-6}	5
Chicken	1.968×10^{-6}	7.570×10^{-7}	8
Elk	6.262×10^{-6}	2.230×10^{-6}	3

Phagocytosis			
Species	ED_{50} (M)	SD	n
Musk ox	1.005×10^{-5}	1.116×10^{-5}	4
Dog	2.549×10^{-6}	3.310×10^{-7}	4
Llama	1.803×10^{-5}	2.137×10^{-5}	3
Sheep	3.528×10^{-5}	4.461×10^{-5}	5
Horse	1.437×10^{-5}	5.143×10^{-6}	9
Cat	2.639×10^{-5}	8.280×10^{-6}	5
Chicken	1.073×10^{-5}	6.383×10^{-6}	8
Elk	NA	NA	–

Notes: Optimal concentrations of Con-A were predetermined for each species tested. Estimated inhibitory concentration that results in decreased response in 50% of the samples tested (IC_{50}) for methylmercury chloride on blastogenesis and phagocytosis for domestic animals and wildlife species. The IC_{50} values were estimated by logit plot analysis. Results are expressed as the mean concentrations that induce 50% inhibition of blastogenesis or phagocytosis.

ED_{50}, effective concentration inducing 50% mortality of exposed organisms; IC_{50}, inhibitory concentration for 50% of samples; NA, value not determined because either 50% inhibition fell outside the highest concentration of metals used in this study (10^{-4} M), or because of poor correlation of the dose–response curve.

graphical plot analysis for each endpoint in each species. The IC_{50} values for phagocytic and blastogenic activities following methylmercury exposure are presented in Table 10.5. The IC_{50} values for CH_3HgCl on blastogenesis confirm a small difference in sensitivity among species with IC_{50} spreading only over one log range, with the horse being the most sensitive species (1.7×10^{-7} M) and the most resistant being the elk (6.3×10^{-6} M). In contrast to blastogenesis, the IC_{50} values for phagocytosis are generally higher, suggesting that phagocytic cells are more resistant to the immunotoxic effects of metals than are lymphocytes (see Chapter 25 in this book).

For other groups of animals tested, including birds, fish, bivalve mollusks, and worms, the amount of heterogeneity in the IC_{50} values for phagocytosis is much lower than in mammals (Figure 10.1). While it may be argued that for fish and birds the number of species tested was limited (n = 3), the studies of worms (n = 5) and bivalves are convincing. Phylogenetic differences are not likely to provide the explanation, considering that within bivalves alone the differences among species are probably as great as those among mammals.

This type of comparative study demonstrates the differences in sensitivity of immune cells to chemical contaminants, and in general, invertebrates and aquatic species are more sensitive than mammals. This work has shown that species-specific susceptibility to immunotoxic chemicals can be established and quantified. This allows for comparison among species and extrapolation of results from species to species. This much needed background information will aid in the selection

Table 10.6   Study Numbers of Acute Toxicity Tests on Various Organisms by Pesticide Class

Acute Toxicity Test	Herbicides	Fungicides	Insecticides/Acaricides	Other	Total
EC_{50}-96 h on algae	104	62	68	8	242
EC_{50}-96 h on *Daphnia*	171	88	124	24	407
LC_{50}-96 h on rainbow trout	200	94	130	23	447
LC_{50}-96 h on bluegill sunfish	127	51	74	15	267
LC_{50}-96 h on fathead minnow	10	1	6	1	18
LC_{50}-96 h on catfish	24	6	11	2	43
LC_{50}-96 h on carp	41	34	30	4	109
LC_{50}-96 h on golden orfe	7	10	27	3	47
Total	714	352	486	81	1633

Note: D_{50}, effective concentration inducing 50% mortality of exposed organisms; IC_{50}, inhibitory concentration for 50% of samples.

From Tremolada, P., et al., *Aquatic Toxicol.*, 2004.

of sentinel species that would be most valuable or relevant for ecotoxicologic assessment. The most suitable species may then be identified for contaminants of concern in particular ecosystems of interest.

10.3.2   Pesticides

Pesticides are an extremely heterogeneous group of toxicants used to control the growth of unwanted plants, fungi, or insects and other invertebrates in terrestrial and aquatic ecosystems. More than 800 chemicals are marketed as multiple formulations for use in various field applications. Pesticides are classified broadly into functional or chemical classes (Table 10.6) (Tremolada et al., 2004). The prescribed use of pesticides is meant to follow application guidelines to minimize damage to nontarget species. Restrictions of use exist with respect to times, amounts, and climatic conditions when pesticides can be used. Although the harmful properties of chemical or biological (natural or seminatural) insecticide products are designed against particular target organisms, close observation of treated ecosystems reveals the inevitable negative impact on nontarget terrestrial and aquatic wildlife. Pesticides are common pollutants to which organisms in many different ecosystems are exposed, since pesticide concentrations in the biosphere have dramatically increased over the past few decades. In a degraded ecosystem, particularly where pesticides persist at sublethal concentrations, structural and functional changes in basal metabolism occur in many species. Some of these changes include the modification of immunological responses.

Oligochaetes (worms), which are widely distributed in the soil and in fresh water, can be used as bioindicators for both terrestrial and fresh water ecosystems. Due to their ubiquitous and sedentary nature, the low capital investment required to study them, the ease of acclimatizing them to the laboratory (Cooper and Parrinello, 1996), and the relatively minor concerns of animal welfare with such species, ecotoxicologic investigations can be conducted under controlled laboratory conditions as well as in field studies. The capacity of worms to accumulate toxic molecules in their caudal region (chloragogene cells), plus their unique, efficient detoxification mechanisms such as mucus excretion and immunological responses, makes them an attractive nonmammalian model. Worms ingest large amounts of soil and are therefore exposed to a wide range of toxicants through their skin and gastrointestinal tracts, resulting in the concentration of toxicants in their body. The bioavailability of pesticides and their breakdown products depend on soil characteristics that must be considered when interpreting the nonspecific immune responses of worms against soil-borne contaminants.

Pesticides tend to work through similar metabolic pathways in vertebrate or invertebrate species, and thus induce similar immunological responses. The partitioning of pesticides into different compartments within ecosystems is described by soil and water chemists and toxicologists. Ecotoxicologic studies evaluating the impact of pesticides on aquatic wildlife have focused on fish populations because of their wide distribution in affected areas and the relatively low cost and ease of laboratory exposure *in situ*. In marine or fresh water ecosystems, fish can accumulate pesticides through external contact surfaces, gills, and dermis (skin), as well as through diet. Internal defense mechanisms are then required for detoxification and elimination of these toxicants. The innate or nonspecific immune system of fish is vulnerable to many xenobiotics including pesticides (Betoulle et al., 2002). Immunosuppression has been used as an early indicator or biomarker of toxicant exposure (Dean et al., 1982; Fournier et al., 2000a). Several immunological variables are potentially useful as bioindicators of effects of xenobiotic exposure in fish. These include white blood cell (leukocyte) and lymphocyte status (measured as total or differential white blood cell counts); nonspecific defense factors (such as lysosomal activity and levels of acute-phase proteins in body fluids); weight and morphology of leukocyte-producing organs (such as thymus, kidney, and spleen); melanomacrophage centers (number, size, and histopathological appearance); macrophage function (chemotaxis, phagocytosis, pinocytosis, and chemiluminescence); and most significantly, increased susceptibility to bacterial infections (van der Oost et al., 2003). Changes in these variables have been documented at sublethal concentrations of pesticides, which are considerably lower than concentrations inducing acute toxicity and the death of nontarget organisms (Fournier et al., 2000b). Rainbow trout (*Oncorhynchus mykiss*), bluegill sunfish (*Lepomis macrochirus*), catfish (*Ictalurus* spp.) and carp (*Cyprinus* spp.) have been used as research subjects for acute toxicity studies of a large range of toxicants. Fish are also important for the numerous studies providing information on the 96-h LC_{50} of various classes of pesticides (Table 10.6).

Immunosuppression occurring with sublethal or chronic pesticide contamination in fish is considered a valuable biomarker of environmental quality. The large number of studies containing toxicological information about the different categories of pesticides is reported in Table 10.6 (Tremolada et al., 2004). Impairment of innate immunity appears to be more biologically significant in fish than in mammals, in part because mounting an adaptive or acquired immune response takes longer in fish compared to higher vertebrates. From an immunological point of view, the use of marine mammals as monitors of environmental health appears to be less sensitive than fish as early warning sentinels, in part due to their more robust immune system (Alexander and Ingram, 1992).

Pesticides, owing to their pest-destroying properties, are also important in food preservation. They may be added to food products during production, processing, storage and/or packaging. Pesticides are often present as residues in food of both vegetable and animal origins meaning that humans at the top of the food chain are also exposed. Pesticides can induce dramatic environmental changes that have a profound impact on the survival of many species comprising an ecosystem, and in this way can affect the equilibrium of ecosystems.

10.3.3 Polycyclic Aromatic Hydrocarbons

Polycyclic aromatic hydrocarbons (PAHs) are ubiquitous environmental contaminants derived primarily from incomplete combustion of organic matter. Since the onset of industrialization, the environmental load of PAHs has increased several orders of magnitude due to the increasing use of fossil fuels. They are found in air, soil, water, and food. While some PAHs are relatively innocuous, others, such as benzo(a)pyrene (BaP), dibenzanthracenes, and certain dibenzopyrenes and nitropyrenes, are compounds with cytotoxic, genotoxic, teratogenic, and carcinogenic effects (Varansi, 1989). PAHs undergo metabolic activation by mammalian tissues responsible for biotransformation. They form highly reactive, toxic, intermediate metabolites that can irreversibly damage cellular macromolecules (DNA, proteins, lipids). This process has been most thoroughly studied for BaP. Formation of toxic BaP metabolites results from cytochrome P450 metabolism initiated via the aryl hyrdoxylase (Ah) receptor. The resultant 7,8,9,10-diol epoxide of BaP forms protein and nucleic acid adducts that promote mutagenic, carcinogenic, and immunotoxic effects.

The cytochrome P450 system for biotransformation has also been identified in earthworms (Achazi et al., 1998), although, to our knowledge, there exist no reports that PAHs may modulate immune function in earthworms. However, earthworms have a high likelihood of exposure to PAHs in soils. When they have been exposed to PAH-contaminated habitats, increased DNA adducts have been reported (van Schooten et al., 1995). In bivalves, dose-dependent suppression of phagocytosis follows PAH-exposure (Anderson et al., 1981; Grundy et al., 1996; Wootton et al., 2003). Highly variable effects of phenanthrene on the immune responses of various bivalve species has been reported (Wootton, et al., 2003). In mammals, exposure to PAHs is known to cause immunomodulation. T lymphocyte–mediated function appears to be particulary susceptible in fish (Faisal and Huggett, 1993), reptiles, birds, and mammals following exposure to PAHs.

10.3.4 Halogenated Aromatic Hydrocarbons

This class of toxic chemicals includes a variety of persistent chlorinated and brominated aromatic compounds with high potential for bioaccumulation. Polychlorinated biphenyls (PCBs) are industrial chemicals present in dielectric fluids of transformers and were common as paint additives. Chlorinated dioxins, furans, and phenols are unintentional by-products of industrial processes. The toxicity of these compounds have been described as falling into two categories: (1) congeners that share structural similarities with TCDD (PCB congeners with coplanar structures) and (2) nonplanar congeners with activities mimicking endogenous hormones, or having calcium ionophoric or membrane-perturbing properties (Rice, 2001). Tetrachlorodibenzo-p-dioxin (TCDD) is extremely immunotoxic in mice (Holsapple et al., 1991; Vos et al., 1991). Like PAHs, these compounds have a high affinity for the Ah receptor.

Due to the stability and consequently long persistence of organochlorines in the environment, they accumulate in fish and shellfish (Smith and Gangolli, 2002). Therefore, they are of great concern as contaminants in marine ecosystems. Determining tissue levels of PCBs, particularly in bivalves, is an important means of monitoring environmental PCB contamination (Connor et al., 2001; Herve et al., 2002; Villeneuve et al., 1999). Immunotoxic effects of PCBs have been described in mussels and appear to be congener specific (Canesi et al., 2003). Also, earthworms readily bioaccumulate these compounds. The commercial PCB mix Aroclor 1254 has been shown to have immunotoxic effects on earthworm coelomocytes (Burch et al., 1999).

Harbor seals, like other piscivorous marine mammals, are at the top of the aquatic food chain, and accumulate high levels of lipophilic contaminants including PCP and PAHs. Seals fed with Baltic Sea herring containing high levels of halogenated aromatic hydrocarbons have decreased NK-cell activity and T-cell mitogen-induced proliferation, as well as suppressed DTH and antibody responses, when compared to control seals fed on Atlantic herring with much lower contaminant levels (De Swart et al., 1994; Ross et al., 1995; Ross et al., 1996). American kestrels (*Falco sparverius*), small birds of prey at the top of the terrestrial food web, have shown prolonged dysfunction of both B-cell (Smits and Bortolotti, 2001) and T-cell (Smits et al., 2002) mediated immune function when exposed to dietary PCBs.

10.4 SUMMARY AND CONCLUSIONS

Ecological risk assessment can be achieved based upon investigations at different levels of organization. Suborganism biomarkers at the level of biochemical and physiological processes are studied to detect deviations from the normal state of health, and are measured using biochemical and immunological endpoints. At the organism's level, survival, growth, and reproduction of individuals provide the basis of ecotoxicologic studies and are well-established endpoints of classic laboratory tests in ecotoxicology.

The most relevant level may be population stability in which decreases in numbers, or modification of the genetic structure of populations would lead to disequilibrium in the disrupted ecosystem. At the level of ecosystems, changes in species composition, numbers, and diversity are studied to describe negative impacts of pollutants on communities (van der Oost et al., 2003).

The discipline of ecological risk assessment has been developed to characterize the adverse effects of human activities or natural catastrophes on ecosystems. In the last decade, much research was devoted to ecological risk assessment by international bodies such as the World Health Organization, Organization for Economic Cooperation and Development, and European Centre for Ecotoxicology and Toxicology of Chemicals. The objective of risk-based environmental regulation is to balance the degree of allowable risk against the cost of risk reduction and weighing those against competing risks (van der Oost et al., 2003). Many tools have been developed to predict the environmental impact of toxicants, the assessment of immune system markers being one of those tools. Because the immune system is well conserved through animal taxa and is sensitive to environmental change, it can serve as a relevant bioindicator of environmental stress. The main challenge in ecotoxicologic studies has been to determine appropriate sentinels and to define specific schemes for ecosystems based on the various communities that identify them, which will lead to an overall understanding of the health status of ecosystems within the biosphere.

Many factors influence the susceptibility of vertebrate and invertebrate species to pollutants, including the route of entry; type of toxicant; and whether exposure is made to naive animals or populations with previous exposure to certain classes or groups of toxicants, viruses, bacteria, and parasites. Due to the diversity and subclassifications of each class of vertebrate or invertebrate animals, ecotoxicologic studies have promoted the use of particular representative species in order to assess environmental risk. A common research strategy has focused on a physiological endpoint represented across all phyla, the immune system. Efforts have been made to identify immunological responses that are sensitive to a large range of toxicants or potentially toxic substances that may be present in the aquatic or terrestrial environment. As human activity or natural catastrophes affect an ecosystem, tools for evaluating biological impacts need objectivity to help provide a global vision of our biosphere and to preserve it for future generations of humans, domestic animals and wildlife.

ACKNOWLEDGMENTS

H.S. was partly supported by grant n:o 104263 from Academy of Finland.

REFERENCES

Achazi, R.K., et al., Cytochrome P450 and dependent activities in unexposed and PAH-exposed terrestrial annelids, *Comp. Biochem. Physiol.*, 121, 339–350, 1998.
Alexander, J.B., and Ingram, G.A., Noncellular nonspecific defence mechanisms of fish, *Ann. Rev. Fish Dis.*, 2, 249–279, 1992.
Andersen, V., Maage, A., and Johannessen, P.J., Heavy metals in blue mussels (*Mytilus edulis*) in the Bergen Harbor area, western Norway, *Bull. Environ. Contam. Toxicol.*, 57, 589–596, 1996.
Anderson, R., et al., Effects of environmental pollutants on immunological competency of the clam *Mercenaria mercenaria*: Impaired bacterial clearance, *Aquatic Toxicol.*, 1, 187–195, 1981.
Attrill, M.J., and Depledge, M.H., Community and population indicators of ecosystem health: targeting links between levels of biological organisation, *Aquatic Toxicol.*, 38, 183–197, 1997.
Bernier, J., et al., Great Lakes health effects: immunotoxicity of heavy metals, *Environ. Health Perspect.*, 103, 23–24, 1995.
Betoulle, S., Etienne, J.C., and Vernet, G., Acute immunotoxicity of gallium to carp (*Cyprinus carpio* L.), *Bull. Environ. Contam. Toxicol.*, 68, 817–823, 2002.
Bols, N.C., et al., Ecotoxicology and innate immunity in fish, *Dev. Comp. Immunol.*, 25, 853–873, 2001.
Borella, P., Manni, S., and Giardino, A., Cadmium, nickel, chromium and nickel accumulate in human lymphocytes and interfere with PHA-induced proliferation, *J. Trace Elements Electrolytes Health Dis.*, 4, 87–95, 1990.

Brousseau, P., et al., Flow cytometry as a tool to monitor the disturbance of phagocytosis in the clam *Mya arenaria* hemocytes following *in vitro* exposure to heavy metals, *Toxicology*, 142, 145–156, 1999.

Burch, S.W., et al., *In vitro* earthworm *Lumbricus terrestris* coelomocyte assay for use in terrestrial toxicity identification evaluation, *Bull. Environ. Contam. Toxicol.*, 62, 547–554, 1999.

Canesi, L., et al., Effects of PCB congeners on the immune function of *Mytilus hemocytes*: alterations of tyrosine kinase-mediated cell signaling, *Aquatic Toxicol.*, 63, 293–306, 2003.

Cifone, M., et al., Effects of cadmium on lymphocyte activation, *Biochim. Biophys. Acta*, 1011, 25–32, 1989.

Clements, W.H., and Kiffney, P.M., Assessing contaminant effects at higher levels of biological organization, *Vet. Immunol. Immunopathol.*, 13, 357–360, 1994.

Connor, L., et al., Recent trends in organochlorine residues in mussels (*Mytilus edulis*) from the Mersey Estuary, *Mar. Environ. Res.*, 52, 397–411, 2001.

Cooper, E.L., and Parrinello, N., Comparative immunologic models can enhance analyses of environmental immunotoxicity, *Ann. Rev. Fish Dis.*, 6, 179–191, 1996.

Cossa, D., Cadmium in *Mytilus* spp.: Worldwide survey and relationship between seawater and mussel content, *Mar. Environ. Res.*, 26, 265–284, 1988.

Dallinger, R., and Rainbow, P., *Ecotoxicology of Metals in Invertebrates*. SETAC Special Publications, Lewis, Chelsea, MI, 1993.

De Swart, R.L., et al., Impairment of immune function in harbour seals (*Phoca vitulina*) feeding of fish from polluted waters, *Ambio*, 23, 155–159, 1994.

Dean, J.H., et al., Procedures available to examine the immunotoxicity of chemicals and drugs, *Pharmacol. Rev.*, 34, 137–148, 1982.

Faisal, M., and Huggett, R.J., Effects of polycyclic aromatic hydrocarbons on the lymphocyte mitogenic responses in spot, *Leiostomus xanthurus*, *Mar. Environ. Res.*, 35, 121–124, 1993.

Fournier, M., et al., Phagocytosis as a biomarker of immunotoxicity in wildlife species exposed to environmental xenobiotics, *Am. Zool.* 40, 412–420, 2000a.

Fournier, M., et al., Immunosuppression in mice fed on diets containing beluga whale blubber from the St Lawrence Estuary and the Arctic populations, *Toxicol. Lett.*, 112–113, 311–317, 2000b.

George, S.G., et al., Detoxication of metals by marine bivalves: an ultrastructural study of the compartmentation of copper and zinc in the oyster *Ostrea edulis*, *Mar. Biol.*, 45, 147–156, 1978.

Grundy, M.M., et al., Phagocytic reduction and effects on lysosomal membranes by polycyclic aromatic hydrocarbons, in haemocytes of *Mytilus edulis*, *Aquatic Toxicol.*, 34, 273–290, 1996.

Haley, P.J., Species differences in the structure and function of the immune system, *Toxicology*, 188, 49–71, 2003.

Herve, S., Heinonen, P., and Paasivirta, J., Survey of organochlorines in Finnish watercourses by caged mussel method, *Resour. Conserv. Recycling*, 35, 105–115, 2002.

Hoffmann, J.A., et al., Phylogenetic perspectives in innate immunity, *Science*, 284, 1313–1318, 1999.

Holsapple, M.P., et al., A review of 2, 3, 7, 8-tetrachlorodibenzo-p-dioxin-induced changes in immunocompetence: 1991 update, *Toxicology* 69, 219–255, 1991.

Lagadic, L., and Caquet, T., Invertebrates in testing of environmental chemicals: are they alternatives? *Environ. Health Perspect.*, 106 (Suppl. 2), 593–611, 1998.

Lukkari, T., et al., Biomarker responses of the earthworm *Aporrectodea tuberculata* to copper and zinc exposure: differences between populations with and without earlier metal exposure, *Environ. Pollut.*, 129, 377–386, 2004.

Macdonald, R.W., and Bewers, J.M., Contaminants in the arctic marine environment: priorities for protection, *ICES J. Mar. Sci.*, 53, 537–563, 1996.

Martin, M., and Richardson, B.J., A paradigm for integrated marine toxicity research? Further views from the Pacific Rim, *Mar. Pollut. Bull.*, 30, 8–13, 1995.

Pipe, R.K., et al., Copper induced immunomodulation in the marine mussel, *Mytilus edulis*, *Aquatic Toxicol.*, 46, 43–54, 1999.

Reinecke, S.A., Helling, B., and Reinecke, A.J., Lysosomal response of earthworm (*Eisenia fetida*) coelomocytes to the fungicide copper oxychloride and relation to life-cycle parameters, *Vet. Immunol. Immunopathol.*, 21, 1026–1031, 2002.

Rice, C.R., Fish immunotoxicology, in *Target Organ Toxicity in Marine and Freshwater Teleosts*, Schlenk, D., and Benson, W.H., Taylor & Francis, London, 2001, 96–138.

Rodriguez-Ariza, A., et al., Metal, mutagenicity, and biochemical studies on bivalve molluscs from Spanish coasts, *Environ. Mol. Mutagen.*, 19, 112–124, 1992.

Roitt, I.M., Brostoff, J., and Male, D., *Immunology*, Mosby, Edinburgh, 2001.

Rojas de Astudillo, L., et al., Heavy metals in green mussel (*Perna viridis*) and oysters (*Crassostrea* sp.) from Trinidad and Venezuela, *Arch. Environ. Contam. Toxicol.*, 42, 410–415, 2002.

Ross, P.S., et al., Contaminant-related suppression of delayed-type hypersensitivity (DTH) and antibody responses in harbor seals (*Phoca vitulina*) fed herring from the Baltic Sea, *Environ. Health Perspect.*, 103, 162–167, 1995.

Ross, P.S., et al., Suppression of natural killer cell activity in harbour seals (*Phoca vitulina*) fed Baltic Sea herring, *Aquatic Toxicol.*, 34, 71–84, 1996.

Sauve, S., et al., Phagocytic activity of marine and freshwater bivalves: *in vitro* exposure of hemocytes to metals (Ag, Cd, Hg and Zn), *Aquatic Toxicol.*, 58, 189–200, 2002a.

Sauve, S., and Fournier, M., Age-specific immunocompetence of the earthworm *Eisenia andrei*: exposure to methylmercury chloride, *Ecotoxicol. Environ. Saf.*, 60, 67–72, 2005.

Sauve, S., et al., Phagocytic response of terrestrial and aquatic invertebrates following *in vitro* exposure to trace elements, *Ecotoxicol. Environ. Saf.* 52, 21–29, 2002b.

Scanes, P., "Oyster watch:" Monitoring trace metal and organochlorine concentrations in Sydney's coastal waters, *Mar. Pollut. Bull.*, 33, 226–238, 1996.

Smith, A.G., and Gangolli, S.D., Organochlorine chemicals in seafood: occurrence and health concerns, *Food Chem. Toxicol.*, 40, 767–779, 2002.

Smits, J.E., and Bortolotti, G.R., Antibody mediated immunotoxicity in American Kestrels (*Falco sparverius*) exposed to polychlorinated biphenyls, *J. Toxicol. Environ. Health*, 62, 217–226, 2001.

Smits, J.E., et al., Thyroid hormone suppression and cell-mediated immunomodulation in American Kestrels (*Falco sparverius*) exposed to PCBs, *Arch. Environ. Contam. Toxicol.*, 43, 338–344, 2002.

Spurgeon, D.J., Hopkin, S.P., and Jones, D.T., Effects of cadmium, copper, lead and zinc on growth, reproduction and survival of the earthworm *Eisenia fetida* (*Savigny*): assessing the environmental impact of point-source metal contamination in terrestrial ecosystems, *Environ. Pollut.*, 84, 123–130, 1994.

Steinberg, C.E.W., Geyer, H.J., and Kettrup, A.A.F., Evaluation of xenobiotic effects by ecological techniques, *Chemosphere*, 28, 357–374, 1994.

Szefer, P., et al., Distribution and relationships of trace metals in soft tissue, byssus and shells of *Mytilus edulis trossulus* from the southern Baltic, *Environ. Pollut.*, 120, 423–444, 2002.

Tremolada, P., et al., Quantitative inter-specific chemical activity relationships of pesticides in the aquatic environment, *Aquatic Toxicol.*, 67, 87–103, 2004.

Turner, R.J., *Immunology: A Comparative Approach*, John Wiley & Sons, Chichester, 1994.

Ulevitch, R.J., Molecular mechanisms of innate immunity, *Immunol. Res.*, 21, 49–54, 2000.

van der Oost, R., Beyer, J., and Vermeulen, N.P.E., Fish bioaccumulation and biomarkers in environmental risk assessment: a review, *Environ. Toxicol. Pharmacol.*, 13, 57–149, 2003.

van Schooten, F.J., et al., DNA dosimetry in biological indicator species living on PAH-contaminated soils and sediments, *Ecotoxicol. Environ. Saf.*, 30, 171–179, 1995.

Varansi, U., *Metabolism of Polycyclic Aromatic Hydrocarbons in the Aquatic Environment*, CRC Press, Boca Raton, FL, 1989.

Villeneuve, J.P., et al., Levels and trends of PCBs, chlorinated pesticides and petroleum hydrocarbons in mussels from the NW Mediterranean coast: comparison of concentrations in 1973/1974 and 1988/1989, *Sci. Total Environ.*, 237–238, 57–65, 1999.

Vos, J.G., Van Loveren, H., and Schuurman, H.J., Immunotoxicity of dioxin: immune function and host resistance in laboratory animals and humans, in *Biological Basis for Risk Assessment of Dioxins and Related Combounds*, Banbury Report 35, Cold Spring Harbor Laboratory Press, Cold Spring Harbor, NY, 1991, pp. 79–93.

Warwick, C.J., Mumford, J.D., and Norton, G.A., Environmental management expert systems, *J. Environ. Manage.*, 39, 251–270, 1993.

Wootton, E.C., et al., Comparisons of PAH-induced immunomodulation in three bivalve molluscs, *Aquatic Toxicol.*, 65, 13–25, 2003.

Zelikoff, J.T., Immunotoxicology across species lines, *Dev. Comp. Immunol.*, 21, 121, 1997.

Zelikoff, J.T., and Thomas, P.T., *Immunotoxicology of Environmental and Occupational Metals*, Taylor & Francis, London, 1998.

Approaches and Models Relevant to the Assessment of the Impact of Chemical-Induced Immunotoxicity on Human Health

CHAPTER 11

Are Changes in the Immune System Predictive of Clinical Diseases?*

Michael I. Luster, Dori R. Germolec, Christine G. Parks, Laura Blanciforti, Michael Kashon, and Robert W. Luebke

CONTENTS

11.1 INTRODUCTION

Although immunosuppression can lead to an increased incidence and severity of infectious and neoplastic diseases, interpreting data from experimental immunotoxicology studies, or even epidemiologic studies, for quantitative risk assessment purposes has been problematic. This is particularly true when the immunological effects, as may be expected from inadvertent exposures in human populations, are slight in nature. In order to accurately predict the risk of immunotoxic exposures in human populations, a scientifically sound framework needs to be established that will allow for the accurate and quantitative interpretation of experimental or clinical immune test data to human health effects. This may require, for example, development of models to equate moderate changes in the numbers of circulating lymphocyte populations or serum immunoglobulin levels, tests that

---

* This report has been reviewed by the Environmental Protection Agency's Office of Research and Development and approved for publication. Approval does not signify that the contents reflect the views of the Agency.

can readily be performed in humans, to potential changes in the incidence or severity of infectious diseases. As an integral step in the development of such a framework, studies on the qualitative and quantitative relationships between immune parameters and disease are reviewed. Initially, the most likely clinical consequences that may occur from chronic mild to moderate immunosuppression are described as well as physiological factors and study design issues that may modify these disease outcomes. Clinical and experimental animal studies that address relationships between immune function and disease development are also discussed in detail and quantitative relationships are described. The most comprehensive databases that address immunodeficiency disease relationships, specifically primary immunodeficiency diseases and AIDS, are not discussed, as these represent extreme examples of immunosuppression, and neither the specific clinical diseases that result nor the eventual outcomes have much in common to that which occurs in individuals with chronic mild to moderate immunosuppression.

It is useful to provide clarification of certain terminology. "Immunosuppression," "immunodeficiency," and "immunocompromised" are nonquantitative terms that reflect a reduced capacity of the immune system to respond to antigens, and are often used interchangeably in immunotoxicology. For the purpose of risk assessment, immunosuppression can be defined as a loss in the ability of the immune system to respond to a challenge at a level that is considered normal, regardless of whether clinical disease ensues. Immunodeficiency often represents an alteration in the immune system that can potentially lead to clinical disease, whether primary (i.e., genetic etiology) or secondary (epigenetic) in nature. The term immunocompromised, like immunosuppression, indicates a deficient immune response, independent of whether it is maladaptive. Immunotoxicity encompasses each of these terms, but specifies that the effect on the immune system originates from xenobiotic exposure.

11.2 DISEASES ASSOCIATED WITH IMMUNOSUPPRESSION

As immunotoxicology testing is increasingly becoming incorporated into toxicological evaluations (House, 2003), there is added impetus to more accurately predict immune system changes detected with these tests to clinical outcomes. Infectious disease is the most obvious consequence of maladaptive immunity, although the etiology, progression, and/or severity of a much broader range of disorders, including certain cancers and autoimmune diseases can also be affected. Identifying the quantitative relationships between altered immune responses and frequency or severity of these diseases in populations is challenging, as many factors may contribute (Moris and Potter, 1997). This is summarized schematically in Figure 11.1, where the appearance, progression, and outcome of infectious disease is viewed as an interrelationship between the virulence of the organism, infectious dose (number of organisms required to produce illness), the integrity of the host's anatomical and functional barriers, and the overall immunocompetence of an individual. The latter, in turn, is affected by genetics as well as age, gender, use of certain medications, drug or alcohol abuse, smoking history, stress, and nutritional status. These factors probably account for most of the variability reported in the values of common immune tests that may, in some cases, exceed two standard deviations. Another factor that affects the quantitative associations between immune function and disease is functional overlap (i.e., redundancy). This reflects the fact that multiple immunologic cell types and effector mechanisms are evoked in response to disease, and have been mistakenly considered as immune reserve. As with the function of other organ-systems, such as the liver or central nervous system, immune reserve cannot exist if infectious diseases occur in individuals with presumably fully intact immune systems. In contrast, immune redundancy is scientifically supported and can be empirically examined (Halloran, 1996). In this respect, the effect of redundancy on the interpretation of immunotoxicology studies was recently addressed using factor analysis and multiple logistic regression (Keil et al., 2001).

CHANGES IN ONSET, COURSE AND OUTCOME OF INFECTIOUS DISEASE

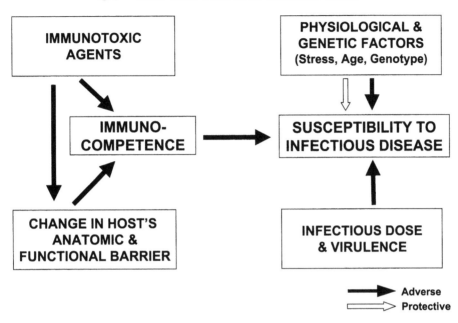

Figure 11.1 Schematic showing factors that influence infectious disease susceptibility.

While both infectious and neoplastic diseases are associated with immunodeficiency, infectious disease incidence is usually the focus of epidemiologic studies, as it represents the most rapid consequence. The particular microorganism responsible for an infection may assist in identifying the qualitative and quantitative nature of the immunodeficiency. For example, extracellular pathogens, such as *Streptococcus pneumoniae* and *Haemophilus influenza*, only multiply outside phagocytic cells, producing disease when they resist phagocytosis. Facultative intracellular pathogens (e.g., *Mycobacterium tuberculosis*) are generally phagocytized, but resist intracellular killing. Thus, infections with extracellular or facultative intracellular organisms will be more frequent in individuals where impaired phagocytic mechanisms exist, such as neutropenia, or when humoral (i.e., antibody) deficiencies are present. Obligate intracellular pathogens, which include all viruses, cannot multiply unless they are within a host cell, and are more commonly observed in individuals with defects in cellular (T-cell) immunity.

Microbial agents associated with immunodeficiency disorders can also be classified into common, opportunistic, or latent pathogens. Common pathogens, such as influenza, occur in the general population at frequencies associated with their infectious nature (e.g., virulence, ease of transmission). The respiratory system is the most vulnerable target for common pathogens, as it is directly exposed to the external environment and has a large surface area, four times the combined total surface areas of the gastrointestinal tract and skin (Gardner, 2001). Upper respiratory infections occur in all age groups, but are most severe in the very young or very old because of age-related immunodeficiencies. Although influenza is responsible for more morbidity and mortality than any other infectious agent in recorded history (Patriarca, 1994), the low individual rates of common infections in the general population (only one or two episodes in an individual per year), combined with underreporting, make it difficult to detect changes in infection rates. While infections with common pathogens occur routinely in the healthy population, opportunistic infections are typically seen in individuals with more severe immunosuppression, such as AIDS patients, and cause disease in the general population at very low incidences. These microorganisms are commonly encountered in food, water, dust, or soil, and include certain protozoans, such as *Toxoplasma gondii*, which causes cerebral infections and intractable diarrhea, the fungi *Candida albicans* and *Pneumocystis* carinii, and bacteria in the *Mycobacterium avium* complex (MAC) (Morris and Potter, 1997). Other pathogenic microorganisms are responsible for latent infections. Cytomegalovirus (CMV), herpes simplex virus (HSV), and Epstein–Barr virus (EBV), all members of the herpes virus family, can remain in the tissue following primary infection for the duration of the host's life without causing disease. In healthy individuals, the immune system usually maintains viral latency, with cellular immunity playing a major role. When the immune response is compromised, viral replication can ensue and potentially cause severe complications or death. Preceding viral activation, a vigorous immune response to viral-specific antigens occurs in response to replication. As discussed later, changes in virus-specific immune response or activation of latent viruses have been observed in individuals with secondary immunodeficiency disorders, and may reflect mild to moderate immunosuppression.

Immunodeficiency is also associated with an increased incidence of certain virus-induced tumors, such as non-Hodgkin's lymphomas and skin tumors (Penn, 2000). In contrast to cancers of internal organs, such as the lung and liver, which are often induced by chemical carcinogens, virus-induced cancers are more immunogenic and, therefore, are more likely influenced by immunological factors. Examples of cancers that are common in immunosuppressed individuals include leukemia and lymphoproliferative disorders, as well as cancers of the skin, seen in transplant patients, and Kaposi's sarcoma and EBV-associated B-cell lymphomas, seen in AIDS patients, due to T-cell deficiency. Natural killer (NK) cells are also involved in resistance to neoplastic diseases, but more likely play a role in resisting the progression and metastatic spread of tumors once they develop, rather than preventing initiation (Herberman, 2001). Studies of individuals with NK cell deficiency states, most of which are associated with single gene mutations, have helped identify a role for NK cells in defense against human infectious disease (Orange, 2002). A common theme in NK cell deficiencies is susceptibility to herpes viruses, suggesting that unexplained severe herpes viral infections should raise the possibility of an NK cell deficit.

11.3 SOCIAL AND ECONOMIC IMPACT OF INFECTIOUS DISEASES AND THE RISK ASSESSMENT PROCESS

In many instances, it is necessary to include the social and economic consequences as part of the risk assessment/management processes. For immunotoxicology, this may involve, for example, estimating the social and economic impact of a change in infectious disease incidence/severity from background. A basic assumption in immunotoxicology is that at the population level, the incidence of infections will increase as immunocompetence decreases. Although the percentage of affected individuals may be small relative to the population, a significant number of individuals will nevertheless bear the costs of illness. Although the exact social or economic impacts of infectious diseases in the general population are not known, data from several sources indicate these to be significant and that even small changes in frequency will have a major impact. The impacts associated with mortality, and to a lesser extent morbidity, from common pathogens such as influenza and pneumonia, have been determined and can serve as a basis in the risk management process. Deaths have the most costly impact on society. In 2000, the age-adjusted death rate for influenza and pneumonia was 23.0 and 0.6 per 100,000, respectively, based on the tenth revision of the *International Statistical Classification of Diseases and Related Health Problems* (World Health Organization, 1992), coded J10–J18, and together these infections were ranked as the seventh leading cause of death in the U.S. for all ages (Anderson, 2002). In 2000, the mortality rates for all infants from influenza and pneumonia were 7.5 deaths per 100,000 live births, a decline from 8.4 in 1999 (Minino et al., 2002; Hoyert et al., 2001). In both of these years, this number was dominated by pneumonia. Other conditions secondarily related to these illnesses, such as disorders related to low birth weights, respiratory distress, or bacterial sepsis, accounted for higher infant deaths (Minino et al., 2002), including neonates (i.e., less than 28 days of age). However, influenza and pneumonia still rank seventh for postneonatal deaths (Anderson, 2002). For the group aged 65 years and over, chronic lower respiratory disease and influenza-pneumonia were the fourth and fifth ranked leading causes of death in 2000, respectively (Anderson, 2002). However, pneumonitis due to aspirating solids and liquids into the lung is becoming a more common cause of death among the elderly, and is now ranked fifteenth (Minino et al., 2002).

Economic impacts resulting from infectious diseases are captured by determining the number of deaths, hospitalizations, and outpatient or emergency room visits for specific illnesses, usually collected in national surveys, and applying formulas to convert these to dollars. Cost of illness methodology can handle, with some degree of confidence, the valuing of medical costs and productivity losses in an attempt to capture the burden of infectious disease mortality and morbidity. However, it should be noted that most estimates of this burden do not account for reduced functional abilities, losses from pain and suffering, or the cost to the individual, family member, or co-worker of psychological or emotional stress. There are many other fundamentally unobservable quantities, such as the value of output that is lost as a result of an employee having an infectious disease episode. Valuing lost workdays does not explain the entire productivity loss but provides a comparison indicator. A National Institutes of Health–sponsored effort used methods from previous disease-specific results to estimate costs and applied inflation factors for the time period under consideration to estimate the total cost of influenza and pneumonia using *International Classification of Diseases Clinical Modification* (ICD-9-CM) codes 480–487 (U.S. Department of Health and Human Services, 1991), in 2000 at $25.6 billion (Kirstein, 2000). This amount included $18.6 billion for medical costs and $7 billion for productivity losses. Leigh et al. (2003) focused on 14 occupational illnesses to determine annual medical costs of occupational illnesses in the U.S. Within this population, the estimate for pneumonia, codes 480–482 and 484 of the ICD-9-CM, which included only the 25- to 64-year-old group, was $24.7 million, with males accounting for $19.9 million of the total. Otitis media infection, the most common cause of hearing loss in children, occurs in 80% of children under 3 years of age, and is the major reason for doctor or emergency room visits in this age group. According to the Agency for Health Care Research and Quality, formerly the Agency for Health Care Policy and Research, the 1991 annual cost for treating 2 year olds for otitis media was $1 billion. However, estimates for 2000 place the figure at $5 billion, with $2.9 billion in direct costs and $2.1 billion in indirect costs (Kirstein, 2000). Langley et al. (1997) estimated the annual cost of respiratory syncytial virus (RSV) infection, using ICD-9-CM code 466.1, at $17 million in children younger than 4 years of age. The largest cost was associated with hospital services for the approximately 0.7% of infected children requiring admission. The average medical care expenditure for all children 2 years of age or younger was $22 per child.

11.4 ISSUES IN USING HUMAN DATA IN IMMUNOTOXICOLOGY RISK ASSESSMENT

There are many advantages of using human data over experimental animal studies in quantitative risk assessment, especially as it avoids the difficulties in interspecies extrapolation and provides data on lower doses that are of interest to public health policy makers (Hertz-Picciotto, 1995). Human studies offer realistic exposure scenarios, including multiple exposure routes, and include a much more diverse range of genetic backgrounds than experimental models. The limitations and challenges of human studies, however, can be considerable and differ depending on whether they represent controlled clinical trials or population-based observational studies. Clinical studies offer advantages in that exposure parameters of interest can often be controlled (e.g., chamber studies of inhaled toxicants, challenge infection with adenovirus), and outcomes can be prospectively monitored. However, there are also disadvantages as ethical considerations prevent human studies involving deliberate exposure to toxic chemicals. Furthermore, studies with extensive biological monitoring and functional immune tests are expensive, and exposures, as well as outcomes of interest, may be difficult to study in the available time frame, as study participants are not typically available for long-term exposures or extended follow-up. For the purpose of obtaining data for an immunotoxicologic risk assessment, clinical studies are particularly useful as they can provide data on frequency of infections or vaccine response under controlled conditions.

Other types of human studies employed in immunotoxicology are typically classified as observational or epidemiologic. Observational studies can be of varying size, and be cross-sectional (one point in time), retrospective, or prospective in nature, each design having advantages and disadvantages. The initial means of control in observational studies is introduced through the study design. The quality and validity of results can be greatly affected by the methods used to select the study sample, and the rigor with which exposures and outcomes are measured. In addition to high costs, observational studies can be challenging for many reasons, including potential confounding by host (age, gender, and lifestyle) and environmental (frequency of exposure to chemicals and infectious agents) factors. A secondary measure of control in observational studies involves the use of multivariable analysis techniques (e.g., regression modeling), providing there is sufficient sample size and information on potential confounders. Overall, well-designed epidemiologic studies (e.g., absence of selection bias, exposure, or outcome misclassification, and control of confounding) can contribute valuable information to the assessment of risk due to immunotoxic exposures.

Existing immunotoxicology studies in humans have generally been based on either fairly small sample sizes, often in individuals with transient high-level occupational exposures, or large groups with chronic low-level exposures. In some instances body burdens of chemicals have been determined, while in other studies exposure has relied on subject recall or rough estimates of the duration and intensity of exposure. Furthermore, in contrast to experimental animals, functional assessment is considerably more difficult in humans as it requires antigen challenge, which involves some risk to the individual. When undertaken, subjects have been provided commercial vaccines, such as hepatitis antigen (van Loveren et al., 2001; Weisglas-Kuperus et al., 2000; Yucesoy et al., 2001; Sleijffers et al., 2003). In this respect, the cellular and humoral immune response to vaccination is thought to be a sensitive indicator of immunosuppression (Glaser et al., 1993), and the vigor of the response an indicator of infectious disease susceptibility (van Loveren et al., 2001; Deseda-Tous et al., 1978). In most epidemiologic studies, testing in humans has been limited to blood collection where peripheral cell counts and differentials, serum immunoglobulin levels, and immunophenotyping are performed. While certainly of value, it is generally agreed that these are less sensitive indicators of immunocompetence, making it difficult to detect low to moderate levels of immunosuppression (Immunotoxicity Testing Committee, 1999).

11.5 IMMUNODEFICIENCY AND RELATIONSHIP TO INFECTIOUS DISEASE

11.5.1 Environmental Chemicals

The need to extend data obtained in experimental studies to humans has been recently reviewed (Tryphonas, 2001). Although a large number of human studies have evaluated immune system endpoints in occupationally and environmentally exposed cohorts, immune function and infectious outcomes generally have not been reported for the same cohort. Some of the more complete immunotoxicology studies have focused on persistent organochlorine compounds, formerly found in pesticides and industrial chemicals (e.g., polychlorinated biphenyls [PCBs]), in children following prenatal or postnatal exposure (via maternal diet and breast milk). Studies of accidentally exposed populations in Japan (Yusho) and China (Yu-Cheng) suggested an association of PCBs, their thermal breakdown products (quaterphenyls), and polychlorinated dibenzofurans with immune abnormalities and increased infections. Children born to exposed mothers between 1978 and 1987 in the Yu-Cheng study group had lower levels of serum IgA and IgM and a higher frequency of respiratory infections and otitis media compared to matched, unexposed controls (Lu and Wu, 1985; Nakanishi et al., 1985; Yu et al., 1998). Similar results have been observed in the Yusho study population (Nakanishi et al., 1985).

An association between PCBs and increased frequency of otitis media in children has also been described in other populations. A study of 343 children in the United States (Michigan), while not showing a general association between organochlorine levels and prevalence of infections, revealed a positive association between polychlorinated biphenyls (PCBs) and DDE (the primary metabolite of DDT) or PCBs and hexachlorobenzene with otitis media (Karmaus et al., 2001). In a study of Inuit infants in Arctic Quebec, Canada (Dewailly et al., 2000), the relative risk of recurrent episodes (at least three per year) of otitis media was higher in breastfed infants in the second and third highest percentile of organochlorine exposure, compared to the lowest. At 3 months of age, breastfed infants with higher exposure levels had lower numbers of white blood cells and lymphocytes, and lower serum IgA levels at ages 7 and 12 months compared to bottle-fed infants. In Dutch preschool children (Weisglas-Kuperus et al., 2000), PCB levels in breast milk (nonortho and coplanar PCBs) were also associated with increased recurrent otitis media and other symptoms of respiratory infection. In this sample, the body burden of PCBs at age 42 months was associated with a higher prevalence of recurrent otitis media and chicken pox. PCB body burden was not associated with differences in lymphocyte markers outside the normal range for age-matched children, although levels in breast milk and cord blood were positively correlated with lymphocyte counts and various T-cell subsets. While these findings linking otitis media with PCB exposure are consistent across a number of studies, it is not possible to determine whether immunotoxicity mediated this association or simply reflected parallel findings.

The immunotoxicity of pesticides following human exposure has been reviewed by several authors (Thomas et al., 1995; Voccia et al., 1999; Luebke, 2002; Vial et al., 1996). Although some studies have described associations among pesticide exposure, altered immune function, and increased rates of infection, sample sizes were generally small and, in some cases, the subjects were self-selected, based on symptoms rather than exposure. Furthermore, the frequency of infections was typically estimated by recall over several years, and immune function data were scarce. Not all studies suffer from these shortcomings. For example, a relatively large (n = 1600) and well-defined population living in and around Aberdeen, North Carolina, near a pesticide dump site (a priority Superfund site containing organochlorine pesticides, volatile organic compounds, and metals), was evaluated for immune function and frequency of viral infections. Compared to a neighboring community, residents of Aberdeen, ages 18 to 40, were found to have a higher incidence of herpes zoster (reactivated herpes infection causing shingles) (Arndt et al., 1999). In a substudy of 302 individuals, those living in Aberdeen had significantly higher age-adjusted levels of plasma DDE than those living in neighboring communities. Furthermore, higher levels of plasma DDE were related to lower lymphocyte responses to mitogens, and higher absolute lymphocyte counts and serum IgA levels (Vine et al., 2001). In a separate analysis, residents living nearer to the pesticide dump site had both a lower lymphocyte response to mitogen stimulation and a greater likelihood of having a lower percentage of CD16+ (NK) cells (< 8%, the lower limit of the normal reference range) (Vine et al., 2000). The association seen with reactivated herpes infection is plausible in light of these changes, given that NK cells play an important role in the generation of cytotoxic T-cells required to help control viral infections (Orange, 2002).

11.5.2 Chronic Stress

Chronic psychological factors (i.e., stressors), such as separation and divorce, caregiving for Alzheimer's patients or bereavement, produce low to moderate degrees of immunosuppression, and increase infectious disease incidences (Biondi and Zannino, 1997; Cohen, 1995; Yang and Glaser, 2000; Kiecolt-Glaser et al., 2002). In one of the few examples of human challenge studies, 394 healthy individuals were assessed for psychological stress and subsequently administered nasal droplets containing RSV or coronavirus (Cohen et al., 1991). The rate of respiratory infections ($p < 0.005$) and clinical colds ($p < 0.02$), as determined by virus-specific antibody levels and viral isolation, increased in a dose-responsive manner with increasing degrees of psychological stress. Although usually conducted in small cohorts, immune testing in chronically stressed individuals has also provided insights into the relationship between mild to moderate immunosuppression and disease (Kiecolt-Glaser et al., 1986, 1987). In chronic stress populations showing an increased rate of infections, total circulating T-cell numbers can be reduced to as much as 20% below mean control values, while the number of circulating B cells remains unaffected. Furthermore, CD4:CD8 ratios can be reduced as much as 40%, and NK cell activity by 10 to 25% below mean control values. Measurement of mitogen-stimulated T-lymphocyte proliferation, although not generally considered a sensitive indicator of immune function, was reduced by approximately 10% from control values in the stressed population. However, as with a number of immunotoxicology studies, these changes were still within the range of normal values.

Associations have also been observed between chronic stress and reactivation of latent viruses, such as CMV, HSV-1, or EBV, as measured either by clinical disease or elevations in specific antibody titers (Glaser et al., 1993; Biondi and Zannino, 1997; Cohen, 1995; Yang and Glaser, 2000; Kasl et al., 1979; Esterling et al., 1993; Glaser et al., 1987). Elevations in antiviral antibody titer (i.e., seroconversion), a reflection of viral activation and replication, precede disease onset, although only about 20% of those with elevated titers actually develop clinical disease. Studies have also been conducted to examine associations between psychological stress and the immune response using hepatitis B, influenza virus, or pneumococcal vaccine responses (Kiecolt-Glaser et al., 2002). In studies of students under defined academic stress, the ability to seroconvert following first and second immunizations with hepatitis B vaccine was inversely associated with the outcome of tests that measure stress levels. In studies involving influenza vaccinations, Alzheimer's disease caregivers responded less often to vaccination, with only 12 (38%) experiencing a fourfold increase in antibody titer (the minimum response considered to be protective) following immunization, compared to 21 controls (66%) (Kiecolt-Glaser et al., 1996). As shown in Figure 11.2, the effects on pneumococcal vaccine responses in caregivers were even more striking, where a significantly blunted antibody response occurred in current caregivers compared to controls over the 6-month period following immunization (F[5.82,142.46] = 2.56, $p < 0.03$) (Glaser et al., 2000).

11.5.3 Hematopoietic Stem Cell Transplantation

Hematopoietic stem cell transplantation, which came into general practice in the 1980s, is employed in the treatment of certain hematological malignancies, aplastic anemia, and inborn genetic errors of cells originating in hematopoietic stem cells. Following cell grafting, immunodeficiency can persist for well over a year due to pregrafting radiation treatment. This is manifested as decreased antibody responses, decreased delayed hypersensitivity responses, low CD4+ cell numbers, and low serum IgG2, IgG4, and IgA levels (Ochs et al., 1995). Thus, prospective studies can help identify quantitative relationships between immune function and disease as the immune system recovers. The incidence of infections exceeds 80% during the first 2 years post-engraftment with 50% of the patients having three or more infections. Opportunistic infections predominate, with fungi being the most common type of organism causing disease, followed by bacteria and viruses (Ochs et al., 1995; Atkinson, 2000). Incidence data for upper-respiratory infections are generally unavailable for these patients, since these infections are seldom monitored in allogeneic bone marrow recipients. Although infections that occur in the first month following transplant are most likely due to severe deficiencies in granulocytes, later infections appear to be due to deficiencies in CD4+ T cells and B cells.

In a prospective study involving 108 transplant patients that were followed between days 100 and 365 post-engraftment, decreases in B, CD4+, and CD8+ lymphocytes, and total mononuclear cells were associated with infectious disease incidence ($p < 0.05$) (Storek et al., 2000). A smaller,

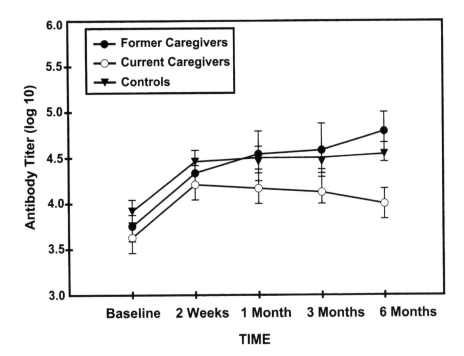

Figure 11.2   Pneumococcal vaccine responses in elderly caregivers, shown as antibody titer over the 6-month period following immunization. Controls are age-matched noncaregivers. From Glaser, R., et al., *Psychosom. Med.*, 62, 804, 2000. With permission.

but more detailed study by Storek et al. (1997) that evaluated 29 patients for 180 days preceding the 1-year post-transplant exam, showed a highly significant inverse correlation between activated CD4+ T-cell counts and total infection score (Figure 11.3), which included frequency and severity ($p = 0.005$ in univariate analysis), but not with CD8+ T-cell numbers, B-cell numbers, serum immunoglobulin levels, or delayed hypersensitivity responses. In comparing the efficacy of allogeneic marrow transplantation to blood stem cell transplantation (Storek et al., 2000), it was demonstrated that a 1.7-fold lower rate of infections in blood stem cell transplants corresponded to about a fourfold higher CD45RA^{high} CD4+ T-cell counts and about a twofold higher count for CD45RA^{low} CD4+ T cells. In studies conducted by Small et al. (1999), which monitored immune cell recovery following bone marrow cell transplantation, the incidence of infection also was inversely correlated with CD4+ cell counts. However, only opportunistic infections were monitored, and were almost exclusively present in patients considered severely immunosuppressed (i.e., with CD4+ T-cell counts of < 200 cells/mm^3). The relationship between CD4+ cell numbers and respiratory virus infections was examined in a small group of T-cell–depleted (using anti-CD52 antibody treatment), stem cell recipients over a 3- to 6-month period following transplantation (Chakrabarti et al., 2001). The relationship between CD4+ T-cell counts and respiratory virus infection was relatively linear; however, the population studied was small and CD4+ T cells in the treated group did not progress above 180 cells/mm^3, compared to normal values, which ranged between 700 to 1100 cells/mm^3.

11.5.4   Organ Transplants

Studies in renal organ transplant patients have also provided insights into the long-term consequences of moderate immunosuppression. While immunosuppressive therapies have greatly improved over the past 40 years, transplant patients are still predisposed to high rates of malignancies and infections. Infection rates range between 65 to 70% during the first 6 months post-transplantation, with CMV representing anywhere from 18 to 67% of the reported infections (Sia

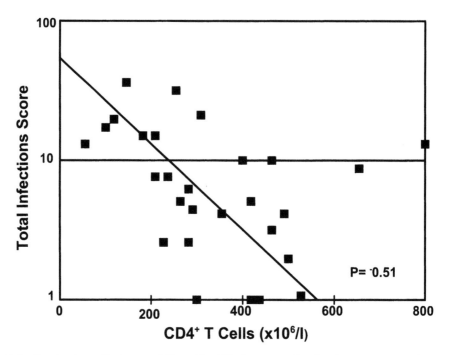

Figure 11.3 Twenty-nine patients were followed for 180 days preceding the 1-year post-transplant exam for CD4[+] T-cell counts, and total infection score, which includes frequency and severity. From Storek, J., et al., *Am. J. Hematol.*, 54, 131, 1997. With permission.

and Paya, 1998). Increased skin cancers have also been noted in patients on long-term immuno-suppressive therapy. For example, the risk of developing skin tumors following renal transplantation is 10% after 10 years and 40% after 20 years, while the incidence of squamous and basal cell carcinomas is tenfold and 250-fold higher, respectively, than in the general population (Hartevelt et al., 1990). Generally, the initial immunosuppressive therapy for renal transplant consists of a combination cyclosporin (CsA), azathioprine, and steroid cocktail. In examining 478 renal trans-plant patients, it was shown that the risk of lymphomas during the first 6 months post-transplantation increased proportionally with the intensity of immunosuppressive therapy (Jamil et al., 1999). The risk of infections was increased 1.5-fold after treatment with steroids and almost threefold after treatment with steroids plus antibodies to deplete CD3[+] T cells. As a result of the surgical procedure, urinary tract infections are commonly observed in all renal transplants, while severe bacterial infections (pneumonia and septicemia) and systemic/invasive fungal infections were almost exclu-sively associated with the most immunosuppressed group. A high incidence of anti-CMV antibodies occurred in all three treatment groups with 9% seroconverting compared to 29% in those patients that also received steroids and 53% in the group that received steroids plus CD3[+] T-cell depletion. Wieneke et al. (1996) also examining renal transplant patients, noted that reduced IgG1 subclass levels and CD4 T-cell counts were the best predictors for infections (frequency of infections increased from 9% in patients with normal values to 38% with lower values). In a small cohort, Clark et al. (1993) noted a reduction in the number of serious viral infections occurred ($p < 0.04$) in transplant patients when the level of CD3[+] lymphocytes were maintained above 500 cells/mm^3.

11.6 EXPERIMENTAL ANIMAL MODELS

Immunotoxicology data most often available for use in the risk assessment process originate from experimental animal studies. Although animal models provide an opportunity to establish more

Table 11.1 **Commonly Employed Experimental Disease Resistance Models**

Challenge Agent	Endpoint Measured
Listeria monocytogenes	Liver CFU[a], spleen CFU, morbidity
Streptococcus pneumoniae	Morbidity
Plasmodium yoelli	Parasitemia
Influenza virus	Morbidity, viral titer/tissue burden
Cytomegalovirus	Morbidity, viral titer/tissue burden
Trichinella spiralis	Muscle larvae, parasite numbers
PYB6 sarcoma	Tumor incidence (subcutaneous)
B16F10 melanoma	Tumor burden (lung nodules)

Note: For details see Burleson, G.R., *Immunopharmacology*, 48, 315, 2000; van Loveren, H., et al., in *Methods in Immunotoxicology*, vol. 2, Burleson, G.R., Dean, J.H., and Munson, A.E., Eds., Wiley-Liss, New York, 1995, p. 243; Bradley, S.G., in *Methods in Immunotoxicology*, vol. 2, Burleson, G.R., Dean, J.H., and Munson, A.E., Eds., Wiley-Liss, New York, 1995a, p. 135; and Selgrade, M.K., *Toxicology*, 133, 59, 1999.

[a] Each bacterial colony growing on artificial culture medium is assumed to arise from a single organism. CFU values therefore reflect the number of viable organisms recovered.

CFU, colony forming unit.

reliable exposure estimates and conduct more informative immune tests than human studies, the accuracy level that can be achieved using such data in extrapolating to humans is often a matter of debate. In immunotoxicology testing, a set of tests, usually referred to as "host resistance assays," has evolved in which groups of experimental rodents are challenged with either an infectious agent or transplantable tumor at a challenge level sufficient to produce either a low incidence or minimal infectivity in the control group (Table 11.1). As the endpoints in these tests have evolved from relatively nonspecific (e.g., animal morbidity and mortality) to continuous measures, such as number or size of tumor foci, viral titers, or bacterial cell counts, the sensitivity of these models has increased, although they are still limited by the number of animals that can be realistically devoted to a study. While there have been considerable efforts to establish interlaboratory variability and the robustness of tests to measure specific immunological endpoints, such as antibody responses (Temple et al., 1993), histopathology (International Collaborative Immunotoxicity Study, 1998; Kuper et al., 2000), quantitation of cell-surface markers by flow cytometry (Zenger et al., 1998; Burchiel et al., 1997), and cytokine production (Langezaal et al., 2002; Hermann et al., 2003), there have been only two programs that have evaluated the sensitivity and predictive value of individual measures of immune outcomes with host resistance tests. In 1979, under the auspices of the U.S. National Toxicology Program (NTP), a panel of experts gathered to prioritize a list of immunological and host resistance assays that would be suitable for use in mouse studies, and a formal validation was initiated (Luster et al., 1988). A smaller effort was undertaken at the National Institute of Public Health and the Environment, which focused on the rat and was based on the Organization of Economic Cooperation and Development Guideline 407 (van Loveren and Vos, 1989; Vos, 1977, 1980). In both programs, host resistance tests were usually considered in a second or third testing level (i.e., tier) of evaluation, and were only performed when there were indications of alterations in a previous tier. Data subsequently obtained from these validation programs indicated, for the most part, that host resistance assays were highly correlated with immune tests, but were unlikely to detect subtle immunosuppression due to relative differences in sensitivity in the test models.

A number of studies have addressed individual relationships between specific immune responses and host resistance in rodent studies. While it is rare for a single component of the immune system to be solely responsible for resistance to a specific infectious agent, certain immune measures showed a significant correlation with the outcome of a host resistance assay. For example, reduction in NK cell activity correlated with increased susceptibility to challenge with PYB6 sarcoma cells, B16F10 melanoma cells, and murine CMV (Luster et al., 1988, 1993; Selgrade et al., 1992). Suppression of cell-mediated immunity, complement deficiency, and depressed macrophage and neutrophil function have been associated with decreased resistance to *Listeria* monocytogenes (Luster et al., 1988; Petit, 1980; Bradley, 1995b). Clearance of parasitic infections, such as *Plasmodium yoelii* and *Trichinella spiralis*, which have both a cellular and humoral component, are associated with depression of both arms of the immune system (van Loveren et al., 1995; Luebke, 1995). In the comprehensive studies conducted by the NTP, concordance between individual immune tests was compared to host resistance tests and found to range from relatively good (e.g., antibody plaque-forming cell assay, 73%; NK cell activity, 73%; and delayed-type hypersensitivity response, 82%) to poor (e.g., lymphoproliferative response to liposaccharide is < 50%). These studies have recently been reviewed in more detail (Germolec, 2004).

Deletion, or "functional blocking," of specific immune components in experimental animals has also been used to elucidate the relative contributions of specific molecules, signaling pathways and cells to disease resistance (Hickman-Davis, 2001). This has been achieved via targeted gene disruption resulting in animals deficient in a specific cell population or soluble mediator that contributes to host defense (e.g., CD4[+] T-cell knockouts), treatment of normal animals with selective toxic agents (e.g., the use of gadolinium chloride to block macrophage function), or administration of neutralizing antibodies against critical cell-specific surface receptors. A study by Wilson et al. (2001) was specifically designed to determine the magnitude of NK cell suppression, thought to be important in preventing metastasis, which would translate into altered resistance in three disease models. These studies were conducted following depletion of NK cells with an antibody to the cell-surface molecule, asialo GM1, using a treatment regimen that did not alter other standard immune function tests. These authors demonstrated that at low levels of tumor challenge, a reduction of approximately 50% or more in NK cell activity was required before significant effects on resistance to NK-sensitive tumors were observed. These studies also demonstrated that the level of suppression needed to alter host resistance was related to the challenge level of the tumor cells. Conversely, studies that have used monoclonal antibodies to effectively deplete CD4[+] and CD8[+] T-lymphocytes have found little evidence of altered resistance to challenge with PYB6 sarcoma cells, a model that was thought to be dependent on cell-mediated immunity (Weaver et al., 2002).

Studies designed to address the contribution of a single immune parameter in host resistance have obvious limitations. In studies designed to specifically address these limitations, Keil et al. (2001) demonstrated that monitoring several immunological parameters concurrently provides information that might not be evident from studies using single tests. Using the prototypical immunosuppressive agent, dexamethasone, these authors demonstrated that contrary to what might be expected based on the compound's suppressive effects on cytokine production, T-cell function and NK cell activity, relatively high levels of dexamethasone were required to decrease resistance to *Listeria monocytogenes*. At doses that suppressed many immune parameters, an increase in neutrophil numbers and nitrite production by peritoneal macrophages was observed. It was suggested that at these doses of dexamethasone, the significant increase in the number of peripheral neutrophils, in conjunction with increased production of nitric oxide, compensated for the decrements in other immune parameters so that overall resistance to the pathogen was not compromised (Keil et al., 2001). Herzyk et al. (1997) have developed a testing paradigm that evaluates immune function within the context of resistance to a specific infection. Following infection with *Candida albicans*, a four-parameter model was used that included survival; numbers of organisms recoverable from the spleen; numbers recoverable from muscle and antibody titers, which allows evaluation of immune responses; and host resistance in the same animal. This approach has proved successful in identifying both immunosuppressive and immunostimulatory compounds, but has yet to undergo validation studies, and is not widely used outside of the pharmaceutical industry.

Unlike human exposures, variables such as virulence and dose of the infectious agent that impact the ability to clear the challenge organism generally remain a constant in laboratory investigations. To address this issue in terms of implications in risk assessment, Luster et al. (1993) showed, using the PYB6 sarcoma cell model, that even immunologically normal animals provided a sufficient number of tumor cells to develop a high frequency of transplanted PYB6 tumors and the number of injected tumor cells required to produce a tumor decreased proportionally to the degree of immunosuppression. The implication of these data in terms of risk assessment would be that, given all other factors are constant, the incidence of disease would increase in a linear fashion as the level of immunosuppression increases. However, the slope of this response curve would vary depending on a number of factors such as infectivity (e.g., virulence) of the specific infectious agent, with the slope increasing as a function of increasing infectivity.

11.7 CONCLUSIONS

Adequate clinical data (i.e., exposure level and disease incidence) are rarely available to accurately determine safe exposure levels to immunotoxic agent, and thus, results from experimental animal models or human biomarkers studies are often employed. Hence, it is important that a scientifically sound framework be established that allows for the accurate and quantitative interpretation of experimental or biomarker data in the risk assessment process. For immunotoxicology data, this may require, for example, development of models to equate changes in leukocyte counts, $CD4^+$ T-cell numbers, and serum immunoglobulin levels that can be readily performed in human populations, to changes in the incidence or severity of infectious diseases. Although experimental animal models provide an opportunity to perform more informative immune tests and establish reliable exposure estimates, extrapolating these findings across species also introduces considerable uncertainty. While this review does not provide a specific framework to perform these extrapolations, it does present background information on qualitative and quantitative relationships between immune parameters and disease that would be an integral part of such an effort. The following general conclusions can be surmised:

1. The major clinical, or at least most readily discernible, consequence of mild to moderate chronic immunosuppression is an increase in the incidence of infectious diseases. Only a few studies have addressed infectious disease severity or neoplastic diseases. In addition to immunosuppression, many nonimmune factors can affect infectious disease incidences, and should be considered in data interpretation. This is particularly evident in infection with common pathogens, such as influenza and pneumonia. The microorganism responsible for the infection is often dependent on the specific arm of the immune system that is affected, and such information can be useful in observational studies. Thus, increased infections with obligate intracellular pathogens, such as viruses, will most likely occur from suppression of cell-mediated immunity, while defects in phagocytic activity, such as neutropenia, will more likely increase susceptibility to facultative intracellular or extracellular microbes.

2. Increases in infectious disease incidence following immunosuppression can be caused by common pathogens, opportunistic microbes, or activation of latent viruses (most often from the herpes family). For reasons described above, the ability to detect changes in the frequency of infections from common pathogens (e.g., increased respiratory infections from influenza) has proved difficult and may require well-controlled, large population studies that are closely monitored. Increases in the incidence of infections due to opportunistic organisms have been observed in immunosuppressed individuals. However, for the most part, these infections are present at a high incidence only in individuals with severe immunodeficiency, such as AIDS, in which $CD4^+$ T-cell numbers are typically reduced by greater than 50% from control values. Increased incidences of latent virus reactivation (e.g., HSV infection) have been commonly observed in populations that are chronically immunosuppressed and may be more closely associated with mild to moderate immunosuppression.

3. The major gap in clarifying the shape of the dose–response curve (i.e., between immune response and disease) is a lack of large-scale epidemiological studies in populations with mild to moderate immunodeficiency that have been monitored simultaneously for immune system parameters and

clinical disease. Assessment of immunocompetence disease relationships in large numbers of patients with AIDS-defining illnesses, such as *Pnuemocyctis carinii* pneumonia (PCP), CMV, and *Mycobaterium avium* complex (MAC) by the Multicohort AIDS centers, for example, are of limited value since participation was limited to individuals with CD4+ T-cell counts of $< 500 \times 10^6$L (Margolick et al., 1998; Pauli and Kopferschmitt-Kubler, 1991; Amornkul et al., 1999). Interestingly, the 5-year cumulative probabilities for AIDS and infectious disease deaths in the latter stages show a relatively linear response from 0 to 76% occurrence in patients with CD4+ cell counts between 500×10^6L and 200×10^6L, respectively (Vlahov et al., 1998). Furthermore, in one study where 864 patients failed to meet the study criteria (CD4+ cell counts $> 500 \times 10^6$L), 50 developed PCP, 4 developed CMV, and 3 developed MAC, indicating that opportunistic infection can occur in less immunosuppressed individuals but at a lower incidence (Lyles et al., 1999). Clinical studies in patients with severe immunodeficiency tend to show a threshold relationship between the incidences of opportunistic infections and CD4+ T-cell counts ($< 50\%$ below normal values). However, threshold relationships are not evident in the limited studies of patients with less severe immunodeficiency, such as chronic stress or stem cell transplant, as infection or activation of latent viruses can clearly occur in populations with small changes in immune parameters. Available clinical data are insufficient to determine whether the relationship between immunosuppression and increases in infectious disease follows a linear or threshold relationship in humans; experimental animal studies support a linear relationship when multiple immune parameters are examined. However, threshold relationships are observed when examining single parameters (e.g., NK cell activity).

ACKNOWLEDGMENTS

This review was prepared in conjunction with the Immunotoxicology Workgroup sponsored by the Environmental Protection Agency (EPA) Office of Research and Development (ORD) (National Center for Environmental Assessment [NCEA] and Health Effects Research Laboratories), EPA Office of Children's Health Protection, National Institute of Environmental Health Sciences (National Toxicology Program) and National Institute for Occupational Safety and Health (Health Effects Laboratory Division). Members of the workgroup not included as authors are Drs. David Chen (EPA/OCPH), Marquea King (EPA/ORD/NCEA) and Yung Yang (EPA, OPPTS). Special thanks to Dr. Bob Sonawane (EPA/ORD/NCEA) for helping to organize this effort. This review has been modified from a recent review by the same authors entitled "Associating changes in the immune system with clinical diseases for interpretation in risk assessment," in *Current Protocols in Toxicology*, Maines, M., Costa, L., Reed, D., et al., Eds., John Wiley & Sons, New York, 2004, in press.

REFERENCES

Amornkul, P.N., et al., Clinical disease associated with HIV-1 subtype B' and E infection among 2104 patients in Thailand, *AIDS*, 13, 1963, 1999.

Anderson, R.N., Deaths: leading causes for 2000, *Natl. Vital Stat. Rep.,* 50, 1, 2002.

Arndt, V., Vine, M.F., and Weigle, K., Environmental chemical exposures and risk of herpes zoster, *Environ. Health Perspect.*, 107, 835, 1999.

Atkinson, K., *Clinical Bone Marrow and Blood Stem Cell Transplantation*, 2nd ed., Cambridge University Press, Cambridge, MA, 2000.

Biondi, M., and Zannino, L.G., Psychological stress, neuroimmunomodulation, and susceptibility to infectious diseases in animals and man: a review, *Psychother. Psychosom.*, 66, 3, 1997.

Bradley, S.G., Introduction to animal models in immunotoxicology: host resistance, in *Methods in Immunotoxicology*, vol. 2, Burleson, G.R., Dean, J.H., and Munson, A.E., Eds., Wiley-Liss, New York, 1995a, p. 135.

Bradley, S.G., Listeria host resistance model, in *Methods in Immunotoxicology*, vol. 2, Burleson, G.R., Dean, J.H., and Munson, A.E., Eds., Wiley-Liss, New York, 1995b, p. 169.

Burchiel, S.W., et al., Assessment of immunotoxicity by multiparameter flow cytometry, *Fundam. Appl. Toxicol.*, 38, 38, 1997.

Burleson, G.R., Models of respiratory immunotoxicology and host resistance, *Immunopharmacology*, 48, 315, 2000.

Chakrabarti, S., et al., Respiratory virus infections in adult T cell-depleted transplant recipients: the role of cellular immunity, *Transplantation*, 72, 1460, 2001.

Clark, K.R., et al., Administration of ATG according to the absolute T lymphocyte count during therapy for steroid-resistant rejection, *Transpl. Int.*, 6, 18, 1993.

Cohen, S., Psychological stress and susceptibility to upper respiratory infections, *Am. J. Respir. Crit. Care Med.*, 152, S53, 1995.

Cohen, S., Tyrrell, D.A., and Smith, A.P., Psychological stress and susceptibility to the common cold, *N. Engl. J. Med.*, 325, 606, 1991.

Deseda-Tous, J., et al., Measles revaccination: persistence and degree of antibody titer by type of immune response, *Am. J. Dis. Child.*, 132, 287, 1978.

Dewailly, E., et al., Susceptibility to infections and immune status in Inuit infants exposed to organochlorines, *Environ. Health Perspect.*, 108, 205, 2000.

Esterling, B.A., et al., Defensiveness, trait anxiety, and Epstein–Barr viral capsid antigen antibody titers in healthy college students, *Health Psychol.*, 12, 132, 1993.

Gardner, D.E., Bioaerosols and disease, in *Patty's Industrial Hygiene and Toxicology*, vol. 1, 5th ed., Bingham, E., Cohrssen, B., and Powell, C.H., Eds., Wiley Publishing, New York, 2001, p. 679.

Germolec, D.R., Sensitivity and predictivity in immunotoxicity testing: Immune endpoints and disease resistance, *Toxcol. Lett.*, 149, 109–114, 2004.

Glaser, R., et al., Stress-related immune suppression: health implications, *Brain. Behav. Immun.*, 1, 7, 1987.

Glaser, R., et al., Stress and the memory T-cell response to the Epstein–Barr virus in healthy medical students, *Health Psychol.*, 12, 435, 1993.

Glaser, R., et al., Chronic stress modulates the immune response to a pneumococcal pneumonia vaccine, *Psychosom. Med.*, 62, 804, 2000.

Halloran, P.F., Rethinking immunosuppression in terms of the redundant and nonredundant steps in the immune response, *Transplant. Proc.*, 28, 11, 1996.

Hartevelt, M.M., et al., Incidence of skin cancer after renal transplantation in The Netherlands, *Transplantation*, 49, 506, 1990.

Herberman, R.B., Immunotherapy, in *Clinical Oncology*, Lenhard Jr., R.E., Osteen, R.T., and Gansler, T., Eds., American Cancer Society, Atlanta, 2001, p. 215.

Hermann, C., et al., A model of human whole blood lymphokine release for in vitro and ex vivo use, *J. Immunol. Methods*, 275, 69, 2003.

Hertz-Picciotto, I., Epidemiology and quantitative risk assessment: a bridge from science to policy, *Am. J. Public Health*, 85, 484, 1995.

Herzyk, D.J., et al., Single-organism model of host defense against infection: a novel immunotoxicologic approach to evaluate immunomodulatory drugs, *Toxicol. Pathol.*, 25, 351, 1997.

Hickman-Davis, J.M., Implications of mouse genotype for phenotype, *News Physiol. Sci.*, 16, 19, 2001.

House, R.V., A survey of immunotoxicology regulatory guidance, in *Encyclopedia of Immunotoxicology*, Vohr, H.-W., Ed., Springer, Heidelberg, in press, 2004.

Hoyert, D.L., et al., Deaths: final data for 1999, *Natl. Vital Stat. Rep.*, 49, 1, 2001.

Immunotoxicity Testing Committee, Application of flow cytometry to immunotoxicity testing: summary of a workshop report, International Life Sciences Institute/Health and Environmental Sciences Institute, Washington, DC, 1999.

International Collaborative Immunotoxicity Study, Report of validation study of assessment of direct immunotoxicity in the rat, *Toxicology*, 125, 183, 1998.

Jamil, B., et al., Impact of acute rejection therapy on infections and malignancies in renal transplant recipients, *Transplantation*, 68, 1597, 1999.

Karmaus, W., Kuehr, J., and Kruse, H., Infections and atopic disorders in childhood and organochlorine exposure, *Arch. Environ. Health*, 56, 485, 2001.

Kasl, S.V., Evans, A.S., and Niederman, J.C., Psychosocial risk factors in the development of infectious mononucleosis, *Psychosom. Med.*, 41, 445, 1979.

Keil, D., Luebke, R.W., and Pruett, S.B., Quantifying the relationship between multiple immunological parameters and host resistance: probing the limits of reductionism, *J. Immunol.*, 167, 4543, 2001.

Kiecolt-Glaser, J.K., et al., Chronic stress alters the immune response to influenza virus vaccine in older adults, *Proc. Natl. Acad. Sci. U.S.A.*, 93, 3043, 1996.

Kiecolt-Glaser, J.K., et al., Psychoneuroimmunology: psychological influences on immune function and health, *J. Consult. Clin. Psychol.*, 70, 537, 2002.

Kiecolt-Glaser, J.K., et al., Modulation of cellular immunity in medical students, *J. Behav. Med.*, 9, 5, 1986.

Kiecolt-Glaser, J.K., et al., Chronic stress and immunity in family caregivers of Alzheimer's disease victims, *Psychosom. Med.*, 49, 523, 1987.

Kirstein, R., Disease-Specific Estimates of Direct and Indirect Costs of Illness and NIH Support: Fiscal Year 2000 Update, U.S. Department of Health and Human Services, Washington, DC, 2000.

Kuper, C.F., et al., Histopathologic approaches to detect changes indicative of immunotoxicity, *Toxicol. Pathol.*, 28, 454, 2000.

Langezaal, I., et al., Evaluation and prevalidation of an immunotoxicity test based on human whole-blood cytokine release, *Alternatives Lab Anim.*, 30, 581, 2002.

Langley, J.M., et al., Economic evaluation of respiratory syncytial virus infection in Canadian children: a Pediatric Investigators Collaborative Network on Infections in Canada (PICNIC) study, *J. Pediatr.*, 131, 113, 1997.

Leigh, J.P., Yasmeen, S., and Miller, T.R., Medical costs of fourteen occupational illnesses in the United States in 1999, *Scand. J. Work. Environ. Health*, 29, 304, 2003.

Lu, Y.C., and Wu, Y.C., Clinical findings and immunological abnormalities in Yu-Cheng patients, *Environ. Health Perspect.*, 59, 17, 1985.

Luebke, R.W., Pesticide-induced immunotoxicity: Are humans at risk? *Hum. Ecol. Risk Assessment*, 8, 293, 2002.

Luebke, R.W., Assessment of host resistance to infection with rodent malaria, in *Methods in Immunotoxicology*, vol. 2, Burleson, G.R., Dean, J.H., and Munson, A.E., Eds., Wiley-Liss, New York, 1995, p. 221.

Luster, M.I., et al., Development of a testing battery to assess chemical-induced immunotoxicity: National Toxicology Program's guidelines for immunotoxicity evaluation in mice, *Fundam. Appl. Toxicol.*, 10, 2, 1988.

Luster, M.I., et al., Risk assessment in immunotoxicology. II. Relationships between immune and host resistance tests, *Fundam. Appl. Toxicol.*, 21, 71, 1993.

Lyles, R.H., et al., Prognostic value of plasma HIV RNA in the natural history of Pneumocystis carinii pneumonia, cytomegalovirus and Mycobacterium avium complex: Multicenter AIDS Cohort Study, *AIDS*, 13, 341, 1999.

Margolick, J.B., et al., Decline in total T cell count is associated with onset of AIDS, independent of CD4(+) lymphocyte count: implications for AIDS pathogenesis, *Clin. Immunol. Immunopathol.*, 88, 256, 1998.

Meltzer, M.I., Cox, N.J., and Fukuda, K., The economic impact of pandemic influenza in the United States: priorities for intervention, *Emerg. Infect. Dis.*, 5, 659, 1999.

Minino, A.M., et al., Deaths: final data for 2000, *Natl. Vital Stat. Rep.*, 50, 1, 2002.

Morris, J.G., Jr., and Potter, M., Emergence of new pathogens as a function of changes in host susceptibility, *Emerg. Infect. Dis.*, 3, 435, 1997.

Nakanishi, Y., et al., Respiratory involvement and immune status in Yusho patients, *Environ. Health Perspect.*, 59, 31, 1985.

Ochs, L., et al., Late infections after allogeneic bone marrow transplantations: comparison of incidence in related and unrelated donor transplant recipients, *Blood*, 86, 3979, 1995.

Orange, J.S., Human natural killer cell deficiencies and susceptibility to infection, *Microbes Infect.*, 4, 1545, 2002.

Patriarca, P.A., A randomized controlled trial of influenza vaccine in the elderly: scientific scrutiny and ethical responsibility, *JAMA*, 272, 1700, 1994.

Pauli, G., and Kopferschmitt-Kubler, M.C., Isocyanates and asthma, in Progress in Allergy and Clinical Immunology, Proceedings of the 14th International Congress for Allergy and Clinical Immunology, 1991, p. 152.

Penn, I., Post-transplant malignancy: the role of immunosuppression, *Drug Saf.*, 23, 101, 2000.

Petit, J.C., Resistance to listeriosis in mice that are deficient in the fifth component of complement, *Infect. Immun.*, 27, 61, 1980.

Selgrade, M.K., Use of immunotoxicity data in health risk assessments: uncertainties and research to improve the process, *Toxicology*, 133, 59, 1999.

Selgrade, M.K., Daniels, M.J., and Dean, J.H., Correlation between chemical suppression of natural killer cell activity in mice and susceptibility to cytomegalovirus: rationale for applying murine cytomegalovirus as a host resistance model and for interpreting immunotoxicity testing in terms of risk of disease, *J. Toxicol. Environ. Health*, 37, 123, 1992.

Sia, I.G., and Paya, C.V., Infectious complications following renal transplantation, *Surg. Clin. North Am.*, 78, 95, 1998.

Sleijffers, A., et al., Cytokine polymorphisms play a role in susceptibility to ultraviolet B-induced modulation of immune responses after hepatitis B vaccination, *J. Immunol.*, 170, 3423, 2003.

Small, T.N., et al., Comparison of immune reconstitution after unrelated and related T-cell–depleted bone marrow transplantation: effect of patient age and donor leukocyte infusions, *Blood*, 93, 467, 1999.

Storek, J., et al., Infectious morbidity in long-term survivors of allogeneic marrow transplantation is associated with low CD4 T cell counts, *Am. J. Hematol.*, 54, 131, 1997.

Storek, J., et al., Low B-cell and monocyte counts on day 80 are associated with high infection rates between days 100 and 365 after allogeneic marrow transplantation, *Blood*, 96, 3290, 2000.

Temple, L., et al., Comparison of ELISA and plaque-forming cell assays for measuring the humoral immune response to SRBC in rats and mice treated with benzo[a]pyrene or cyclophosphamide, *Fundam. Appl. Toxicol.*, 21, 412, 1993.

Thomas, P.S., Yates, D.H., and Barnes, P.J., Tumor necrosis factor-α increases airway responsiveness and sputum neutrophilia in normal human subjects, *Am. J. Respir. Crit. Care Med.*, 152, 76, 1995.

Tryphonas, H., Approaches to detecting immunotoxic effects of environmental contaminants in humans, *Environ. Health Perspect.*, 109 Suppl. 6, 877, 2001.

U.S. Department of Health and Human Services, *International Classification of Diseases Clinical Modification* (IDC-9-CM), vol. 1, 9th rev., 4th ed., U.S. Department of Health and Human Services, Washington, DC, 1991 (PHS–91–1260).

van Loveren, H., et al., Vaccine-induced antibody responses as parameters of the influence of endogenous and environmental factors, *Environ. Health Perspect.*, 109, 757, 2001.

van Loveren, H., Luebke, R.W., and Vos, J.G., Assessment of immunotoxicity with the parasitic infection model Trichinella spiralis, in *Methods in Immunotoxicology*, vol. 2, Burleson, G.R., Dean, J.H., and Munson, A.E., Eds., Wiley-Liss, New York, 1995, p. 243.

van Loveren, H., and Vos, J.G., Immunotoxicological considerations: a practical approach to immunotoxicity testing in the rat, in *Advances in Applied Toxicology*, Dayan, A.D., and Paine, A.J., Eds., Taylor and Francis, London, 1989, p. 143.

Vial, T., Nicolas, B., and Descotes, J., Clinical immunotoxicity of pesticides, *J. Toxicol. Environ. Health*, 48, 215, 1996.

Vine, M.F., et al., Plasma 1, 1-dichloro–2, 2-bis(p-chlorophenyl)ethylene (DDE) levels and immune response, *Am. J. Epidemiol.*, 153, 53, 2001.

Vine, M.F., et al., Effects on the immune system associated with living near a pesticide dump site, *Environ. Health Perspect.*, 108, 1113, 2000.

Vlahov, D., et al., Prognostic indicators for AIDS and infectious disease death in HIV-infected injection drug users: plasma viral load and CD4+ cell count, *JAMA*, 279, 35, 1998.

Voccia, I., et al., Immunotoxicity of pesticides: a review, *Toxicol. Ind. Health*, 15, 119, 1999.

Vos, J.G., Immune suppression as related to toxicology, *CRC Crit. Rev. Toxicol.*, 5, 67, 1977.

Vos, J.G., Immunotoxicity assessment: Screening and function studies, *Arch. Toxicol.*, S4, 95, 1980.

Weaver, J.L., et al., Serial phenotypic analysis of mouse peripheral blood leukocytes, *Toxicol. Mech. Methods*, 12, 95, 2002.

Weisglas-Kuperus, N., et al., Immunologic effects of background exposure to polychlorinated biphenyls and dioxins in Dutch preschool children, *Environ. Health Perspect.*, 108, 1203, 2000.

Wieneke, H., et al., Predictive value of IgG subclass levels for infectious complications in renal transplant recipients, *Clin. Nephrol.*, 45, 22, 1996.

Wilson, S.D., et al., Correlation of suppressed natural killer cell activity with altered host resistance models in B6C3F1 mice, *Toxicol. Appl. Pharmacol.*, 177, 208, 2001.

World Health Organization, *International Statistical Classification of Diseases and Related Health Problems*, 1989 Revision, Geneva, 1992.

Yang, E.V., and Glaser, R., Stress-induced immunomodulation: impact on immune defenses against infectious disease, *Biomed. Pharmacother.*, 54, 245, 2000.

Yu, M.L., et al., The immunologic evaluation of the Yucheng children, *Chemosphere*, 37, 1855, 1998.

Yucesoy, B., et al., Association of tumor necrosis factor-alpha and interleukin-1 gene polymorphisms with silicosis, *Toxicol. Appl. Pharmacol.*, 172, 75, 2001.

Zenger, V.E., et al., Quantitative flow cytometry: inter-laboratory variation, *Cytometry*, 33, 138, 1998.

CHAPTER 12

Developmental Immunotoxicology in Rodent Species

Barry R. Blakley and Patricia M. Blakley

CONTENTS

12.1 INTRODUCTION

The ability to detect manifestations of toxicity has evolved considerably in recent years. Traditional approaches evaluated toxicants associated with acute or chronic high-level exposure. Endpoints reflecting toxic damage focused on organ damage, cell death, and gross anatomical or histopathological changes. Functional endpoints such as behavior or immunocompetence were rarely evaluated. However, in many instances, these functional endpoints have proven to be more sensitive indicators of exposure or toxicity. This may be particularly relevant in situations where the manifestations of toxicity remain subclinical for extended periods of time. Noteworthy examples may include carcinogenesis or teratogenesis where the clinical syndromes may not be observed for months or years. Autoimmunity, hypersensitivity, and allergy are examples of immune dysfunction that may be triggered by immunotoxicants. Subtle functional alterations of the immune system may be useful to identify chemical exposure prior to gross anatomical or biochemical change. If functional disturbances are present without evidence of toxicity as identified by more classical testing, the existing threshold limit values or no effect levels for many xenobiotics may require reevaluation.

0-415-30854-2/05/$0.00+$1.50

Teratological studies have demonstrated that the fetus is at considerable risk for abnormal development and often more susceptible to xenobiotic exposure as compared to adults under similar circumstances. Based on this premise, new approaches for toxicity testing have been developed. These approaches have been extended to include developmental immunotoxicants. Fetal development, whether morphological or functional development, is genetically programmed. Exposure to immunotoxicants *in utero* may alter this programmed development, resulting in sustained postnatal immune alteration (Holladay and Luster, 1996). The immunoteratogens may act by a variety of mechanisms. Altered fetal gene expression represents one potential mechanism. Many teratogens are known to be immunotoxic (Holladay et al., 2002). Considering the dynamic nature and susceptibility of the developing immune system to immunotoxicants, and its interaction with the endocrine and central nervous systems through cytokines, it is not surprising that developmental immunotoxicants may cause endocrine disruption or reproductive disorders (Holsapple, 2003; Van Loveren et al., 2003). Consequently, the developing immune system may be uniquely susceptible to xenobiotic exposure. Alterations in immune function observed postnatally may reflect abnormalities in various systems — including the endocrine, nervous, reproductive, and immune systems — which may have permanent health implications.

12.2 FETAL SUSCEPTIBILITY TO TERATOGENS

The stage of embryonic development influences the susceptibility of the fetus to teratogens. Rapidly dividing and differentiating cells develop into tissues and organs in an ordered fashion. Teratogens disrupt this ordered development. A number of ordered, dynamic changes occur during development of the immune system. Consequently, like many other organ systems, the immune system is susceptible to chemical perturbation during its development. Traditionally, teratology has focused on gross anatomical and biochemical abnormalities. This approach has been modified to include functional endpoints associated with behavior (Andersen et al., 2000) or immune alteration (Holsapple, 2003).

In each instance analogous to traditional teratological evaluations, the stage of development influences the susceptibility of the fetus to the teratogen. During development both the immune and nervous systems are structurally, chemically, and functionally immature. Many cell types lack specific receptors necessary for critical cell–cell interactions. The immunochemical and neurochemical regulation and maintenance within the fetus may be easily disrupted resulting in a loss of function or ordered development. The fetus is uniquely susceptible to xenobiotic exposure. Placental barriers may provide little protection for the fetus. A poorly developed blood–brain barrier further compromises fetal viability and physiological function. Impaired mechanisms for excretion and metabolism that are evident during fetal development may result in greater accumulation of the teratogen during gestation. Consequently, the impact of a xenobiotic or teratogen on the fetus and fetal development may be more significant as compared to the adult. Anatomical and functional development in most species may extend beyond parturition. The window of vulnerability may extend into postnatal life, including lactation when significant exposure to lipid-soluble xenobiotics may occur (Holsapple, 2003). Postnatally, after organogenesis is complete, only functional alterations may persist. Using a functional approach to teratological testing, evidence of exposure and toxicity may become apparent at lower dosages.

In contrast to classical teratological studies, functional abnormalities must be evaluated in viable offspring. However, in many instances, functional disturbances may not manifest until maturity. In man, both the central nervous and immune systems continue to develop anatomically and functionally, postnatally. Consequently, depending on the specific behavioral or immunological function, the time of development and maturation may vary considerably. Corresponding functional disturbances will also be delayed in a comparable fashion. Unlike behavioral or immunotoxicologic studies in mature subjects, the exposure conditions in teratological studies must be related to the period of organ development. Following exposure, the endpoints for functional evaluation in the offspring or the adult subjects are similar. This approach has led to the establishment of two subdisciplines known as neurobehavioral teratology and developmental immunotoxicology.

The assessment of immune function is now considered to be an important component of many reproductive studies (Van Loveren et al., 2003). Endocrine disruption, a major avenue of investigation in reproductive studies, may be associated with immune dysfunction (Holsapple, 2003). Ordered development in the fetus is genetically controlled. Disruption of this control during pregnancy may result in functional or morphological alterations in the fetus. Many teratogens are considered to be immunotoxic. Maternal immunoproteins may be altered by immunotoxicants during pregnancy. These alterations may disrupt genetically programmed development and fetal gene expression. The consequences may become evident postnatally. It has been demonstrated that immunostimulation during pregnancy reduces the impact of certain teratogens (Holladay et al., 2002; Sharova et al., 2000) on the fetus. Although the impact may be triggered by an immunotoxicant, the manifestations may be associated with various organ systems that may ultimately alter gene expression in the fetus.

12.3 SIGNIFICANCE WITH RESPECT TO IMMUNE DEVELOPMENT

The immune system is uniquely sensitive as a "target organ" for xenobiotics both prenatally and postnatally. There are a variety of explanations that account for this observation. Table 12.1 summarizes many of the features that are often examined using functional assays in immunotoxicology. The immune system is complex, metabolically active, and regulated extensively by other systems. Consequently, the immune system is vulnerable to dysfunction that may manifest in a variety of ways (Table 12.1).

Alterations in immune function have been associated with prenatal and postnatal exposure. Direct comparisons of immune dysfunction associated with exposure during pregnancy and postnatally are difficult to make. The nature of the exposure circumstances is markedly different. Consequently, the actual dose of the agent reaching the target cell type or receptor cannot be readily determined. In spite of this limitation, considerable evidence suggests that the developing immune system is uniquely susceptible to chemical perturbation. Immune alterations following exposure during pregnancy are often more dramatic and persistent as compared to comparable effects observed subsequent to exposure in adults (Holsapple, 2003; Holladay and Smialowicz, 2000).

During gestation the immune system, like other organ systems, develops in a systematic and programmed fashion. Precursor cells from the thymus and liver differentiate and mature in an organized manner to form functional cell types such as CD_4 and CD_8 lymphocytes with specific cell-surface antigens that form distinct lymphocyte subpopulations with varied effector or regulatory functions (Van Loveren et al., 2003; Holladay and Smialowicz, 2000; Holladay and Luster, 1996). The combination of a dynamic and metabolically active system undergoing a series of programmed

Table 12.1 Characteristics of Immune System that Contribute to Unique Susceptibility as "Target Organ" for Xenobiotics

Dynamic nature
Rapid response capacity
Involvement of many cell types
Diversity of immune functions
Complex system with many sites for dysfunction
Considerable neuroendocrine regulation
High metabolic requirements for proliferation and differentiation
High degree of cellular and biochemical specificity
Development is under genetic control

developmental changes places the immune system in an extremely vulnerable situation. Impaired metabolism and excretion of xenobiotics and the impact of endocrine disrupters may add further insult to the developing fetus. Depending on the nature of the toxicant, the window of vulnerability may be extended into lactation or early postnatal life (Dietert et al., 2000; Holsapple, 2003). It has been suggested by several investigators that the impairment or destruction of pluripotent stem cells or specialized cells *in utero* may result in more devastating consequences later in life (Holsapple 2003; Holladay and Smialowicz, 2000; Dietert et al., 2000; Holladay and Luster, 1996).

In addition to immunosuppression, immune dysfunction associated with a loss of self-tolerance or autoimmunity (Holladay, 1999) may develop. The effects as observed later in life may be more severe, more persistent, or permanent. Compensatory mechanisms that may be present to attenuate extreme imbalances in immune regulation or control may also be disrupted during development. Therefore, the consequences of immune alteration associated with fetal development may be quantitatively and qualitatively more devastating in postnatal life.

12.4 MODEL SYSTEMS FOR DEVELOPMENTAL IMMUNOTOXICOLOGY

Rodent species have been the preferred choice as an animal model to conduct developmental immunotoxicologic studies. The immune system of the rodent, particularly the mouse, has been extensively investigated. There are considerable similarities with the human immune system to allow for interspecies comparisons. In addition to the economic factors and the availability of inbred strains of mice with specific immunodeficiency predispositions, many mouse-specific reagents such as monoclonal antibodies or cytokines are available commercially for routine use. Reproductive characteristics of mice (short estrous and gestational periods) and an in-depth understanding of embryological development and organogenesis of the fetus make this species a logical choice for developmental immunotoxicologic investigation. Nonhuman primate species have also been promoted as good animal models to assess developmental immunotoxicology (Buse et al., 2003). The developmental patterns of lymphatic tissues are much more comparable to humans. In addition, many human-specific antibodies or vaccines may not interact with rodent receptors, but may interact with corresponding primate receptors. In spite of these factors, rodent species remain the model of choice.

12.4.1 Time of Exposure

The time of fetal exposure to a xenobiotic may vary depending on the properties of the agent or the nature of the study. Table 12.2 outlines several potential milestones that may be considered in a developmental immunotoxicologic investigation (Table 12.2). The embryonic origin of the lymphoid tissues is ectoendodermal in nature. In the murine species, thymic (T-lymphocyte) organogenesis occurs on days 9 to 10 of gestation. T-lymphocyte migration from the liver and subsequent development and maturation may be extended to day 17 of gestation (Verlarde and Cooper, 1984; Owen and Raff, 1972; Dean et al., 1994; Adkins et al., 1987). The bursa-equivalent (B-lymphocyte) organogenesis occurs on days 10 to 13 (Dean et al. 1994). B-lymphopoiesis and precursor development occur from days 12 to 14 (Hayakawa et al., 1994). The most susceptible period for abnormal development of the immune system in the mouse is associated with days 9 to 13 of gestation, although specific patterns of immune dysfunction that may reflect abnormal development may not be evident functionally until 15 days postnatally (Morin et al., 1992). Organogenesis in the mouse for most organ systems occurs from days 6 to 16 of gestation. Organogenesis associated with the immune system occurs predominantly during the midportion of the susceptible period. Exposure during lactation may also result in immune dysfunction. Since maturation and differentiation of cells may still be occurring to a limited extent, persistent alterations may continue to develop at doses that do not affect adult cell populations (Dietert et al., 2000).

Table 12.2   Developmental Milestones Relevant for Immunotoxicologic Investigations
              In Utero in Mice

Time Period	Comment
Prior to pregnancy	May prevent pregnancy.
	May induce metabolic enzymes.
Day 0–6	May prevent implantation.
	May prevent fertilization.
	May induce metabolic enzymes.
Day 9–10	Period of maximum susceptibility for thymic organogenesis.
	Period of thymic epithelium formation.
Day 10–11	Precursor T-lymphocytes migrate from liver to thymus.
Day 10–13	Period of maximum susceptibility for bursa-equivalent organogenesis.
Day 12–13	B-lymphopoiesis begins in liver.
Day 14	B-lymphocyte precursors present in liver.
Day 14–17	Development and maturation of T-lymphocytes expressing surface antigens such as $CD3^+$, $CD4^+$, and $CD8^+$.
Day 9–13	Period of maximum susceptibility for immune system organogenesis.
Day 6–16	Period of organogenesis (all systems).
Day 16–17	B-lymphocytes present.
Day 16–20	Period of low susceptibility (immune system).
Postpregnancy	Exposure to xenobiotic may cause immune dysfunction (lactation).

If the investigator wishes to minimize the impact of enzyme induction, impaired implantation or fertilization or lactational exposure on the development of immune-mediated functional disturbances, treatment should be restricted to days 9 to 13 of gestation in the mouse. Based on the physical and biological properties of the xenobiotic, the "window" of susceptibility may shift marginally. If metabolic activation of the xenobiotic by fetal tissues is essential for immune dysfunction, this becomes an unlikely outcome since this metabolic capability in rodents occurs late in pregnancy after organogenesis has been completed.

12.4.2 Route of Exposure

The route of exposure frequently influences the nature and magnitude of the adverse response. These differences may be attributed to altered toxicokinetic properties of the toxicant. It is therefore critical to expose dams by the route that is most frequently encountered in the environment. Dermal, inhalation, or oral routes of exposure are most appropriate. Since the dams are pregnant and predisposed to stress and subsequent abortion, it is important to minimize handling and invasive treatment protocols. If properties of the toxicant permit (palatability and water solubility), the agent may be placed in the drinking water for the appropriate period of time.

12.4.3 Animal Monitoring During Pregnancy

The objective of most developmental immunotoxicologic studies is to determine the primary functional immune alterations induced by the immunotoxicants following *in utero* exposure. To achieve this objective in classical teratological or immunological oriented developmental studies, it is an essential prerequisite that maternal health be unaffected by exposure to the agent during pregnancy. Maternal toxicity or disease may result in nonspecific abnormal fetal development (Chernoff et al., 1989). With particular reference to immunotoxicologic studies, adverse health effects in the dam may produce secondary manifestations in the fetus that may induce immune dysfunction unrelated to the primary mechanism of action of the xenobiotic.

Weight gain, water or feed consumption, and general physical appearance should be recorded at regular intervals during pregnancy. Particular attention should be directed toward the period of administration of the immunotoxicants. Reduced weight gain, water consumption, or feed consumption in treated dams may be general indicators of maternal toxicity and pending nonspecific immune alterations in the offspring. Caution must be used during evaluation of these data. Small litter sizes that may have no direct association with the treatment may also produce similar observations, but have little immunological consequence in the offspring.

12.4.4 Animal Monitoring After Pregnancy

Post-pregnancy, both the offspring and the dam require observation. The period of lactation is demanding and stressful for both. Exposure to the toxicant during lactation as a consequence of treatment or sustained secretion into the milk from previous exposure during pregnancy may result in immune alterations (Holsapple, 2003; Snyder et al., 2000; Ilback et al., 1991; Thuvander et al., 1996). Cross-fostering of the offspring with unexposed dams during lactation will eliminate exposure to the immunotoxicants in the milk subsequent to treatment during pregnancy. High metabolic demands and susceptibility to disease that are unrelated to the immunotoxicants are important sources of stress and potential immune dysfunction during lactation. To minimize these effects or to improve uniformity among litters, it is desirable to cull the litters to a similar size. For example, a maximum of eight offspring (four male and four female) should be retained in each litter. Culling should occur after sex, litter size, or other reproductive variables related to the litter have been recorded. In the offspring, weights at birth, day 7, day 21 (weaning date), and day 35, should be recorded to reflect other delayed health effects associated with the toxicant that may have occurred during lactation or the postweaning period.

In the dams, body weight, and water and feed consumption should be recorded during the period of lactation since the offspring begin to drink water and consume regular feed during the latter stages of lactation. Subsequent to the weaning period, the dams are no longer essential for the study. For comparative purposes, some studies may evaluate similar immunological endpoints in the offspring and the dam, although most studies focus entirely on the offspring. The dams are frequently euthanized and examined for implantation sites in the uterus and corpora lutea on the ovaries. These observations are indicators of fetal resorption, early embryonic death or ovulation. Such information may provide useful insight in terms of fetotoxicity or immunotoxicity.

12.5 EVALUATION OF IMMUNOTOXICOLOGIC DATA IN A DEVELOPMENTAL MODEL SYSTEM

The immune system is complex and dynamic. It involves many cell types with a diversity of functions. Immune function is regulated at many different levels through cellular interactions within the immune system and by external mechanisms such as the neuroendocrine axes. Standard batteries of immunological tests have been developed to assess humoral immunity, cell-mediated immunity, and a variety of nonspecific immune functions (Luster et al., 1988). Depending on the objectives of the study, a variety of immunological assays can be run. Frequently, a two-tier approach involving both screening and comprehensive testing of immune function is utilized (Luster et al., 1989). Maturation of the immune system in the mouse is usually complete by approximately 6 weeks of age. Routine evaluation prior to this time may reflect the extent of maturation rather than immune dysfunction. Unless delayed maturation is the desired endpoint, the immune system should not be evaluated in the offspring until they are 6 weeks of age.

For each litter, it is possible to evaluate sex-related differences and a variety of immune functions. It must be kept in mind from a statistical perspective that the litter is the experimental unit. Individual offspring in the litter are normally considered replicates within the unit. Litter effects unrelated to the immunotoxicants can therefore be avoided.

Exposure to immunotoxic chemicals during pregnancy provides an added dimension to the discipline of immunotoxicology. Physiological interactions, developmental aspects, and unique

Table 12.3   Immunotoxicologic Concepts that May Be Explored Using a Developmental Model System

Nature of Immune Alteration	Comment
Permanent	Reflects abnormal development of a precursor cell.
Permanent	Reflects persistence of immunotoxicants.
Delayed maturation	One or more cell types.
Relative organ susceptibility	Thymus, spleen, lymph node.
Relative precursor cell susceptibility	
Period of high immunosusceptibility	
Specific or generalized immune alteration	

Mechanism of Action	
Loss of specific cell population	Example: T-helper lymphocyte.
Induction of a specific cell population	Example: T-suppressor lymphocyte.
Loss of a specific function	Cell proliferation, protein synthesis, antigen processing.

Dose–Response Relationships	
Relevance of low dose exposure and residues	
Comparison between adult and fetal effects	Susceptibility, nature of immune alteration, mechanism of action.
Predictive ability	Adult vs. fetus.

Other Considerations	
Significance of lactational exposure	
Significance of placental exposure	
Endocrine–immune system interaction	
Reproductive–immune system interaction	

exposure circumstances allow this approach to explore horizons in immunotoxicology from new perspectives. Consequently, new hypotheses and dose–response relationships can be evaluated. Table 12.3 describes some of the new opportunities and potential immunotoxicologic concepts that may be examined.

The underlying question that ultimately needs to be answered is whether the fetus responds to immunotoxicants in a manner quantitatively and qualitatively similar to adults. A number of investigators have demonstrated that endocrine disruption, stage of development, time of exposure (lactation vs. gestation), and level of exposure have a significant impact on the nature and extent of the immune alteration (Holsapple, 2003; Snyder et al., 2000; Gehrs et al., 1997a, 1997b; Santoni et al., 1997; Holladay and Luster, 1996; Thuvander et al., 1996; Schuurman et al., 1992; Ilback et al., 1991).

12.6 IMMUNOTOXICANTS ASSOCIATED WITH PRENATAL EXPOSURE

A limited number of studies have documented immune dysfunction subsequent to exposure to xenobiotics during pregnancy or the early postnatal period. Many xenobiotics that cause immune alterations in the adult may yield an entirely different immunotoxicologic profile in the offspring subsequent to *in utero* exposure. Analogous to the thalidomide experience several decades ago, certain agents that specifically act as immunoteratogens may have little effect on immunocompetence when administered to an adult. Yet, these agents when administered during pregnancy may produce permanent immune alterations in the offspring that may impact directly on the quality of life.

Alterations of immunocompetence following prenatal exposure to xenobiotics have been documented with a variety of agents. In many instances, continued exposure during lactation makes it difficult to determine whether the damage to the immune system occurred during the prenatal or

Table 12.4   Metals Associated with Immune Alteration Following Exposure *In Utero*

Metal	Effect	Reference
Cadmium	Impaired thymocyte development.	Van Loveren et al. (2003)
Lead	Increased plasma IgE; reduced splenic white cell count.	Snyder et al. (2000)
	Impaired T-lymphocyte and macrophage function.	Luster et al. (1978a)
	Impaired T-lymphocyte mitogen responses.	Faith et al. (1979)
	Persistent immune alterations.	Miller et al. (1998)
	Altered Th1–Th2 function.	Bunn et al. (1998)
Methylmercury	Impaired humoral immunity (*Brucella* antigen).	Spyker and Fernandes (1973)
	Increased thymocyte number; increased B-lymphocyte mitogen response; increased CD4+ and CD8+ subpopulations; increased humoral immunity (viral antigen).	Thuvander et al. (1996)
	Impaired natural killer cell activity.	Ilback et al. (1991)
Bis(tri-*n*-butyltin) oxide	Impaired thymocyte development.	Van Loveren et al. (2003)
Di-*n*-butyltin chloride	Impaired thymocyte development.	Van Loveren et al. (2003)

Table 12.5   Therapeutic Agents, Hormones, and Drugs of Abuse Associated with Immune Alteration Following Exposure *In Utero*

Agent	Effect	Reference
Acyclovir	Impaired thymocyte development.	Van Loveren et al. (2003)
	Reduced antibody production.	Stahlmann et al. (1991)
Ampicillin/cloxacillin	Enhanced humoral immunity (sheep red blood cell antigen).	Dostal et al. (1994)
Azahioprine	Impaired thymocyte development.	Van Loveren et al. (2003)
Cannabinoid	Reduced T-helper lymphocyte.	del Arco et al. (2000)
Cyclosporine	Impaired thymocyte development.	Van Loveren et al. (2003)
Dexamethazone	Increased autoimmune disease.	Bakker et al. (2000)
Diazepam	Altered interleukin profiles.	Schlumpf et al. (1993)
	Reduced humoral immunity (sheep red blood cell antigen).	Butikofer et al. (1993)
Diethylstilbestrol	Thymic atrophy; altered CD4 and CD8 antigen expression.	Holladay et al. (1993a, 1993b)
	Impaired antibody production.	Luster et al. (1978b)
	Impaired delayed type hypersensitivity; reduced T-lymphocyte blastogenesis.	Luster et al. (1979)
	Thymic atrophy.	Greenman et al. (1977)
	Reduced natural killer cell activity.	Kalland (1980)
Ethanol	Reduced T-lymphocyte blastogenesis; reduced cell mediated immunity.	Gottesfeld and Ullrich (1995)
	Delayed B-lymphocyte maturation.	Biber et al. (1998)
	Impaired B- and T-lymphopoiesis.	Wolcott et al. (1995)
Glucocorticoids	Impaired thymocyte development.	Van Loveren et al. (2003)
Popanil	Impaired thymocyte development.	Van Loveren et al. (2003)
Rapamycin	Impaired thymocyte development.	Van Loveren et al. (2003)

postnatal periods. Table 12.4 provides examples of immunotoxicants that have been associated with metals following *in utero* exposure. A wide variety of immune alterations have been observed. Similarly, pesticides, environmental contaminants, hormones, and various therapeutic agents have also produced a diverse array of immune dysfunction following *in utero* exposure. Tables 12.5, 12.6, and 12.7 provide summaries of the reported immune alterations. The effects of prenatal exposure to immunotoxicants have been well documented in a review by Holladay (1999). These partial lists of developmental immunotoxicants or immunoteratogens, in most instances, are immunotoxicants following exposure in adults. Immunotoxicants that produced comparable evidence of immune dysfunction in both the adult and the offspring were generally evaluated using similar

Table 12.6   Environmental Contaminants Associated with Immune Alteration Following Exposure In Utero

Agent	Effect	Reference
Benzo[a]pyrene	Impaired humoral immunity (sheep red blood cell antigen).	Urso and Gengozian, (1980); Urso and Johnson (1987)
	Thymic atrophy; altered CD4 and CD8 antigen expression.	Holladay and Smith (1994)
7,12-dimethylbenz[a]anthracene	Reduced thymic cellularity; reduced hematopoietic cell subpopulations in liver.	Holladay and Smith (1995)
2,3,7,8-tetrachlorodibenzo-p-dioxin	Altered CD4 and CD8 antigen expression.	Holladay et al. (1991)
	Thymic atrophy.	Fine et al. (1989)
	Thymocyte maturation.	Blaylock et al. (1992)
	Impaired lymphocyte blastogenesis; impaired graft vs. host reaction; reduced T-lymphocyte precursors;	Dietert et al. (2000)
	reduced spleen and thymus size. Reduced CD3+/CD4-CD8- ratio.	Gehrs et al. (l997a, 1997b)
	Impaired delayed-type hypersensitivity response.	Gehrs and Smialowicz (1999); Faith and Moore (1997)
	Reduced T-lymphocyte blastogenesis.	Luster et al. (1979)
T-2 toxin	Thymic atrophy; altered CD4 and CD8 antigen expression.	Holladay et al. (1993a, 1993b)
	Impaired thymocyte development.	Van Loveren et al. (2003)
	B-lymphocyte precursor alteration.	Holladay et al. (1995)

Table 12.7   Pesticides Associated with Immune Alteration Following Exposure In Utero

Agent	Effect	Reference
Chlordane	Impaired mixed lymphocyte reaction.	Spyker-Cranmer et al. (1982)
	Impaired delayed-type hypersensitivity.	Barnett et al. (1987a)
	Altered bone marrow colony formation.	Barnett et al. (1990)
Chlorpyrifos	Reduced T-lymphocyte blastogenesis.	Navarro et al. (2001)
Cypermethrin	Enhanced natural killer cell and antibody-dependent cytotoxic activity.	Santoni et al. (1997)
2,4-dichlorophenoxyacetic acid	Impaired lymphocyte blastogenesis.	Blakley and Blakley (1986)
	Altered tumor production.	Lee et al. (2000)
	Increased B-lymphocyte numbers; reduced T-cytotoxic/suppressor lymphocyte numbers.	Lee et al. (2001)
Diethyldithiocarbamate	Enhanced humoral immunity (sheep red blood cell antigen).	Binder et al. (1985)
Heptachlor	Reduced humoral immunity (male) (sheep red blood cell antigen); increased percentage of B-lymphocytes (male).	Smialowicz et al. (2001)
Hexachlorobenzene	Impaired delayed-type hypersensitivity; reduced B-lymphocytes; impaired mixed lymphocyte reaction.	Barnett et al. (1987b)
Methoxychlor	Reduced humoral immunity (sheep red blood cell antigen).	Chapin et al. (1997)

testing protocols. Differences related to absorption, distribution, and excretion of the immunotoxicants are reflected in the quantitative discrepancies between the responses of the adult and offspring. An immunotoxicant that is highly persistent will be retained by the fetus postnatally, resulting in altered immune function similar to the profile observed subsequent to exposure in the adult. This profile will not be similar if the immunotoxicants disrupt development of the immune system during pregnancy.

In some instances, immunotoxicants that are potent agents in the adult may not produce comparable effects in the offspring. For example, the mycotoxin, T-2, is an extremely potent immunotoxicant in adult mice (Tomar, et al., 1988), causing suppression of T-lymphocyte–dependent humoral immunity. In contrast, a comparable study with T-2 toxin in pregnant mice demonstrated that the offspring exhibited a normal T-lymphocyte–dependent humoral immune response (Blakley et al., 1987), although substantial embryolethality was evident. This suggests that the immunosuppressive action of T-2 toxin may only persist if exposure to T-2 toxin is maintained. The rapid elimination of T-2 toxin may be an important consideration. Thymic atrophy and altered expression of lymphocyte antigens following *in utero* exposure to T-2 toxin have been reported (Holladay et al., 1993b). These alterations, which may reflect abnormal development rather than direct transitory cytotoxicity, may persist after the T-2 toxin has been eliminated. The impact of these persistent effects may be of greater significance from a long-term health perspective. The dioxin 2,3,7,8-tetrachlorodibenzo-*p*-dioxin has been associated with persistent functional alterations following prenatal exposure (Gehrs et al., 1997a, 1997b; Faith and Moore, 1977). Several investigators have suggested that the functional disturbances in the immune system are more severe or persistent following *in utero* exposure (Holladay and Smialowicz, 2000; Holladay and Luster, 1996; Schuurman et al., 1992).

The herbicide 2,4-dichlorophenoxyacetic acid (2,4-D) has been shown to suppress humoral immunity following acute exposure in mice (Blakley and Schiefer, 1986). Lymphocyte blastogenesis of T- and B-lymphocytes remained normal. A similar study in pregnant mice (Blakley and Blakley, 1986) demonstrated that the humoral immunity was unaffected in the offspring exposed prenatally to 2,4-D, whereas a generalized suppression of lymphocyte blastogenesis was evident. The delay between 2,4-D exposure and the assessment of immune function was approximately 50 days. Since 2,4-D has a relatively short half-life, it is unlikely that significant residues of 2,4-D remained at the time of immune function assessment. This suggests that permanent damage to a precursor cell type rather than a direct transitory alteration may have occurred. A more recent study, Lee et al. (2001), also reported altered lymphocyte blastogenesis. In a comparable situation, the insecticide cypermethrin, a synthetic pyrethroid that is considered to be relatively nontoxic in mammalian species, has been shown to impair humoral and cell-mediated immune responses in adult animals (Desi et al., 1986). In contrast, prenatal exposure to cypermethrin in the rat resulted in enhanced natural killer cell- and antibody-dependent cytotoxic activity in the offspring. The dams exhibited no evidence of these changes (Santoni et al., 1997). Exposure to lead during pregnancy has also been reported to produce more persistent immune alterations in the offspring as compared to the dams (Miller et al., 1998). Occasionally, the offspring may be less susceptible to the immunotoxicants. Cyclophosphamide, which produces profound suppression of B-lymphocyte activity and T-lymphocyte–dependent antibody production, has limited effects following *in utero* exposure in the offspring (Holladay and Smialowicz, 2000).

Although it is not a simple matter to make direct comparisons between adult and prenatal exposure situations, it is apparent that significant differences associated with immune alterations are evident that are not dose related. The response profile and nature of the immune alterations are distinctly different qualitatively where comparisons can be made. Comparisons between the adult and fetal response profiles may only be similar if the agent in question has a long half-life, and produces predominantly direct cytotoxicity in both the adults and the offspring. This may be the case in a few instances. The high probability of differences associated with immune alterations induced during adult or fetal periods would suggest that developmental immunotoxicologic assessment should be a critical consideration with many classes of immunotoxicants.

12.7 SUMMARY

Abnormal development classically has been associated with anatomical or biochemical alterations. Recently functional disturbances such as immunosuppression have been reported subsequent to *in utero* exposure. The developing immune system appears to be uniquely susceptible to xenobiotic exposure. The developing fetus is highly susceptible to immunological perturbations if the chemical exposure occurs during organogenesis of the lymphoid tissues. Prenatal immune alterations have been associated with a variety of immunotoxicants. Quantitatively and qualitatively, the effects in the offspring differ from the comparative alterations observed in the adult. In many instances, the effects in the offspring exposed during pregnancy are more persistent and severe in nature. Prediction of immune alterations in the offspring exposed *in utero* based on adult exposure studies is likely to be unreliable. Specific developmental immunotoxicology studies are essential if gestational exposure is a likely scenario.

REFERENCES

Adkins, B., et al., Early events in T-cell maturation, *Ann. Rev. Immunol.*, 5, 325–365, 1987.

Andersen, H.R., Neilsen, J.B., and Grandjean, P., Toxicologic evidence of developmental neurotoxicity of environmental chemicals, *Toxicology*, 144, 121–127, 2000.

Bakker, J.M., et al., Neonatal dexamethasone treatment increases susceptibility to experimental autoimmune disease in adult rats, *J. Immunol.*, 165, 5932–5937, 2000.

Barnett, J.B., Soderberg, L.S.F., and Menna, J.H., The effect of prenatal chlordane exposure on the delayed hypersensitivity response of BALB/c mice, *Toxicol. Lett.*, 25, 173–183, 1987a.

Barnett, J.B., et al., The effect of in utero exposure to hexachlorobenzene on the developing immune response of BALB/c mice, *Toxicol. Lett.*, 39, 263–274, 1987b.

Barnett, J.B., et al., Long-term alteration of adult bone marrow colony formation by prenatal chlordane, *Fundam. Appl. Toxicol.*, 14, 688–695, 1990.

Biber, K.L., et al., Effects of in utero alcohol exposure on B-cell development in the murine fetal liver, *Clin. Exp. Res.*, 22, 1706–1712, 1998.

Binder, P., et al., Immunomodulation in offspring mice after neonatal immunostimulation of mothers or newborn mice, *Tohoku. J. Exp. Med.*, 146, 379–383, 1985.

Blakley, B.R., and Blakley, P.M., The effect of prenatal exposure to the *n*-butylester of 2, 4-dichlorophenoxyacetic acid (2, 4-D) on the immune response in mice, *Teratology*, 33, 15–20, 1986.

Blakley, B.R., Hancock, D.S., and Rousseaux, C.G., Embryotoxic effects of prenatal T-2 toxin exposure in mice, *Can. J. Vet. Res.*, 51, 399–403, 1987.

Blakley, B.R., and Schiefer, H.B., The effect of topically applied *n*-butylester of 2, 4-dichlorophenoxyacetic acid on the immune response in mice, *J. Appl. Toxicol.*, 6, 291–295, 1986.

Blaylock, B.L., et al., Exposure to tetrachlorodibenzo-p-dioxin (TCDD) alters thymocyte maturation, *Toxicol. Appl. Pharmacol.*, 112, 207–213, 1992.

Bunn, T.L., Golemboski, K.A., and Dietert, R.R., In utero exposure to lead modulates Th1/Th2 associated functions and is influenced by gender, *Toxicol. Sci.*, 42 (Abstr.), 206, 1998.

Buse, E.B., et al., Reproductive/developmental toxicity and immunotoxicity assessment in the nonhuman primate model, *Toxicology*, 185, 221–227, 2003.

Butikofer, E.E., Lichtensteiger, W., and Schlumpf, M., Prenatal exposure to diazepam causes sex-dependent changes of the sympathetic control of rat spleen, *Neurotoxicol. Teratol.*, 15, 377–382, 1993.

Chapin, R.E., et al., The effects of perinatal/juvenile methoxychlor exposure on adult rat nervous, immune and reproductive system function, *Fundam. Appl. Toxicol.*, 40, 138–157, 1997.

Chernoff, N., Rogers, J.M., and Kavlock, R.J., An overview of maternal toxicity and prenatal development: Considerations for developmental toxicity hazard assessments, *Toxicology*, 59, 111–125, 1989.

Dean, J.H., et al., Immune system: Evaluation of injury, in *Principles and Methods of Toxicology*, 3rd ed., Hayes, A.W., Ed., Raven Press Ltd., New York, 1994, chap. 30.

del Arco, I., et al., Maternal exposure to the synthetic cannabinoid HU–210: effects on the endocrine and immune systems of the adult male offspring, *Neuroimmunomodulation*, 7, 16–26, 2000.

Desi, I., Dobronyi, I., and Varga, L., Immuno-, neuro-, and general toxicologic animal studies on a synthetic pyrethroid: cypermethrin, *Ecotoxicol. Environ. Saf.*, 12, 220–232, 1986.

Dietert, R.R., et al., Workshop to identify critical windows of exposure for children's health: immune and respiratory systems work group summary, *Environ. Health Perspect.*, 108 (Suppl. 3), 483–490, 2000.

Dostal, M., Soukupova, D., and Horka, I., Treatment of pregnant mice with antibiotics modulates the humoral response of the offspring: Role of prenatal and postnatal factors, *Int. J. Immunopharmacol.*, 16, 1035–1042, 1994.

Faith, R.E., Luster, M.I., and Kimmel, C.A., Effects of combined pre- and postnatal lead exposure on cell mediated immune functions, *Clin. Exp. Immunol.*, 35, 413–420, 1979.

Faith, R.E., and Moore, J.A., Impairment of thymus-dependent immune functions by exposure of the developing immune system to 2, 3, 7, 8-tetrachlorodibenzo-p-dioxin (TCDD), *J. Toxicol. Environ. Health*, 3, 451–464, 1977.

Fine, J.S., Gasiewicz, T.A., and Silverspoon, A.E., Lymphocyte stem cell alterations following perinatal exposure to 2, 3, 7, 8-tetrachlorodibenzo-p-dioxin, *Mol. Pharmacol.*, 35, 18–25, 1989.

Gehrs, B.C., et al., Alterations in the developing immune system of the F344 rat after perinatal exposure to 2, 3, 7, 8-tetrachlorodibenzo-p-dioxin. I. Effects on the fetus and the neonate, *Toxicology*, 122, 219–228, 1997a.

Gehrs, B.C., et al., Alterations in the developing immune system of the F344 rat after perinatal exposure to 2, 3, 7, 8-tetrachlorodibenzo-p-dioxin. II. Effects on the pup and the adult, *Toxicology*, 122, 229–240, 1997b.

Gehrs, B.C., and Smialowicz, R.J., Persistent suppression of delayed-type hypersensitivity in adult F344 rats after perinatal exposure to 2, 3, 7, 8-tetrachlorodibenzo-p-dioxin, *Toxicology*, 134, 79–88, 1999.

Gottesfeld, Z., and Ullrich, S.E., Prenatal alcohol exposure selectively suppresses cell-mediated but not humoral immune responsiveness, *Int. J. Immunopharmacol.*, 17, 247–254, 1995.

Greenman, D.L., Dooley, K., and Breeden, C.R., Strain differences in the response of the mouse to diethylstilbestrol, *J. Toxicol. Environ. Health*, 3, 589–595, 1977.

Hayakawa, K., Tarlinton, D., and Hardy, R.R., Absence of MHC class II expression distinguishes fetal from adult B lymphopoiesis in mice, *J. Immunol.*, 152, 4801–4807, 1994.

Holladay, S.D., Prenatal immunotoxicant exposure and postnatal autoimmune disease, *Environ. Health Perspect.*, 107 (Suppl. 5), 687–691, 1999.

Holladay, S.D., and Luster, M.I., Alterations in fetal thymic and liver hematopoietic cells as indicators of exposure to developmental immunotoxicants, *Environ. Health Perspect.*, 104 (Suppl. 4), 809–813, 1996.

Holladay, S.D., and Smialowicz, R.J., Development of the murine and human immune system: differential effects of immunotoxicants depend on time of exposure, *Environ. Health. Perspect.*, 108 (Suppl. 3), 463–473, 2000.

Holladay, S.D., and Smith, B.J., Fetal hematopoietic alterations after maternal exposure to benzo[a]pyrene: a cytometric evaluation, *J. Toxicol. Environ. Health*, 42, 259–273, 1994.

Holladay, S.D., and Smith, B.J., Alterations in murine fetal thymus and liver hematopoietic cell populations following developmental exposure to 7, 12-dimethylbenz[a] anthracene, *Environ. Res.*, 68, 106–113, 1995.

Holladay, S.D., Smith, B.J., and Luster, M.I., B lymphocyte precursor cells represent sensitive targets of T2 mycotoxin exposure, *Toxicol. Appl. Pharmacol.*, 131, 309–315, 1995.

Holladay, S.D., et al., Selective prothymocyte targeting by prenatal diethylstilbestrol exposure, *Cell. Immunol.*, 152, 131–142, 1993a.

Holladay, S.D., et al., Fetal thymic atrophy after exposure to T-2 toxin: Selectivity for lymphoid progenitor cells, *Toxicol. Appl. Pharmacol.*, 121, 8–14, 1993b.

Holladay, S.D., et al., Perinatal thymocyte antigen expression and postnatal development altered by gestational exposure to tetrachlorodibenzo-p-dioxin (TCDD), *Teratology*, 44, 385–393, 1991.

Holladay, S.D., et al., Maternal immune stimulation in mice decreases fetal malformations caused by teratogens, *Internat. Immunopharmacol.*, 2, 325–332, 2002.

Holsapple, M.P., Developmental immunotoxicity testing: a review, *Toxicology*, 185, 193–203, 2003.

Ilback, N.G., Sundberg, J., and Oskarsson, A., Methylmercury exposure via placenta and milk impairs natural killer (NK) cell function in newborn rats, *Toxicol. Lett.*, 58, 149–158, 1991.

Kalland, T., Reduced natural killer activity in female mice after neonatal exposure to diethylstilbestrol, *J. Immunol.*, 124, 1297–1302, 1980.

Lee, K., Johnson, V.J., and Blakley, B.R., The effect of exposure to a commercial 2, 4-D herbicide during gestation on urethan-induced lung adenoma formation in CD-1 mice, *Vet. Hum. Toxicol.*, 42, 129–132, 2000.

Lee, K., Johnson, V.J., and Blakley, B.R., The effect of exposure to a commercial 2, 4-D formulation during gestation on the immune response in CD-1 mice, *Toxicology*, 165, 39–49, 2001.

Luster, M.I., Faith, R.E., and Clark, G., Laboratory studies on the immune effects of halogenated aromatics, *Ann. N.Y. Acad. Sci.*, 320, 473–485, 1979.

Luster, M.I., Faith, R.E., and Kimmel, C.A., Depression of humoral immunity in rats following chronic developmental lead exposure, *J. Environ. Pathol. Toxicol.*, 1, 397–401, 1978a.

Luster, M.I., Faith, R.E., and McLaghlan, J.A., Alterations of the antibody response following in utero exposure to diethylstilbestrol, *Bull. Environ. Contam. Toxicol.*, 20, 433–437, 1978b.

Luster, M.I., et al., Effect of in utero exposure to diethylstilbestrol on the immune response in mice, *Toxicol. Appl. Pharmacol.*, 47, 279–285, 1979.

Luster, M.I., et al., Development of a testing battery to assess chemical-induced immunotoxicology: national toxicology program's guidelines for immunotoxicity evaluation in mice, *Fundam. Appl. Toxicol.*, 10, 2–19, 1988.

Luster, M.I., et al., Perturbations of the immune system by xenobiotics, *Environ. Health Perspect.*, 81, 157–162, 1989.

Miller, T.E., et al., Developmental exposure to lead causes persistent immunotoxicity in Fischer 344 rats, *Toxicol. Sci.*, 42, 129–135, 1998.

Morin, C., Jotereau, F., and Augustin, A., Patterns of responsiveness of T cell lines and thymocytes reveal waves of specific activity in the post-natal murine thymus, *Int. Immunol.*, 4, 1091–1101, 1992.

Navarro, H.A., et al., Neonatal chlorpyrifos administration elicits deficits in immune function in adulthood: a neural effect? *Dev. Brain Res.*, 130, 249–252, 2001.

Owen, J.J.T., and Raff, M.C., Studies on the differentiation of thymus-derived lymphocytes, *J. Exp. Med.*, 132, 1216–1223, 1972.

Santoni, G., et al., Prenatal exposure to cypermethrin modulates rat NK cell cytotoxic functions, *Toxicology*, 120, 231–242, 1997.

Schlumpf, M., Lichtensteiger, W., and Ramseier, H., Diazepam treatment of pregnant rats differentially affects interleukin-1 and interleukin-2 secretion in their offspring during different phases of postnatal development, *Pharmacol. Toxicol.*, 73, 335–340, 1993.

Schuurman, H.J., et al., Chemicals trophic from the thymus: risk for immunodeficiency and autoimmunity, *Int. J. Immunopharmacol.*, 14, 369–375, 1992.

Sharova, L., et al., Nonspecific stimulation of the maternal immune system. II. Effects on gene expression in the fetus, *Teratology*, 62, 420–428, 2000.

Smialowicz, R.J., et al., The effects of perinatal/juvenile heptachlor exposure on adult immune and reproductive system function in rats, *Toxicol. Sci.*, 61, 164–175, 2001.

Snyder, J.E., et al., The efficiency of maternal transfer of lead and its influence on plasma IgE and splenic cellularity of mice, *Toxicol. Sci.*, 57, 87–94, 2000.

Spyker, J.M., and Fernandes, G., Impaired immune function in offspring form mercury-treated mice, *Teratology*, 7, 28A, 1973.

Spyker-Cranmer, J.M., et al., Immunoteratology of chlordane: cell-mediated and humoral immune response in adult mice exposed in utero, *Toxicol. Appl. Pharmacol.*, 62, 402–408, 1982.

Stahlmann, R., et al., Structural anomalies of thymus and other organs and impaired resistance to trichinella spiralis infection in rats after prenatal exposure to acyclovir, *Thymus Update*, 4, 129–155, 1991.

Thuvander, A., Sundberg, J., and Oskarsson, A., Immunomodulating effects after perinatal exposure to methylmercury in mice, *Toxicology*, 114, 163–175, 1996.

Tomar, R.S., Blakley, B.R., and DeCoteau, W.E., Antibody producing ability of mouse spleen cells after subacute dietary exposure to T-2 toxin, *Int. J. Immunopharmacol.*, 10, 145–151, 1988.

Urso, P., and Gengozian, N., Depressed humoral immunity and increased tumor incidence in mice following in utero exposure to benzo[a]pyrene, *J. Toxicol. Environ. Health*, 6, 569–576, 1980.

Urso, P., and Johnson, R.A., Early changes in T lymphocytes and subsets of mouse progeny defective as adults in controlling growth of a syngeneic tumor after in utero insult with benzo[a]pyrene, *Immunopharmacology*, 14, 1–10, 1987.

Van Loveren, H., et al., Immunotoxicological consequences of perinatal chemical exposures: a plea for inclusion of immune parameters in reproduction studies, *Toxicology*, 185, 185–191, 2003.

Verlarde, A., and Cooper, M.D., An immunofluorescence analysis of the ontogeny of myeloid, T, and B lineage cells in mouse hemopoietic tissues, *J. Immunol.*, 133, 672–677, 1984.

Wolcott, R.M., Jennings, S.R., and Chervenak, R., Intrauterine exposure to ethanol affects postnatal development of T and B lymphocytes but not natural killer cells, *Clin. Exp. Res.*, 19, 170–176, 1995.

CHAPTER 13

Gut Mucosal Immunotoxicology in Rodents

Genevieve S. Bondy and James J. Pestka

CONTENTS

13.1 INTRODUCTION

The mucosal immune system is made up of surface and associated lymphoid tissues found in the nasal and oral cavities, gastrointestinal (GI) tract, upper respiratory tract, mammary glands, and urogenital tract. Immunization with a foreign antigen at one mucosal surface was found to generate both local antibodies and also similar antibodies at other mucosal sites, leading to the postulation of a common mucosal immune system (CMIS) (Mayer, 2000). Tissues of the CMIS, including epithelial cells that collaborate with underlying immune tissues, maintain a balance between tolerance and responsiveness at mucosal surfaces which are constantly exposed to microorganisms, foreign antigens, and other potentially harmful substances from food or the environment (Neutra et al., 2001; Neurath et al., 2002). The potential for ingested or inhaled xenobiotics to upset this balance is an area of immunotoxicology that has not been fully explored. This is, in part, because mucosal immunology is still a comparatively young subdiscipline of immunology. In the area of systemic immunotoxicology, potential immunotoxins are screened using a tiered approach that employs morphological and immune function assays. These assays were selected to account for the complexity of immune responses, as well as the relevance and sensitivity of endpoints in predicting disease (Luster et al., 1992). Successful development of a comparable set of assays for assessing mucosal immunity will depend on advances in our understanding of mucosal immune function and the relationship between mucosal and systemic immune responses.

Many methods have been devised to elucidate and clarify the structure and function of the mucosal immune system. Much of this work has been comprehensively summarized by Ogra et al. (1999). While assays developed for assessing gut mucosal immune responses will undoubtedly evolve from basic research, many of these methods will require modification and validation in order to be practical for measuring immunotoxicologic hazards. Kawabata et al. (1995) articulated this concern and summarized the status of this process in a past review. The purpose of the present chapter is to provide an updated overview of the immunoassays that are currently available for assessing the effects of ingested xenobiotics on gut mucosal immune function in rodent models, and to identify areas of research that may improve strategies for assessing gut mucosal immuno-toxicology in the future.

13.2 BRIEF OVERVIEW OF THE GUT MUCOSAL IMMUNE SYSTEM

The mucosal surfaces of the GI tract are lined with a continuous monolayer of epithelial cells that forms a protective and interactive barrier between the body and the external environment. This monolayer forms the first line of defense against microorganisms by restricting their passage from the gut into surrounding tissues. The intestinal lumen is protected from microbial invasion by the barrier properties of the mucosal epithelium and by nonspecific, extrinsic factors which act outside the monolayer to limit microbial entry (reviewed by Pitman and Blumberg, 2000). These extrinsic factors include gastric and intestinal secretions as well as mucus, which creates a viscous barrier and provides a vehicle for antiviral and microbicidal substances such as lysozyme, lactoferrin, complement components, defensins, and secretory immunoglobulin A (sIgA) (Pitman and Blum-berg, 2000). When microbial invaders cross the epithelium, they induce overlying epithelial cells to produce chemokines that attract immune effector cells (Hornef et al., 2002). These nonadaptive immune responses, including physical barriers, secretions, phagocytic cell attraction and activation are referred to as innate immunity because they are not highly specific for a particular pathogen. Foreign antigens are transported, processed, and presented at gut mucosal sites containing organized lymphoid tissues and specialized overlying epithelial cells, thereby leading to adaptive responses. Adaptive immune responses are antigen specific and include specific immunoglobulin production by plasma B cells and cytotoxicity by T cells. Lymphoid follicles are present throughout the intestine. In the distal ileum lymphoid follicles are grouped in large patches known as Peyer's patches (Neutra et al., 2001). These follicles, along with diffuse lymphoid tissue which includes lymphocytes, dendritic cells, macrophages and mast cells in the lamina propria underlying the epithelium, as well as intraepithelial lymphocytes (IELs) in the epithelial basement membrane, make up the gastrointestinal lymphoid tissue (GALT). The importance of gut lamina propria lymphocytes is exemplified by the facts that 80% of all plasma cells are found in this tissue and that these cells produce more immunoglobulin A (IgA) than all other immunoglobulin isotypes combined (Fagarasan and Honjo, 2003).

Antigens from the gut lumen are endocytosed by intestinal epithelial cells or taken up by M cells in the epithelial monolayer overlying the Peyer's patches where active immune responses are generated (Pestka, 1993). The dome-shaped Peyer's patches are GALT inductive sites (Figure 13.1). Under the follicle-associated epithelium covering the Peyer's patch are B-cell zones containing

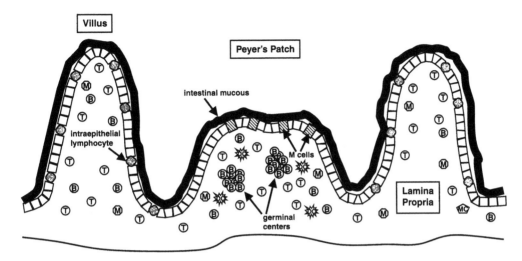

Figure 13.1   Simplified diagram of gut mucosal lymphoid tissue. B, B-lymphocytes; DC, dendritic cells; M, macrophages; MC, mast cells; T, T-lymphocytes.

germinal centers which are sites of B-cell proliferation. Near these B-cell zones are T-cell areas that contain regulatory and effector subsets. Accessory cells such as macrophages and dendritic cells are also present in Peyer's patches. This complement of immunocompetent cells facilitates gut humoral and cell-mediated responses. In response to antigens from the gut lumen, mature T cells and IgA-bearing B cells induced in the Peyer's patches migrate through the mesenteric lymph node into the circulation and finally to mucosal effector sites such as the intestinal lamina propria, which lies under the mucosal epithelium (Fagarasan and Honjo, 2003; Pestka, 1993). Intraepithelial lymphocytes, a heterogenous population of T cells located on the basolateral side of the mucosal epithelium, are also mucosal effector cells. Studies indicate that IEL populations are composed of immunoregulatory and cytotoxic T cells that participate in mucosal defense against infections, in the modulation of epithelial integrity, and in the regulation of mucosal immune responses (reviewed by Aranda et al., 1999).

13.3 BIOMARKERS OF CHANGES IN GUT-ASSOCIATED LYMPHOID TISSUES

Existing immunotoxicology testing guidelines for xenobiotic exposure using rodent models recommend a tiered approach that incorporates morphological as well as functional endpoints (Hinton, 2000; Luster et al., 1992; International Programme on Chemical Safety [IPCS], 1996). First-tier endpoints generally include hematology, histopathology of immune tissues, lymphocyte blastogenesis assays, and natural killer cell assays, all of which can be incorporated into "standard" toxicology studies that are designed to identify multiple target organs and effects. Enumeration of antibody (IgM) plaque-forming cells for sheep red blood cells is sometimes included as a first-tier assay, although it involves the incorporation of satellite control and treatment groups. The next tier of assays is undertaken to define immunotoxic effects for xenobiotics that are positive for changes in first-tier endpoints, and often includes tests for changes in host resistance and cellular responses. Although a similarly systematic approach has not been formulated for the identifying potential mucosal immune modulators, the following section indicates that many of the assays used to assess systemic immune function could be applied to mucosal tissues and responses.

13.3.1 Nonfunctional Assessment of Changes in GALT

13.3.1.1 Histopathology and Immunohistopathology

Immunotoxicologic evaluation requires both functional and morphologic assessment of immune organs and cells, of which lymphoid tissue histopathology is an essential component. Recommendations for first-tier immunotoxicologic or toxicologic testing in rodents include evaluation of hematoxylin-eosin stained, paraffin-embedded slides of spleen, thymus, small intestine (Figure 13.2), large intestine, and lymph node sections (Hinton, 2000; IPCS, 1996; Organization for Economic Cooperation and Development [OECD], 1995). Recently a more comprehensive pathology evaluation was found to detect azathioprine and cyclosporin-induced immunotoxicity more effectively than the limited immune tissue histopathology suggested in OECD Guideline 407 (International Collaborative Immunotoxicity Study [ICIS], 1998). It is likely that this enhanced pathology protocol would flag changes in GALT as it includes microscopic examination of all lymphoid tissues that are frequently examined in gut mucosal immunology studies, including ileum (with Peyer's patches), colon, and mesenteric lymph nodes. Treatment-related histopathologic changes were graded by several criteria, including percentage deviation from morphologically "normal" tissues and cellularity by compartment within each tissue (ICIS, 1998). To facilitate histologic examination of the entire length of the GI tract, rodent intestinal tissues are most effectively collected using the Swiss roll technique, in which the intestine is slit open, cleaned, and rolled with the mucosa side outward prior to fixation (Moolenbeek and Ruitenberg, 1981). In practice, the histological assessment of mucosal tissues in the context of a general toxicological study can be used to identify suspected targets of mucosal immunomodulation. For example, acute exposure to high doses of the mycotoxins deoxynivalenol and other trichothecene mycotoxins was found to cause treatment-related depletion of Peyer's patches lymphocytes (Forsell et al., 1987). In contrast, prolonged oral exposure of mice to deoxynivalenol caused enlargement of Peyer's patches with increased numbers of germinal centers (Pestka et al., 1990b). These disparate dose-related effects are consistent with immunosuppression and immune stimulation observed following high- and low-dose trichothecene exposure, respectively (Bondy and Pestka, 2000).

The toxicologic pathology of the immune system, both systemic and GALT, can be further expanded by immunochemical staining. These methods allow for semiquantitative assessment of changes in specific lymphocyte subsets, taking into account morphological considerations such as compartmentalization of B- and T-lymphocytes within lymphoid tissues (Kuper et al., 2000). Numerous studies have employed immunohistochemical or immunofluorescent staining techniques

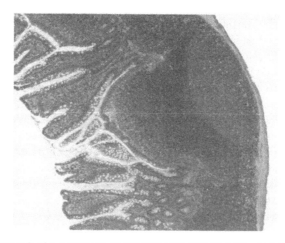

Figure 13.2   Photomicrograph of a section of rat intestine with Peyer's patch; hematoxylin and eosin stain.

to visualize xenobiotic-induced changes in mucosal tissues. Oral exposure of mice to the environmental pollutant acrylonitrile significantly decreased numbers of duodenal, ileal and jejunal IgA$^+$ B cells (Hamada et al., 1998). Immunohistochemical methods have also been used to visualize changes in rodent intestinal epithelium that directly contacts intestinal lymphoid patches. In cyclosporin-A–treated rats, monoclonal antibodies against epithelial cell keratins indicated that as lymphoid tissues were diminished by treatment, glandular structures and single cells containing epithelial cell markers became visible (Bandaletova, 1992). While most mucosal immunotoxicology studies focus on changes in lymphoid tissues, intestinal lymphoepithelial interactions also play a key role in immune responses. In the absence of standardized methods for assessing epithelial changes, both enhanced pathology and immunohistochemistry provide a means of identifying changes in intestinal epithelial cells that may impact mucosal immune function.

Lymphocyte phenotyping is a sensitive measure of immunotoxicity. Analyses of immunotoxicology data for over 50 test chemicals indicate that cell-surface marker analyses, in addition to the plaque-forming cell assay, were the two most sensitive and predictive tests for immunotoxicity in rodents (Luster et al., 1992). The results were based mainly on comparisons of splenic total T- and B-lymphocytes (Thy 1.2 and sIg$^+$, respectively), which were enumerated in most studies. Comparisons indicated that changes in total T-lymphocyte numbers were a more sensitive index of immunotoxicity than changes in total B-lymphocytes. Based on a smaller number of studies, changes in splenic CD4$^+$ and CD8$^+$ T-lymphocytes may also be sensitive to immunotoxins (Luster et al., 1992). For example, immunofluorescent microscopy revealed that deoxynivalenol feeding elevates CD4$^+$ and IgA$^+$ populations in Peyer's patches and spleen (Pestka et al., 1990b), and this corresponded with elevated IgA production *ex vivo* (Bondy and Pestka, 1991) and *in vivo* (Dong et al., 1991; Dong and Pestka, 1993). For flow cytometry, lymphocyte suspensions must be prepared from appropriate GALT, usually Peyer's patches, mesenteric lymph nodes, and small intestine. Lymphocyte suspensions from lymph nodes are relatively easily prepared by mechanical tissue dispersion followed by filtration through sterile gauze or mesh (Bondy and Pestka, 1991). Preparation of rat or mouse IEL and lamina propria lymphocyte suspensions is more labor intensive and requires successive incubations to release IELs from intestinal segments followed by collagenase digestion to release lamina propria lymphocytes, rendering absolute quantification difficult (Bruder et al., 1999; Li et al., 1995). Cell populations most frequently enumerated in GALT include total T cells, CD4$^+$ and CD8$^+$ T cells, TCRαβ+ and TCRγδ+ T cells, total and IgA$^+$ B cells, and macrophages. Several immunotoxic fungal toxins have been shown to alter relative numbers of lymphocyte subsets in GALT. In rats, concurrent oral exposure to aflatoxin B$_1$ and ovalbumin significantly reduced CD4$^+$/CD8$^+$ ratios in mesenteric lymph nodes (Watzl et al., 1999). There were significantly fewer CD4$^+$ and CD8$^+$ cells in mesenteric lymph nodes from mice treated with an acute high dose of the trichothecene T-2 toxin, along with a significant decrease in IgA$^+$ B cells in Peyer's patches (Nagata et al., 2001).

While immunohistochemistry and flow cytometry can detect changes in gut lymphoid tissues, there are limitations to each approach. In aging Fischer 344 rats, flow cytometric analyses did not detect a significant age-related change in Peyer's patch CD8$^+$ cell numbers. However, immunohistochemical analyses indicated that CD8$^+$ cells were redistributed from the interfollicular areas of the Peyer's patches in adult rats to a wider distribution within the lymphoid follicles of aging rats (Daniels et al., 1993). This underscores the importance of recognizing that flow cytometry detects changes in numbers of specific lymphoid cell subsets but not changes in tissue distribution. The two methods were further compared by Bruder et al. (1999) to detect changes in intraepithelial and lamina propria T-lymphocytes in various rat strains exposed to known immunotoxins. Of the three strains examined, internal elastic layer composition was more variable in Wistar than in Fischer 344 or Lewis rats. The study indicated that flow cytometry of isolated lymphocytes was a more sensitive means of detecting labeled cells, while lymphocytes with low receptor numbers may not be easily visible in stained sections. However in lymphocyte suspensions from intestinal epithelium and lamina propria, purity is a concern for flow cytometry, particularly contamination of IELs with lamina propria lymphocytes (Bruder et al., 1999). While both methods detected changes in intestinal T-lymphocytes in rats exposed to immunomodulators, it remains to be clarified which of the two offers the best choice for a standardized approach to assessing mucosal immune function.

13.3.2 Functional Assessment of GALT

13.3.2.1 Basal Circulating IgA Levels

IgA predominates in mucosal secretions and plays a central role in mucosal humoral responses. At mucosal surfaces, IgA is primarily in the form of secretory IgA, which is a dimer linked by a J (joining) chain and a secretory component that enhances its functionality in mucosal tissues and its survivability in the intestinal lumen (Goldblum, 1990). In Peyer's patches, which are the GALT IgA inductive sites, antigen uptake occurs via specialized epithelial cells (M cells) that facilitate antigen delivery to underlying B and T cells (Figure 13.3). In Peyer's patch follicles, or B-cell zones with germinal centers, B cells proliferate and switch to an IgA+ phenotype; however, final differentiation into IgA plasma cells occurs after these cells leave the Peyer's patches via the lymphatic system and take up residence in mucosal tissues (Fagarasan and Honjo, 2003). As a result, exposure to xenobiotics could affect IgA responses at any point from the inductive phase in the Peyer's patches to the effector phase at distant mucosal sites (Kawabata et al., 1995).

One obvious indirect way of assessing effects of chemical treatment on mucosal humoral responses is to measure serum IgA concentrations. It has been suggested that sandwich ELISAs for total IgM and IgG in rat serum may be useful for detecting nonspecific humoral changes in subchronic or chronic exposure scenarios (IPCS, 1996). Similarly, ELISA methods have been used to measure changes in total serum IgA in mice and rats (Pestka, et al., 1990a; Poon et al., 1997). One issue with this approach relates to the question of whether a change in mucosal IgA production might be similarly reflected at the systemic level. Another concern relates to the possibility that the extrapolation of serum IgA data from rodent studies to putative human effects might be

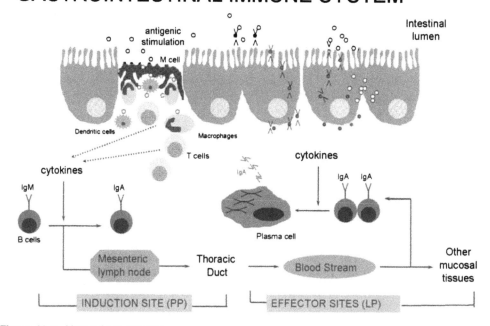

Figure 13.3 Mucosal IgA induction.

complicated by species differences in the form and origin of circulating IgA. While sIgA is produced locally in mucosal tissues, circulating IgA in humans originates primarily from the bone marrow. Peppard and Russell (1999) reported that human plasma typically contains 2 to 3 mg/mL of circulating IgA, of which 85 to 90% is monomeric in form. In contrast, circulating IgA in rodents and other nonprimate animals contains variable ratios of monomeric and polymeric IgA that can change upon immunization (Delacroix et al., 1985; Mascart-Lemone et al., 1987). Serum IgA in these species is intestinal in origin, and present at levels around 0.3 mg/mL or less (Peppard and Russell, 1999). Another major difference is that human IgA occurs as two subclasses, IgA_1 and IgA_2, whereas rodent IgA consists of only one isotype. Thus, the biological function of circulating IgA remains to be clarified as do the physiological consequences of species differences between humans and laboratory rodents.

In spite of the aforementioned uncertainties, changes in serum total IgA have been shown to be indicative of xenobiotic-induced mucosal immunomodulation in mice exposed to deoxynivalenol in food. Here, the observation was made that feeding deoxynivalenol at the 2- to 25-ppm concentration elicited isotype-specific increases in serum IgA (primarily polymeric), whereas levels of IgG and IgM were decreased or unaffected (Forsell et al., 1986; Pestka et al., 1989). Subsequently, it was determined that these serum IgA increases were produced by or coincided with increases in IgA^+-bearing cells, and IgA-secreting cells in Peyer's patches and spleen (Bondy et al., 1991; Pestka et al., 1990a). The resultant antibody IgA was polyclonal, had weak affinity, and could bind to various microbial and tissue antigens (Rasooly and Pestka, 1992, 1994; Rasooly et al., 1994). Prolonged deoxynivalenol feeding results in elevated IgA immune complexes and IgA deposition in kidney that mimics human IgA nephropathy, which is the most common glomerulonephritis worldwide (Pestka et al., 1989; Dong et al., 1991; Dong and Pestka, 1993). Recent studies indicate that deoxynivalenol-induced IgA effects are mediated by elevated IL-6 expression (Pestka, 2003).

13.3.2.2 Antigen-Specific Humoral Responses

In systemic immunotoxicologic assessment, antigen-specific antibody responses are generally regarded to be a more sensitive measure of humoral responses than total immunoglobulin levels. Antibodies specific for the T-cell–dependent antigens tetanus toxin, ovalbumin, or keyhole limpet hemocyanin have been measured in rodents as a test of systemic humoral responses (Bunn et al., 2001; IPCS, 1996). The most widely used humoral response assay is the T-cell–dependent, sheep red blood cell (SRBC)-specific IgM antibody response, which can be quantified using an ELISA or, more commonly, using the hemolytic plaque assay (Temple et al., 1993).

For mucosal responses, a suitable humoral response assay would ideally focus on antigen-specific sIgA in mucosal secretions (Schmucker, 2002). Recent intensive efforts to enhance mucosal IgA responses for prevention of infection by HIV and other mucosal pathogens have generated a plethora of opportunities for modeling this function in an immunotoxicologic assessment. While parenteral or intravenous antigen exposure is typically used for systemic humoral assessment, mucosal immunization is a more effective means of inducing antigen-specific sIgA at mucosal sites. Nonetheless, oral vaccine research indicates that the choice of an immunogen for mucosal humoral challenge using oral exposure would be complicated by factors such as low efficacy, induction of immunotolerance instead of humoral responses, or antigen degradation from exposure to proteases and acids in the GI tract and stomach (Shalaby, 1995).

The *Vibrio cholerae* enterotoxin, or cholera toxin (CT), strongly elicits sIgA and serum antibodies. Its potent adjuvant qualities when co-administered with unrelated proteins have been explored in oral vaccine development (Fujihashi et al., 2001). Although the clinical effects associated with CT exposure preclude its use as an oral vaccine adjuvant in humans, CT may have value as a mucosal humoral challenge agent or adjuvant in experimental animal models. In rodents treated with immunomodulating xenobiotics, cholera toxin administered enterally or duodenally has been successfully employed as a mucosal immune challenge (Karacic and Cowdery, 1988; Vargas et al.,

1998). Should this agent be used as a functional challenge, an important consideration will be to include a suboptimal dose since optimal or supraoptimal doses might yield a maximum response and override subtle down- or up-regulatory effects generated by a putative immunotoxicant.

Although parenteral immunization is a less effective means of inducing antigen-specific IgA at mucosal surfaces, clinically significant mucosal responses can be obtained via routes other than oral exposure. As a result, rodent models of mucosal humoral responses need not be restricted to oral antigen challenge. For example, an antigenic preparation of extracellular products from a strain of *Aeromonas hydrophilia* associated with human gastroenteritis elicited *Aeromonas*-specific IgA and IgG in intestinal secretions upon intraperitoneal administration of the antigen preparation in rats. This model was used to measure changes in antibody secretion induced by administration of the neuromodulator bombesin (Jin et al., 1989).

The most straightforward and noninvasive approach to measure changes in secretory IgA is to use fecal pellets (deVos and Dick, 1991). A distinct advantage of the approach is that the kinetics of IgA responses can be monitored without animal stress or sacrifice. Secretory sIgA can also be measured in intestinal secretions collected at necropsy by washing standardized sections of intestinal lumen (Cozon et al., 1991). Methods have also been developed for repeated sampling of both gut lavage fluids and saliva for sIgA analyses (Elson et al., 1984; Hau et al., 2001; Keren et al., 1989). However, these are stressful, and factors such as matrix, sample viscosity, and volume and level of mucus and proteases can introduce variability in IgA analyses. Antigen-specific IgA or IgA-producing cells have also been measured by ELISA and ELISPOT, respectively, in lymphocyte cultures from Peyer's patches, mesenteric lymph nodes, and intestinal lamina propria from rodents treated with enteric reovirus or CT (Cuff et al., 1998; Karacic and Cowdery, 1988; Vargas et al., 1998).

13.3.2.3 Lymphocyte Proliferation in Response to B- and T-Cell Mitogens

Mitogen-induced lymphocyte blastogenesis and proliferation assays provide a means of evaluating either B- or T-cell function depending on mitogen specificity (White, 1992). In rodent assays, both lipopolysaccharide (LPS) and *Salmonella typhimurium* mitogen (STM) have been used to assess humoral changes by stimulating B-lymphocyte proliferation. Mouse B-lymphocytes will proliferate and differentiate in the presence of either mitogen. In rat lymphocyte cultures, STM is a more potent mitogen than LPS, and both mitogens induce proliferation without differentiation (IPCS, 1996). Changes in T-lymphocyte function are generally assessed with concanavalin A, phytohema-glutinin, or anti-CD3 antibodies. Regardless of mitogen specificity, lymphocyte blastogenesis assays follow a similar protocol in which uptake of tritiated thymidine or an appropriate analogue is measured in rodent spleen or lymph node cultures after a standard incubation period. Mitogen responsiveness has been used to assess rodent lymphocyte function in GALT, particularly mesenteric lymph nodes or Peyer's patches, after exposure to xenobiotics in the diet or by oral gavage (Burchiel et al., 1990; Davis et al., 1994; Watzl et al., 1999). The specificity of responses to some test chemicals indicates that these assays can provide useful data on immunomodulation in gut lymphoid tissues. For example, Peyer's patch T-cell responses were suppressed to a greater extent than Peyer's patch B-cell responses or spleen and mesenteric lymph node B- or T-cell responses in mice treated with a heterocyclic food mutagen (Davis et al., 1994). Likewise, LPS responses in murine Peyer's patch and mesenteric lymph node cultures were more sensitive than LPS responses in spleen lymphocyte cultures to oral 7,12-dimethylbenz[*a*]anthracene exposure (Davis et al., 1991).

13.3.2.4 Cellular Responses

Other than mitogen-stimulated T-lymphocyte proliferation assays, there are no standardized methods for assessing mucosal cell-mediated responses. Mesenteric lymph nodes and Peyer's patches provide ready access to mucosal T cells for functional studies; lamina propria lymphocytes and IELs are less accessible but because of their crucial roles in local responses, functional studies of these subpopulations would be valuable. The development of a colorimetric assay to measure mouse IEL-mediated cytotoxicity to epithelial cells indicates that widely used cellular response assays can be adapted for mucosal cells (Ni and Hollander, 1996). Systemic responses to gut-specific pathogens can offer a means of exploring cellular responses linked to mucosal infection. For example, enteric reovirus-specific cytotoxic T-cell responses in murine spleen cells were inhibited by cyclophosphamide treatment. Since the virus induces precursor cytotoxic T cells in the gut similar studies could be done with Peyer's patch or IEL lymphocytes (Cuff et al., 1998).

13.3.2.5 Cytokines

Effects of xenobiotics on cytokine production by Peyer's patch lymphocytes can be addressed by measuring mRNA expression. For example, acute deoxynivalenol exposure induces IL-6, TNF-α, IL-1, IFN-γ, IL-2, IL-10, and IL-12 (Azcona-Olivera et al., 1995; Zhou et al., 1997, 1999) mRNA expression *in vivo*. The observation that similar effects occurred in the spleen suggests that the mycotoxins were not only targeting the mucosal immune system but systemic immunity as well. Another approach to measure cytokines is to culture Peyer's patch lymphocytes *ex vivo* after toxin exposure. This strategy employed cell isolation from toxin-exposed mice and reconstitution in cell culture to dissect the role of macrophages and T cells in IL-2, IL-5, and IL-6 expression in deoxynivalenol-induced IgA (Yan et al., 1997, 1998). Similar strategies can be used for measuring other immune-related genes such chemokines and selectins.

13.3.2.6 Host Resistance Assays

An optimal host resistance assay for examining the effects of xenobiotics on gut mucosal immune function should most appropriately be based on infectivity and immune responses to microbes that specifically infect the intestinal tract. For systemic infectivity assays, a wide range of bacterial, viral, and parasitic infectious agents have been used, including *Listeria monocytogenes*, *Streptococcus* spp., cytomegalovirus, influenza virus, *Trichinella spiralis*, and *Plasmodium* spp. (IPCS, 1996). For gut mucosal immunity, no single host resistance model has been validated or is widely used, although advances in the understanding of specific intestinal pathogens may eventually provide a practical model.

   One bacterium that might be a good model is *Yersinia enterocolitica*, which specifically infects Peyer's patches and mesenteric lymph nodes, but not liver and spleen in 3 to 6 days (Dube et al., 2003; Petersen and Young, 2002). Another possibility could be based on *Helicobacter*, which causes infections limited to the gastroduodenal regions of the digestive tract and are characterized histopathologically by chronic, active inflammation. Innate responses are triggered by chemokine production in gastric epithelial cells that draw neutrophils to the site of infection. Although B and T cells contribute to *Helicobacter*-induced inflammation, Th1 cell responses play a prominent role (Ernst and Pappo, 2001). Practical application of advances in *Helicobacter* immunology for developing an infectivity assay that would be useful for assessing mucosal responses in an immunotoxicologic setting may be hampered by the time required postinoculation for clinical manifestations of gastritis to develop in rodents, usually longer than 1 week (Lee et al., 1997).

   Relative to viral pathogens, enteric infection with respiratory enteric orphan virus (reovirus) has been proposed as a robust gut mucosal infectivity model (Cuff et al., 1998). In immunocompetent adult mice, reovirus infection is limited to the gut. Mice treated with cyclophosphamide exhibit impaired viral elimination following challenge. In conjunction with increased intestinal viral titers, intestinal humoral and cellular responses such as virus-specific IgA production by Peyer's patches and lamina propria cultures and reovirus-specific cytotoxic T-cell activity were suppressed in cyclophosphamide-treated mice.

   Flagellated protozoans of the genus *Giardia* cause gastroenteritis in humans and in other mammals, birds, and reptiles. The infection is extracellular and is confined to the small intestine.

The *Giardia muris* mouse model provides a useful tool for the study of giardiasis, and may provide a host resistance model for assessing mucosal immune function. Both expulsion of *G. muris* from the intestines of infected mice and the concomitant appearance of anti-*G. muris* IgA in intestinal secretions would be convenient infectivity endpoints. Changes in T-cell subsets in intraepithelial and lamina propria lymphocytes have also been observed in infected mice, specifically increased numbers of CD8$^+$ cells that peak during the acute phase, leading to a peak in CD4$^+$ cells during the elimination phase (Faubert, 2000).

Rodents infected with *Trichinella spiralis* and *Nippospongylus brasiliensis* have been used as models to study immune responses to parasitic nematodes. In particular, *Trichinella spiralis* has been used in immunotoxicology studies to assess xenobiotic-induced changes in host resistance. Although these nematodes invade other organs from the gut, few mucosal responses to this agent have been measured in an immunotoxicologic setting. Parasite expulsion from the intestine is a standard measure of infectivity with this organism, and mesenteric lymph node (MLN) cell-proliferative responses, IL-2 production and changes in numbers of lymphocyte subgroups have been measured in *T. spiralis*-infected mice and rats exposed to cyclophosphamide (Luebke et al., 1992). It is known that CD4$^+$ T cells play an important role in protective immune responses to these nematodes and that IL-4 can be required for host protection and can limit the severity of infection (Finkelman et al., 1997); however, a more comprehensive understanding of mucosal immune responses to parasite models would be required to facilitate their application in mucosal host resistance assays.

13.3.3 Specialized Rodent Models

Based on the value of studies using transgenic and gene-targeted or "knockout" mice in under-standing the pathogenesis of inflammatory bowel disease (IBD), similar models may prove useful for clarifying the mechanisms of xenobiotic-induced immune modulation in GALT. It has been hypothesized that IBD develops as a result of an uncontrolled mucosal immune response to one or more normal gut constituents, which upsets the homeostasis between immune responsiveness to putative gut pathogens and unresponsiveness to foods and commensal gut microflora (Kiyono et al., 2001). Transgenic rats expressing human HLA-B27 and β2-microglobulin are susceptible to severe intestinal inflammatory disease, which is dependent on expression of HLA-B27 on immune cells. The presence of gut bacterial flora is required for disease development, as germ-free animals remain relatively disease-free (James and Klapproth, 1996). Gene-targeted mutant mice models of IBD include knockouts for IL-2, IL-10 T-cell receptors, (TCR)-α, and TCR-β (Elson et al., 1995). In each model, intestinal inflammation develops as a result of abnormal immune responses, allowing for insight into the role of specific immune pathways in the pathogenesis of IBD. Relatedly, knockout mice have been used to clarify the role of cytokine upregulation of IgA upregulation by deoxynivalenol. Here it was determined that mice deficient in IL-6 expression were recalcitrant to IgA induction by this mycotoxin (Pestka and Zhou, 2000). Similarly, tumor necrosis factor receptor 1 has been found to play a contributory role in IgA dysregulation (Pestka and Zhou, 2002).

13.4 CONCLUSIONS AND DIRECTIONS FOR FUTURE RESEARCH

The previous discussion highlights potential strategies and limitations related to gut mucosal immuno-toxicologic assessment. A primary issue related to quantitative assessment of mucosal immunotoxicity is that although it is functionally useful to separate mucosal and systemic compartments, immuno-competent cells constantly move back and forth across tissues delineated in these two groupings. With similar indifference, many pathogens bidirectionally cross the mucosal–systemic interface. Thus, existing tiered immunotoxicologic assessments might be adequate to detect the potential of those chemicals that modulate mucosal immunity as part of a generalized immune effect. However, the possibility exists that because a mucosal site such as the gut is barraged by a massive number of microbial and food-derived immune challenges, it may be much more susceptible to down- or up-regulation of immunocompetetent cell responses by natural toxins, phytochemicals, food additives, environmental chemicals, processing-induced toxicants, or drugs than would cells of the systemic immune compartment. In addition, the gut mucosal immune compartment may be selectively sensitive to food- or water-borne chemicals or macromolecules that are not readily absorbed but nevertheless may negatively affect a resident immunocompetent cell.

Taken together, further research is needed in several areas relative to development of reliable assays for gut mucosal immunotoxicity assessment. First, new methods for cell isolation, multicolor flow cytometry, confocal microscopy, and intracellular cytokine measurement should be incorporated in improved immunophenotyping protocols for analyzing mucosal tissues. Second, newly evolving strategies for mucosal vaccines relative to novel oral antigens, adjuvants, and dosing protocols should be used as templates for assessing humoral and, potentially, cell-mediated responses to antigens. Third, promising bacterial, viral, and parasitic mucosal infection models need to be further evaluated and modified toward the goal of reproducible, quantitative, mucosal immune response endpoints. Fourth, antigen and host-resistance models can be extended by using DNA and protein microarrays delimited for immune-related genes as well as transgenic and knockout mice. The robustness of these approaches will require testing using graded doses of prototypical immunotoxicants, such as cyclophosphamide, as well as xenobiotics, such as carageenan, that may be poorly absorbed but have extended residency in the intestine. Gut mucosal immunotoxicity data will ultimately have to be reconciled to possible effects on nasal lymphoid tissues as well as the impact of food allergenicity.

ACKNOWLEDGMENTS

The authors thank Paul Rowsell for valuable discussions and advice, Chris Oberg for assistance preparing Figure 13.1, Don Caldwell for supplying the photomicrograph for Figure 13.2, and Maria Victoria Tejada-Simon for the artwork in Figure 13.3.

REFERENCES

Aranda, R., Sydora, B.C., and Kronenberg, M., Intraepithelial lymphocytes: function, in *Mucosal Immunology*, 2nd ed., Ogra, P.L., Mestecky, J., Lamm, M.E., Strober, W., Bienenstock, J., and McGhee, J.R., Eds., Academic Press, Toronto, 1999, p. 429.

Azcona-Olivera, J.I., et al., Induction of cytokine mRNAs in mice after oral exposure to the trichothecene vomitoxin (deoxynivalenol): relationship to toxin distribution and protein synthesis inhibition, *Toxicol. Appl. Pharmacol.*, 133, 109, 1995.

Bandaletova, T.Y., Morphogenesis of epithelial lesions in the lymphoid patches of Fischer 344 rats during cyclosporine-A-induced prolonged immunosuppression, *Tumori*, 78, 366, 1992.

Bondy, G.S., and Pestka, J.J., Dietary exposure to the trichothecene vomitoxin (deoxynivalenol) stimulates terminal differentiation of Peyer's patch B cells to IgA-secreting plasma cells, *Toxicol. Appl. Pharmacol.*, 108, 520, 1991.

Bondy, G.S., and Pestka, J.J., Immunomodulation by fungal toxins, *J. Toxicol. Environ. Health Part B*, 3, 109, 2000.

Bruder, M.C., et al., Intestinal T lymphocytes of different rat strains in immunotoxicity, *Toxicol. Pathol.*, 27, 171, 1999.

Bunn, T.L., et al., Developmental immunotoxicity assessment in the rat: age, gender, and strain comparisons after exposure to lead, *Toxicol. Methods*, 11, 41, 2001.

Burchiel, S.W., et al., Inhibition of lymphocyte activation in splenic and gut-associated lymphoid tissues following oral exposure of mice to 7, 12-dimethylbenz[a]anthracene, *Toxicol. Appl. Pharmacol.*, 105, 434, 1990.

Cozon, G., et al., Transient secretory IgA deficiency in mice after cyclophosphamide treatment, *Clin. Immunol. Immunopathol.*, 61, 93, 1991.

Cuff, C.F., et al., Enteric reovirus infection as a probe to study immunotoxicity of the gastrointestinal tract, *Toxicol. Sci.*, 42, 99, 1998.

Daniels, C.K., Perez, P., and Schmucker, D.L., Alterations in CD8+ cell distribution in gut-associated lymphoid tissues (GALT) of the aging Fischer 344 rat: a correlated immunohistochemical and flow cytometric analysis, *Exp. Gerontol.*, 28, 549, 1993.

Davis, D.A.P., et al., Inhibition of humoral immunity and mitogen responsiveness of lymphoid cells following oral administration of the heterocyclic food mutagen 2-amino-1-methyl-6-phenylimidazol[4, 5-b]pyridine (PhIP) to B6C3F$_1$ mice, *Fundam. Appl. Toxicol.*, 23, 81, 1994.

Davis, D.P., et al., Suppression of local gut-associated and splenic mitogen responsiveness of lymphoid cells following oral exposure of B6C3F1 mice to 7, 12-dimethylbenz[a]anthracene, *Fundam. Appl. Toxicol.*, 17, 429, 1991.

Delacroix, D.L., Malburny, G.N., and Vaerman, J.P., Hepatobiliary transport of plasma IgA in the mouse: contribution to clearance of intravascular IgA, *Eur. J. Immunol.*, 15, 893, 1985.

deVos, T., and Dick, T.A., A rapid method to determine the isotype and specificity of coproantibodies in mice infected with Trichinella or fed cholera toxin, *J. Immunol. Methods*, 141, 285, 1991.

Di Genaro, M.S., et al., Intranasal immunization with Yersinia enterocolitica O:8 cellular extract protects against local challenge infection, *Microbiol. Immunol.*, 42, 781, 1998.

Dong, W., Sell, J.E., and Pestka, J.J., Quantitative assessment of mesangial immunoglobulin A (IgA) accumulation, elevated circulating IgA immune complexes, and hematuria during vomitoxin-induced IgA nephropathy, *Fundam. Appl. Toxicol.*, 17, 197, 1991.

Dong, W., and Pestka, J.J. Persistent dysregulation of IgA production and IgA nephropathy in the B6C3F1 mouse following withdrawal of dietary vomitoxin (deoxynivalenol), *Fundam. Appl. Toxicol.*, 20, 38, 1993.

Dube, P.H., et al., The rovA mutant of Yersinia enterocolitica displays differential degrees of virulence depending on the route of infection, *Infect. Immun.*, 71, 3512, 2003.

Elson, C.O., Ealding, W., and Lefkowitz, J., A lavage technique allowing repeated measurement of IgA antibody in mouse intestinal secretions, *J. Immunol. Methods*, 67, 101, 1984.

Elson, C.O., et al., Experimental models of inflammatory bowel disease, *Gastroenterology*, 109, 1344, 1995.

Ernst, P.B., and Pappo, J., T-cell-mediated mucosal immunity in the absence of antibody: lessons from Helicobacter pylori infection, *Acta Odontol. Scand.*, 59, 216, 2001.

Fagarasan, S., and Honjo, T., Intestinal IgA synthesis: regulation of front-line body defences, *Nat. Rev. Immunol.*, 3, 63, 2003.

Faubert, G., Immune response to Giardia duodenalis, *Clin. Microbiol. Rev.*, 13, 35, 2000.

Finkelman, F.D., et al., Cytokine regulation of host defense against parasitic gastrointestinal nematodes: lessons from studies with rodent models, *Ann. Rev. Immunol.*, 15, 505, 1997.

Forsell, J.H., et al., Effects of 8-week exposure of the B6C3F1 mouse to dietary deoxynivalenol (vomitoxin) and zearalenone, *Food Chem. Toxicol.*, 24, 213, 1986.

Forsell, J.H., et al., Comparison of acute toxicities of deoxynivalenol (vomitoxin) and 15-acetyldeoxynivalenol in the B6C3F1 mouse, *Food Chem. Toxicol.*, 25, 155, 1987.

Fujihashi, K., et al., A revisit of mucosal IgA immunity and oral tolerance, *Acta Odontol. Scand.*, 59, 301, 2001.

Goldblum, R.M., The role of IgA in local immune protection, *J. Clin. Immunol.*, 10, 64S, 1990.

Hamada, F.M., et al., Possible functional immunotoxicity of acrylonitrile, *Pharmacol. Res.*, 37, 123, 1998.

Hau, J., Andersson, E., and Carlsson, H.-E., Development and validation of a sensitive ELISA for quantification of secretory IgA in rat saliva and feces, *Lab. Anim.*, 35, 301, 2001.

Hinton, D.M., "Redbook II" immunotoxicity testing guidelines and research in immunotoxicity evaluations of food chemicals and new food proteins, *Toxicol. Pathol.*, 28, 467, 2000.

Hornef, M.W., et al., Bacterial strategies for overcoming host innate and adaptive immune responses, *Nat. Immunol.*, 3, 1033, 2002.

International Collaborative Immunotoxicity Study, Report of validation study of assessment of direct immunotoxicity in the rat, *Toxicology*, 125, 183, 1998.

International Programme on Chemical Safety, Environmental Health Criteria 180, Principles and methods for assessing direct immunotoxicity associated with exposure to chemicals, Geneva, World Health Organization, 1996.

James, S.P., and Klapproth, J.-M., Major pathways of mucosal immunity and inflammation: cell activation, cytokine production and role of bacterial factors, *Aliment. Pharmacol. Ther.*, 10, 1, 1996.

Jin, G.-F., Guo, Y.-S., and Houston, C.W., Bombesin: an activator of specific Aeromonas antibody secretion in rat intestine, *Digest Dis. Sci.*, 34, 1708, 1989.

Karacic, J.J., and Cowdery, J.S., The effect of single dose, intravenous cyclophosphamide on the mouse intestinal IgA response to cholera toxin, *Immunopharmacology*, 16, 53, 1988.

Kawabata, T.T., et al., Immunotoxicology of regional lymphoid tissue: the respiratory and gastrointestinal tracts and skin, *Fundam. Appl. Toxicol.*, 26, 8, 1995.

Keren, D.F., et al., The enteric immune response to Shigella antigens, *Curr. Top. Microbiol. Immunol.*, 146, 213, 1989.

Kiyono, H., et al., The mucosal immune system: from specialized immune defense to inflammation and allergy, *Acta Odontol. Scand.*, 59, 145, 2001.

Kuper, C., et al., Histopathologic approaches to detect changes indicative of immunotoxicity, *Toxicol. Pathol.*, 28, 454, 2000.

Lee, A., et al., A standardized mouse model of Helicobacter pylori infection: introducing the Sydney strain, *Gastroenterology*, 112, 1386, 1997.

Li, J., et al., Bombesin affects mucosal immunity and gut-associated lymphoid tissue in intravenously fed mice, *Arch. Surg.*, 130, 1164, 1995.

Luebke, R.W., et al., Host resistance to Trichinella spiralis infection in rats and mice: species-dependent effects of cyclophosphamide exposure, *Toxicology*, 73, 305, 1992.

Luster, M.I., et al., Risk assessment in immunotoxicology. I. Sensitivity and predictability of immune tests, *Fundam. Appl. Toxicol.*, 18, 200, 1992.

Mascart-Lemone, F., et al., A polymeric IgA response in serum can be produced by parenteral immunization, *Immunology*, 61, 409, 1987.

Mayer, L., Mucosal immunity and gastrointestinal antigen processing. *J. Pediatr. Gastroenterol. Nutr.*, 30, S4, 2000.

Moolenbeek, C., and Ruitenberg, E.J., The "Swiss roll": a simple technique for histological studies of the rodent intestine, *Lab. Anim.*, 15, 57, 1981.

Nagata, T., et al., Development of apoptosis and changes in lymphocyte subsets in thymus, mesenteric lymph nodes and Peyer's patches of mice orally inoculated with T-2 toxin, *Exp. Toxicol. Pathol.*, 53, 309, 2001.

Neurath, M.F., Finotto, S., and Glimcher, L.H., The role of Th1/Th2 polarization in mucosal immunity, *Nat. Med.*, 8, 567, 2002.

Neutra, M.R., Mantis, N.J., and Kraehenbuh, J.-P., Collaboration of epithelial cells with organized mucosal lymphoid tissues, *Nat. Immunol.*, 2, 1004, 2001.

Ni, J., and Hollander, D., Applications of the MTT assay to functional studies of mouse intestinal intraepithelial lymphocytes, *J. Clin. Lab. Anal.*, 10, 42, 1996.

Ogra, P.L., et al., Eds., *Mucosal Immunology*, Academic Press, Toronto, 1999.

Organization for Economic Cooperation and Development, Guideline for the Testing of Chemicals 407: Repeated Dose 28-Day Oral Toxicity Study in Rodents, OECD, Paris, 1995.

Peppard, J.V., and Russell, M.W., Phylogenetic development and comparative physiology of IgA, in *Mucosal Immunology*, 2nd ed., Ogra, P.L., et al., Eds., Academic Press, Toronto, 1999, p. 163.

Pestka, J., Food, diet, and gastrointestinal immune function, *Adv. Food Nutr. Res.*, 37, 1, 1993.

Pestka, J.J., Deoxynivalenol-induced IgA production and IgA nephropathy-aberrant mucosal immune response with systemic repercussions, *Toxicol. Lett.*, 140, 287, 2003.

Pestka, J.J., et al., Effect of dietary administration of the trichothecene vomitoxin (deoxynivalenol) on IgA and IgG secretion by Peyer's patch and splenic lymphocytes, *Food Chem. Toxicol.*, 28, 693, 1990a.

Pestka, J.J., et al., Elevated membrane IgA+ and CD4+ (T helper) populations in murine Peyer's patch and splenic lymphocytes during dietary administration of the trichothecene vomitoxin (deoxynivalenol), *Food Chem. Toxicol.*, 28, 409, 1990b.

Pestka, J.J., Moorman, M.A., and Warner, R.L., Dysregulation of IgA production and IgA nephropathy induced by the trichothecene vomitoxin, *Food Chem. Toxicol.*, 27, 361, 1989.

Pestka, J.J., and Zhou, H.R., Interleukin-6–deficient mice refractory to IgA dysregulation but not anorexia induction by vomitoxin (deoxynivalenol) ingestion, *Food Chem. Toxicol.*, 38, 565, 2000.

Pestka, J.J., and Zhou, H.R., Effects of tumor necrosis factor type 1 and 2 receptor deficiencies on anorexia, growth and IgA dysregulation in mice exposed to the trichothecene vomitoxin, *Food Chem. Toxicol.*, 40, 1623, 2002.

Petersen, S., and Young, G.M., Essential role for cyclic AMP and its receptor protein in Yersinia enterocolitica virulence, *Infect. Immun.*, 70, 3665, 2002.

Pitman, R.S., and Blumberg, R.S., First line of defense: the role of the intestinal epithelium as an active component of the mucosal immune system, *J. Gastroenterol.*, 35, 805, 2000.

Poon, R., et al., Effects of subchronic exposure of monochloramine in drinking water on male rats, *Regul. Toxicol. Pharmacol.*, 25, 166, 1997.

Rasooly, L., and Pestka, J.J., Vomitoxin-induced dysregulation of serum IgA, IgM and IgG reactive with gut bacterial and self antigens, *Food Chem. Toxicol.*, 30, 499, 1992.

Rasooly, L., and Pestka, J.J., Polyclonal autoreactive IgA increase and mesangial deposition during vomitoxin-induced IgA nephropathy in the BALB/c mouse, *Food Chem. Toxicol.*, 32, 329, 1994.

Rasooly, L., et al., Polyspecific and autoreactive IgA secreted by hybridomas derived from Peyer's patches of vomitoxin-fed mice: characterization and possible pathogenic role in IgA nephropathy, *Food Chem. Toxicol.*, 32, 337, 1994.

Schmucker, D.L., Intestinal mucosal immunosenescence in rats, *Exp. Gerontol.*, 37, 197, 2002.

Shalaby, W.S., Development of oral vaccines to stimulate mucosal and systemic immunity: barriers and novel strategies, *Clin. Immunol. Immunopathol.*, 74, 127, 1995.

Temple, L., et al., Comparison of ELISA and plaque-forming cell assays for measuring the humoral response to SRBC in rats and mice treated with benzo[a]pyrene or cyclophosphamide, *Fundam. Appl. Toxicol.*, 21, 412, 1993.

Vargas, J.A., Vessey, D.A., and Schmucker, D.L. Effect of dehydroepiandrosterone (DHEA) on intestinal mucosal immunity in young adult and ageing rats, *Exp. Gerontol.*, 33, 499, 1998.

Watzl, B., et al., Short-term moderate aflatoxin B_1 exposure has only minor effects on the gut-associated lymphoid tissue of Brown Norway rats, *Toxicology*, 138, 93, 1999.

White, K.A. Jr., Specific immune function assays, in *Principles and Practise of Immunotoxicology*, K. Miller, K., Turk, J., and Nicklin, S., Eds., Blackwell Scientific Publications, Boston, 1992, p. 304.

Yan, D., et al., Potential role for IL-5 and IL-6 in enhanced IgA secretion by Peyer's patch cells isolated from mice acutely exposed to vomitoxin, *Toxicology*, 122, 145, 1997.

Yan, D., et al., Role of macrophages in elevated IgA and IL-6 production by Peyer's patch cultures following acute oral vomitoxin exposure, *Toxicol. Appl. Pharmacol.*, 148, 261, 1998.

Zhou, H.R., Yan, D., and Pestka, J.J., Differential cytokine mRNA expression in mice after oral exposure to the trichothecene vomitoxin (deoxynivalenol): dose response and time course, *Toxicol. Appl. Pharmacol.*, 144, 294, 1997.

Zhou, H.R., et al., Amplified proinflammatory cytokine expression and toxicity in mice coexposed to lipopolysaccharide and the trichothecene vomitoxin (deoxynivalenol), *J. Toxicol. Environ. Health*, A57, 115, 1999.

CHAPTER 14

The Use of Old World Nonhuman Primates as Models for Evaluating Chemical-Induced Alterations of the Immune System

Helen Tryphonas and Douglas L. Arnold

CONTENTS

14.1 INTRODUCTION

It is increasingly recognized that immunotoxicology — the investigation of chemical-induced alterations on the structure and function of the immune system — should be an inherent component in any multidisciplinary toxicology research endeavor. This type of investigation is a necessary component of the risk assessment process, which seeks to determine whether exposure to a chemical will result in an adverse health effect. While a few databases exist regarding the direct exposure of humans to potentially harmful chemicals, the majority of data supporting human health risk evaluations are currently generated by extrapolation of data from experimental animal models such as rodents and canines. However, it is known that considerable differences exist in the metabolism of chemicals and the function of the immune system between humans and experimental animal models. For this reason, evaluators have implemented the use of uncertainty factors when using experimental animal data to determine human health risks.

While data derived from direct human exposure situations would negate species–species uncertainty, for obvious ethical reasons the use of humans as experimental research models is prohibited. While some epidemiologic data may be available from industrial exposure studies, industrial accidents, and inadvertent environmental exposure, such data have a number of shortcomings because they are limited in scope. Consequently, there is a need to establish experimental animal models that would closely mimic human responses to xenobiotic chemicals. To fulfill this need, nonhuman primate centers are actively engaged in the development and validation of immunologic methodologies for several nonhuman primate species. The need to develop a nonhuman primate model for immunological testing is becoming increasingly important in safety evaluations as new and potentially immunotoxic pharmacologic agents and biotechnology products are being developed. The major focus of this chapter is devoted to the description and discussion of several immunologic assays that have been developed, adapted, and successfully applied to the study of chemical-induced immunotoxicity in nonhuman primates. However, the necessity for and the ability to maintain a high-quality experimental animal cannot be overlooked. Therefore, the discussion will commence with issues pertaining to the husbandry, breeding, dosing, and clinical evaluation considerations inherent when a research program strives to maintain the highest-quality animal model possible, which in turn will go a long way toward ensuring the acquisition of valid data (Canadian Council on Animal Care, 2004).

14.2 TOXICOLOGIC CONSIDERATIONS

14.2.1 General Comments

Nonhuman primates present an experimental model that is unique among laboratory species because of its phylogenetic proximity to humans. However, it should not be assumed that monkeys will respond pharmacokinetically or toxicokinetically to xenobiotics in a manner analogous to humans (Dixon, 1976; Ruelius, 1987; Fuller et al., 1992; Brown, 1994; Mes et al., 1995a). For example, some species of monkeys appear to be more sensitive to the dermal toxicity of polychlorinated biphenyls (PCBs) than humans (Safe, 1994). That aside, the larger species of monkeys do afford an experimenter the luxury of using each monkey as its own control; that is, one can acquire blood, fat, and semen specimens prior to study initiation to ascertain the animal's pretreatment values and health status. Similarly, one can keep previously dosed monkeys on test for an indefinite period to collect additional blood and fat specimens for purposes of determining the half-lives of the test chemical and/or its metabolites (Mes et al., 1995a, 1995b). Concurrently, immunological testing can also be conducted during pre-dosing, dosing, and post-dosing. Another advantage when working with monkeys is the fact that monkey colonies are relatively new and, consequently, the genetic makeup of a study population will be more heterogeneous than would be found with purpose-bred species. In some experimental situations though, such variability could result in a requirement for a larger study population. As an aside, the Health Canada cynomolgus monkey colony offers some potential assistance in this regard since their lineage for the past two decades is known. In addition, these monkeys are now considered specific pathogen free (SPF). In the North American context, the term SPF is defined as the monkeys being free (i.e., antibody negative) for herpes B virus; simian immunodeficiency virus (siv); type D simian retrovirus (srv); and simian T-lymphotropic virus (stlv-1). Furthermore, ageing monkeys provide unique experimental models regarding immunological evaluation, since latent onset diabetes does occur within our cynomolgus monkey colony and the rhesus monkeys are more prone to develop endometriosis than our colony's cynomolgus monkeys. Furthermore, immune senescence in rhesus monkeys appears to occur around 20 years of age (Coe, 2004), which presents another unique experimental.

14.2.2 Experimental Considerations

14.2.2.1 Dosing

One of the major advantages of using cynomolgus or rhesus monkeys for toxicological studies is that they can be trained to consume gelatin capsules containing the test chemical. This consideration is particularly helpful when the test chemical is volatile or distasteful to consume. To preclude consumption problems, we have found that it is best to use a masking agent, such as glycerol, during the training and experimental period. While monkeys will often play with their food, occasionally throwing it out of their cages, we have found that they will readily consume a gelatin capsule containing the glycerol and test chemical. In one 10-year study, only 0.027% of the gelatin capsules were discarded by the monkeys. However, changing the contents of the gelatin capsule from a liquid to a solid may result in the monkey pulling the capsule apart and dumping its contents into the litter pan. Our experience has been that the inclusion of the test agent in the drinking water or diet severely compromises accurate consumption data and could result in contamination of the animals' facility (Sansone et al., 1977; Sansone and Fox, 1977; Keene and Sansone, 1984). Consumption data for test entities delivered in the drinking water were particularly frustrating since monkeys readily shake their cages, with a concurrent loss of water from their drinking vesicle.

While book chapters have been written on dose selection, suffice it to say that normally experimental protocols for immunotoxicity studies use a control (nontreated vehicle control) and three dose groups (low, medium, and high). As with all toxicological studies, at least one of the doses should result in frank toxicity. However, since immunotoxicologic studies are designed to detect somewhat subtle toxicological effects, the use of a dose level that results in significant body weight loss compared to the controls (i.e., 5 to 10%) or debilitating clinical manifestations are clearly inappropriate for immunotoxicologic studies. A significant induction of endoplasmic enzymatic activity in the treated versus control is the type of toxicological effect desired for an immunological study.

The last two points that should be mentioned in regard to dosing pertains to the test chemical and pilot studies. The test chemical should always be analyzed prior to the start of the study, as well as each subsequent batch, even when the vendor has provided a certificate of analysis. A pilot study should also be undertaken when one is starting out a research program with a new chemical, since there are a number of environmental and other factors that can alter the relative toxicity of a test chemical (Gad, 1992). Pilot studies may also alert one to an unexpected toxicological response, thereby facilitating subsequent data interpretation.

14.2.2.2 Specimens

As previously indicated, numerous samples of blood, fat, and semen can be obtained from monkeys if sufficient time is allocated between sampling periods. For example, the accepted rule of thumb regarding a single blood sampling event is up to 10% of the circulating blood volume; for a rhesus monkey, the circulating blood volume is approximately 54 ml/kg of body weight (b.w.), while it is 65 ml/kg b.w. for the cynomolgus monkey. This volume of blood can be repeated every 3 to 4 weeks. For repeated blood sampling (i.e., every 24 hours), the amount that can be obtained is $0.01 \times$ circulating blood volume or roughly 0.6 ml/kg b.w./day. This level of frequency did not induce any degree of detectable anamia in our monkeys (Joint Working Group, 1993; Hawk and Leary, 1995; Diehl et al., 2001).

We usually obtained fat samples from the scapular area — which minimized the monkey's ability to pick at the incision — using a local anesthetic. This procedure requires an experienced technician to conduct the sampling in order to separate the connective tissue from the fat. Appropriate spacing of fat and blood collection allows for the calculation of accumulation rates and elimination of half-lives. To collect blood and fat biopsies from female rhesus or cynomolgus monkeys, the monkey would be placed in a head-only stockade device attached to a table. The blood sample can be readily acquired from the femoral artery or vein (Fernie et al., 1994). Upon conclusion of sample acquisition, the monkey received a fruit or vegetable reward. Some laboratories have reported that they trained their monkeys to stick their arms through the cage bars while the blood sample was obtained (Reinhardt, 1992, 1996). We did not attempt to undertake this method of blood sampling. Occasionally, when obtaining blood from infant monkeys, it was necessary to anesthetize them, as they often were exceedingly excitable upon separation from their mother. However, such treatment did not appear to affect their feed/water consumption. Just about any manipulation of the males required that they be anesthetized (ketamine). The use of an anesthetic was necessary when obtaining blood from a male monkey, and it often resulted in the male going off his feed and water for 24 to 48 hours, which is not an ideal result. Consequently, male specimens were not obtained with any periodicity. Similarly, sperm was only obtained pre-dosing, pre-breeding, and post-breeding by electro-ejaculation using a procedure that was developed in-house (Maleck et al., 1996).

14.2.2.3 Housing

For most toxicological studies, individual housing of test animals is desirable for a number of reasons, including but not limited to data acquisition (feed and water consumption), enhanced quantification of test substance ingestion/exposure, transmission of disease, and fighting/wounding/injury/death. The major drawback is that laboratory species have some type of "societal" structure in their native habitat, especially monkeys who are known to have a hierarchical "society." Even with pair housing during a toxicological study, there is still a concern regarding compatibility between potential partners, which can be exacerbated when one is conducting a reproduction study. We were successful in minimizing potential incompatibilities by placing the female's transfer box on the front of the male's cage for a few minutes, and if nothing untoward happened, the female was allowed to enter the male's cage. However, an animal attendant remained in the area for several minutes since the occasional male would initially appear to be compatible, only to exhibit subsequent aggressive behavior. In this situation, the male would be removed from the study.

One of the greatest concerns with nonhuman primates that are housed in animal laboratories is the development of stereotypies and/or autoaggression. Various types of laboratory enrichment stategies have been developed in an attempt to minimize these abnormal behaviors (Bryant et al., 1988; Beaver, 1989; Crockett et al., 1989; Fajzi et al., 1989; Reinhardt et al., 1995).

14.2.2.4 Clinical Parameters

As most monkey colonies are genetically heterogeneous, it is often difficult to compare one colony's hematological and serum biochemistry values with those reported in the literature. This situation is somewhat exacerbated by the lack of sufficient detail regarding experimental procedures. Even with the improvements in animal care procedures over the years, interlaboratory comparisons of clinical data are due to such experimental variables as the time of day that the sample was acquired, whether an anesthetic was used, instrumentation, age and sex of the monkey, sample handling, and housing arragements (Fernie et al., 1994; Flow and Jaques, 1997). We found that one of the more important experimental variables when acquiring blood sample is the order in which the monkeys were sampled. In short, we found it necessary to sample the monkeys within any one room in a totally random order, particularly when one was obtaining blood samples for hormone analysis. However, our experience has shown that the more extensive the database for the experimental group, the more confidence one has when interpreting the subsequent experimental data obtained.

14.3 IMMUNE PARAMETERS AVAILABLE FOR USE IN NONHUMAN PRIMATES

14.3.1 General

The systematic development and cross-laboratory validation of immunologic methods applicable to nonhuman primate models is a desirable but difficult task to achieve for many reasons including maintenance costs and scarcity of primate facilities. Consequently, the existing limited methodology is the product of research carried out in a handful of institutions that support an in-house nonhuman primate facility. Through such efforts, a number of immunologic parameters have been developed/adapted and validated for application to nonhuman primates. In addition to hematologic profiles (total white blood cell counts and differentials) and immunohistopathologic techniques, many other assays are typically used in a clinical laboratory and have been successfully applied to monkeys. These are grouped as shown in Table 14.1.

Table 14.1   Immune Parameters Developed/Adapted for Use in Nonhuman Primates

A. General parameters of immune system
- Total white blood cell counts
- Immunophenotyping of peripheral blood lymphocytes

B. Assays to study effects on humoral immunity
- Total serum immunoglobulin (IgG, IgA, and IgM) levels
- Challenge with specific antigens
- Sheep red blood cells, tetanus toxoid, pneumococcal antigens, and determination of antigen-specific antibody levels in serum

C. Assays to study cell-mediated immunity
- Lymphocyte transformation (^{3}H-thymidine incorporation) in response to the mitogens phytohemagglutinin-P (PHA-P), concanavalin A, and pokeweed mitogen (PWM), or specific antigens such as tetanus toxoid
- Mixed lymphocyte cultures using allogeneic lymphocytes
- Delayed-type hypersensitivity using dinitrochlorobenzene as the sensitizing agent

D. Assays to study nonspecific immunity
- Monocyte function (activation, phagocytosis, and respiratory burst activity)
- Natural killer cell numbers and function
- Serum complement levels
- Cytokine levels basal and in lectin-activated cultures
- Hydrocortisone levels

Detailed descriptions of these methodologies are found elsewhere (Karpinski et al., 1987; Arnold et al., 1993; Tryphonas et al., 1989, 1991a, 1991b, 1996, 1999; Loo et al., 1989). In this chapter, key issues to be taken into consideration when using these methodologies are presented and discussed.

14.3.1.1 Total White Blood Cell Counts

Quantitative and morphologic investigations of total white blood cells (WBC) and differential counts of peripheral blood (PB) are basic investigations, and have been included in most of the immuno-toxicity studies involving nonhuman primates. In nonhuman primate studies, WBC counts have proven useful in signaling chemical-induced alterations on the immune system (Arnold et al., 1993). Both relative and absolute numbers of WBCs are quantified. However, the consensus is that absolute numbers provide biologically more relevant information, as the use of percentages of cell types may mask some cytopenias or excessive numbers of a cell type, which would lead to falsely high/low numbers of a particular cell (Perkins et al., 1999).

14.3.1.2 Immunophenotyping of Peripheral Blood Leukocytes

Immunophenotyping of PB leukocytes, using monoclonal antibodies directed to cell-surface markers, and flow cytometric techniques, have become an important tool in the diagnosis of hematologic and immunologic disorders (Perkins, 1999). Cross-reactivity between mouse–antihuman monoclonal antibodies (mAbs) and monkey leukocyte surface antigens has been demonstrated for several human mAbs using the whole-blood lysis technique in two-color, fluorescein isothiocyanate, and phycoerythrin (PE), flow cytometric analysis (Ahmed-Ansari et al., 1989; Tryphonas et al., 1996). Reference values have been established for infants and adult *Macaque fascicularis* (cynomolgus) (Tryphonas et al., 1996), the *Macaque mulatta* (rhesus) (Tryphonas et al., 1991), the *Macacca nemestrina* (pig-tailed macaque) (Ahmed-Ansari et al., 1989), the *Cercocebus atys* (sooty mang-abeys) (Ahmed-Ansari et al., 1989), the *Callithrix jacchus* (marmoset), and the *Pan troglodytes* (chimpanzees) (Eichberg et al., 1988).

Valuable data can be gained from repeated measures of lymphocyte subpopulations during the course of a study. Such data can be useful in elucidating the mode of action of the chemical in question. For example, infant cynomolgus monkeys exposed to toxaphene *in utero*, via lactation and postweaning to adulthood, presented with significant shifts in the CD4+ and CD8+ T-cell subpopulations at weaning and approximately a year later (Tryphonas et al., CPT Consultants in Pathology and Toxicology Inc., Nepean, Ontario, Canada, manuscript in preparation).

14.3.1.3 Quantification of Total Serum Immunoglobulin Levels

Total serum immunoglobulin (Ig) levels (IgG, IgM, and IgA) can be quantified in nonhuman primates using the ELISA technique and appropriate statistics to evaluate the results (Karpinski et al., 1987; Tryphonas et al., 1989). Mean Ig levels determined in 80 healthy adult rhesus monkeys are listed in Table 14.2. However, the determination of total serum Ig levels in experimental animals has not proven useful, since significant effects on immune function are required before any effect

Table 14.2 Serum Immunoglobulin Levels in Naïve Adult Healthy Rhesus *(Macaca mulatta)* Monkeys

Ig Class	N	Mean mg/dL	Standard Error	Range
IgG	80	1581.09	32.26	879.57–3467.71
IgM	80	230.89	55.17	71.26–528.47
IgA	80	399.96	11.92	81.09–2120.71

on total serum Ig levels can be observed. For example, no effects on serum Ig levels were observed in rhesus monkeys exposed to low levels of PCBs in a chronic study, although significant effects were observed regarding the ability of the same animals to respond to a foreign antigen (Tryphonas et al., 1991).

14.3.2 Humoral Immunity

14.3.2.1 Challenge with Specific Antigens

The immune system is endowed with a large functional reserve capacity, and the changes in WBC numbers or shifts in lymphocyte subsets observed in many studies may not be accompanied by changes in immune function (Tryphonas and Feeley, 2001). Thus, the functional capacity of the immune system must be established. This can be accomplished by challenging the control and treated monkeys with foreign antigens, followed by the determination of antigen-specific antibody levels in serum samples collected prior to antigenic challenge (baseline titers) and at weekly intervals postimmunization. Unlike total serum immunoglobulins and T-lymphocyte subsets that are determined sequentially, the challenge with foreign antigens is performed once during or at the end of treatment. For chemicals such as PCBs and toxaphene that are known to accumulate in the fat of the experimental animal model, it will be biologically more meaningful if an antigenic challenge was performed when the chemical in question reached detectable fat/blood equilibrium. In nonhuman primates challenge with specific antigens has been highly predictive of effects on humoral immunity (reviewed by Tryphonas and Feeley, 2001).

The response to an antigenic challenge involves the sequential and tightly orchestrated interactions of competent immune cells, including the macrophage/monocyte (antigen-processing and antigen-presenting cells), and the activated T- and B-lymphocytes. Consequently, much information can be derived from challenging the host with foreign antigens. Clinically relevant information includes the ability of the host to respond to a foreign antigen (primary response) and to establish memory (secondary or anamnestic) response. Establishing memory endows the host with the ability to respond to a second insult by the same antigen in a much shorter time relative to the primary response. The analysis of pre- and post-immunization titers of antigen-specific antibody (IgM and IgG) in the serum collected at weekly intervals can also be used to study the catabolic rate of the specific antibody and the ability of the immune system to switch from IgM to IgG (i.e., isotypic class switching), both of which influence the level of detectable antibody to a given antigen (Virella, 1998).

For antigenic challenge to be meaningful, it is important to have documented histories of the experimental monkeys since previous immunization can lead to erroneous results and should be avoided. Antigens that are routinely used in rhesus and cynomolgus monkeys include sheep red blood cells (SRBC), tetanus toxoid (tt), and pneumococcus (pneu) antigens. A typical immunization schedule would be as follows. First, an intravenous dose of 1×10^9 SRBC in physiologic saline is injected iv for a primary response. This dose is repeated 4 weeks later for a secondary response. SRBC-specific titers (IgM and IgG) are determined in aliquots of sera collected at preimmunization and at weekly intervals for 8 weeks thereafter using the hemagglutination-complement lysis (IgM) or the indirect hemagglutination (IgG) methods (Tryphonas et al., 1991) (Figures 14.1 and 14.2). Peak antibody titers are observed 7 days postimmunization, and gradually return to almost baseline level by week 4 postimmunization. In view of the observed differences in immunogenicity among different batches of SRBC, it is advisable to use the same source of SRBC for immunization and for the antibody detection systems throughout the study. We routinely maintain two sheep for the purpose of collecting SRBC, and standardization of procedures is performed at the initiation of a study using SRBC from each of these sheep. The sheep's health status is closely monitored during the sampling period.

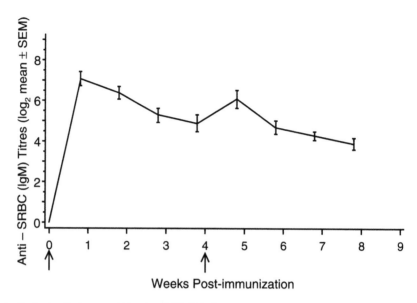

Figure 14.1 Typical anti-sheep red blood cell (SRBC) (IgM) titer curves observed in adult female cynomolgus (*Macaque fascicularis*) monkeys following immunization with SRBC for a primary and secondary (anamnestic) response. An immunizing dose of 1×10^9 SRBC in physiologic saline is injected intravenously at week 0 (primary response) and at week 4 (secondary response). Aliquots of serum are collected prior to immunization and at weekly intervals thereafter, and anti-SRBC titers are determined using the hemagglutination-complement assay. (Modified from Tryphonas, H., *Environ. Health Perspect.*, 109(Suppl. 6), 877–884, 2001.)

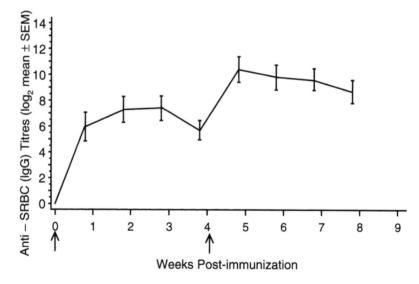

Figure 14.2 Typical anti-sheep red blood cell (SRBC) (IgG) titer curves observed in healthy adult female cynomolgus (*Macaque fascicularis)* monkeys following immunization with SRBC for a primary and secondary (anamnestic) response. An immunizing dose of 1×10^9 SRBC in physiologic saline is injected intravenously at week 0 (primary response) and at week 4 (secondary response). Aliquots of serum are collected prior to immunization and at weekly intervals thereafter, and anti-SRBC titers are determined using the indirect hemagglutination assay. (Modified from Tryphonas, H., *Environ. Health Perspect.*, 109(Suppl. 6), 877–884, 2001.)

Immunization with tt and pneu antigens has been incorporated during the last 2 weeks of the SRBC immunization schedule. At this point, 0.5 ml pneu (pneumovax-23, Merck, Sharp & Dohme, Canada; Division of Merck Frost Canada Inc., Kirkland, Quebec, Canada) or 0.5 ml (5 Lf units)

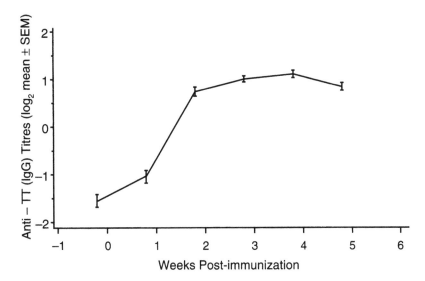

Figure 14.3   Typical anti-tetanus toxoid (tt) (IgG) titer curves observed in healthy adult female cynomolgus (*Macaque fascicularis)* monkeys following immunization with 0.5 ml (5 Lf units) of tt intramuscularly. Aliquots of serum are collected prior to immunization and at weekly intervals thereafter, and anti-tt titers are determined using the ELISA technique. Anti-tt titers ($\log_{2\ mean} \pm$ SEM) are expressed as IgG antibody nitrogen/mL by multiplying the calculated micrograms of IgG per milliliter value in reference to the standard by 0.16 (16% protein nitrogen) assay. (Modified from Tryphonas, H., *Environ. Health Perspect.*, 109(Suppl. 6), 877–884, 2001.)

of tt (Bureau of Biologics, Health Products and Food Branch, Ottawa, Ontario, Canada) are administered intramuscularly at different sites. The development of anti-pneu– or anti-tt–specific antibodies (IgG) are determined in sera collected at preimmunization and at weekly intervals thereafter using ELISA techniques (Tryphonas et al., 2000). Results of the two antigens are indicative of a secondary (IgG) response, and normally a high preimmunization titer exists. However, a drop in antibody titer has been observed during the first week following immunization. This is presumably due to the *in vivo* binding of the incoming antigen to the preformed antibody.

Subsequently, antibody titers to both antigens increase and reach a plateau at 2 weeks post-immunization and remain at this level for at least 5 weeks (Figures 14.3 and 14.4). As is the case with the SRBC, a detailed history of experimental animals, particularly with respect to previous immunization, is imperative.

14.3.3 Cell-Mediated Immunity

Cell-mediated immunity (CMI) has been studied in nonhuman primates using three methods: the lymphocyte transformation (³H-thymidine incorporation) (LT) in response to the T-cell mitogens phytohemagglutinin-P (PHA-P), concanavalin A (Con-A) and the B-cell pokeweed mitogen (PWM) or to specific antigens such as tetanus toxoid; the mixed lymphocyte culture assay using allogeneic cells; and the delayed-type hypersensitivity response using dinitro-chlorobenzene (DNCB) as the sensitizing agent. It is advisable to run these assays prior to any immunization of the monkeys. The LT and mixed lymphocyte culture (MLC) assays can be run concurrently, and are followed by the DTH assay.

14.3.3.1 Lymphocyte Transformation

The LT assay is considered an *in vitro* clinical correlate of DTH. Plant mitogens such as the PHA and Con-A, which are T-cell mitogens, and a B-cell mitogen such as the PWM, are frequently used

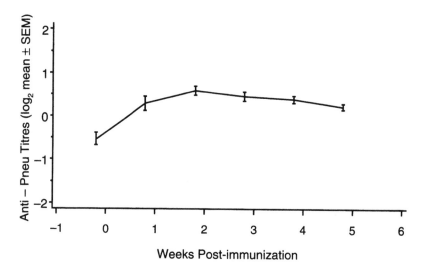

Figure 14.4 Typical anti-pneumococcal (IgG) titer curves observed in healthy adult female cynomolgus (*Macaque fascicularis*) monkeys following immunization with 0.5 mL pneumococcal (pneumovax-23) vaccine. Aliquots of serum are collected prior to immunization and at weekly intervals thereafter, and anti-pneumococcal titers are determined using the ELISA technique. Anti-pneumococcal titers ($log_{2 mean} \pm$ SEM) are expressed as IgG antibody nitrogen/mL by multiplying the calculated micro-grams of IgG per milliliter value in reference to the standard by 0.16 (16% protein nitrogen) assay. (Modified from Tryphonas, H., *Environ. Health Perspect.*, 109(Suppl. 6), 877–884, 2001.)

in this assay. Variations among different lots of mitogens have been observed. Therefore, it is recommended that the lot of mitogens remain constant within a given study, and that each lot of mitogens be titrated prior to initiation of a study to determine optimum levels to be used in the experimental protocols. The LT assay used in this context has provided useful information regarding the mechanism of action for some chemicals but its predictive value for immune functional impairment is low. A useful extension of the LT assay is the quantitative analysis of Igs in culture supernatants.

14.3.3.2 Mixed-Lymphocyte Cell Cultures

The MLC assay uses responder cells (peripheral blood mononuclear cells) from control and treated monkeys isolated by Ficoll–Hypaque gradient, and stimulator cells (pools of peripheral blood mononuclear cells) isolated by Ficoll–Hypaque gradients. Three serum pools (four monkeys/pool) are prepared and frozen using the Programmable Freezing Controller Model 701 (CRYO-MED, Mt. Clemens, MI). It is important that the source of monkeys used as stimulator cells be, as far as possible, unrelated to the experimental monkeys.

14.3.3.3 Delayed-Type Hypersensitivity Response

The DTH assay measures the response of monkeys to dinitrochlorobenzene (DNCB, Aldrich Chemical Company, St. Louis, MO). It utilizes the sensitization and challenge protocols previously detailed by Bugelski et al. (1990). Induration of the challenge sites is evaluated by measuring skin fold thickness in millimeters (average of three measures) with a constant pressure calliper (Starret, Athol, MA) prior to applying DNCB or acetone (control), and at 24 and 48 hours postapplication. The challenge site is also scored for macroscopic appearance. Erythema, contour, induration, texture, and appearance of superficial layers are scored by comparing these to adjacent skin. A clinical score is calculated by assigning a value of 1 for each affected parameter. The scores for affected parameters are added to give the clinical score. The highest possible clinical score was 5. In addition, each of the five parameters is scored for severity with a score of 0 = absent, 1 = low, 2 = slight, 3 = moderate, 4 = marked, and 5 = severe. The total clinical score is calculated by adding the individual scores. All evaluations are performed in a blind fashion, without prior knowledge of dose.

14.3.4 Nonspecific Immunity

14.3.4.1 Natural Killer Cell Activity

A number of peripheral blood cells, including cytotoxic T-lymphocytes, natural killer (NK) cells, and mononuclear phagocytic cells, are endowed with cytotoxic abilities, and are thus very efficient in immunosurveillance mechanisms against neoplastic cells and viral infections. We have used the NK cell function assay because, like the macrophage/monocyte lineage of cells, its role as the first line of immune-mediated defense against viral and bacterial infections has been conclusively established. In immunocompromised hosts, a correlation has been observed between low NK cell activity and morbidity (Levy et al., 1991) or the incidence and severity of upper respiratory infections (Whiteside and Herberman, 1989). NK cells are identified by the phenotype CD3-CD16[+] and/or CD56[+] (Trinchiery, 1989). NK cell function is measured in a 4-hour [51]chromium (^{51}Cr)-release assay, whereby freshly isolated PB leukocytes (effector cells) from which monocytes have been removed by adherence to plastic culture flasks, and ^{51}Cr-labeled K562 cells (target cells) are co-cultured, and the release of label in culture supernatants is quantified. At least three concentrations of effector cells are added to a fixed number of target cells. The amount of ^{51}Cr-released is directly proportional to the level of NK cytotoxicity (Trinchiery, 1989). Although the number of circulating NK cells is small (7 to 15% of circulating lymphocytes), NK cells are functionally very efficient cells, as unlike other cells whose function requires association with the major histocompatibility complex (MHC), NK cell action against certain malignant and virus-infected cells is MHC unrestricted. Furthermore, NK activity is nonspecific with respect to the type of cell targeted (Trinchiery, 1989). In addition, NK cells produce numerous cytokines such as tumour necrosis factors α (alpha) and β (beta), interferons α and β, granulocyte-macrophage colony-stimulating factor, and interleukin-3 (IL-3) upon immune stimulation, all of which have a profound effect on immune reactivity (Trinchiery, 1989). Appropriate controls including a spontaneous ^{51}Cr release (SR) control and a total ^{51}Cr release (TR) control must be included with every assay plate and these are entered into the calculation of the percent cytotoxicity using the following formula:

$$\% \text{ Cytotoxicity} = \frac{\left[\text{Experimental}\left(\text{mean counts/minute}\right) - \text{SR}\left(\text{mean counts/minute}\right)\right]}{\left[\text{TR}\left(\text{mean counts/minute}\right)/\text{SR}\left(\text{mean counts/minute}\right)\right]} \times 100$$

A cut-off value of ≤ 10 for SR is recommended. Experiments with SR >10 should be discarded.

14.3.4.2 Phagocytosis

In addition to their cytotoxic properties, the monocyte/macrophage lineage of cells is important in antigen recognition, processing, and presentation to T-lymphocytes. These cells are potential targets for chemical-induced immunotoxicity, and assays to study functional aspects should be included in the experimental design. One of the assays used successfully in nonhuman primates is the activation of peripheral blood monocytes by zymosan and the potent activator phorbol myristate acetate (PMA) (Tryphonas et al., 1991).

Table 14.3   Serum Hydrocortisone Levels in Adult Healthy Female Rhesus
(*Macaca mulatta*) Monkeys[a]

Observation Period in Months	Hydrocortisone Mean (ng/mL)	Observation Period in Months	Hydrocortisone Mean (ng/mL)
1	321.39	9	313.96
2	286.01	10	306.01
3	304.21	11	336.20
4	331.37	14	319.91
5	341.54	17	275.71
6	311.51	20	328.94
7	294.92	23	305.99
Mean (standard deviation)			313.54 (18.6)

[a] N= 16.

14.3.5 Measurement of Serum Hydrocortisone Levels

Other assays that measure serum complement levels and cytokine levels basal and in lectin-activated cultures are useful parameters in the interpretation of immunologic data. Another factor that needs to be considered in any immunotoxicology study is the influence of hormones, that is, corticosteroids, on the experimental results. To exclude the possibility that the observed effects on the immune system are not the result of an indirect effect on hormone levels, investigators have systematically determined serum levels of hydrocortisone prior to and at intervals during the study. As it has been emphasized before due to the effect of stress and the observed diurnal variation of corticosteroids, the collection of serum specimens need to be carried out by adhering to a strict randomized order of all monkeys in the study for each blood collection. Sequential determination (over 23 months) of hydrocortisone levels in the sera of naïve adult rhesus monkeys have been determined using previously detailed methods (Table 14.3) (Loo et al., 1989).

The majority of these assays have been developed/adapted for use in cynomolgus, rhesus, pig-tailed macaque, and marmoset monkeys. In comparison, fewer assays have been developed/adapted for use in baboon, squirrel, and chimpanzee monkeys. While validation of these assays across laboratories is an issue that needs to be addressed, assays for CMI, challenge with foreign antigens, NK cell assay, and CSMA have been reproduced in several laboratories. Of these assays, challenge with foreign antigens, CSMA, and NK have repeatedly proven to be the most sensitive in detecting chemical-induced immunotoxicity.

14.4 NONHUMAN PRIMATES USED IN IMMUNOTOXICITY STUDIES: EXAMPLES

14.4.1 PCBs and Toxaphene

Nonhuman primates have been used extensively in studies designed to investigate the potential immunotoxic effects of chemicals, particularly those of environmental concern such as PCBs (Tryphonas et al., 1989, 1991a, 1991b); 2,3,7,8-TCDD (dioxins) (Hong et al., 1989); and toxaphene (Tryphonas et al., 2000, 2001). For example, the immunotoxic effects of the commercial PCB mixture, Aroclor 1254, has been extensively studied in adult and infant rhesus monkeys. This chronic, multidose, one-generation toxicity/reproductive/immunotoxicity study generated a great deal of data on effects of PCBs, not only in adults but also in infant monkeys (Tryphonas et al., 1989, 1991a, 1991b). A large number of parameters were affected. In particular, the response to SRBC antigens was significantly affected at levels of Aroclor 1254 as low as 5 µg/kg of body weight per day.

In addition to PCBs, recent studies have shown that the pesticide toxaphene, a complex mixture of chlorinated bornanes with more than 13,000 individual isomers, is also immunotoxic in cynomolgus monkeys (Tryphonas et al., 2000, 2001). In these studies, young adult female monkeys (10 monkeys per group), were administered doses of 0.00, 0.1, 0.4, or 0.8 mg/kg of body weight per day for 75 weeks, while 5 male monkeys per group were administered toxaphene at a dose of 0.8 or 0.0 mg/kg of body weight per day (Tryphonas et al., 2000, 2001). A striking feature of this study was the statistically significant reduction in antibody titers in response to immunization with SRBC and tt antigens without any significant effect on the antibody response to pneu antigens indicating that the T-cell–dependent humoral immune response was compromised. This effect was highly significant at the 0.8 mg/kg dose. Studies on the infants of the same monkeys at weaning indicated that while the immune response to SRBC was not statistically different from the control, statistically significant shifts had occurred in the T-cell subsets, which may have accounted for no effects on the anti-SRBC titer. Specifically, a trend toward increased % CD4$^+$ (T gate) cell numbers (p = 0.0016 for trend) was observed, and in comparison to the control this increase was statistically significant at the 0.4 and 0.8 mg/kg doses (p = 0.0254 and p = 0.0384 for the medium and high dose, respectively). In contrast, the % CD2$^+$CD8$^+$ and CD8$^+$ (T gate) were decreased (statistically significant for trend p = 0.004 and p = 0.0005, respectively). The CD4$^+$:CD8$^+$ cell ratio was significantly increased (p = 0.0004 for trend, and p = 0.0517 and p = 0.0137 for the middle and high dose, respectively, compared to control). This trend for both cell types and the T subset ratio was also observed when infants were retested at 40 weeks postweaning. Specifically, the CD2$^+$CD4$^+$ were statistically significantly increased for all doses (p = 0.0124, p < 0.0001, and p = 0.0417, for control vs. low, control vs. medium, and control vs. high-dose comparisons, respectively); the % CD2$^+$CD8$^+$ and CD8$^+$ (T gate) cells were statistically significantly reduced in the high dose (p = 0.0349 and p = 0.0466, respectively). The CD4$^+$:CD8$^+$ ratio was decreased (p = 0.0199 for trend, and p = 0.0515 and p = 0.0466 for the middle and high dose compared to control, respectively) (Tryphonas et al., CPT Consultants in Pathology and Toxicology Inc., Nepean, Ontario, Canada, manuscript in preparation). These data indicated that the regulatory cells of the immune system were affected by treatment, and emphasized the need for multiple immune parameter assessment.

14.5 RELEVANCE TO HUMANS

The relevance of immunotoxicity data generated in monkeys in relation to the human population remains unresolved. However, the available data for humans accidentally or occupationally exposed to various agents of environmental concern strongly indicate that the human immune system is a target for chemical-induced immunotoxic effects (Tryphonas et al., 2001). Examples of these include populations exposed to PCBs, polychlorodibenzofurans (PCDFs), and polychloroquater-phenyls (PCQ) via contaminated rice oil (Wu et al., 1985; Chang et al., 1981; Lu and Wu, 1985); humans consuming fatty fish species from the Baltic Sea, and studies on the Inuit (northern Quebec) populations consuming large amounts of fish fat (Dewailly et al., 2000). Studies in newborn and children exposed *in utero* to ambient levels of PCBs and dioxins suggest that this population may be particularly sensitive to the immunotoxic effects of environmental chemicals. This is due to the fact that certain chemicals including the PCBs are known to cross the human placenta and to be secreted in mother's milk (Weisglas-Kuperus, 1995, 2000). All of these studies report effects on several parameters of the immune system. Many affected parameters are similar to those for which effects were shown in nonhuman primates exposed to PCBs or dioxins (Tryphonas and Feeley, 2001; Hong et al., 1989).

14.6 APPLICATION OF DATA TO REGULATORY ISSUES

The ultimate purpose of conducting immunotoxicity studies is to enable the regulatory agencies to determine "safe" levels of unwanted chemicals in the environment and in the food chain. To facilitate this process, several countries have issued guidelines for immunotoxicity testing in rodents, and the majority of the studies performed during the last decade followed these guidelines. Although no such guidelines exist for nonhuman primate models, several of the assays used in monkeys correspond to those listed in the proposed guidelines for rodents, which makes the process of cross-species comparisons possible (National Research Council, 1992).

Studies in experimental animals such as guinea pigs, rabbits, and rodents have been helpful in identifying a no observed adverse effect level (NOAEL) for several chemicals of environmental concern. The best example of such studies would be those performed using several of the commercially available PCBs. The calculated NOAELs in these animals were high in comparison to those calculated from similar data generated in monkeys (reviewed in Tryphonas and Feeley, 2001). This was attributed to the higher rate of PCBs eliminated in mice and rats compared to the rate observed in monkeys (reviewed in Tryphonas and Feeley, 2001). The Aroclor 1254 immunotoxicity data in monkeys have been used extensively by regulatory and advisory agencies. The U.S. Agency for Toxic Substances and Disease Registry (2001) has derived a minimal risk level (MRL) of 0.02 µg/kg/day for chronic-duration oral exposure to Aroclor 1254. The chronic oral MRL is based on a lowest observed adverse effect level (LOAEL) of 0.005 mg/kg/day for immunological effects in adult monkeys that were evaluated after 23 and 55 months of exposure to Aroclor 1254 (Tryphonas et al., 1991). Typically, these calculations apply an uncertainty factor of 300 (10 for extrapolating from a LOAEL to a NOAEL, 3 for extrapolating from monkeys to humans, and 10 for compensating for the observed variability among humans). Similarly, the U.S. Environmental Protection Agency has calculated an oral reference dose (RfD) of 0.02 µg/kg/day for Aroclor 1254 based on the evaluation of dermal/ocular and immunologic effects in monkeys, and an oral RfD of 0.07 µg/kg/day based on reduced birth weight in monkeys (U.S. Environmental Protection Agency, 2000).

14.7 SUMMARY

In this chapter, we presented and discussed key toxicologic and immunotoxicologic approaches pertaining to the husbandry of and experimentation using nonhuman primates. To date, most immunotoxicologic studies have been carried out within the context of a toxicologic study, and this trend is likely to continue. Several immunologic parameters have been developed and applied to the study of environmental chemicals that are of concern to regulators and to the public at large. While there are obvious advantages to using nonhuman primates as experimental animals in immunotoxocology, the limitations inherent to the model, including availability, the requirement for special housing, and maintenance costs are serious deterrents in the widespread use of nonhuman primates in research. Furthermore, challenges with multiple antigens for humoral and DTH responses require long periods of observation, which increases maintenance costs. Presently, three antigens — SRBC, tt, and pneumococcus have been used to immunize monkeys. The protocols for these immunizations span several weeks of the study. In addition to difficulties encountered in incorporating these tests within a given experimental protocol, such procedures result in a substantial increase in maintenance costs. It may be possible to combine such immunization regimens and shorten the length of this procedure by immunizing monkeys with all three antigens at a single time point of the study. This, however, would require additional developmental work to rule out any potential interaction among the antigens used for immunization. A further drawback in using the monkey model is the difficulty in developing infectivity assays. In rodents, infectivity procedures have been established, and are considered valid indicators of a compromised immune system. In monkeys, the need for an additional group of animals that increases study costs, and the establishment of special facilities to accommodate infectivity assays have discouraged researchers from developing such models in nonhuman primates. It is hoped that with the increasing need for monkeys in research these deterrents will be minimized.

REFERENCES

Ahmed-Ansari, A., et al., Flow microfluorometric analysis of peripheral blood mononuclear cells from non-human primates: correlation of phenotype with immune function, *Am. J. Primatol.*, 17, 107–131, 1989.

Arnold, D.L., et al., Toxicological consequences of aroclor 1254 ingestion by female rhesus (Macaque mulatta) monkeys. Part 1B. Prebreeding phase: clinical and analytical laboratory findings, *Fundam. Chem. Toxicol.*, 31, 811–824, 1993.

Bach, B.A., Normal values: definition of a reference range for lymphocyte subsets of healthy adults, in *Becton Dickinson Immunocytometry Systems*, Clinical Monograph 1, Bach, B.A., Ed., San Jose, CA, 1989.

Barton, R., Thrall, R., Neubauer R., Binding of lymphocyte specific monoclonal antibodies to common marmoset lymphoid cells, *Cell. Immunol.*, 84, 446–451, 1984.

Beaver, B.V., Environmental enrichment for laboratory animals, *Inst. Lab. Anim. Res. News*, 31, 5–13, 1989.

Brown, J.F. Jr., Unusual congener selection patterns for PCB metabolism and distribution in the rhesus monkey, *Organohalogen Compounds*, 21, 29–31, 1994.

Bryant, C.E., Rupniak, N.M.J., and Iversen, S.D., Effects of different environmental enrichment devices on cage sterotypies and autoaggression in captive cynomolgus monkeys, *J. Med. Primatol.*, 17, 257–269, 1988.

Canadian Council on Animal Care, available at: www.ccac.ca, 2004.

Chang, K.J., et al., Immunologic evaluation of patients with polychlorinated biphenyl poisoning: determination of lymphocyte subpopulations, *Toxicol. Appl. Pharmacol.*, 61, 58–63, 1981.

Coe, C.L., Biological and social predictors of immune senescence in the aged primate, *Mech. Ageing Dev.*, 125, 95–98, 2004.

Crockett, C., Bielitzki, J., Carey, A., et al., Kong toys as enrichment devices for singly-housed macaques, *Lab. Primate Newslett.*, 28, 21–22, 1989.

Dewailly, E., et al., Susceptibility to infections and immune status in Inuit infants exposed to organochlorines, *Environ. Health Perspect.*, 108, 205–211, 2000.

Diehl, K.-H., et al., A good practice guide to the administration of substance and removal of blood, including routes and volumes, *J. Appl. Toxicol.*, 21, 15–23, 2001.

Dixon, R.L., Problems in extrapolating toxicity for laboratory animals to man, *Environ. Health Perspect.*, 13, 43–50, 1976.

Eichberg, J.W., et al., Lymphocyte subsets in chimpanzees, *Lab. Anim. Sci.*, 38, 197–198, 1988.

Ellingsworth, L.R., et al., Characterization of rhesus macaque peripheral blood T-lymphocyte subpopulations, *Vet. Immunol. Immunopathol.*, 4, 517–532, 1983.

Fajzi, K., Reinhardt, V., and Smith, M.D., A review of environmental enrichment strategies for singly caged nonhuman primates, *Lab. Anim.*, 18, 23–29, 31, 33, 35, 1989.

Fernie, S., et al., Normative hematologic and serum biochemical values for adult and infant rhesus monkeys (Macaca mulatta) in a controlled laboratory environment, *J. Toxicol. Environ. Health*, 42, 53–72, 1994.

Flow, B.L., and Jaques, J.T., Effect of room arrangement and blood sample collection sequence on serum thyroid hormone and cortisol concentrations in cynomolgus macaques (Macaca fascicularis), *Am. Assoc. Lab. Anim. Sci.*, 36, 65–68, 1997.

Fuller, G.B., Hobson, W.C., Renquist, D.M., et al., Nonhuman primates, in *Animal Models in Toxicology*, Gad, S.C., and Chengelis, C.P., Eds., Marcel Dekker, New York, 1992, pp. 675–735.

Gad, S.C., *Drug Chem. Toxicol.*, 8, 841–859, 1992.

Hawk, C.T., and Leary, S.L., *Formulary for Laboratory Animals*, Iowa State University Press, Ames, 1995.

Hong, R., Taylor, K., and Abonour, R., Immune abnormalities associated with chronic TCDD exposure in rhesus, *Chemosphere*, 18, 313–321, 1989.

Joint Working Group, Removal of blood from laboratory mammals and birds, *Lab. Anim.*, 27, 1–22, 1993.

Karpinski, K.F., Hayward, S., and Tryphonas, H. Statistical considerations in the quantitation of serum immunoglobulin levels using the enzyme-linked immunosorbent assay (ELISA), *J. Immun. Methods*, 103, 189–194, 1987.

Keene, J.H., and Sansone, E.B., Airbone transfer of contaminants in ventilated spaces, *Lab. Anim. Sci.*, 34, 453–457, 1984.

Levy, S.M., et al., Persistently low natural killer cell activity, age, and environmental stress as predictors of infectious morbidity, *Nat. Immun. Cell. Growth Regul.*, 10, 289–307, 1991.

Loo, J.C.K., et al., Effects of Aroclor 1254 on hydrocortisone levels in adult rhesus monkeys (Macaca mulatta), *Bull. Environ. Contam. Toxicol.*, 43, 667–669, 1989.

Lu, Y.-C., and Wu, Y.-C., Clinical findings and immunological abnormalities in Yu-Cheng patients, *Environ. Health Perspect.*, 59, 17–29, 1985.

Luster, M.I., et al., Risk assessment in immunotoxicology. I. Sensitivity and predictability of immune tests, *Fundam. Appl. Toxicol.*, 18, 200–210, 1992.

Maleck, H., Serra, A., and Florence, G., Setup of a rectal probe and a stimulation procedure to obtain electroejaculation in rhesus monkeys, *Folia Primatol.* 67, 98–99, 1996.

Mes, J., Arnold, D.L., and Bryce, F., The elimination and estimated half-lives of specific polychlorinated biphenyl congeners from the blood of female monkeys after discontinuation of daily dosing with aroclor 1254, *Chemosphere*, 30, 789–800, 1995a.

Mes, J., Arnold, D.L., and Bryce, F., Postmortem tissue levels of polychlorianted biphenyls in female rhesus monkeys after more than six years of daily dosing with aroclor 1254 and in their non-dosed offspring, *Arch. Environ. Contam. Toxicol.*, 29, 69–76, 1995b.

National Research Council, *Biologic Markers in Immunotoxicology*, U.S. National Academy of Science, Subcommittee on Immunotoxicology, Committee on Biologic Markers, Board on Environmental Studies and Toxicology, Commission on Life Sciences, National Academy Press, Washington, DC, 1992.

Neubert, R., Helge, H., and Neubert, D., Nonhuman primates as models for evaluating substance-induced changes in the immune system with relevance for man, in *Experimental Immunotoxicology*, Smialowicz, R.J., Holsapple, M.P., Eds., CRC Press, Boca Raton, FL, 1996, pp. 63–117.

Perkins, S.L., Examination of the blood and bone marrow, in *Wintrobe's Clinical Hematology*, 10th ed., vol. 1, Lee, G.R., et al., Eds., Williams & Wilkins, Baltimore, 1999, pp. 9–35.

Reinhardt, V., Improved handling of experimental rhesus monkeys, in *The Inevitable Bond: Examining Scientist–Animal Interactions*, Davis, H., and Balfour, D., Eds., Cambridge University Press, Cambridge, 1992, pp. 171–177.

Reinhardt, V., Refining the blood collection procedure for macaques, *Lab. Anim.*, 25, 32–35, 1996.

Reinhardt, V., Liss, C., and Stevens, C., Social housing of previously single-caged macaques: what are the options and the risks? *Anim. Welfare*, 4, 307–328, 1995.

Ruelius, H.W., Extrapolation from animals to man: predictions, pitfalls and perspectives, *Xenobiotica*, 17, 255–265, 1987.

Safe, S.H., Polychlorinated biphenyls (PCBs): Enviromental impact, biochemical and toxic responses, and implications for risk assessment, *Crit. Rev. Toxicol.*, 24, 87–149, 1994.

Sansone, E.B., and Fox, J.G., Potential chemical contamination in animal feeding studies: evaluation of wire and solid bottom caging systems and gelled feed, *Lab. Anim. Sci.*, 27, 457–465, 1977.

Sansone, E.B., Losikoff, A.M., and Pendleton, R.A., Potential hazards from feeding test chemicals in carcingone bioassay research, *Toxicol. Appl. Toxicol.*, 39, 435–450, 1977.

Trinchiery, G., Biology of natural killer cells, *Adv. Immunol.*, 47, 187–376, 1989.

Tryphonas, H., Approaches to detecting immunotoxic effects of environmental contaminants in humans, *Environ. Health Perspect.*, 109(Suppl. 6), 877–884, 2001.

Tryphonas, H., and Feeley, M., Polychlorinated biphenyl-induced immunomodulation and human health effects, in *PCBs: Recent Advances in Environmental Toxicology and Health Effects*, Robertson, L.W., and Hansen, L.G., Eds., University Press of Kentucky, Lexington, 2001, pp. 194–209.

Tryphonas, H., et al., Cell surface marker evaluation of infant Macaca monkey leukocytes in peripheral whole blood using simultaneous dual-color immunophenotypic analysis, *J. Med. Primatol.*, 25, 89–105, 1996.

Tryphonas, H., et al., The effect of butylated hydroxytoluene on selected immune surveillance parameters in rats bearing enzyme-altered hepatic preneoplastic lesions, *Food Chem., Toxicol.*, 37, 671–681, 1999.

Tryphonas, H., et al., Effect of chronic exposure of PCB (Aroclor 1254) on specific and nonspecific immune parameters in the rhesus (Macaca mulatta) monkey, *Fundam. Appl. Toxicol.*, 16, 773–786, 1991a.

Tryphonas, H., et al., Effects of PCB (aroclor 1254) on non-specific immune parameters in rhesus (*Macaca mulatta*) monkeys, *Int. J. Immunopharmacol.*, 13, 639–648, 1991b.

Tryphonas, H., et al., Effects of toxaphene on the immune system of cynomolgus (Macaca fascicularis) monkeys: a pilot study, *Food Chem. Toxicol.*, 38, 25–33, 2000.

Tryphonas, H., et al., Effects of toxaphene on the immune system of cynomolgus (Macaca fascicularis) monkeys, *Food Chem. Toxicol.*, 39, 947–958, 2001.

Tryphonas, H., et al., Immunotoxicity studies of PCB (aroclor 1254) in the adult rhesus (Macaca mulatta) monkey B preliminary report, *Int. J. Immunopharmacol.*, 11, 199–206, 1989.

U.S. Agency for Toxic Substances and Disease Registry, Toxicological Profile for Polychlorinated Biphenyls (Update), U.S. Department of Health and Human Services, Agency for Toxic Substances and Disease Registry, Atlanta, GA, 2001.

U.S. Environmental Protection Agency, Integrated Risk Information System, polychlorinated biphenyls (PCBs), 2000, available at: www.epa.goc/ngispgm3/iris/subst/0294.htm.

Virella, G., Ed., Introduction, in *Medical Immunology*, 4th ed., Marcel Dekker, New York, 1998.

Weisglas-Kuperus, N., Immunologic effects of polychlorinated biphenyl (PCB) and dioxin exposure in Dutch toddlers, *Toxicologist* 54 (Abstr.), 1047, 2000.

Weisglas-Kuperus, N., et al., Immunologic effects of background prenatal and postnatal exposure to dioxins and polychlorinated biphenyls in Dutch infants, *Pediatr. Res.* 38, 404–410, 1995.

Whiteside, T.L., and Herberman, R.B., The role of natural killer cells in human disease, *Clin. Immunol. Immunopathol.*, 53, 1–23, 1989.

Wu, Y.-C., et al., Cell-mediated imunity of patients with polychlorinated biphenyl poisoning, *J. Formosan Med. Assoc.*, 83, 419–429, 1985.

CHAPTER 15

Practical Considerations for Toxicologic Pathology Assessment of the Immune System in Rodents and Nonhuman Primates

Olga Pulido, Don Caldwell, Meghan Kavanagh, and Santokh Gill

CONTENTS

15.1 INTRODUCTION

Immunotoxicology is the study of direct or indirect adverse effects on the immune system induced by exposure to environmental chemicals, drugs, or radiation. It includes conditions of immune regulation (suppression and rarely enhancement), allergies, and autoimmunity (Descotes et al., 2000; European Agency for the Evaluation of Medicinal Products, 2000; Kimber and Dearman, 2002; Kuper et al., 2002; Luster et al., 1998). Immunotoxicology studies using experimental animals can be conducted to detect the detrimental effects of xenobiotics on the immune system. The data generated are an important component of human risk assessment (European Agency for the Evaluation of Medicinal Products, 2000; Kuper et al., 2002).

Rodent and nonhuman primate models have been used for safety evaluation of chemicals including pharmaceuticals, industrial intermediates, agricultural compounds, and food additives. The usefulness and suitability of nonhuman primates for assessing immunotoxicology has been the subject of recent reviews and publications (Luster et al., 1998; Buse et al., 2003; Bleavins and de la Iglesia, 1995; Hendrick et al., 2000; House and Thomas, 2002; Lappin and Black, 2003). Various tests are currently available for the assessment of immunotoxicity in nonhuman primate models (Luster et al., 1998; Hendrick et al., 2000). It is evident that both animal models have applications and can be used for defining immunotoxicity.

The pathology evaluation of lymphoid tissues is necessary to understand the immunotoxic effects of chemicals. The rationale is that chemically induced alterations may manifest as qualitative or quantitative changes in the histology of the lymphoid organs (Vos, 1987; Vos and van Loveren, 1987). Daily insults, ageing, and toxins can alter the primary function of lymphoid organs. Hence, it is important to distinguish and differentiate lesions induced by xenobiotics as compared to those lesions that occur naturally during development and aging (Ward, 1993). This means that knowledge of the morphological and functional maturation of the immune system relevant to the animal model used is important for the design and interpretation of toxicological studies (Buse et al., 2003).

This chapter emphasizes current methodologies used to conduct enhanced toxicologic pathology assessments of lymphoid tissues in the rat and nonhuman primate models. We review the histopathology protocols recommended in the tiered system, and the general approach used in our laboratory for pathology evaluation. Consideration is given to the immunophenotyping of lymphoid tissues. We describe the current methods used in our laboratory, and provide examples of the practical applications to the rat as a rodent model and to the cynomolgus (*Macaca fascicularis*) monkey as a nonhuman primate model. Preliminary data on the immunohistochemical characterization of glutamate receptors (GluRs) and neural biomarkers are also included. The practical application of the methods and procedures described are illustrated in Figures 15.1, 15.2, and 15.3, and Tables 15.1 and 15.2.

15.2 ENHANCED TOXICOLOGIC PATHOLOGY OF THE IMMUNE SYSTEM

15.2.1 Background

The pathology assessment of lymphoid tissues adopts, and is an integral component of, the tiered system used for immunotoxicology assessment. The historical perspectives leading to the tiered system are the subject of many publications and book chapters (European Agency for the Evaluation of Medicinal Products, 2000; Kuper et al., 2002; Basketter et al., 1995; De Waal and van Loveren, 1997; Gopinath, 1996; Institoris et al., 1998; Kuper et al., 1995; Kuper et al., 2000; Schuurman et al., 1992; van Loveren and Vos, 1989; Vos, 1980; Vos, 1987; Hinton, 2000). The International Collaborative Immunotoxicity Study (ICIS) (1998) is an example of the efforts made to validate and standardize the operating procedures used among various laboratories for the universal implementation of this tiered approach (ICIS, 1998). Here we review and describe selected aspects of the tiered system that are relevant to the pathology evaluation of the immune system and to the implementation of the recommended methodology in our laboratory.

15.2.2 Methods and Procedures

It is beyond the scope of this chapter to review the available literature describing methodologies relevant to the toxicologic pathology assessment of lymphoid tissues. Therefore, we describe the methods and procedures that are currently used in our laboratory. However, selected key publications are cited for further reading and support of our presentation.

15.2.2.1 Tissue Sampling

At the end of the experimental period, animals are sacrificed by exsanguination via the abdominal aorta under isofluorane anesthesia (Arnold et al., 2001; Tryphonas et al., 2004; Cooke et al., 2004). A practical and reliable pathology assessment of the immune system can be made by sampling the primary (thymus, bone marrow) and secondary (lymph nodes, spleen, mucosa-associated lymphoid tissue [MALT]) lymphoid organs from each experimental group (Gopinath, 1996; Tryphonas et al., 2004). The weight is recorded for unfixed lymphoid organs with their surrounding fat removed. Bone marrow may be collected from the femur or sternum and processed for cellularity assessment by making fixed sections, smears, or suspensions. Samples of MALT include Peyer's patches using the Swiss-roll technique (Tryphonas et al., 2004; Moolenbeek and Ruitenberg, 1981; Chapter 14 of this book). Other tissues can also be included, such as nasal-associated lymphoid tissue (NALT) and bronchial-associated lymphoid tissue (BALT) (De Waal and van Loveren, 1997; Kawabata et al., 1995). It is recommended that lymphoid tissues contacting or draining the site of administration are examined (Vos and van Loveren, 1987; Kuper et al., 1995; Kawabata et al., 1995). For compounds administered intravenously, the spleen is considered the draining site (U.S. Food and Drug Administration, 2001). After a detailed gross inspection, tissues are fixed by immersion in 10% neutral buffered formalin and processed for Paraplast paraffin embedding (Tryphonas et al., 2004). Fixation by perfusion and the use of other fixatives may be appropriate in some studies. Sampling of organs and tissues is done in a standardized way to facilitate proper tissue embedding, allowing examination of all organ compartments and providing consistency in the evaluation (Institoris et al., 1998).

15.2.2.2 Histopathology Evaluation

Paraffin-embedded sections (4 to 5μ thick) stained with hematoxylin and eosin (H&E) provide good morphological resolution and allow the viewing of large areas for the grading of lesions (Kuper et al., 1995; Tryphonas et al., 2004). In the case of lymph nodes and Peyer's patches, orientation is crucial to permit assessment of all tissue compartments (Kuper et al., 1995; Chapter 13 of this book). In our laboratory, tissues are examined and photographed with an Axiophot Zeiss microscope (Germany) equipped with a digital camera linked to an image analysis system (Axio-Cam Zeiss camera and Axiovision image analysis and archiving system).

The morphological details obtained with paraffin sections stained with H&E are illustrated in Figure 15.1B, E and F and Figure 15.2A through D. Reticulin stain is used to assess organ architecture. Figure 15.1A shows a reticulin stain of a rat lymph node; finely interwoven reticulin fibers emphasize the overall tissue architecture and compartmentalization of a lymph node. Reticulin fibers are more evident in the medullary region and surrounding blood vessels.

For histopathological evaluation, it is important to appreciate the variation in normal lymphoid organs (Basketter et al., 1995; Kuper et al., 1995; Kuper et al., 2000). Grading of the lesions is conducted in accordance with the Organization for Economic Cooperation and Development Guideline 407/enhanced pathology protocol as outlined by the ICIS (ICIS, 1998). This protocol recommends dividing the lymphoid organs into compartments for assessing the effects of treatment on cellularity and area, as follows:

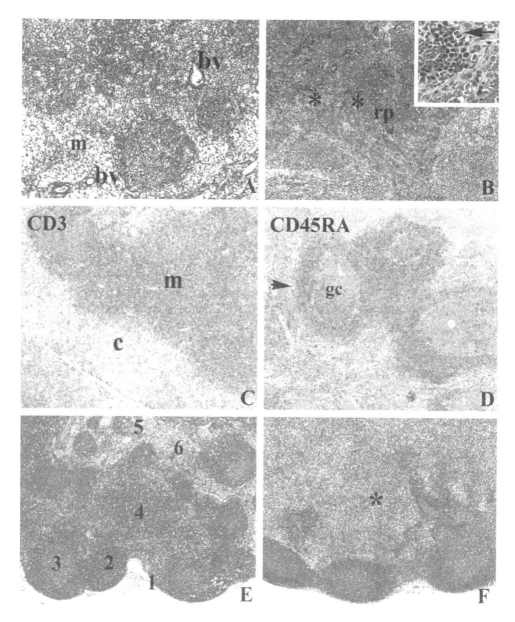

Figure 15.1 **(See color insert following page 236)** Histopathology of rat lymphoid tissues. All tissues were fixed by immersion in 10% buffered neutral formalin and embedded in paraffin. Photographs show sections (4 to 5 µ) of tissues from adult Sprague–Dawley (SD) (A, C through F) and from aged Fisher (FIS) rats (B). Grading of lesions in E and F was conducted in accordance with the Organization for Economic Cooperation and Development Guideline 407/enhanced pathology protocol as outlined by the International Collaborative Immunotoxicity Study. (A) Reticulin stain of a lymph node from adult control SD rat showing interwoven reticulin fibers. This is more evident in the medullary region (m) and surrounding blood vessels (bv); 20X objective. (B) Spleen of an aged FIS rat showing clusters of leukemic cell infiltrates (asterisk) in the red pulp (rp); 20X objective. The cellular pleomorphism and a mitotic figure (arrow) are shown at higher magnification in the inset; 40X objective, H&E stain. (C) Thymus of a control SD rat stained by immunohistochemistry using the streptavidin-biotin complex indirect method. It depicts the distribution of CD3 immunoreactive cells (T-lymphocytes) with stronger intensity of the staining in the medulla (m) as compared to the cortex (c); 10X objective. (D) Spleen of control SD rat stained by immunohistochemistry using the biotin-free secondary amplification indirect method. It depicts the distribution of CD45RA (B-lymphocytes) in the follicles of the white pulp. The intensity of the stain is stronger within the tightly packed zone of B cells in "the mantle" (arrow) that surrounds the germinal centers (GC) of the secondary follicles; 10X objective. (E) Mesenteric lymph node of 90-day-old control SD rat assessed as grade 1 or normal. The various compartments are easily recognized including (1) capsule, (2) primary follicle, (3) germinal center of secondary follicle, (4) paracortex, (5) medullary cord, (6) medullary sinus, H&E stain; 5X objective. (F) Mesenteric lymph node of a 90-day-old SD rat exposed to tributyltin (0.25-mg/kg/day) *in utero* and postnatally until 90 days of age. It shows a grade 3 decrease (hypocellular) in paracortical lymphocyte density (asterisk), as compared to the age-matched control shown in E, H&E stain; 5X objective.

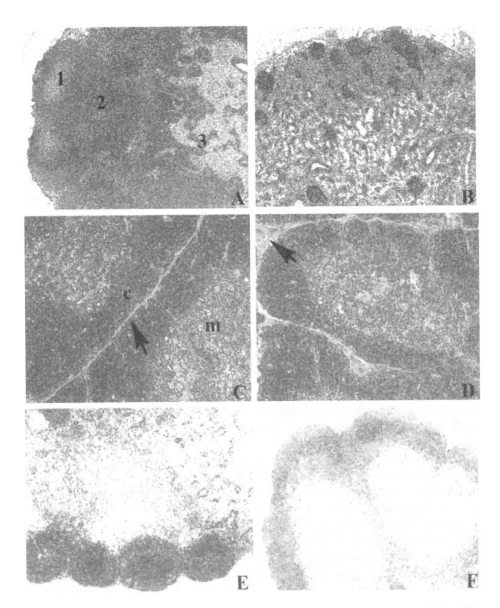

Figure 15.2 (See color insert) Histopathology of lymphoid tissues of adult (A, B) and 78-week-old infant (C through F) cynomolgus (*Macaca fascicularis*) monkey (*Mf*). All tissues were fixed by immersion in 10% buffered neutral formalin and embedded in paraffin. Photographs show sections (4 to 5 μ) of mesenteric lymph nodes of adult monkey stained with H&E (A, B), thymus of infant animals stained with H&E (C, D), and lymph node of infant monkey stained by inmunohistochemistry (E, F). Grading of the lesions in A to D was conducted in accordance with the Organization for Economic Cooperation and Development Guideline 407/enhanced pathology protocol as outlined by the International Collaborative Immunotoxicity Study. (A) Mesenteric lymph node of 7-year-old adult control monkey showing a grade 1 normal histology. The various compartments are easily recognized including large, densely populated secondary follicle (1), paracortex (2), and medullary cords (3); 5X objective, H&E stain. (B) Mesenteric lymph node of an age-matched monkey exposed for 125 weeks to daily oral toxaphene (1.0 mg/kg/day). It depicts a generalized B- and T-lymphoid involution assessed as grade 4 compared to the control shown in A. There is a decrease in follicular lymphoid cell density, follicular area, paracortical lymphoid cell density, and paracortical area; 5X objective, H&E stain. (C) Thymus of a 78-week-old infant control monkey showing grade 1 normal histology. The cortex (c) and medulla (m) are well developed with dense cell populations and large lobules separated by interlobular connective tissue (arrow); 10X objective, H&E stain. (D) Thymus of an age-matched infant animal exposed for 11 months to oral toxaphene (1.0 mg/kg/day). A grade 2 deviation from normal as compared to the age-matched control in C. There is a decrease in cortical lymphoid density and area with smaller lobules and increased interlobular connective tissue (arrow); 10X objective, H&E stain. (E) Mesenteric lymph node from 78-week-old infant control monkey stained by immunohistochemistry using the streptavidin-biotin complex indirect method. It depicts the distribution of CD20 labeled B-lymphocytes within the well-populated primary follicles in the cortex; 5X objective. (F) Mesenteric lymph node of an age-matched infant monkey exposed for 11 months to oral toxaphene (1.0 mg/kg/day). The section was stained using the same method as in E. It shows a decrease in the stain intensity, cell population, and distribution of CD20 labeled B-lymphocytes. It was assessed as a grade 3 decrease in follicular lymphoid cell density, and grade 2 decreases in follicular area; 5X objective.

Spleen — Periarteriolar lymphocyte sheath (PALS), follicles, marginal zone, and red pulp
Thymus — Cortex, medulla, and cortex-medulla ratio
Lymph node — Follicles, interfollicular area, and medullary cords
Peyer's patches — Follicle and interfollicular area
Bone marrow — Myeloid and erythroid cells populations

The various microenvironments or compartments each have their own characteristic lymphoid and nonlymphoid cell populations (Kuper et al., 1995). Standardization of diagnostic terms is beneficial, and a typical system has been previously reviewed by Kuper et al. (2000). In the assessment of the variable morphology of lymphoid organs, consideration should be given to stress, steroid hormones, nutritional status, antigenic load, age, and spontaneous lesions for each strain of animal (van Loveren and Vos, 1989). An example of a spontaneous, strain-specific rat lesion is illustrated in Figure 15.1B. It shows the spleen of an aged Fisher rat with clusters of leukemic cell infiltrates in the red pulp. Marked cellular pleomorphism and frequent mitotic figures are visualized within the leukemic infiltrate, and are better viewed in the inset.

15.2.2.3 Grading of Lesions

The recommended grading system (ICIS, 1998) is a semiquantitative estimation conducted by the pathologist as follows: grade 1, normal; grade 2, up to 25% different from normal (increase or decrease); grade 3, a change of 25% to 50% from normal; and grade 4, > 50% different from normal. The grading scheme applies to cell density and compartmental area where the B- and T-lymphocyte–dependent regions are individually assessed (Tryphonas et al., 2004; U.S. Food and Drug Administration, 2001). Changes in cell numbers are described as increased or decreased cellularity rather than hyperplasia or atrophy, to be descriptive and not interpretive. The stroma, stationary leukocytes, and blood vessels should be carefully examined, as these components interact intensively with migrating-passenger lymphocytes and macrophages (Kuper et al., 2000). We have been applying this grading system to evaluate the toxic effects of environmental contaminants in rats and nonhuman primates using H&E-stained histological preparations. To facilitate discussion of lesions, grade 2 is equated to mild, grade 3 to moderate, and grade 4 to severe.

Secondary lymphoid organs are organized into thymus-dependent and thymus-independent regions. An indication of the relative effect of the xenobiotic on T- and B-cell compartments can be obtained (Vos, 1987; Tryphonas et al., 2004). Tissues from treated animals are compared to their control counterparts in order to grade the severity of the lesion, including the overall architecture, cell density, and distribution of cell populations. Figure 15.1E shows a section from a mesenteric lymph node of a 90-day-old Sprague–Dawley rat assessed as grade 1 or normal. This is used as a base pattern to compare with a lymph node from the same region collected from an animal of the same age exposed to the test compound tributyltin (0.25 mg/kg of body weight/day). A similar system is used for nonhuman primates. Figure 15.2A compares the mesenteric lymph node (grade 1) from a 7-year-old control monkey to the node in Figure 15.2B from an age-matched animal treated daily for 125 days with toxaphene (1.0 mg/kg/day). Similarly, Figure 15.2C illustrates the grading of the thymus of a 78-week-old infant control monkey (grade 1) as compared to the thymus (grade 2) in Figure 15.2D of an age-matched monkey treated for 11 months with oral toxaphene (1.0 mg/kg/day).

Suppression or stimulation of cell-mediated immunity may be reflected microscopically as hypocellularity or hypercellularity of the paracortical area in lymph nodes and PALS in the spleen. The number of lymphoid follicles, germinal centers, and plasma cells in the lymph nodes (Figure 15.1E through F and Figure 15.2A and B) and spleen, as well as the size and cellularity of the marginal zone in the spleen are indicators of B-cell involvement (Figure 15.1D). Cytotoxicity may also be reflected as cellular degeneration. Following antigenic stimulation to B cells in lymph nodes, primary follicles may enlarge into secondary follicles that have pale-staining central areas with macrophages and proliferating lymphocytes. In these germinal centers, B cells differentiate into plasmablasts and become plasma cells in medullary cords. Following exposure to an antigen stimulating a T-cell response, the paracortex enlarges and lymphoblasts become prominent. Similar to lymph nodes, compartmentalization in the spleen and Peyer's patches allow the assessment of B- and T-cell regions.

The histopathology examination of photomicrographs of similar magnifications can detect changes in tissue architecture and overall estimation of organ size (Figures 15.2C and D). The thymus from the toxaphene-treated monkey (Figure 15.2D) shows smaller lobules and thicker interlobar connective tissues as compared to the age-matched control (Figure 15.2C). These changes in size correlate with changes in weight. However, the ability to detect changes in cell numbers may be limited if all cell numbers are altered proportionally (Basketter et al., 1995). A provisional hypothesis about the mechanism of toxicity may also be made based on the nature of lymphoid lesions. However, different mechanisms of tissue injury can lead to similar lesions (International Programme on Chemical Safety, 1996; Hashek et al., 2002). The criteria of the grading system used should be defined and all lesions recorded (Kuper et al., 1995; Tryphonas et al., 2004; Dayan et al., 1998; Harleman, 2000). The sensitivity of the histopathology assessment in flagging immuno-toxicity may depend in part on the dose of the xenobiotic employed in the study; at low doses, the histopathology may be insensitive (Schuurman et al., 1994).

The ability to differentiate subpopulations of lymphocytes is limited in conventional histology (Kuper et al., 1995; Hashek et al., 2002). Lesions detected during histopathology evaluation using H&E stained slides can be further assessed by immunohistochemistry (Section 15.3, Figure 15.1C and D, and Figure 15.2E and F) or flow cytometry. Quantitative histomorphometry of lymphoid tissues (Table 15.1) can help to define the affected cell populations and compartmental changes (Tryphonas et al., 2004; Hashek et al., 2002; Gossett et al., 1999).

Other aspects to consider follow:

1. Immunologic processes such as autoimmune disease and hypersensitivity/allergy can lead to tissue damage, protein (immune complex) deposits, and inflammatory cell infiltrates predominantly in nonlymphoid organs (Kuper et al., 1995).
2. Well-known nonlymphoid target sites include vasculature, kidneys, synovial membranes, thyroid, liver, and lungs (Rose and Caturegli, 1997). An immunotoxicologic pathology evaluation should also include examination of these tissues.
3. The morphological hallmark of autoimmune disease and allergy is inflammation. The lesions of inflammation mediated by nonimmune factors may be indistinguishable from immune-mediated phenomena (Sell, 2001).

15.2.2.4 *Histomorphometry*

Histomorphometry of lymphoid tissues is currently being conducted in our laboratory using the image analysis software, Northern Eclipse (Empix Imaging Inc., Mississauga, Ontario, Canada). For this, H&E-stained sections of the selected tissues are scanned (Nikon Supercool scan 4000) to produce digital images of the full cross-section of the tissue. Area values in square millimeters are computed and compared for control and treated groups as in Table 15.1. This software can also be used for a quantitative comparison of lymphocyte populations highlighted by immunohistochemical staining for specific clusters of differentiation (CD) markers (Tryphonas et al., 2004).

15.3 IMMUNOHISTOCHEMICAL PHENOTYPING OF LYMPHOID TISSUES

Immunohistochemistry is a simple and reliable staining technique, providing that the antigen in the tissues is kept intact and available for antibody binding. The labeling of specific components in individual cells can then be related to physiological or pathological events. Formaldehyde fixation

Table 15.1 Histomorphometric Evaluation of Thymus from _Sprague–Dawley_ Rats Exposed Orally (Gavage) to Tributyltin Chloride _In Utero_ and Postnatally Using Northern Eclipse Software on H&E Slides

	30 Days			60 Days			90 Days		
	Dose Mean ± Standard Error[a]	P value[b]	Control	Dose Mean ± Standard Error[a]	P value[b]	Control	Dose Mean ± Standard Error[a]	P value[b]	Control
Males									
Thymus									
Cortex (area in mm^2)	67.89 ± 7.77 (5)	47.64 ± 7.46	>0.05	N/A	N/A	N/A	53.56 ± 5.04	31.64 ± 3.97[c]	0.0066
Medulla (area in mm^2)	18.57 ± 4.50 (5)	8.59 ± 0.91 (S)	>0.05	N/A	N/A	N/A	12.27 ± 1.10	9.00 ± 0.82[c]	0.0381
C + M	86.46 ± 11.14 (5)	56.22 ± 7.79[c]	0.0483	N/A	N/A	N/A	65.83 ± 5.67	40.64 ± 4.63[c]	0.0063
C/M ratio	4.69 ± 1.16 (5)	5.83 ± 0.97	>0.05	N/A	N/A	N/A	4.44 ± 0.42	3.52 ± 0.26	>0.05
C/C and M	0.80 ± 0.03	0.83 ± 0.03	>0.05	N/A	N/A	N/A	0.81 ± 0.01	0.77 ± 0.02	>0.05
Females									
Thymus									
Cortex (area in mm^2)	51.22 ± 5.70	40.48 ± 3.25	>0.05	60.56 ± 7.11	46.16 ± 8.86	>0.05	N/A	N/A	N/A
Medulla (area in mm^2)	9.86 ± 1.86	8.03 ± 1.580	>0.05	15.65 ± 1.97	9.70 ± 1.63[c]	0.0425	N/A	N/A	N/A
C + M	61.07 ± 5.74	48.51 ± 4.52	>0.05	76.21 ± 1.97	55.86 ± 10.42	>0.05	N/A	N/A	N/A
C/M Ratio	6.34 ± 1.36	5.64 ± 0.68	>0.05	3.93 ± 0.28	4.62 ± 0.37	>0.05	N/A	N/A	N/A
C/C and M	0.83 ± 0.03	0.84 ± 0.02	>0.05	0.79 ± 0.01	0.82 ± 0.01	>0.05	N/A	N/A	N/A

[a] Doses are milligrams per kilogram of body weight per day. Values are based on six animals per group, unless otherwise indicated in parentheses.

[b] Values are the t-test pooled for equal variance, unless Satterwaite for unequal variance indicated by (S) where $p \leq 0.05$.

[c] Statistically significant from control ($p \leq 0.05$).

Source: Modified from Tryphonas, H., et al., _Food Chem. Toxicol.,_ 42, 221, 2004.

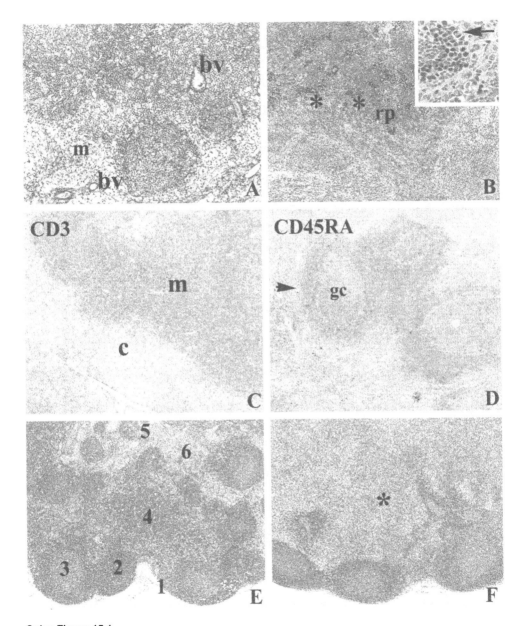

Color Figure 15.1

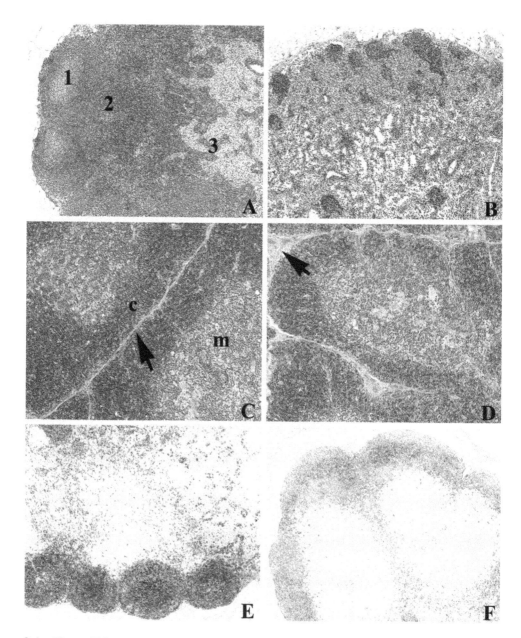

Color Figure 15.2

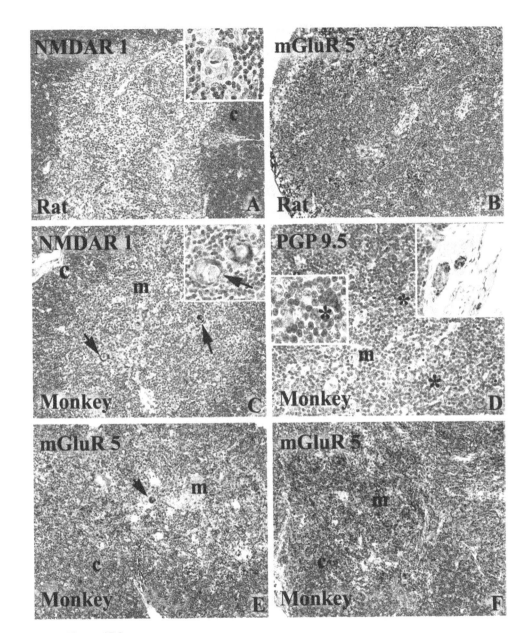

Color Figure 15.3

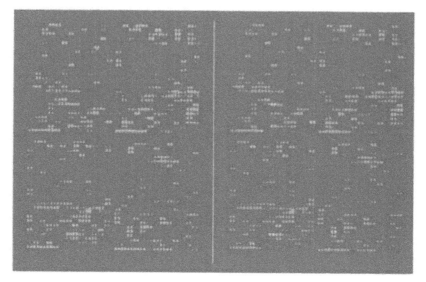

Color Figure 23.3

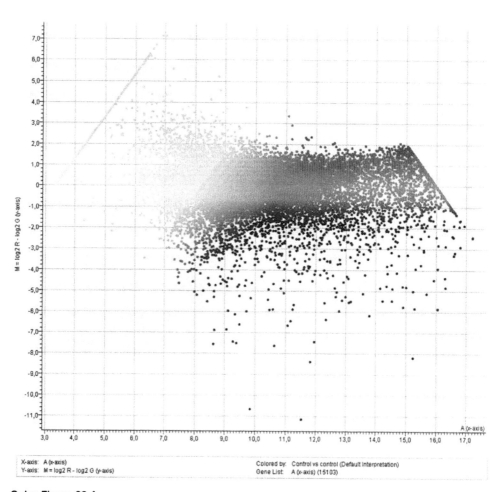

X-axis: A (x-axis)
Y-axis: M = log2 R - log2 G (y-axis)

Colored by: Control vs control (Default Interpretation)
Gene List: A (x-axis) (15103)

Color Figure 23.4

and histological processing often alter the three-dimensional structure of the antigen by forming methylene bridges between reactive sites within the same or adjacent proteins causing "masking" of the antigen (Cattoretti et al., 1993). Although the use of frozen sections circumvents this problem, this is a labor-intensive procedure when applied to the large number of samples in a toxicological study.

For immunotoxicology assessment staining of formalin-fixed, paraffin-embedded sections is the method of choice. It provides the morphological resolution required for the detailed histology evaluation recommended in the tiered system. The advent of efficient antigen-retrieval methods (Section 15.3.1.1) has allowed the visualization and assessment of specific cell populations using CD markers by immunohistochemistry in the same paraffin-embedded tissues used for histopathology and grading of lesions (Figures 15.1 and 15.2). An additional advantage of paraffin sections is the feasibility of conducting restrospective evaluation of archived tissues (Gill and Pulido, 2001; Mueller et al., 2003; Imam et al., 1995; McQuaid et al., 1995). However, since several commercially available antibodies only react in frozen tissue sections, sample collection of lymphoid tissues should include frozen and formalin fixed tissues.

15.3.1 Methods

15.3.1.1 Microwave Antigen Recovery Method

Formalin fixation is known to cause major chemical changes in tissues, thereby masking the antigen sites. Several techniques such as protease digestion are reported in the literature for retrieving antigen sites for immunohistochemistry (Cattoretti et al., 1993). After testing several of these methods, we found that the microwave antigen recovery method was the most reliable and efficient. It allows optimal antigen retrieval of a wide range of biomarkers on standard formaldehyde-fixed, paraffin-embedded tissues, including archived specimens (Gill and Pulido, 2001; Mueller et al., 2003; Imam et al., 1995; McQuaid et al., 1995).

We use paraffin sections mounted on silanated slides. Sections are deparaffinized and passed through a series of 100% ethanol. Endogenous peroxidase sites in the tissue are blocked in a 0.5% hydrogen peroxide/100% ethanol solution, and rinsed in 95% ethanol and double distilled water. The slides are transferred into coplin jars of 10 mM sodium citrate buffer (BDH), pH 6.0, and then microwaved for two 3-minute periods at 450W with gentle agitation. After cooling to room temperature to prevent the drying of sections, the slides are processed for immunohistochemistry using one of the following methods.

15.3.1.2 Streptavidin-Biotin Complex Indirect Method

Although several methods for immunohistochemical staining exist, our preferred choice is the avidin-biotin indirect method. Avidin is a basic glycoprotein with four high-affinity binding sites for the vitamin biotin. Streptavidin, a protein isolated from the bacterium *Streptomyces avidinii*, is substituted for avidin in the immunohistochemical method because it has the same high affinity for the biotin, while its physical properties avert the nonspecific binding of avidin. The conjugation of biotin to the secondary antibody forms the highly specific and efficient streptavidin-biotin complex used in the colorometric detection of the antigen (Cattoretti et al., 1993).

The procedure used in our laboratory begins with the deparaffinization, blocking of peroxidase sites, and microwave antigen retrieval as described above. Then the slides are blocked for avidin and biotin sites to ensure specific staining. They are then incubated with the primary antibody at a previously determined concentration overnight at 4°C in 15% normal serum specific to the type of antibody. A biotinylated F (antibody') 2, secondary antibody directed against the primary antibody is diluted (1:100) in 15% antibody-specific normal serum. This is used to incubate the slides for 1 hour at room temperature, followed by 30 minutes incubation at room temperature with streptavidin against the biotinylation on the secondary antibody at (1:200) dilution. The chromogen most often used is 3,3'-diaminobenzidine tetrahydrochloride (DAB, 40 mg/36 µl of 30% H_2O_2 in 200 ml of Tris buffer). Other chromogens such as aminoethylcarbazole are also available (Mueller et al., 2003). To allow the histomorphometry analysis of CD markers, the tissues are not counterstained (Figure 15.1C; Figure 15.2E and F; Tryphonas et al., 2004). For other biomarkers, tissues are counterstained with instant hematoxylin (Gill and Pulido, 2001).

15.3.1.3 Biotin-Free Secondary Amplification Indirect Method

The most recent indirect method is the biotin-free secondary amplification, which has several advantages over the traditional streptavidin-biotin complex indirect method. This procedure uses a unique enzyme-conjugated polymer backbone that is biotin-free and carries multiple secondary antibody and peroxidase (HRP) sites (DAKO Envision Plus System). This eliminates required blocking step for endogenous biotin in the tissue, and hence eliminates the possibility of nonspecific staining. The polymer consists of a strand of multiple secondary antibody and HRP, which amplifies the signal, thereby making it beneficial especially for antigens with low tissue expression. Another advantage of this procedure is that it can be completed within the same day.

After deceration, blocking of peroxidase sites, and microwave antigen retrieval, the slides are incubated with the optimal concentration of primary antibody in dilution buffer for 30 minutes at room temperature. The secondary antibody/HRP polymer backbones (DAKO Envision Plus System) are added undiluted to each slide for 30 minutes at room temperature. The chromogen and counter-stain used are similar to that described above. Examples of the application of this method are depicted in Figure 15.1D and Figure 15.3A through F.

15.3.1.4 Immunotyping Using Clusters of Differentiation Markers

The expression of CD markers can be used to characterize and differentiate lymphocyte subtypes in lymphoid tissues of rat and nonhuman primates. Paraffin sections from monkey and rat thymus, lymph node, and spleen were used. A variety of commercially available polyclonal and monoclonal antibodies for CD markers were tested using the methods described above. Table 15.2 summarizes the results and sources of the antibodies showing positive reaction in paraffin sections. The table includes the localization and distribution observed for each marker within the tissues. These results were consistent with the specifications given by the manufacturers for CD3 labeled T-lymphocytes, CD8 labeled cytotoxic T cells, CD45RA and CD20 labeled B cells, and ED1 labeled tissue mac-rophages in formalin-fixed, paraffin-embedded tissues. We also tested other antibodies from various manufacturers that did not react on paraffin sections, and hence required cryostat sections: CD3 (DakoCytomation, Mississauga, Ontario, Canada); CD11b (Chemicon International, Temecula, CA); and CD45 (Cymbus Biotechnology, Chandlers Ford, Hantes, United Kingdom).

Sell (2001) reviewed in detail the steps of lymphoid cell differentiation and CD acquisition. Briefly, pre-T cells from the bone marrow enter the thymus, develop a large variety of specific receptors, and express both CD4 and CD8 markers. CD4 is one of the first identifiable markers in the thymic cortex. As the thymocytes mature, the thymic medulla expresses the CD3 marker. As the final differentiation step, CD4+ and CD8+ cells segregate into two separate populations. CD4 designates the T-helper population and CD8 designates the T-cytotoxic population. After a selection process, the cells in the thymus that survive and differentiate into CD3/CD4+ or CD3/CD8+ leave the thymus for further development in the peripheral lymphoid organs.

Immunohistochemistry employing CD markers allows the visualization of specific lymphocyte cell populations in thymus and other lymphoid tissues. Figure 15.1C shows CD3 positive T cells in the medulla of the rat thymus. Figure 15.1D shows B-cell–dependent regions in the rat spleen. The follicles of the white pulp are highlighted by CD45RA immunostain. The intensity of the stain is stronger within the tightly packed zone of B cells — "the mantle" — that surrounds the germinal

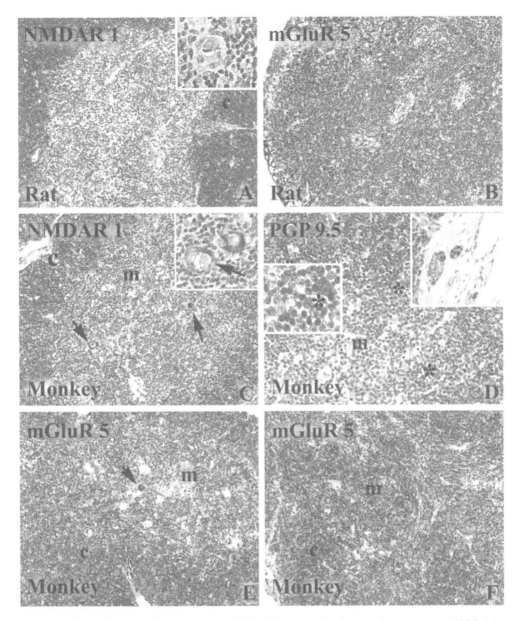

Figure 15.3 (See color insert) Differential distribution of the neural marker protein gene product (PGP 9.5) and the glutamate receptors (GluRs) subtypes NMDAR1 and mGluR5 in lymphoid tissues of rat and nonhuman primate. Paraffin sections (4 to 5 μ) of lymphoid tissues from Sprague–Dawley (SD) adult rats (A, B) and from 78-week infant *Macaca fascicularis* monkeys (C through F) were stained by immunohistochemistry using the biotin-free secondary amplification indirect method. (A) Thymus of an adult control SD rat showing cortex (c) and a very pale NMDAR1 immunostain scattered within the medulla in cells that appear to be epithelial cells; 20X objective. These cells are better viewed at higher magnification in the inset; 40X objective. (B) Mesenteric lymph node of an adult control SD rat showing strong mGluR5 immunostain particularly in the subcapsular and medullary sinuses; 20X objective. (C) Thymus of a control infant monkey showing NMDAR1 immunostain particularly prominent within the medulla (m) and rare in the cortex (c). The overall intensity and labeling of epithelial cells/Hassall's corpuscles (arrows and inset) are stronger in the infant monkey than in the rat thymus shown in A. (D) Thymus of a control infant monkey showing PGP 9.5 in cells that appear to be dendritic cells/macrophages (asterisk); 20X objective. These cells are better viewed at higher magnification (40X objective) in the middle inset (asterisk). The top right enclosure shows immunostained nerve fibers embedded within the capsule connective tissue (40X objective), confirming the well-known specific affinity of this marker for neural tissues. (E) Thymus of a control infant monkey showing mGluR5 immunostain particularly within the medulla (m) as compared to the cortex (c). Labeled epithelial cells/Hassall's corpuscles (arrow) are easily recognized; 20X objective. (F) Thymus of an infant monkey exposed for 11 months to oral toxaphene (1.0 mg/kg/day). The section shows mGluR5 immunostain with a preferential distribution in the medulla (m) as compared to the cortex (c); 20X objective.

Table 15.2   Immunohistochemical Phenotyping of Rat and Monkey Lymphoid Tissues[a]

Biomarker	Source	Main Cellular Expression	Species and Lymphoid Tissue Tested
Clusters of Differentiation (CD)			
CD45RA (monoclonal)	Biosource International, Montreal	Leucocyte common antigen (LCA), B cells	Rat: LN, S,T Monkey: T
CD20cy (monoclonal)	DakoCytomation, Missaussaga, Ontario, Canada	B-cell precursors and mature B cells	Rat: LN, S,T Monkey: LN, S, T
CD8 (monoclonal)	Biosource International, Montreal	Majority of NK cells Majority of thymocytes	Rat: LN, S,T Monkey: T
CD3 (polyclonal)	Sigma-Aldrich Canada Ltd.	T cells	Monkey: S, LN
ED1 (monoclonal)	Sertec, Raleigh, NC	Tissue macrophages	Rat: LN, S,T Monkey: T
Neuronal Markers			
PGP 9.5 (polyclonal)	Chemicon International, Temecula, CA	Nerve fibers Dendritic cells and macrophages, particularly monkey thymus	Rat: LN, S,T Monkey: T
NMDAR1 (polyclonal)	Chemicon International, Temecula, CA	NMDAR receptor Endocrine epithelium, dendritic cells, macrophages, Hassall's corpuscles of monkey thymus	Rat: LN, S,T Monkey: T
GluR2/3 (polyclonal)	Chemicon International, Temecula, CA	GluR2 and GluR3 receptors Monkey thymus T-cell medulla>cortex, macrophages, Hassall's corpuscles	Rat: LN, S,T Monkey: T
mGluR5 (polyclonal)	Chemicon International, Temecula, CA	mGluR receptor Endocrine epithelium, dendritic cells, macrophages, Hassall's corpuscles of monkey thymus>rat thymus	Rat: LN, S,T Monkey: T

Note: LN, lymph node; S, spleen; T, thymus.

[a] Paraffin sections (4 to 5 μ) were microwaved for antigen recovery and stained by immunohistochemistry using commercially available antibodies for clusters of differentiation (CD), glutamate receptors (GluRs), and the neural marker protein gene product (PGP 9.5). All antibodies listed showed positive immunostain in paraffin sections. Distribution and intensity varied with the specie and tissue tested. The commercial source of the antibodies and the main cellular expression are also included. These immunohistochemical procedures (Section 15.3) can be used as an extension of routine toxicity/pathology studies, and as an investigational tool for immunotoxicology.

centers. On the other hand, Figure 15.2E and F illustrate the distribution of CD20-positive B cells within the the primary follicules of the cortex of a mesenteric lymph node from an infant control monkey (Figure 15.2E), as compared to a toxaphene-treated animal (Figure 15.2F). The stain intensity, cell population, and distribution of CD20-labeled B-lymphocytes are decreased in the treated animal as compared to the control.

In addition to the lymphocytes, macrophages and Langerhans-type dendritic cells can also be visualized by immunohistochemistry. We tested the ED1 marker for tissue macrophages with acceptable results in paraffin sections.

15.3.1.5   *Immunohistochemistry of Glutamate Receptors and Neural Markers*

During the past few years, we have been interested in excitatory signaling and the use of neural biomarkers in peripheral tissues. We have successfully used these neural biomarkers in several tissues of monkey and rat (Gill and Pulido, 2001; Mueller et al., 2003). Interaction between the nervous and immune systems has attracted considerable attention during recent years. Therefore, we decided to conduct a pilot study to investigate the presence and distribution of GluRs and specific neural biomarkers in the lymphoid tissues of rats and monkeys. Although the results of this study are the subject of a separate publication (Gill et al., submitted), we have included here a summary of our findings. Our results suggest that GluRs and the neural biomarker, protein gene product (PGP 9.5), may be useful for cell specific phenotyping of lymphoid tissues (Figure 15.3, Table 15.2). The data illustrate that the antibody to PGP 9.5 has specific affinity for dendritic cells and macrophages, particularly in monkeys (Figure 15.3D, middle inset). The well-known preferential affinity of this marker for neural tissues is seen in the top inset of Figure 15.3D (Thompson et al., 1983; Schofield et al., 1995; Wilkinson et al., 1989). Antibodies to the GluRs subtypes, NMDAR 1 and mGluR 5, also appear to have specific affinity for the endocrine epithelium, dendritic cells, and macrophages, particularly in monkey tissues (Figure 15.3B, C, E, and F). Hassall's corpuscles show strong NMDAR1 and mGluR5 immunostaining in monkey thymus, but not in rat thymus. Whether the differences between rats and monkeys are age or species specific needs to be determined.

15.4 PRACTICAL APPLICATION TO RODENTS AND NONHUMAN PRIMATES STUDIES

Studies have been conducted at Health Canada to assess the toxic effects of the environmental contaminants tributyltin chloride (TBTC) and toxaphene in rats and nonhuman primates. Recent publications (Arnold et al., 2001; Tryphonas et al., 2004) describe the experimental protocols. Tissue samples from some of these animals were used to develop and implement the above methodologies. Figure 15.1E and F and Figure 15.2A through D illustrate the histopathology of lymphoid tissues from rats and monkeys exposed to these environmental contaminants.

15.4.1 Rat

Tryphonas et al. (2004) studied the effects of tributyltin chloride on the immune system of *in utero* and postnatally exposed Sprague–Dawley rats. In this study, the pups were exposed *in utero* from day 8 of gestation, through lactation, and then postweaning until the pups were necropsied. Dams were dosed orally by gavage with TBTC in olive oil at 0, 0.025, 0.25, and 2.5 mg/kg of body weight per day from day 8 of pregnancy. Dosing continued at the same level through lactation. At weaning (21 days of age), male and female pups within each treatment group were randomly assigned to three subsets and continued to receive the same dose of TBTC by gavage as their respective dam until age 30 days (subset 1, males and females), 60 days (subset 2, females), or 90 days (subset 3, males). The lymphoid tissues were used for immunotoxicologic evaluation and enhanced histopathologic assessment. The thymus showed mild to moderate cortical atrophy characterized by decreased numbers of cortical lymphocytes, particularly in the 30-day high-dose group. Other lymphoid tissues were affected including lymph nodes, spleen, and ileum (Figure 15.1E and F).

The semiquantitative histopathology results were substantiated using histomorphometry. The thymus cortical and medullary area values in square millimeters were computed and compared for control and treated groups (Table 15.1). There was a significant reduction in the combined cortex plus medulla area of the 30-day high-dose male rats ($p = 0.048$), and in the medullary area of the 60-day high-dose female rats ($p = 0.042$). In the 90-day high-dose male rats, there was a significant reduction in the area of the cortex, the medulla, and the cortex plus medulla ($p = 0.0066$, $p = 0.0381$, and $p = 0.0063$, respectively). This methodology was also used for comparison of the number of T cells highlighted by a CD3 immunohistochemical stain in the thymus, spleen, ileum, and mesenteric and popliteal lymph nodes in the control and high-dose groups (Figure 15.1C and

D). Five readings using a 40X power objective were recorded for each tissue. Data were analyzed using trend analysis and pairwise comparisons. Trends were considered significant when $p < 0.05$. Data from control and treated rats were compared using one-way analysis of variance for multiple comparisons, followed by Dunnett's test for pairwise comparisons if necessary. There were no statistically significant effects on the T-cell population in any of the tissues examined of treated animals compared to controls when slides were stained with the CD3 marker ($p > 0.05$, data not shown). Although there were no significant changes, the trends suggest that B-cell–dependent regions were most affected.

15.4.2 Nonhuman Primates

Nonhuman primates have been used as a nonrodent animal model for preclinical toxicology and safety assessment trials. This is based on similarity and comparability between nonhuman primates and humans (Li et al., 1993). The validity of the nonhuman primate models applies to many aspects of toxicology testing. Buse et al. (2003) recently reviewed the validity of these models in reproductive/developmental toxicity and immunotoxicity assessments. The authors point out that many immunotherapeutics do not interact with rodent receptors, but frequently cross-react with primate tissues. Knowledge of the morphology and functional maturation of immune system components is relevant to the design and interpretation of toxicological studies. There is very limited information available on the phenotypes and development of the lymphoid organ in nonhuman primate models. The cynomolgus (*M. fascicularis*) monkey is the primate model used in our laboratory. We initiated the evaluation and application of the enhanced histopathology and immunohistochemistry procedures using tissues obtained from a recent study that examined the effects of toxaphene (Arnold et al., 2001). A synopsis of the preliminary data is included here to illustrate the application of the methods to this nonhuman primate model.

Cynomolgus monkeys obtained from the Health Canada monkey breeding colony were used in these studies (Arnold et al., 2001). They were housed individually during the 5-month acclimatization period in accordance with the guidelines of the Canadian Council of Animal Care (1984). At the end of the 5 months, the female animals were randomized to four treatment groups. The monkeys received gelatin capsules containing doses of 0, 0.1, 0.4, or 0.8 mg (groups A, B, C, or D, respectively) of technical-grade toxaphene per kilogram of body weight per day. During the 75-week prebreeding phase, a pharmacokinetic steady state was achieved for toxaphene in the monkeys' fat and blood. Ten untreated male cynomolgus monkeys were randomly housed with the females for breeding. The methods and procedures used in monitoring the resulting infants' development from parturition through weaning were described (Arnold et al., 2001). Prior to weaning, the infants were exposed to toxaphene via their mothers' milk. From weaning at 22 weeks of age until 30 weeks of age they did not ingest toxaphene, but were taught to ingest gelatin capsules. From 30 weeks of age until necropsy at 78 weeks of age, they ingested toxaphene-filled capsules in the same four dose groups as their mothers.

An enhanced histopathology assessment of the lymphoid tissues from this study was carried out using the methodology described in Sections 15.2 and 15.3. Figure 15.2A and B depict the cross-section of a mesenteric lymph node from an adult control monkey and from an animal treated with 0.8 mg/kg/day of toxaphene. The lymph node from the treated animal shows a grade 4 generalized decrease in the B- and T-cell–dependent areas and nonspecific cell populations when compared to an age-matched control. Figure 15.2D depicts the thymus from a 78-week-old infant monkey exposed to 0.8 mg/kg/day of toxaphene with a grade 2 decrease in cortical area and cell density, and an apparent increase in interlobular connective tissue. These changes represent isolated responses from individual monkeys, since no major differences were observed in given treatment groups when compared to controls. Thymuses in both adult groups showed expected age-related involution. Since semiquantitative grading systems are known to be limited, we are currently assessing these tissues using combined histomorphometry and immunohistochemistry. These studies may provide valuable additional information on the effects of toxaphene in lymphoid tissues.

15.5 SUMMARY AND CONCLUSION

The pathology assessment of lymphoid tissues is an integral component of the methods employed to detect immunotoxicity of chemicals and is routinely employed in Tier 1 testing. The focus of this chapter (Section 15.2) is to highlight the assessment of the lymphoid tissues using conventional and enhanced pathology methods. The incorporation of more detailed histopathology and grading of lesions by semiquantitative methods has greatly enhanced the toxicologic pathology evaluation of these tissues. Since phenotypic characterization of lymphoid tissues reveals a complexity of cell types that are not apparent by conventional stains, we tested the application of two immunohistochemical procedures to paraffin-embedded sections (Section 15.2). A battery of selected CD immunohistochemical markers were used to identify subpopulations of lymphocytes to complement the enhanced histopathology evaluation. Application of the software program Northern Eclipse was also tested and validated as a morphometric procedure (Section 15.2). Although it proved to be a valuable tool to generate numerical data, it does not replace the semiquantitave evaluation conducted by a trained pathologist. Furthermore, software programs tend to be expensive and require computer hardware specifications that may not be available in all laboratories around the world. In summary, histomorphometry and immunohistochemical phenotyping of lymphoid cells using CD markers can be included as sensitive tools in the pathology evaluation of immunotoxicity. However, it is still the challenge of the toxicologic pathologists to interpret the pathology data within the complete clinical evaluation of the entire animal. This includes making contributions to more sensitive detection of the immunotoxic potential of compounds and to understanding the involved mechanisms.

Because the relevance to this chapter and uniqueness of findings, we included a brief summary and illustrations of our investigations in neuroexcitatory signaling pathways (Gill and Pulido, 2001; Mueller et al., 2003). A complete report is the subject of a separate publication (Gill et al., submitted). This study demonstrates for the first time that there are structures in lymphoid tissues such as dendritic cells, macrophages, and Hassall's corpuscles that have specific affinity for the neural biomarkers PGP 9.5 and GluR receptor subtypes NMDAR1, GluR 2/3, and mGluR5. These findings reflect the structural complexity of the lymphoid tissues and open new avenues for the investigation of the function and role that these structures may have in health and disease. We hope that this chapter will foster further research and advances in toxicologic pathology of the immune system.

ACKNOWLEDGMENTS

The authors are grateful to Drs. Helen Tryphonas, Gerry Cooke, and Doug Arnold for allowing the use of tissues collected in the course of toxicological studies. The contributions of Dr. Rudi Mueller in the preparation of the photographs, and Dr. Michael Barker in the reading of the document and collection of the tissues during necropsy are greatly appreciated. The technical assistance of Peter Smyth, Ian Greer, and James Elwin is also acknowledged.

REFERENCES

Arnold, D.L., et al., Toxicological consequences of toxaphene ingestion by cynomolgus (Macaca fascicularis) monkeys. Part 1: pre-mating phase, *Food. Chem. Toxicol.*, 39, 67, 2001.

Basketter, D.A., et al., The identification and classification of skin irritation hazard by a human patch test, *Food Chem. Toxicol.*, 32, 769, 1994.

Basketter, D.A., et al., Pathology consideration for and subsequent risk assessment of chemicals identified as immunosuppressive in routine toxicology, *Food Chem. Toxicol.*, 33, 239, 1995.

Bleavins, M.R., and de la Iglesia, F.A., Cynomolgus monkeys (Macaca fascicularis) in preclinical immune function safety, *Toxicology*, 95, 103, 1995.

Buse, E., et al., Reproductive/developmental toxicity and immunotoxicity assessment in the nonhuman primate model, *Toxicology*, 221, 221, 2003.

Cattoretti, G., et al., Antigen unmasking on formalin-fixed, paraffin-embedded tissue sections, *J. Pathol.*, 171, 83, 1993.

Cooke, G.M., et al., Oral (gavage) in utero and postnatal exposure of Sprague–Dawley rats to low doses of tributyltin chloride. Part I: Toxicology, histopathology and clinical chemistry, *Food Chem. Toxicol.*, 42, 211, 2004.

Dayan, A.D., et al., Report of validation study of assessments of direct immunotoxicity in the rat, *Toxicology*, 125, 183, 1998.

Descotes, J., et al., Responses of the immune system to injury, *Toxicol. Pathol.*, 283, 479, 2000.

De Waal, E. J., and van Loveren, H., Practice of tiered testing for immunosuppression in rodents, *Drug Infect. J.*, 31, 1317, 1997.

European Agency for the Evaluation of Medicinal Products, Evaluation of Medicines for Human Use: Note for Guidance on Repeated Dose Toxicity, 2000, London, England.

Gill, S., and Pulido, O., Glutamate receptors in peripheral tissues: Current knowledge, future research and implications, *Toxicol. Pathol.*, 29, 208, 2001.

Gill, S., et al., Immunophenotyping of lymphoid tissues using cluster of differentiation (CD), the Neural biomarker PGP 9.5 and glutamate receptors, *Toxicol. Pathol.*, submitted.

Gopinath, C., Pathology of the toxic effects on the immune system, *Inflammation Res.* 45, S74, 1996.

Gossett, K.A., et al., Flow cytometry in the preclinical development of biopharmaceuticals, *Toxicol. Pathol.*, 27, 32, 1999.

Harleman, J.H., Approaches to the identification and recording of findings in the lymphoreticular organs indicative for immunotoxicity in regulatory type toxicity studies, *Toxicology*, 142, 213, 2000.

Hashek, W.M, Rousseaux, C.G., and Wallig, M.A., Eds., *Handbook of Toxicology Pathology*, Academic Press, San Diego, 2002.

Hendrick, A.G., Makori, N., and Peterson, P., Nonhuman primates: their role in assessing developmental effects of immunomodulatory agents, *Hum. Exp. Toxicol.*, 19, 219, 2000.

Hinton, D.M., "Redbook II" immunotoxicity testing guidelines and research in immunotoxicity evaluations of food chemicals and new food proteins, *Toxicol. Pathol.*, 28, 467, 2000.

House, R.V., and Thomas, P.T., Immunotoxicology: fundamentals of preclinical assessment, in *Handbook of Toxicology*, Derelanko, M.J., and Hollinger, M.A., Eds., CRC Press, Boca Raton, FL, 2002, p. 401.

Imam, S.A., et al., Comparison of two microwave based antigen-retrieval solutions in unmasking epitopes in formalin-fixed tissue for immunostaining, *Anticancer Res.*, 15, 1153, 1995.

Institoris, L., et al., Extension of the protocol of OECD guideline 407 (28-day repeated dose oral toxicity test in the rat) to detect potential immunotoxicity of chemicals, *Hum. Exp. Toxicol.*, 17, 206, 1998.

International Collaborative Immunotoxicity Study, Report of validation study assessment of direct immunotoxicity in the rat, *Toxicology*, 125, 183, 1998.

International Programme on Chemical Safety, Principles and Methods for Assessing Direct Immunotoxicity Associated with Exposure to Chemicals, Environmental Health Criteria 180, World Health Organization, Geneva, 1996.

Kawabata, T.T., et al., Immunotoxicology of regional lymphoid tissue: the respiratory and gastrointestinal tracts and skin, *Fundam. Appl. Toxicol.*, 26, 8, 1995.

Kimber, I., and Dearman R.J., Immune Responses: adverse versus non-adverse effects, *Toxicol. Pathol.*, 30, 54, 2002.

Kuper, C.F., Schuurman, H.J., and Vos, J.G., Pathology in immunotoxicology, in *Methods in Immunotoxicology*, Burleson, G.R., Dean, J.H., and Munson, A.E., Eds., Wiley-Liss, New York, 1995.

Kuper, C.F., et al., Histopathologic approaches to detect changes indicative of immunotoxicity, *Toxicol. Pathol.*, 28, 454, 2000.

Kuper, C.F., et al., Immune System, in *Handbook of Toxicology Pathology*, Haschek, W.M., Rousseaux, C.G., Wallig, M.A., Eds., San Diego, Academic Press, 2002.

Lappin, P.B., and Black, L.E., Immune modulator studies in primates: the utility of flow cytometry and immunohistochemistry in the identification and characterization of immunotoxicity in the identification and characterization of immunotoxicity, *Toxicol. Pathol.*, 31, 111, 2003.

Li, S.H., et al., Immunohistochemical distribution of leucocyte antigens in lymphoid tissues of cynomolous monkey (Macaca fascicularis), *J. Med. Primatol.*, 22, 285, 1993.

Luster, M.I., et al., Development of a testing battery to assess chemical-induced immunotoxicity: national toxicology program's guidelines for immunotoxicity evaluation in mice, *Fundam. Appl. Toxicol.*, 10, 2, 1998.

McQuaid, S., et al., Microwave antigen retrieval for immunocytochemistry on formalin-fixed, paraffin-embedded post-mortem CNS tissue, *J. Pathol.*, 176, 207, 1995.

Moolenbeek, C., and Ruitenberg, E.J., The Swiss roll: a simple technique for histological studies of the rodent intestine, *Lab. Anim.*, 15, 57, 1981.

Mueller, R., Gill, S., and Pulido, O., The monkey *(Macaca fascicularis)* heart neural structures and conducting system: an immunochemical study of selected neural biomarkers and glutamate receptors, *Toxicol. Pathol.*, 31, 227, 2003.

Neubert, R., Helge, H., and Neubert, D., Nonhuman primate model for evaluating substance-induced changes in the immune system with relevance for man, in *Experimental Immunotoxicology*, Smialowicz R.J., and Holsapple, M.P., Eds., CRC Press, Boca Raton, FL, 2002.

Rose, N.R., and Caturegli, P.P., Environmental and drug-induced autoimmune diseases of humans, in *Comprehensive Toxicology*, Lawrence, D.A., Ed., Pergamon, Oxford, 1997, p. 5.

Schofield, J.N., et al., PGP9.5, a ubiquitin C-terminal hydrolase: pattern of mRNA and protein expression during neural development in the mouse, *Brain Res. Dev.*, 85, 229, 1995.

Schuurman, H.J., et al., Histopathological approaches, in *Principles and Practice of Immunotoxicology*, Turk, J., Nicklin S., and Miller K., Eds., Blackwell Scientific, Oxford, 1992, p. 279.

Schuurman, H.J., et al., Histopathology of the immune system as a tool to assess immunotoxicology, *Toxicology*, 86, 187, 1994.

Sell, S., Lymphoid organs, in *Immunology, Immunopathology and Immunity*, ASM Press, Washington, DC, 2001, p. 198.

Thompson, R.J., et al., PGP 9.5 — a new marker for vertebrate neurons and neuroendocrine cells, *Brain Res.*, 278, 224, 1983.

Tryphonas, H., et al., Oral (gavage) in utero and postnatal exposure of Sprague-Dawley rats to low doses of tributyltin chloride. Part II: effects on immune system, *Food Chem. Toxicol.*, 42, 221, 2004.

U.S. Food and Drug Administration, Center for Drug Evaluation and Drug Research, Immunotoxicity Evaluation of Investigational New Drugs, U.S. Department of Health and Human Services, Washington, DC, 2001.

van Loveren, H., and Vos, J.G., Immunotoxological considerations: a practical approach to immunotoxicity testing in the rat, in *Advances in Applied Toxicology*, Dayan, A.D., and Paine, A.S., Eds., Taylor & Francis, London, 1989.

Vos, J.G., Immunotoxicology assessment: screening and function studies, *Arch. Toxicol. Suppl.*, 4, 95, 1980.

Vos, J.G., The role of histopathology in the assessment of immunotoxicology, in *Immunotoxicology*, Berlin, A., Dean, J., Drapel, M.H., et al., Eds., Martinus Nijhoff, Dordrecht, 1987, p. 125.

Vos, J.G., and van Loveren H., Immunotoxicity testing in the rat, in *Advances in Modern Environmental Toxicology*, Burger, E.J., Tardiff, R.G., and Bellanti, J.A., Eds., Princeton Scientific Publishing, Princeton, NJ, 1987, p. 167.

Ward, J.M., Uno, H., and Frith, C., Immunohistochemistry and morphology of reactive lesions in lymph nodes and spleen from rats and mice, *Toxicol. Pathol.*, 21, 199, 1993.

Wilkinson, K.D., et al., The neuron-specific protein PGP 9.5 is a ubiquitin carboxyl-terminal hydrolase, *Science*, 246, 670, 1989.

CHAPTER 16

Approaches to Immunotoxicology in Human Population Studies

Frédéric Dallaire, Éric Dewailly, and Pierre Ayotte

CONTENTS

0-415-30854-2/05/$0.00+$1.50

16.1 INTRODUCTION

The assessment of immunotoxicity in humans involves several epidemiologic and ethical challenges. Experimental designs are rarely feasible and investigators have to rely on the identification of exposed and unexposed populations to conduct their studies. These populations differ in regards to disease prevalence, access to medical treatment, and to genetic and environmental factors. Furthermore, the function of the immune system has to be investigated by strictly noninvasive methods that are not always easy to standardize. In this chapter, we discuss epidemiologic designs, factors to consider, and some useful biological and clinical endpoints relevant to human immunotoxicology. Whenever possible, we also provide examples of the methods discussed by referring the reader to published studies on the topic.

16.2 EPIDEMIOLOGIC CONSIDERATIONS

Human immunotoxicology studies are subject to the same rules, challenges, methods, and problems as any other epidemiology study. It is well beyond the scope of this work to discuss the many subtleties of epidemiology and population studies. For a more detailed presentation, the reader is referred to the many excellent publications on epidemiology, including Rothman and Greenland (1998). For the purpose of this discussion, only epidemiologic principles most relevant to immunotoxicology are presented.

16.2.1 Study Design

For obvious ethical reasons, most studies in human immunotoxicology are nonexperimental, that is, the exposure to the xenobiotic under study is not assigned by the investigator for the purpose of a study; it is the participants who, willingly or not, have exposed themselves to the substance. In nonexperimental human studies, investigators have to identify two groups of participants, one exposed and the other unexposed (or less exposed), who are in similar situations. These situations should be comparable enough so that the differences between the two groups will not influence the disease under study more than the exposure itself. In this section, we briefly explore three nonexperimental study designs: cohort, case-control, and cross-sectional. The reader should be aware of the existence of other designs, as well as of the many variations within the three designs mentioned (Rothman and Greenland, 1998).

16.2.1.1 Cohort Studies

To design a cohort study in immunotoxicology, the investigator has to define various groups of participants differing by their exposure to a xenobiotic. The most important aspect of a classical cohort study is that the recruited participants must be free of the disease under investigation at the beginning of the study. A cohort study will follow the participants over time and compare the incidence rates between exposure groups, that is, the rates of negative participants turning positive for a given endpoint. This design is the most straightforward, but is also the most expensive and time consuming. In classical cohort studies, the endpoint is usually the onset of a disease and the analyses are conducted to determine differences of incidence of this disease between the exposure groups. Nevertheless, cohort studies can also be used with continuous biological endpoints, such as the ones often used in immunotoxicology. In this case, the cohort design has the enormous advantage of allowing the investigators to follow the participants in order to record the variations of a given endpoint during the study time period. Finally, it is often possible to incorporate other types of design within a cohort study setting. Such concepts are better understood when specific examples are considered.

Rogan et al. (1987) used a cohort design to evaluate the effects of polychlorinated biphenyls (PCBs) and dichlorodiphenyl dichloroethylene (DDE) on growth, morbidity, and duration of breast-feeding in children. The neonates were grouped according to exposure levels and the incidence of illnesses was recorded during the first year of life. Considering that the recorded illnesses were "new" illnesses, that is, illnesses not present at recruitment, the design would correspond to that of a cohort study. Similarly, Weisglas-Kuperus et al. (2000) investigated the effects of PCBs and dioxins on infections in children. Again, only healthy newborns were recruited and "new" infections were considered. In the latter study, the authors also evaluated T-cell markers. In this case, the design did not technically correspond to a cohort study since T-cell markers were not measured at birth. When only one measurement is made, the relation between exposure and the endpoint at one specific time is evaluated instead of the effect of exposure on the "progression" of this endpoint. This corresponds to a cross-sectional design since the investigator could not ascertain that the difference observed at the end of the study was not present at the beginning (see section cross-sectional design for further discussion on this topic).

When researchers wish to benefit from the advantage of a cohort design using biologic endpoints, which is to evaluate the variation of the endpoint during the follow-up, the endpoints at the beginning (baseline) and at the end of the follow-up must be measured. This allows for the calculation of the difference between the level at the end of the follow-up and the baseline level. This strategy will help in dealing with interindividual variations of the baseline level of the given endpoint. Statistically, this would be achieved in multiple regression by inserting a variable representing the difference between the two measurements into the regression model as the main dependent variable. More sophisticated strategies exist, such as time series and hazard functions, when more than two measurements are made.

16.2.1.2 Case-Control Studies

Case-control studies are generally used when the disease under investigation is rare so that it would be impractical or too expensive to recruit healthy subjects and follow them until a sufficient number is diagnosed with the disease (as in a cohort study). It should be noted, however, that the rarity of the disease is in most cases not an essential criterion for a case-control study to be valid (Rothman and Greenland, 1998). To conduct a case-control study, one has to identify cases, preferably subjects newly diagnosed with the disease, in a source population. One will also have to recruit controls, that is, participants who are free of the disease, within the same source population. The key aspect in case-control studies is that the participants must be selected independently from their exposure status. The comparison of the proportions of exposed and unexposed participants between the cases and the controls will yield an estimate of the relative risk (odds ratio) between the exposed and unexposed subjects.

In the field of immunotoxicology, case-control studies should be used to investigate potential associations between the exposure to xenobiotics and the development of chronic diseases such as cancer or systemic lupus erythematosus. Hardell and Ericksson (1999) used the case-control design to evaluate the association between non-Hodgkin's lymphoma and exposure to pesticides. The authors identified 404 Swedish males diagnosed with non-Hodgkin's lymphoma, and 741 controls. Exposure to pesticides was estimated by the use of a standardized questionnaire and a telephone interview. By doing this, they found an association between the risk of non-Hodgkin's lymphomas and previous exposure to herbicides, insecticides, and fungicides (odds ratios of 1.6, 1.2, and 3.7, respectively).

The case-control design also allows for the comparison of continuous variables between the cases and the controls. For example, one could assess the level of pesticides in the plasma of the participants and compare the mean concentrations between the cases and the controls. This can give an insight on the difference between both groups without having to create categories from a continuous variable (for instance, grouping the participants by quartiles of exposure). This can be done for every assessed variable. However, to remain valid, the recruitment must be done independently from such variables, that is, the factors should not be selection criteria in either cases or controls (Rothman and Greenland, 1998).

16.2.1.3 Cross-Sectional Studies

A cross-sectional study design is defined as a study in which participants are recruited at one time point, irrespective of their status regarding the particular endpoint to be investigated. In this design, the objective is to describe the status of a population at one point in time (a "cross-section" in time). The exposure status and the endpoint are usually investigated simultaneously.

The main limitation of cross-sectional study designs is that only one measurement of the endpoint is performed, without any follow-up. If the endpoint is a disease, the information obtained by the investigator is a disease prevalence, that is, the proportion of participants having the disease at one point in time. When disease prevalence is used in epidemiology, the investigator faces the problem of length bias (Rothman and Greenland, 1998). Disease prevalence is a function of both the incidence and the duration of the disease (the longer the disease is present in an individual, the higher is the probability of recruiting a participant having the disease at a given point in time). Consequently, cross-sectional designs do not allow for the discrimination between an effect on the incidence of an event and an effect on the duration of the same event. Furthermore, the valuable information on the variation or on the progression of an endpoint during the follow-up is impossible to obtain. The strategy where a participant serves as its own control, such as in cohort designs, is therefore, not applicable, and interindividual variations have greater effects on results.

Despite these drawbacks, cross-sectional studies are often used for practical and financial reasons, and a well-executed cross-sectional study may yield valuable and valid information. Svensson et al. (1994) used this design to evaluate the association between fish consumption and several immune parameters in Swedish adult males. These investigators recruited fish eaters and nonfish eaters as a surrogate of organochlorines (OCs) exposure and evaluated, among other endpoints, T-cell markers. They found a lower proportion of natural killer (NK) cells in fish eaters as compared to non-fish eaters. This result is subject to the limitations mentioned above, but is still very useful. The insight on the potential effect of OCs on NK cells would most likely have been much more expensive to gain from a cohort study.

16.2.2 Confounding

Reducing the random error in a study, and therefore increasing precision, is important to obtain statistically significant results. Failing to do so, however, will not bias the study results. A study with high random error will most likely produce nonsignificant results but will still be valid. It is the systematic error, or the differential error, that will bias the results. We do not discuss all possible biases and systematic errors here. However, confounding is a key aspect in the validity of any epidemiologic study, and we feel that this aspect needs to be addressed in some detail.

There are three essential characteristics for a factor to confound an association. First, it must affect the endpoint under study. Second, it must be associated with the exposure, and third, the factor must not be an intermediate step in the causal chain of the postulated mechanism of effect. Of all the factors that could affect the immune system, those that fulfill these three criteria could potentially become confounders.

Let us suppose that an investigator wants to study the effect of occupational exposure to dioxins on the antibody response to vaccination. If the participants exposed to dioxins are also exposed to excessive stress due to their work environment, then stress could possibly confound the association. In this setup, stress is known to affect antibody response, and is also associated with the exposure. Furthermore, stress is not an intermediary step in the postulated mechanism of dioxin toxicity, that is, the hypothesis does not state that the effect of dioxin on humoral response is the result of stress. This means that a potential association between dioxin exposure and response to vaccination could partly be due to the effect of stress, therefore introducing bias in the results.

In the example above, the "stress" factor displays the three essential prerequisites to becoming a confounder. It does not mean that it will necessarily confound the association. It is a common misconception that any factor associated with the endpoint and the exposure must be controlled for in the analysis. Observing a significant association between a factor and the endpoint (p value < 0.05) is not a criterion for the inclusion of this factor in a statistical model when a causal hypothesis is investigated. One must ensure that the factor could be a true confounder. First, the factor must influence the endpoint under study, that is, it must be a risk factor for the endpoint; it is not enough to be merely statistically associated with the endpoint. Second, the inclusion of a variable representing the potential confounder in a statistical model must affect the association between the exposure and the endpoint. If a factor is significantly associated with the endpoint, but does not influence the association of interest, then it does not, by definition, confound the association and should be excluded from the model. Including it would only make the model more complicated, less stable, and produce results that would be harder to interpret.

Many factors can influence immune response. Furthermore, several environmental circumstances, such as hygiene and socioeconomic status, can considerably affect clinical endpoints. When one thinks that one of these factors could also be associated with the exposure to the xenobiotic under study, then one should consider it as a potential confounder. If so, one has to evaluate such a factor, verify if it indeed confounds the association, and if needed, find ways to minimize the confounding effect.

16.2.3 Exposure Assessment and Classification

16.2.3.1 Exposure Assessment

Measurement of exposure and classification of participants in exposure groups are central to human immunotoxicology. The challenges in evaluating the exposure are similar in human studies and in animal studies, although exposure assessment in humans must remain noninvasive. In human immunotoxicology, the key aspect of exposure assessment is a strong and detailed hypothesis because measurement and assigning of exposure will both be planned according to the stated hypothesis. The hypothesis should detail to what extent the dose and length of exposure time are related to the toxic effect. Data collection and analyses are then performed accordingly. If the hypothesis states that it is the dose of exposure that is crucial in determining the toxic effect, and that the duration is less important, then the exposure assignment and statistical analysis should be performed mainly on the dose of exposure. Of course, more than one type of analysis could be carried out. For example, one could analyze the results in terms of the duration of exposure, various expressions of dosage (cumulative, peak, etc.), and all sorts of interaction between them. Specific and thorough knowledge of the xenobiotics under study are central in determining how to deal with exposure assessment.

16.2.3.2 Exposure Classification

Again, the hypothesis is central in the decision of exposure assignment (who is considered exposed, and who is not). One willing to test the hypothesis which states that "cannabis use reduces the macrophage ability to secrete cytokines" will have nothing to work with when the decision of who is considered exposed to cannabis will have to be made. When the hypothesis is clearly established, such as "smoking cannabis at least once a day for more than 1 year reduces cytokine secretion by lung macrophages," the assignment of exposure is clear-cut. In this example, everybody smoking cannabis less than once a day is unexposed. Similarly, somebody smoking cannabis more than once a day but for less than 1 year should also be considered unexposed. The definition of an induction period is important in determining the group to which a subject belongs during the follow-up period. In our example, the hypothesis states that there is a 1-year induction time, that is, that smoking less than a year should not be enough to induce an effect. This induction time should be considered when the contribution to the denominators of the exposed and unexposed groups is determined for each subject. Thus, the development of a clear hypothesis is essential. This is a scientific decision based on existing literature, not merely on statistics.

16.3 HUMAN STUDIES SPECIFICITIES

16.3.1 Factors Influencing Immune Response in Humans

Several factors can influence the immune system response. Although many are comparable between humans and animals, some need to be dealt with distinctively in human studies. Only those factors are addressed specifically in this section.

Among these factors is the large genetic variability of human populations as compared to laboratory animals. The variations of immune responses among human individuals will be proportional to the genetic heterogeneity in the population under study. Examples of this include vaccination responses to measles and to hepatitis B, which are strongly influenced by the human leukocyte antigen (HLA) genotype, and the influence of the promoter region of several cytokine genes on the inflammatory response (Bidwell et al., 1999; Howell et al., 2002; Kay, 1996; van Loveren et al., 2001). It would be impractical and unnecessary to measure such genetic variations in all epidemiologic studies. Although they greatly contribute to interindividual variability, and thus to statistical power, they do not necessarily introduce biases. As with other factors affecting the immune system, it is important to determine if they are potentially associated with the exposure under investigation in the study population. If so, they need to be considered as potential confounders (see Section 16.2.2 on confounding). If not, the random error created by interindividual variability can be minimized by recruiting an appropriately large number of subjects.

Important environmental factors influencing immune response in humans include malnutrition, stress, age, smoking habits, exercise, diet, lifestyle, and radiation. For a study to be valid, it is imperative that the investigator takes into account these factors in study design, and in the analyses and interpretation of results.

Malnutrition is the most common cause of immunodeficiency (Chandra, 2002). Although severe malnutrition is not prevalent in developed countries, the issue must often be addressed since only a small imbalance between energy intake and expenditure, as well as single-nutrient deficiencies, could affect many pathways of the immune response (Chandra, 2002; Marti et al., 2001). Malnutrition itself is delicate to investigate, but a good estimate of socioeconomic status is critical and should always be considered.

Psychologic stress and stress hormones such as glucocorticoids and catecholamines can modulate the immune response. Stress can decrease immune endpoints by affecting white blood cell counts, immunoglobulin (Ig) levels, and antibody response to vaccination (Cohen et al., 2001; Elenkov and Chrousos, 2002; Herbert and Cohen, 1993). Long- and short-term exposures to stress can be tricky to evaluate and one will often rely on random distribution of subjects between the different groups under study. However, stress exposure can be of particular concern in occupational studies, especially when the exposure to a xenobiotic is strongly associated with the type of occupation, which in turn can be associated with occupational stress.

Age is a critical factor, as most immune functions will change with aging. The fetus and the neonate are unable to react to several foreign substances and have less efficient neutrophils compared to healthy adults (Wilson, 1986). Aging adults have impaired T- and B-cell–mediated responses, and a modification of their repertoire could contribute to the development of autoimmunity (Antonaci et al., 1987; Pawelec and Solana, 1997; Urban et al., 2002).

Several lifestyle-related factors have an influence on the immune system. Smoking affects leukocyte counts (Burton et al., 1983), recreational drugs could reduce cell immunity (Roth et al., 2002), and UV radiation affects hypersensitivity reactions and NK cell activity (Sleijffers et al., 2002). On the contrary, exercise slows down immune senescence (Pedersen et al., 2000; Venjatraman and Fernandes, 1997), and vitamin A and fatty acids are important contributors of the immune response (Gil, 2002; Grimm et al., 2002; Semba, 1994). While most of the above mentioned factors can be measured or estimated, an evaluation of all lifestyle factors is financially and technically prohibitive. Depending on the type of outcome that is to be evaluated, one can rely on careful subject selection to minimize their effects. A thorough identification of lifestyle factors potentially associated with the exposure to the xenobiotic under study will also help to reduce the number of factors to evaluate. Importantly, the influence of lifestyle factors on any epidemiologic outcome should not be underestimated. The alteration that they produce can often be greater than that of occupational and environmental exposure to xenobiotics. Unknown and unmeasured factors are unavoidable and could likely bias the results. This always needs to be taken into account when positive or negative results are interpreted.

16.3.2 Clinical Immunodeficiency, Pre-Existing Diseases, and Medication Use

The functional immune system reacts to the incursion of a pathogen in the body, and many dormant immune parameters become activated in the presence of an infection. Extrinsic induction of immune and inflammatory processes by trauma, previous infection, and vaccination are also frequent in human populations. On the other hand, some microorganisms have the ability to suppress the immune system. All these variations in immune function can greatly affect the study results. For example, the prevalence of the BCG vaccine could severely bias an endpoint such as the response to the tuberculin test (van Loveren et al., 2001). A small difference in the prevalence of an immunosuppressing microorganism, such as HIV, could also significantly bias the final results of a study. These are obvious examples, but one must keep in mind that many individuals have asymptomatic acute or chronic infections that can weigh in the immune balance in unpredictable ways.

Particular to human studies is the widespread use of immunomodulating drugs, predominantly the nonsteroidal anti-inflammatory drugs (NSAIDs). Corticosteroids, NSAIDs, cyclophosphamide, and cyclosporine are all substances specifically designed to interfere with the normal immune response. The easiest way to eliminate the effects of pharmaceutical drugs is to ensure that the subjects under study abstain from any medication for a specified period before and during the course of the investigation, if possible. Medication use could be considered in statistical models, but it is not always simple to deal with the doses and the interactions among several drugs. Furthermore, study participants will usually remember the prescription drugs they have taken, but could easily forget sporadic ibuprofen use. This is to be considered especially when acute inflammatory endpoints are measured.

16.4 ASSAYS IN HUMAN IMMUNOTOXICOLOGY

The immune system is functionally and structurally complex. It is an integrated arrangement of several tissues and organs working together. It has great functional reserves and numerous overlaps and backups. One must always keep in mind that when an "abnormal" pathway is present, it may coexist with other "normal" pathways. Therefore, normal findings do not exclude abnormal functions, which can be compensated by other pathways and thus remain clinically silent. It is accepted that on an individual level, the immune system of a healthy adult can support some level of insult without affecting host resistance. Nevertheless, on a population level, it is plausible to assume that even small impairments of the integrity of the immune function could have an effect on some vulnerable individuals. Consequently, the "threshold" relationship observed on the individual level may not hold in a population, in which a more linear relationship between immunotoxicity and health status can be expected (Kimber and Dearman, 2002; Luster et al., 1993). In this context, it has been argued that the so-called immune reserve should not be considered in the interpretation of population immunotoxicologic data (Kimber and Dearman, 2002; Selgrade, 1999).

Unless one seeks to identify an effect on a very specific pathway, the immune system should always be regarded as a whole and its investigation should be done accordingly. We favor testing schemes in which most of the main immune subsystems are investigated simultaneously. Four such schemes elaborated by different groups (reviewed by Tryphonas, 2001) are summarized in Table 16.1. The most complete scheme is the one proposed by the U.S. National Academy of Science (1992). It should be noted that to our knowledge, none of these schemes have been used as a whole. It is therefore difficult to predict their ability to detect immunotoxic effects in human populations.

16.4.1 Total and Differential Blood Counts

The determination of total blood count (TBC) with differential absolute counts constitutes a good, cheap, easily available starting point. Although TBC is a poor predictor of host resistance and has a low concordance with other immune response tests (Luster et al., 1992, 1993), it allows the identification of significant decreases in cells responsible for immune response. TBC tests are usually well standardized but are unfortunately easily influenced by several factors such as age, sex, infections, and lifestyle. Nevertheless, since they will be critical in the interpretation of other functional tests, they should always be performed in all participants. They should not, however, be considered sensitive (Rose and Margolick, 1992; U.S. National Academy of Science, 1992; World Health Organization, 1996).

Table 16.1   Proposed Testing Schemes for Assessing Immunotoxicology in Humans

Assays	WHO	CDC	U.S. NAS[a]			Colosio et al.[a]		
			1st Tier	2nd Tier	3rd Tier	1st Tier	2nd Tier	3rd Tier
A. Blood count and clinical chemistry								
• Complete blood count with differential counts	✔	✔	✔			✔		
• Clinical chemistry	✔	✔						
B. Inflammation and nonspecific immunity								
• C-reactive protein	✔	✔						
• NK cells	✔			✔		✔		
• NK cell function					✔			
• Phagocytosis assay	✔							
C. Cellular immunity								
• Primary delayed-type hypersensitivity reaction	✔			✔				
• Secondary delayed-type hypersensitivity reaction	✔		✔					
• Surface analysis for CD4, CD8, CD3, and CD20	✔	✔	✔			✔		
• Other T- and B-cell markers (CD5, CD11, CD16, CD19, CD23, CD64, class II MHC)				✔				
• Class I and II MHC antigen typing				✔				
• Proliferative response to mitogens					✔			
• Serum levels of cytokine				✔				
• Cytokine production (*in vitro*)								✔
D. Humoral immunity								
• Immunoglobulin concentration in blood	✔	✔	✔			✔		
• Antibody to ubiquitous antigen	✔		✔					
• Primary response to protein antigen	✔		✔	✔				
• Primary response to polysaccharide antigen			✔	✔				
• Secondary response to protein antigen	✔							
• Auto-antibody titers	✔	✔	✔			✔		
• Proliferation to recall antigen	✔						✔	
• IgE to allergen	✔							
• Immunoglobulin subclass					✔			
• Antiviral titers					✔			
• Polyclonal immunoglobulin production *in vitro*							✔	

Note: Testing schemes proposed by World Health Organization (1996); Centers for Disease Control and Prevention and Agency for Toxic Substances and Disease Registry (World Health Organization, 1996); U.S. National Academy of Sciences (1992), and Colosio et al. (1999).

[a] The proposed schemes included a three-tier approach in which the tests in the second and third tiers are done if abnormalities are detected in the first tier or second tier, respectively. These tests could also be performed on the subgroup of the population included in the first tier.

16.4.2   Inflammation and Nonspecific Immunity

16.4.2.1   *Neutrophil and Monocyte Function*

When fresh blood is available, the chemotaxis and respiratory burst of polymorphonuclear leukocytes can be evaluated. The most standardized method is the nitroblue tetrazolium dye reduction used to evaluate the respiratory burst (Fernandes and Queiroz, 1999). More recent methods using flow cytometric techniques can also be used for rapid evaluation of phagocytosis, respiratory burst, activation, and bacterial killing of neutrophils and monocytes (Fruhwirth et al., 1998; Prodan et al.,

1995; Salih et al., 2000). Warnings have been issued against the reproducibility of bacterial killing assays (World Health Organization, 1996).

16.4.2.2 Complement

Alterations of the basal levels of complement proteins can be misleading and hard to interpret. A more useful approach is the measurement of the complement hemolytic activity of the classical (CH50) and alternative pathways (AP50) (Servais et al., 1991; White, et al., 1986). The evaluation of complement components and function is important when complement deficiency is suspected clinically (suspicion of increased incidence of systemic lupus erythematosus or of recurrent infections by *Neisseria* species and pyogenic organisms). Further analyses of complement function should be left to research teams specialized in the complement system.

16.4.2.3 Natural Killer Cells

NK cells contribute to the immune response in the early phase of infection. They are very potent cells acting along nonspecific pathways, independently from the major histocompatibility complex (MHC) (Janeway and Travers, 1996; Trinchieri, 1989). NK cells are lymphocytes expressing the CD3-CD16 and/or CD56 surface markers. The gold standard for the evaluation of their activity is the 51Cr-release assay, in which K562 target cells are cultured with freshly isolated NK cells from peripheral blood (Laso et al., 1997; Trinchieri, 1989). Other methods using fluorescence instead of radioactivity have been developed (Kantakamalakul et al., 2003; Piriou et al., 2000).

16.4.3 Cellular Immunity

16.4.3.1 Immunophenotyping of Lymphocytes

The development of flow cytometric techniques has allowed the rapid phenotyping of lymphocytes for many surface antigens. Most published work on human immunotoxicology recommends a surface analysis of lymphocytes, at least for CD4, CD8, CD3, and CD20. Alterations of lymphocyte subpopulations, whether they are reductions of specific subpopulations or imbalances between two or more subpopulations, can yield valuable results. It was shown that the variation of surface markers had a good predictive value for other immune function tests (Luster et al., 1992), and for biologically relevant *in vivo* effects in mice (Luster et al., 1993). However, investigators should be aware that lymphocyte populations vary greatly with age. Historical controls exist (Babcock et al., 1987; Erkeller-Yuksel et al., 1992), but age must be considered in the design and the analysis when lymphocyte phenotyping is performed. It is noteworthy to mention that most data provided by flow cytometric analysis have a broad distribution, and it has been argued that such analyses are not warranted in human immunotoxicologic studies (van Loveren et al., 1999). The same authors also mentioned the difficulty to predict the biological significance of the observed differences in lymphocyte subpopulations, which is often true. The phenotyping of lymphocytes can provide valuable information, but quantitative thresholds of T-cell subsets that are indicative of clear immune competence remain to be determined (Ward, 1992).

16.4.3.2 Delayed-Type Hypersensitivity Reactions

The overall function of intricate cell-mediated immunity (CMI) can be evaluated *in vivo* by delayed-type hypersensitivity (DTH) skin tests. The development of the CMI Multitest system (Merieux, France) has helped to standardize the antigen potency and the technique of administration (Rosenstreich, 1993). The CMI Multitest allows the simultaneous intradermal injection of seven common antigens and a glycerin control. The measurement of induration 48 hours after injection provides information on the ability of the CMI to respond to previously encountered antigens. This test has been used on several occasions and results for healthy subjects are available in the literature (Corriel et al., 1985; Hickie et al., 1995; Kniker et al., 1985; Moesgaard et al., 1987; Murgueytio and Evans, 1988; Rosenstreich, 1993).

The DTH response is sensitive to the usual factors affecting the immune response. Frequency of previous contacts with the antigens included in the Multitest must also be accounted for (prevalence of tuberculosis, immunization coverage, etc.). Geographic and ethnic variations exist, and results should always be compared to a control group rather than to previously published values. The use of DTH response in children less than 1 year old is not warranted.

While the usefulness of well-performed DTH skin tests is not questioned, the standardization of administration and reading can be quite challenging. Well-trained staff can achieve high reproducibility between observers and retested participants (Frazer et al., 1985). Unfortunately, inter-reader variability can lead to unreliable results, as noted for an important proportion of subjects in a study on the effects of dioxin exposure (Hoffman et al., 1986). Furthermore, it has been argued that since the reactions elicited by the Multitest system are secondary reactions, the uncertainty of prior antigen exposure limits the usefulness of DTH skin reaction (van Loveren et al., 1999). Unless well-trained and experienced staff are available, DTH testing may not be the best choice when subtle immune effects are investigated.

16.4.3.3 Lymphocyte Proliferation

The proliferative response of peripheral blood leukocytes to mitogens or specific antigens is the *in vitro* correlate of the DTH skin test. Substances such as phytohemagglutinin and concanavalin A will induce lymphocyte proliferation, which can be measured by 3H-thymidine incorporation assay (Rose et al., 1992), or by more recent techniques avoiding the use of radioisotopes (Maino et al., 1995; Schoel et al., 1996; Sottong et al., 2000).

Quantification of cytokine production by stimulated peripheral blood cells is useful for pinpointing mechanisms of action. It should not be used for the screening of immunotoxic effects, unless a precise mechanism is proposed.

16.4.4 Humoral Immunity

16.4.4.1 Serum Immunoglobulin Concentrations

Determination of Ig levels is not a sensitive method because immunodeficiencies can be observed in the presence of normal Ig levels. In nonhuman primates, exposure to PCBs produced no effect on serum Ig while the ability to respond to a foreign antigen was impaired (Tryphonas et al., 1991). Determination of serum immunoglobulin concentrations is not warranted unless very specific mechanisms are sought, or when increased incidence of Ig-mediated pathologies, such as hyper-gammaglobulinemia, are suspected (International Union of Immunological Societies/World Health Organization Working Group, 1982, 1988).

16.4.4.2 Antibodies to Ubiquitous Antigen

Evaluation of antibodies against widely occurring antigens allows the identification of profound defects in antigen-specific Ig production in individuals with otherwise normal Ig levels. An example of this is the Wiskott–Aldrich syndrome in which subjects lack isohemagglutinins but have normal Ig levels. Absence of natural antibodies to blood group antigens and to *Escherichia coli* can easily be identified by simple agglutination tests. This type of investigation should be performed only when an immunotoxic mechanism severely affecting the humoral response is suspected. It is not sensitive, and will most likely yield deceiving results when subtle effects are sought.

16.4.4.3   Antibodies to Self-Antigen

The identification of clinically relevant autoimmunity is complex. Autoimmune diseases are diagnosed based on biologic and clinical criteria, and the significance of the presence of autoantibodies in asymptomatic individuals is not clear (Holsapple, 2002; Vial et al., 1996). Other tests exist in animal models but they cannot be used in human studies (Pieters et al., 2002; Vos and van Loveren, 1995). Because autoantibodies are present in healthy individuals, and because the significance of their presence is poorly understood, using them as biomarkers of autoimmunity without other clinical markers is not warranted (Descotes et al., 1995; Holsapple, 2002).

16.4.4.4   Antibody Response to Immunization

The evaluation of antibody development following vaccination is the gold standard of human immunotoxicology. The immune system needs proper antigen processing and presentation, as well as functional B cells and T cells in order to mount an appropriate response to a protein antigen. In rodents, assessment of an antibody response was shown to be the most adequate indicator of immunotoxicity (Luster et al., 1992, 1993). The determination of response to immunization also offers the opportunity to improve public health by increasing the vaccination coverage in the population under study.

Generally, investigators seem to agree to the fact that the primary response is more sensitive than the secondary response for assessing insults to the immune system (Rose and Margolick, 1992; van Loveren et al., 1999; World Health Organization, 1996). However, a review by Cohen et al. (2001) on the effect of stress on response to immunization underlined a more stable effect reported by investigators using the secondary response as the endpoint. Different pathways are at play in primary and secondary responses, and both should be tested whenever possible.

When antibody response to immunization is investigated, it is essential to evaluate serum antigen-specific antibody before and after the challenge. Unknown previous encounters or challenges with the antigen for some participants can greatly influence the results. Sequential assessments would also be useful to determine the catabolic rate of specific antibody, as well as Ig isotypic class switching. Vaccination response is influenced by the type of vaccine, route of administration, and time elapsed between the challenge and assessment of the response (van Loveren et al., 2001). Other factors such as age, genetics, stress, smoking, nutrition, and some infections also affect the response to vaccination. All these factors should be considered as potential confounders when vaccination response is used as an endpoint (van Loveren et al., 2001).

When possible, vaccines offering public health advantage should be preferred. Synchronization with childhood vaccination programs can be particularly effective when immunotoxic effects in children are investigated. In adults, hepatitis B vaccine could be used, but the increasing immunization coverage begins to limit its utilization as a marker of effect for primary response. The use of harmless, but immunogenic antigens foreign to humans, such as the bacteriophage phiX174 (Rubinstein et al., 2000) and the keyhole limpet hemocyanin (Harris and Markl, 1999), can be of great value. These two antigens can also be used to evaluate the anamnestic response by re-challenging participants. Polysaccharide vaccines, such as the pneumococcal vaccine, are useful for the evaluation of T-cell–independent response. Despite the usefulness of this marker of effect, very few investigators have used it in human immunotoxicology studies (van Loveren et al., 2001).

16.5   CLINICAL ENDPOINTS

All methods mentioned above allow the identification of alterations in one or more pathways of the immune system. When such alterations are proven or suspected, one needs to evaluate if these alterations affect the health status of a population. Since the ultimate role of the immune system is to maintain the integrity of the body by dealing adequately with the multitude of threats coming from inside and outside, any alteration of the immune system could potentially result in loss of this integrity. This can happen because the immune system does not fight hard enough, fights too much, or begins to forget what to fight, and what not to fight.

In addition to the factors affecting the immune system discussed above, other risk factors for the clinical endpoint under study, not necessarily related to the immune system, will need to be considered as potential confounders. For example, the incidence of infectious diseases is affected by the frequency of contacts with pathogens (daycare attendance, crowding), education, vaccination coverage, breastfeeding, and access to medical services. Exposure to xenobiotics is often related to the socioeconomic status, which in turn can affect many of the abovementioned factors. A thorough knowledge of all factors directly or indirectly affecting the studied endpoint is critical.

In this section, we discuss some clinical endpoints in relation to immunotoxicology, with a focus on infection susceptibility. We did not attempt to cover all methods and all potential clinical endpoints of interest in immunotoxicology. This section should be regarded as a starting point when one considers the inclusion of such endpoints in a study. The participation of expert epidemiologists, researchers, and clinicians in the area of interest remains essential.

16.5.1 Infections

Contrary to cancer and other chronic diseases, acute infections have a very short latency period. As soon as the immune system is compromised, the risk of having acute infectious episodes increases. This aspect is not to be overlooked. Much effort is dedicated to the design of early biomarkers of effect, but infection incidence should actually be considered as such. Because the evaluation of infectious disease frequency is relatively easy to perform, such an endpoint should be included whenever possible.

Many factors affect infection incidence. In particular, the socioeconomic status of individuals, but also of the population as a whole, strongly influence the risk of having infections. Factors such as education, hygiene, crowding, nutrition, and access to health services are often uneven in populations, and are sometimes strongly correlated with exposure to environmental and occupational xenobiotics. An example of this is the important exposure to dioxin experienced by the inhabitants of the Quail Run Mobile Home Park in Missouri following the spraying of dioxin-contaminated oil near the park (Hoffman et al., 1986). The low socioeconomic status of the exposed participants represented a challenge for the investigation of immunotoxic effects. The recruitment of a control group of participants living in similar homes and conditions enhanced the credibility and validity of study results.

Even in populations for which there is no suspicion of immune deficiencies, there is still a background of acute and chronic infections of all sorts. If immune alterations are suspected in a population, one could expect an increase in the incidence of infections, an increase in the severity of these infections, or both. The central question in measuring infections in human immunotoxicology is the identification of a difference in incidence or severity between an exposed group and an unexposed group. Often, incidence and severity are entangled with each other, as an increase in the severity of an otherwise benign infection will facilitate its identification in a population and, consequently, increase its observed incidence. In this section, we focus on increased incidence.

It is sound to expect that a group with a higher incidence of a given infection will also have a higher prevalence and a higher risk of developing the disease in a given time frame. The evaluation of prevalence (x persons having the disease out of y participants at a given time) and average risk (x persons having one or more episodes out of y participants during a given period of time, also called incidence proportion) are useful shortcuts. However, the most straightforward method is to try, when possible, to directly evaluate the incidence.

16.5.1.1 Incidence of Acute Infections

Ideally, to evaluate the incidence of, say, middle-ear infections, the investigator should follow a group of healthy participants regularly in the hope of diagnosing middle-ear infections as these become identifiable. Of course, it is seldom possible in large cohorts to visit the participants, or to ask them to visit the clinic every week. Less frequent follow-ups during which the participants are asked if they have been diagnosed for middle-ear infections since the last follow-up is a good alternative. By doing so, the investigator has to deal with episodes that are recalled and reported by the participants instead of episodes diagnosed by a physician, therefore opening the door to recall biases. The date of episode onset is also less precise.

For benign infections, a questionnaire or a self-maintained log in which participants record daily or weekly their symptoms related to the infection under study could be quite useful. When the definition of the symptoms is clear and complete, this method could yield valuable results with accurate dates. This method is efficient with common colds and upper respiratory tract symptoms, provided that only self-reported symptoms will be available.

A thorough review of the medical charts can be used to evaluate incidence. The medical chart review is usually easier to perform in remote rural areas where most of the population members attend the same health center. The principal advantage of the medical chart review is that dates of diagnoses are available and several pathologies can be evaluated at the same time. However, only infections for which medical attention was sought will be detected, and diagnoses are usually not standardized. Benign infections could be missed when participants decide not to go to the clinic, a decision that is often related to other risk factors such as socioeconomic status, which potentially introduces a bias. Asking the participants to go to a specific clinic when they have symptoms, and having the physicians of that clinic participate in the study will help to standardize disease definitions and reduce the number of participants who decide not to consult for their symptoms.

The approach of medical chart review was recently used by our group to evaluate the effect of OCs on infection incidence (Dallaire et al., 2004). We reviewed the medical charts of 199 Inuit infants during the first months of life, and used Poisson regression to evaluate the association between the incidence of acute infections and prenatal exposure to OCs. We found that the infants in the higher exposure groups (second, third, and fourth quartiles) had an increased incidence of acute infections such as otitis media, compared to the infants in the lower exposure group (first quartile).

The combination of more than one of the methods mentioned above is warranted. Cross-checking self-reported episodes with medical charts helps in rectifying dates and standardizing diagnoses.

16.5.1.2 Prevalence and Average Risk of Acute Infections

When it is impossible or impractical to evaluate the incidence of infection, other measures of disease frequency, such as prevalence and average risk (incidence proportion), can be used. The biggest advantage of prevalence is the need for only one assessment for each participant, therefore eliminating the infrastructure necessary for follow-up of participants. In this context, it is easier to thoroughly examine each participant in order to make clear standardized diagnoses. This approach was used by Chao et al. (1997) for the evaluation of children prenatally exposed to PCBs in Taiwan. In this study, two otolaryngologists examined each child for ear abnormalities and a diagnosis was made when the two agreed. The cross-sectional nature of the data renders them subject to the warnings discussed in Section 16.3.1 on study designs.

Average risk is usually the measure obtained when questionnaires are used. Questions such as "has a doctor ever given your child a diagnosis of otitis media" (Weisglas-Kuperus et al., 2000) will yield average risk (X% of the participants who ever had otitis media since they were born). Average risk is also the measure obtained by Dewailly et al. (2000) in a study assessing the effect of OC exposure on middle-ear infections. To evaluate the number of episodes in the first year of life, the authors conducted interviews and medical examinations at 3, 7, and 12 months of age, during which they asked the mothers for the occurrence of any previous episodes. Then, they cross-checked the self-reported episodes in the medical charts of the infants and computed the risk of having 1 episode, and 3 episodes. Because this type of evaluation has a dichotomous outcome, the investigator has to determine, sometimes arbitrarily, a threshold value. This has the disadvantage of potentially reducing the statistical power of the study as compared to continuous outcomes.

16.5.1.3 Chronic and Opportunistic Infections

The methods discussed above can be applied to chronic or opportunistic infections. Case-control designs represent also an interesting alternative when the disease studied is less frequent. There exist many other methods for identifying participants with the disease under study. The use of available standardized databases can be of great value in some settings.

16.5.2 Other Clinical Endpoints

16.5.2.1 Allergy

Clinically defined allergy can usually be assessed using the same strategy as infections. As with infection, prevalence and average risk will be easier to obtain than incidence. Whenever a questionnaire is administered, simple questions about allergies such as "has your child ever had eczema or an allergic reaction" should be asked (Weisglas-Kuperus et al., 2000). So far, the search for efficient biomarkers of allergy that could be applied to large cohorts has been deceptive (Odelram et al., 1995).

16.5.2.2 Autoimmunity and Neoplasic Changes

Autoimmunity can be briefly defined as the loss of immune tolerance for autoantigens. The sole presence of autoantibodies does not mean that symptoms of autoimmunity will be present. Since no clear biomarkers of autoimmunity are known, and because the diagnosis of clinical autoimmunity is complex, studies focusing on clinical autoimmunity should be left to teams specialized in the subject.

Although it has long been observed that immunosuppression was linked to increased frequency of cancer (Spector et al., 1978), the hypothesis of a reduced surveillance of tumors by immuno-deficient individuals is controversial (Newcombe, 1992). It is unlikely that cancer will be the first manifestation of immunotoxicity. Screening for increased cancer incidence is not warranted for the purpose of identifying early immunotoxic properties of an exposure. However, a large population exposed to a known immunotoxic agent could be at increased risk, and investigation of links between the development of cancer and exposure might be an interesting research avenue. On the other hand, observation of an unexplained increase of the cancer rate in a population should elicit researchers to consider an immunotoxic etiology.

It is noteworthy that researchers in France have initiated a sentinel program in which autoim-mune diseases and non-Hodgkin's lymphomas are screened in order to flag higher-than-expected incidences associated with chemical exposure (Descotes et al., 1996). The hypothesis that can spur from this approach could be interesting to pursue.

16.6 RESEARCH NEEDS

Useful standardized noninvasive approaches exist for the assessment of immunotoxicity in human populations. However, relatively few studies have been conducted in human settings and the methods varied greatly. Unfortunately, the proposed testing schemes have hardly ever been used and their ability to detect immunotoxic effects has yet to be established. The field of human immunotoxicology would greatly benefit from the widespread use of recognized schemes of assays by several researchers on several suspected immunotoxic agents. Furthermore, the relation between biomarkers and clinical endpoints is still obscure in humans. The design of studies in which both clinical and biologic endpoints are assessed is greatly encouraged. Only these studies, if they are well designed and conducted, will allow the identification of relevant markers with high specificity and sensitivity to predict adverse effects on the immune system in human populations. Finally, the assessment of acute clinical events, such as the incidence of infections, remains a relevant and easily evaluated endpoint that should be used as often as possible in human population studies.

16.7 SUMMARY

Human immunotoxicology poses several epidemiologic challenges. Experimental designs are rarely feasible, and the investigator is forced to rely on careful identification of exposed and control subjects. Valid studies rely on the choice of an adequate study design with an appropriate understanding of the advantages and drawbacks of the selected design. A proper identification and selection of the exposed and unexposed populations, a carefully elaborated hypothesis, and high-quality measurements of the exposures, endpoints, and confounding factors, are also crucial. Many biological endpoints are available to assess immunotoxicity, but not all can be applied to human studies. The most acknowledged methods are the phenotyping of lymphocytes, the evaluation of NK cell activity, skin tests for delayed-type hypersensitivity, and response to vaccination. Complete testing schemes already exist, but they have rarely been used. The inclusion of acute clinical endpoints, such as the incidence of infections, is warranted to better understand the relations between biological markers, the immune system, and health status in human populations.

ACKNOWLEDGMENTS

We are grateful to Daria Pereg for her critical review and useful inputs during preparation of the manuscript.

REFERENCES

Antonaci, S., et al., Immunoregulation in aging, *Diagn. Clin. Immunol.*, 5, 55–61, 1987.
Babcock, G.F., et al., Flow cytometric analysis of lymphocyte subset phenotypes comparing normal children and adults, *Diagn. Clin. Immunol.*, 5, 175–179, 1987.
Bidwell, J., et al., Cytokine gene polymorphism in human disease: on-line databases, *Genes Immun.*, 1, 3–19, 1999.
Burton, R.C., et al., Effects of age, gender, and cigarette smoking on human immunoregulatory T-cell subsets: establishment of normal ranges and comparison with patients with colorectal cancer and multiple sclerosis, *Diagn. Clin. Immunol.*, 1, 216–223, 1983.
Chandra, R.K., Nutrition and the immune system from birth to old age, *Eur. J. Clin. Nutr.*, 56(Suppl. 3), S73–76, 2002.
Chao, W.Y., et al., Middle-ear disease in children exposed prenatally to polychlorinated biphenyls and polychlorinated dibenzofurans, *Arch. Environ. Health*, 52, 257–262, 1997.
Cohen, S., et al., Psychological stress and antibody response to immunization: a critical review of the human literature, *Psychosom. Med.*, 63, 7–18, 2001.
Colosio, C., et al., Immune parameters in biological monitoring of pesticide exposure: current knowledge and perspectives, *Toxicol. Lett.*, 108, 285–295, 1999.

Corriel, R.N., et al., Cell-mediated immunity in schoolchildren assessed by multitest skin testing. Normal values and proposed scoring system for healthy children, *Am. J. Dis. Child.*, 139, 141–146, 1985.

Dallaire, F., et al., Acute infections in Inuit infants in relation to environmental exposure to organochlorines, *Environ. Health Perspect.*, 112, 1359–1364, 2004.

Descotes, J., et al., Assessment of immunotoxic effects in humans, *Clin. Chem.*, 41, 1870–1873, 1995.

Descotes, J., et al., Sentinel screening for human immunotoxicity, *Arch. Toxicol. Suppl.*, 18, 29–33, 1996.

Dewailly, E., et al., Susceptibility to infections and immune status in Inuit infants exposed to organochlorines, *Environ. Health Perspect.*, 108, 205–211, 2000.

Elenkov, I.J., and Chrousos, G.P., Stress hormones, proinflammatory and antiinflammatory cytokines, and autoimmunity, *Ann. N. Y. Acad. Sci.*, 966, 290–303, 2002.

Erkeller-Yuksel, F.M., et al., Age-related changes in human blood lymphocyte subpopulations, *J. Pediatr.*, 120, 216–222, 1992.

Fernandes, M.D., and Queiroz, M.L., Measurement of the respiratory burst and chemotaxis in polymorphonuclear leukocytes from anti-ChE insecticides-exposed workers, *Immunopharmacol. Immunotoxicol.*, 21, 621–633, 1999.

Frazer, I.H., et al., Assessment of delayed-type hypersensitivity in man: a comparison of the "Multitest" and conventional intradermal injection of six antigens, *Clin. Immunol. Immunopathol.*, 35, 182–190, 1985.

Fruhwirth, M., et al., Flow-cytometric evaluation of oxidative burst in phagocytic cells of children with cystic fibrosis, *Int. Arch. Allergy Immunol.*, 117, 270–275, 1998.

Gil, A., Polyunsaturated fatty acids and inflammatory diseases, *Biomed. Pharmacother.*, 56, 388–396, 2002.

Grimm, H., et al., Regulatory potential of n-3 fatty acids in immunological and inflammatory processes, *Br. J. Nutr.*, 87(Suppl. 1), S59–67, 2002.

Hardell, L., and Eriksson, M., A case-control study of non-Hodgkin's lymphoma and exposure to pesticides, *Cancer*, 85, 1353–1360, 1999.

Harris, J.R., and Markl, J., Keyhole limpet hemocyanin (KLH): a biomedical review, *Micron*, 30, 597–623, 1999.

Herbert, T.B., and Cohen, S., Stress and immunity in humans: a meta-analytic review, *Psychosom. Med.*, 55, 364–379, 1993.

Hickie, C., et al., Delayed-type hypersensitivity skin testing: normal values in the Australian population, *Int. J. Immunopharmacol.*, 17, 629–634, 1995.

Hoffman, R.E., et al., Health effects of long-term exposure to 2, 3, 7, 8-tetrachlorodibenzo-p-dioxin, *JAMA*, 255, 2031–2038, 1986.

Holsapple, M.P., Autoimmunity by pesticides: a critical review of the state of the science, *Toxicol. Lett.*, 127, 101–109, 2002.

Howell, W.M., et al., Gene polymorphisms, inflammatory diseases and cancer, *Proc. Nutr. Soc.*, 61, 447–456, 2002.

International Union of Immunological Societies/World Health Organization Working Group, Use and abuse laboratory tests in clinical immunology: critical considerations of eight widely-used diagnostic procedures. Report of an IUIS/WHO working group, *Clin. Immunol. Immunopathol.*, 24, 122–138, 1982.

International Union of Immunological Societies/World Health Organization Working Group, Laboratory investigations in clinical immunology: methods, pitfalls and clinical indications. A second IUIS/WHO report, *Clin. Exp. Immunol.*, 74, 494–503, 1988.

Janeway, C., and Travers, P., *Immunobiology*, Garland Publishing, London, 1996.

Kantakamalakul, W., et al., A novel enhanced green fluorescent protein (EGFP)-K562 flow cytometric method for measuring natural killer (NK) cell cytotoxic activity, *J. Immunol. Methods*, 272, 189–197, 2003.

Kay, R.A., TCR gene polymorphisms and autoimmune disease, *Eur. J. Immunogenet.*, 23, 161–177, 1996.

Kimber, I., and Dearman, R.J., Immune responses: adverse versus non-adverse effects, *Toxicol. Pathol.*, 30, 54–58, 2002.

Kniker, W.T., et al., Cell-mediated immunity assessed by Multitest CMI skin testing in infants and preschool children, *Am. J. Dis. Child.*, 139, 840–845, 1985.

Laso, F.J., et al., Decreased natural killer cytotoxic activity in chronic alcoholism is associated with alcohol liver disease but not active ethanol consumption, *Hepatology*, 25, 1096–1100, 1997.

Luster, M.I., et al., Risk assessment in immunotoxicology. I. Sensitivity and predictability of immune tests, *Fundam. Appl. Toxicol.*, 18, 200–210, 1992.

Luster, M.I., et al., Risk assessment in immunotoxicology. II. Relationships between immune and host resistance tests, *Fundam. Appl. Toxicol.*, 21, 71–82, 1993.

Maino, V.C., et al., Rapid flow cytometric method for measuring lymphocyte subset activation, *Cytometry*, 20, 127–133, 1995.

Marti, A., et al., Obesity and immune function relationships, *Obes. Rev.*, 2, 131–140, 2001.

Moesgaard, F., et al., Cell-mediated immunity assessed by skin testing (Multitest). I. Normal values in healthy Danish adults, *Allergy*, 42, 591–596, 1987.

Murgueytio, P.U., and Evans, R.G., Delayed cutaneous hypersensitivity: multitest CMI reliability assessment in groups of volunteers, *Ann. Allergy*, 61, 463–465, 1988.

Newcombe, D.S., Immunotoxicology: a new challenge, in *Clinical Immunotoxicology*, Newcombe, D.S., Rose, N.R., and Bloom, J.C., Eds., Raven Press, New York, 1992, pp. 1–8.

Odelram, H., et al., Predictors of atopy in newborn babies, *Allergy*, 50, 585–592, 1995.

Pawelec, G., and Solana, R., Immunosenescence, *Immunol. Today*, 18, 514–516, 1997.

Pedersen, B.K., et al., Cytokines in aging and exercise, *Int. J. Sports Med.*, 21(Suppl. 1), S4–9, 2000.

Pieters, R., et al., Predictive testing for autoimmunity, *Toxicol. Lett.*, 127, 83–91, 2002.

Piriou, L., et al., Design of a flow cytometric assay for the determination of natural killer and cytotoxic T-lymphocyte activity in human and in different animal species, *Cytometry*, 41, 289–297, 2000.

Prodan, M., et al., Flow cytometric assay for the evaluation of phagocytosis and oxidative burst of polymorphonuclear leukocytes and monocytes in myelodysplastic disorders, *Haematologica*, 80, 212–218, 1995.

Rogan, W.J., et al., Polychlorinated biphenyls (PCBs) and dichlorodiphenyl dichloroethene (DDE) in human milk: effects on growth, morbidity, and duration of lactation, *Am. J. Public Health*, 77, 1294–1297, 1987.

Rose, N.R., et al., *Manual of Clinical Laboratory Immunology*, 4th ed., American Society for Microbiology, Washington, DC, 1992.

Rose, N.R., and Margolick, J.B., The immunological assessment of immunotoxic effects in man, in *Clinical Immunotoxicology*, Newcombe, D.S., Rose, N.R., and Bloom, J.C., Eds., Raven Press, New York, 1992, pp. 9–25.

Rosenstreich, D.L., Evaluation of delayed hypersensitivity: from PPD to poison ivy, *Allergy Proc.*, 14, 395–400, 1993.

Roth, M.D., et al., Effects of delta–9-tetrahydrocannabinol on human immune function and host defense, *Chem. Phys. Lipids*, 121, 229–239, 2002.

Rothman, K.J., and Greenland, S., *Modern Epidemiology*, 2nd ed., Lippincott-Raven, Philadelphia, 1998.

Rubinstein, A., et al., Progressive specific immune attrition after primary, secondary and tertiary immunizations with bacteriophage phi X174 in asymptomatic HIV- 1 infected patients, *AIDS*, 14, F55–62, 2000.

Salih, H.R., et al., Simultaneous cytofluorometric measurement of phagocytosis, burst production and killing of human phagocytes using Candida albicans and Staphylococcus aureus as target organisms, *Clin. Microbiol. Infect.*, 6, 251–258, 2000.

Schoel, B., et al., Rapid determination of gamma delta T-cell stimulation by microfluorimetry, *Immunol. Lett.*, 53, 135–139, 1996.

Selgrade, M.K., Use of immunotoxicity data in health risk assessments: uncertainties and research to improve the process, *Toxicology*, 133, 59–72, 1999.

Semba, R.D., Vitamin A, immunity, and infection, *Clin. Infect. Dis.*, 19, 489–499, 1994.

Servais, G., et al., Simple quantitative haemolytic microassay for determination of complement alternative pathway activation (AP50), *J. Immunol. Methods*, 140, 93–100, 1991.

Sleijffers, A., et al., Ultraviolet radiation, resistance to infectious diseases, and vaccination responses, *Methods*, 28, 111–121, 2002.

Sottong, P.R., et al., Measurement of T-lymphocyte responses in whole-blood cultures using newly synthesized DNA and ATP, *Clin. Diagn. Lab. Immunol.*, 7, 307–311, 2000.

Spector, B.D., et al., Genetically determined immunodeficiency diseases (GDID) and malignancy: report from the immunodeficiency-cancer registry, *Clin. Immunol. Immunopathol.*, 11, 12–29, 1978.

Svensson, B.G., et al., Parameters of immunological competence in subjects with high consumption of fish contaminated with persistent organochlorine compounds, *Int. Arch. Occup. Environ. Health.*, 65, 351–358, 1994.

Trinchieri, G., Biology of natural killer cells, *Adv. Immunol.*, 47, 187–376, 1989.

Tryphonas, H., Approaches to detecting immunotoxic effects of environmental contaminants in humans, *Environ. Health Perspect.*, 109(Suppl. 6), 877–884, 2001.

Tryphonas, H., et al., Effect of chronic exposure of PCB (Aroclor 1254) on specific and nonspecific immune parameters in the rhesus (Macaca mulatta) monkey, *Fundam. Appl. Toxicol.*, 16, 773–786, 1991.

Urban, L., et al., On the role of aging in the etiology of autoimmunity, *Gerontology*, 48, 179–184, 2002.

U.S. National Academy of Science, Biologic Markers in Immunotoxicology report from the Subcommitte on Immunotoxicology (Committee on Biologic Markers), National Academy Press, Washington, DC, 1992.

van Loveren, H., et al., Report of the Bilthoven Symposium: advancement of epidemiological studies in assessing the human health effects of immunotoxic agents in the environment and the workplace, *Biomarkers*, 4, 135–157, 1999.

van Loveren, H., et al., Vaccine-induced antibody responses as parameters of the influence of endogenous and environmental factors, *Environ. Health Perspect.*, 109, 757–764, 2001.

Venjatraman, J.T., and Fernandes, G., Exercise, immunity and aging, *Aging (Milano)*, 9, 42–56, 1997.

Vial, T., et al., Clinical immunotoxicity of pesticides, *J. Toxicol. Environ. Health*, 48, 215–229, 1996.

Vos, J.G., and van Loveren, H., Markers for immunotoxic effects in rodents and man, *Toxicol. Lett.*, 82–83, 385–394, 1995.

Ward, P.A., *Flow Cytometric Analysis of the Immune and Phagocytic Cells*, Newcombe, D.S., Rose, N.R., and Bloom, J.C., Eds., Raven Press, New York, 1992, pp. 43–47.

Weisglas-Kuperus, N., et al., Immunologic effects of background exposure to polychlorinated biphenyls and dioxins in Dutch preschool children, *Environ. Health Perspect.*, 108, 1203–1207, 2000.

White, K.L., Jr., et al., Modulation of serum complement levels following exposure to polychlorinated dibenzo-p-dioxins, *Toxicol. Appl. Pharmacol.*, 84, 209–219, 1986.

Wilson, C.B., Immunologic basis for increased susceptibility of the neonate to infection, *J. Pediatr.*, 108, 1–12, 1986.

World Health Organization, Principles and Methods for Assessing Direct Immunotoxicity Associated with Exposure to Chemical, World Health Organization, Geneva, 1996.

Models and Approaches to Chemical-Induced Allergenicity

CHAPTER 17

Guinea Pig, Mouse, and Rat Models for Safety Assessment of Protein Allergenicity

Katherine Sarlo, Rebecca J. Dearman, and Ian Kimber

CONTENTS

17.1   INTRODUCTION

The ability to assess the potential for proteins to cause type 1 IgE allergic antibody-mediated hypersensitivity reactions is a daunting task. For safety professionals, being able to predict if a protein can cause this type of immune response under relevant conditions of exposure is key to safety and risk assessments. No *in vitro* tests exist that can predict the inherent potential of a protein to induce type 1 allergy, nor are any *in vitro* tests available that can estimate potential risk under relevant conditions of exposure. For this reason, safety professionals look to animal tests to help answer the basic questions of protein allergenicity.

Practically all foreign proteins can be immunogenic under appropriate conditions of exposure and dose. Some proteins that are derived from the same species may even be immunogenic. Whether the protein can function as an allergen and induce IgE antibody is a different and important question since the outcome of an allergic response can range from rhinitis to life-threatening (and ending) anaphylactic reactions.

Animal models have been used for years to enhance our understanding of basic mechanisms of IgE allergic antibody-mediated disease. Most of the work has revolved around the use of model allergens (frequently ovalbumin) in a rodent strain that generally displays an atopic-like predisposition (e.g., BALB/c mice, Brown Norway rats). This work has greatly advanced our understanding of allergic disease. However, exploiting these models to predict the allergenic potential of proteins delivered to the immune system is not a simple task. One must consider how the choice of species and the genetic makeup of the strain can affect immune recognition of different proteins. The mode of exposure such as route (directly to the target tissue such as lung or gut vs. systemic or intraperitoneal), and regimen (intermittent vs. daily dose) need to be considered. A mouse or rat strain that can mount a good IgE antibody response to an allergen following one route of exposure may not be able to mount an IgE response following a different route of exposure. The need for, and choice of, an adjuvant must be carefully weighed since such materials can affect distribution and uptake of antigen, and possibly confer on inherently nonallergenic proteins the ability to induce type 1 allergy. In addition, the choice of measuring different antibody subclasses (e.g., IgE vs. IgGs), cytokines, and symptoms needs to be standardized to ensure comparison across various types of proteins.

Having identified the issues associated with developing and using animal models to predict protein allergenicity, there have been some successes with a limited range of proteins which indicate that the goal might be achievable. This chapter focuses on models that have been used and/or that are being developed to assess allergenic potential of proteins in the context of safety assessments. These models use guinea pigs, rats, or mice. Specifically, this chapter focuses on the identification and characterization of proteins that may be implicated as causes of respiratory or food allergy.

17.2 ASSESSMENT OF PROTEINS AS RESPIRATORY ALLERGENS

17.2.1 Guinea Pig Models

The guinea pig has been used as an animal model for respiratory allergy and anaphylaxis for many years due to the ease in measuring the respiratory response to an allergen (Ratner et al., 1927). This species displays immediate- and late-onset reactions on exposure to allergens, and exhibits pulmonary responses upon exposure to histamine (Griffith-Johnson et al., 1993). They have also been used to study allergic conjunctivitis (Groneberg, et al., 2003), antigen-induced rhinitis, and airway hyperreactivity (Ishida et al., 1994).

Guinea pigs develop an immune response to a wide variety of proteins and exhibit dose-dependent allergic antibody and pulmonary responses upon exposure to a variety of allergens. Guinea pigs will make small amounts of IgE antibody following allergen exposure. However, the major homocytotropic (mast cell-binding) antibody is IgG1, which has been associated with respiratory symptomatology and eosinophilic inflammatory responses after exposure to allergen. Both antibodies can be detected by the passive cutaneous anaphylaxis (PCA) test. IgG1 can be detected within 48 hours, whereas 4 to 7 days are needed for IgE since it is still associated with mast cell receptors at this time. Both antibody types cause the same type of immediate-onset hypersensitivity responses, while IgE antibodies have been associated with late-onset pulmonary reactions. The relationship between IgG1 antibody and IgE antibody and the development of pulmonary reactivity is not fully understood in this species.

Microbial enzymes used in laundry detergents were first described as a cause of occupational allergy and asthma by Flindt (1969) and Pepys et al. (1969). Since the guinea pig easily exhibits allergen-induced pulmonary responses, this species was chosen to assess immediate hypersensitivity responses to detergent enzymes. Intratracheal (IT) exposure to enzyme protein was used as an alternative to inhalation exposure to deliver enzyme allergen to the airways (Ritz et al., 1993). The antibody response to the protease enzyme from *Bacillus licheniformis* (tradename Alcalase) was very similar in animals exposed to this enzyme via IT instillation, or via inhalation of enzyme aerosol in a closed system. An intradermal injection model was developed as another alternative to inhalation systems. Animals exposed this way also make a dose-dependent antibody response to the protease that is comparable to the responses obtained in the IT model and by inhalation (Blaikie et al., 1994).

Inhalation studies showed an exposure-dependent relationship for antibody production and pulmonary responses to protease enzyme. An inhalation protocol for inducing antibody to proteins was developed in 1980s, and was used to assess exposure response to a subtilisin enzyme (Hillibrand et al., 1987; Thorne et al., 1986). Animals were exposed to subtilisin aerosols for 15 minutes/day for 5 consecutive days. Sera were collected and tested for subtilisin-specific antibody. Also, animals were challenged with a subtilisin aerosol, and the development of immediate- and delayed-onset respiratory reactions were measured via whole body plethysmography. Animals exposed to 8 or 41 µg of subtilisin protein/m^3 developed low levels of anti-subtilisin antibody but no pulmonary symptoms. Animals exposed to 150 µg of subtilisin protein/m^3 or higher did develop enzyme-specific antibody, as well as immediate- and late-onset pulmonary responses when re-exposed to the enzyme.

In the inhalation study by Ritz et al. (1993), guinea pigs were exposed for 6 hours per day, 4 days per week for 10 consecutive weeks, to 1, 3, 11, or 35 µg/m^3 of Alcalase aerosols. The generation of Alcalase-specific IgG1 antibody was exposure dependent, with antibody responses measured at the lowest exposure concentration. The responses of the inhalation-exposed animals were very similar to the responses of animals exposed to weekly IT instillations of enzyme showing that dose–response relationships exist for generation of antibody to respiratory protein allergens. The onset of immediate pulmonary symptoms also showed an exposure (dose)-dependent relationship for both inhalation- and IT-exposed animals. While some exposure (dose) levels were capable of inducing enzyme-specific antibody, these same levels were not sufficient to elicit pulmonary reactions, suggesting a lower threshold for induction of antibody and a higher threshold for elicitation of respiratory symptoms.

Alcalase enzyme became the model allergen in the guinea pig tests since a threshold limit value (TLV) has been established for this protein, and a wealth of human data pertaining to antibody and pulmonary responses to this enzyme in the exposed workforce exists (Schweigert et al., 2000; American Conference of Governmental Industrial Hygienists, 2003). These data provide a basis for comparing allergic responses to different enzymes to the response to Alcalase and using the dose–antibody response data to assign potency values to new proteins as they compare to Alcalase. The allergenic potency of a new enzyme is defined by the dose of protein needed to induce allergic antibody titers comparable to the titers obtained with Alcalase protein. New enzymes can be assessed as more potent, less potent, or equipotent to Alcalase. Several enzymes have been tested with potency ranging from ten-fold less potent to ten-fold more potent than Alcalase. This information has been used to develop operational exposure guidelines to protect workers against developing allergic antibody to new enzymes.

A guinea pig intradermal injection model has also been developed for comparison of the allergenic potency of new enzymes to that of Alcalase (Blaikie et al., 1994). Enzyme-specific serum and tissue-fixed allergic antibody, but not pulmonary symptoms, are assessed in this model. Comparisons are made between a new enzyme and Alcalase, and the difference in the response rate is used to assign potency to the new enzyme with this information being used to develop operational exposure guidelines for manufacturing facilities. The ranking of new enzymes as compared to Alcalase with this injection model was very similar to that obtained with the intratracheal model (L. Blaikie, personal communication, 1997).

The guinea pig has also been used to assess the effects of adjuvants on the allergic antibody response to enzymes. Intratracheal or inhalation exposure to enzyme in a detergent vehicle leads to an allergic antibody response that is greater than the response from animals exposed to the enzyme in saline (Markham and Wilkie, 1976; K. Sarlo, unpublished observations). The surfactants in the detergent have been implicated as adjuvants. The ability to detect the detergent adjuvant effect is more sporadic in the injection model suggesting that this activity may be optimal in the respiratory tract. Active protease enzyme also functions as an adjuvant, enhancing allergic antibody responses to other proteins when there is co-exposure (Sarlo et al., 1997a). Inactivation of the protease leads to total abrogation of the adjuvant activity. Again, the ability to detect the protease adjuvant effect is sporadic in the injection model.

A key question is how do the assessments of enzymes from the guinea pig models compare with experience in man? Prospective evaluation of the detergent workforce has allowed for an assessment of allergic antibody responses to different enzymes as well as a better understanding of exposure–response relationships (Sarlo et al., 1997b; Schweigert et al., 2000; Sarlo and Kirchner, 2002). In the early days of manufacture of enzyme-containing detergents, it was noted that the epidemiology of enzyme allergy was different between the detergent manufacturers and the enzyme manufacturers. The prevailing opinion was that the ingredients in the detergent were adjuvants and enhanced the allergic response to enzymes in the detergent facilities. The guinea pig models have shown that detergent has adjuvant activity, thereby adding support to the interpretation of the epidemiology. The guinea pig models have also shown that different enzymes can have different potencies as compared to Alcalase, so the exposure guidelines for these enzymes are set to be different from Alcalase. Prospective evaluation of skin prick test responses among workers exposed to various enzymes showed that these guidelines were protective, thus supporting the use of the guinea pig model as a predictive tool for type 1 allergy to enzymes. These observations apply to exposures to single enzymes as well as enzyme mixtures, thus providing support for the way the guinea pig assesses potency of single enzymes and detects protease adjuvant effects. The guinea pig has also shown that there are two thresholds for induction of antibody and elicitation of symptoms to enzymes. This same observation can be made from exposed populations in the detergent industry where there appears to be two different thresholds of intensity of exposure for induction of allergic antibody and elicitation of respiratory symptoms.

17.2.2 Mouse Models

The mouse has been used in immunology research for many years, and the original description of T helper (h)1 and Th2 type immune responses was based on investigations conducted in this species (Mosmann et al., 1986, 1989). Much of the published work on type 1 allergy in the mouse has focused on dissecting immune responses to model protein allergens (e.g., ovalbumin); only a few laboratories have used the mouse to assess allergic responses to a variety of proteins in the contexts of hazard and risk assessment. Th2 cells regulate type 1 allergic responses in the mouse and the interleukins (IL) 4 and 13 (IL-4 and IL-13) play pivotal roles in the production of IgE and IgG1 antibodies. In humans, the Th1 and Th2 divide is not as clear as in the mouse, but IgE and IgG4 antibody production are similarly dependent on IL-4 and IL-13 (Romagnani et al., 1997; Wills-Karp and Chiaramonte, 2003). IgE is the primary homocytotropic antibody in humans and mice, although murine IgG1 antibody can in some circumstances drive type 1 allergic/anaphylactic responses (Oettgen et al., 1994; Oshiba et al., 1996). Eosinophils are a hallmark of allergic inflammation in both humans and mice. Mice will develop allergic antibody to a protein following systemic exposure (e.g., intraperitoneal injection), local exposure (e.g., nasal, lung), or topical exposure (via normal skin under occlusion or skin that has been damaged).

The mouse intranasal test (MINT) was developed by Robinson et al. (1996) to assess detergent enzymes as respiratory allergens. The goal was to develop a test system that could rank enzymes in terms of their relative allergenic potency in less time than the 10 to 12 weeks required for the guinea pig models. The BDF1 mouse, a cross between C56B1/6 and DBA/2 mice, is used for this test. Mice receive several doses of enzyme protein via intranasal instillations over a period of 2 to 4 weeks. Serum is collected 5 to 7 days after the last exposure, and protein-specific IgG1 antibody is measured by enzyme linked immunosorbent assay (ELISA). IgE antibody is measured in the rat PCA test. The IgG1 and IgE response to several detergent enzymes is similar to the responses obtained in the guinea pig model, indicating that the BDF1 mouse has the same capability as the guinea pig for predicting relative allergenic potency for this class of proteins (Robinson et al., 1998). These mice will develop eosinophilia in the nasal passages following exposure to allergen (Horn et al., 1999). Similar antibody responses with eosinophilia develop in the lungs following IT instillation of enzyme protein (Kawabata et al., 2000; Clark et al., 2000). Like the guinea pig, the antibody responses to enzymes were enhanced in the presence of detergent showing that the mouse can detect the adjuvant effect of this material. The interlaboratory reproducibility of these responses in this mouse system has been demonstrated when the same batches of enzyme materials were used (Parris et al., 2000). However, a third laboratory was not able to generate concordance with guinea pig data when testing a set of unidentified enzyme proteins (Blaikie and Basketter, 1999). Although numerous technical questions remain to be addressed, these data point to the complexity of developing animal systems for assessing allergenic potential of proteins.

Using BALB/c mice, Ward et al. (1998, 2000) have developed a model of Th2 dominated airway responses to fungal organisms used as biopesticides. In this system, mice received an intraperitoneal injection of fungal extract (*Metarhizium anisopliae*) in alum (as adjuvant) followed by IT challenge with extract 14 days later. Pulmonary reactivity, histopathology, and measurements of cytokines and total IgE were made. Only the fungal-injected mice experienced significant changes in pulmonary function and airway reactivity following challenge with extract. In addition, there was a significant influx of eosinophils and lymphocytes into the airways of these animals. There was elevation of IL-4, IL-5, and total IgE in lavage fluid without a change in IFN-γ indicating that the response to extract of *M. anisopliae* was of the Th2 type. These investigators were also able to measure specific IgE antibody to several proteins in the fungal extract (Ward, personal communication, 2003). Using systemic exposure to induce sensitization followed by respiratory exposure to elicit reactions is a common approach to study the symptomatic phase of type 1 allergy. Whether this approach can be used to assess risk of symptoms on exposure to proteins in a previously sensitized population needs to be explored.

It has been known for many years that the major histocompatibility complex (MHC) class II haplotype of mouse strains can play an important role in immune recognition of proteins, and eventual development of antibody responses. Early work with ovalbumin and ovamucoid showed that recognition of these proteins as immunogens and allergens was strain dependent (Vaz et al., 1970, 1971). Therefore, the choice of strain is important when developing mouse models to assess proteins as allergens. Work with BDF1 and CB6F1 mice, along with the parental strains (C57Bl/6, DBA/2, BALB/c), showed that the F1 mice responded similarly to enzymes and ranked these allergens the same as the guinea pig model (Sarlo et al., 2000). However, the parental strains either did not rank the enzymes in the same way, or could only respond to high doses of enzyme, indicating a link between responsiveness to enzyme and MHC class II molecules. Extension of this work to ovalbumin and bovine serum albumin (BSA) showed that the difference in responsiveness was not restricted to enzymes (Parris et al., 2001). While the IgG1 and IgE antibody responses in the BDF1 mouse "ranked" Alcalase as more potent than ovalbumin, and ovalbumin as more potent than BSA, the antibody responses in the BALB/c mouse did not mirror each other. Using IgG1, BSA was more potent than Alcalase, and ovalbumin was comparable to enzyme. Using IgE, ovalbumin was more potent than Alcalase, and there was no response to the BSA. BSA is recognized as a rare allergen in humans, so the IgE data in both strains would be consistent with the human experience. The data on human responses to ovalbumin as compared to enzymes is less clear. A more in-depth analysis of inhalation exposure to ovalbumin among occupationally exposed cohorts can help to address the question of whether ovalbumin is a more or less potent allergen than Alcalase. Taken together, these data show that when developing mouse models for protein allergy, testing several strains may be required to understand how the MHC class II haplotype plays a role in the allergic response to protein.

17.3 ASSESSMENT OF PROTEINS AS FOOD ALLERGENS

17.3.1 Rat Models

For the characterization of the potential allergenic hazard of food proteins, several groups of investigators have elected to employ the Brown Norway (BN) rat, a strain that has been characterized as mounting strong IgE antibody responses. One of the attractions of this approach is that in addition to the capacity to monitor the kinetics of specific serum antibody (IgE and IgG) responses, it is possible to study oral challenge–induced responses in previously sensitized animals as a function of changes in gut permeability, respiratory functions, and blood pressure. Methods employing either oral or intraperitoneal administration of protein combined with exposure to adjuvant have been described (Atkinson and Miller, 1994; Atkinson et al., 1996; Miller et al., 1999), although such have been limited to the investigation of a handful of proteins. The approach employing BN rats that has attracted the most interest is one in which the test protein is delivered by daily gavage over a 42-day period in the absence of adjuvant. During and following exposure, specific IgE (and IgG) antibody responses are measured (Knippels et al., 1998a, 1998b, 1999a, 1999b; Penninks and Knippels, 2001).

In initial studies, antibody responses to ovalbumin, a major food allergen found in chicken egg white, were examined. The impact of mode (gavage versus ad libitum in the drinking water) and frequency of application (daily, twice a week, once a week, once every two weeks) on antibody responses was investigated (Knippels et al., 1998b). Daily gavage administration of 1 mg of ovalbumin for 42 consecutive days, without the use of adjuvant, resulted in the induction of ovalbumin-specific IgG and IgE responses in the majority of rats. The percentage of IgE responders to ovalbumin was approximately 80%, as measured by either ELISA or homologous PCA test. Optimal ovalbumin-specific IgE antibody responses were observed 28 to 35 days after the initiation of exposure. Using less frequent administration regimes of 1 mg of OVA by gavage failed to provoke marked specific IgG or specific IgE antibody responses. Interestingly, dietary exposure (*ad libitum* in drinking water) to ovalbumin under the same conditions induced no, or only a very low frequency, of IgE antibody responses (Knippels et al., 1998b). It is presumed that, at least in part, the failure to induce strong humoral responses following dietary exposure is secondary to the development of immunological tolerance (Strobel and Mowat, 1998).

Comparative sensitization studies using different strains of rats have been performed (Knippels et al., 1999b). Oral (gavage) exposure of Wistar, PVG, Hooded Lister, and BN rats to ovalbumin resulted in the detection of ovalbumin-specific IgE antibodies in the BN rat only, confirming that this was the most suitable strain for such studies. Immune-mediated effects elicited following oral challenge of sensitized animals have been studied also (Knippels et al., 1999a). Local effects on gut permeability and systemic effects on respiratory function and blood pressure were examined. Oral challenge of ovalbumin-sensitized BN rats resulted in a significant increase in gut permeability, as evidenced by an increased uptake of a bystander protein (β lactoglobulin) although only minor systemic effects were recorded in a minority of animals.

More recently, the relative allergenicity of a range of purified proteins has been examined. These comprised the major peanut allergen, Ara h 1; the major allergen in shrimp, tropomyosin; patatin, the major allergen of the potato, recognized as a relatively minor allergenic food source; and tropomyosin purified from beef, a protein considered to lack significant allergy potential. Two identical oral sensitization studies were performed with these purified proteins in BN rats. In one study, the oral sensitizing potential decreased in the following order: Ara h 1 > shrimp tropomyosin > patatin, with no sensitization to beef tropomyosin. In a second study, the relative responses of the three allergenic materials were different to those observed previously, although beef tropomyosin consistently failed to stimulate detectable IgE antibody production (Kimber et al., 2003).

The sensitizing potential of mixtures of proteins has been investigated. Daily gavage dosing of BN rats with different concentrations of hen egg white (HEW) and cow's milk (CM) proteins induced relatively vigorous antigen-specific IgG responses, but only a limited number of IgE responders was observed as measured by PCA. However, immunoblotting of the rat sera demonstrated the presence of specific IgE antibodies against both HEW proteins and CM proteins that recognized a comparable profile of allergens to those of sera derived from HEW- or CM-allergic patients (Knippels et al., 2000). The reasons for this apparent discrepancy are not clear, although it is possible that the protein mixture used for PCA challenge was not appropriate, or optimal, for elicitation of PCA reactions. In another study conducted by the same investigators, BN rats were sensitized with different doses of either crude raw peanut extract or roasted peanut extract. Dose levels ranged from 0.01 to 10 mg of peanut protein extract by daily gavage. Only a limited number of animals were IgE positive as measured by PCA, although the majority was IgG positive (Kimber et al., 2003). These data suggest that it may be more difficult to assess potential allergenicity as a function of induced IgE responses in BN rats following gavage exposure when complex mixtures of proteins rather than purified protein allergens are utilized. The reasons for this are not immediately clear, but the observation does add another complexity when considering model development.

Although experience to date indicates that the BN rat may be a useful animal model in which to study the potential allergenicity of "novel" food proteins, several issues must be considered. For example, as described above, there appears to be considerable variability in the induction of IgE antibody responses in BN rats. Thus, Knippels et al. (1998b) observed that whereas in most experiments 80% or more of BN rats were IgE responders, in a minority of experiments daily gavage dosing with 1 mg of ovalbumin failed to induce specific IgE antibody in any recipient. Other investigators have reported the lack of a robust IgE response to ovalbumin, and to other proteins, following oral administration to BN rats (Dearman et al., 2001). In these experiments, BN rats received daily gavage doses of 1 mg of protein (ovalbumin or peanut agglutinin, an important peanut allergen) (Burks et al., 1994) for up to 42 days. Both of these test proteins, when administered by gavage, was immunogenic in BN rats, inducing IgG antibody responses. However, there was no evidence in any instance for IgE antibody production as determined by homologous PCA assay. Thus, BN rats failed to mount specific IgE responses to allergenic proteins administered using a standard gavage exposure regime, despite the confirmed immunogenicity of these proteins and the ability of the same rats to produce IgE antibody upon delivery of antigen by a parenteral route (intraperitoneal exposure). Although in theory, the development of oral tolerance is one possibility to accommodate the fact that vigorous IgG responses were induced, it would have to be argued that such tolerance was partial and selective. It appears that the ability of rats, and possibly BN rats in particular, to mount IgE responses is inherently variable and subject to a number of influences including environmental conditions, age, and the presence or absence of subclinical infection (Kemeny, 1994). In addition, it has been demonstrated that unscheduled dietary pre-exposure of the test animals or of the parental generations to the antigen under investigation will impact the ability of rats to respond to gavage treatment with the induction of antibody (Knippels et al., 1998a). Thus, exposure of the parental generation to soy was found to influence the outcome of sensitization studies with the offspring. These studies showed that BN rats bred and raised on a soy protein-containing diet for several generations have soy-specific IgG antibodies. When these rats were fed before breeding with a soy protein–free diet for half a year, soy protein–specific IgG antibodies were still detectable in the parental animals, and also in serum samples collected from the F1 generation of offspring rats fed on soy protein–free diets for periods of 6 to 12 months. In the second, third, and fourth generations of offspring bred on a soy protein–free diet no soy-specific IgG was detected. Oral sensitization to soy could be achieved in these rats. Therefore, when oral sensitization studies with proteins are performed, it may be necessary to ensure that at least two generations of animals have to be bred on a diet free of the antigen under investigation in order to provide for responsive animals.

17.3.2 Mouse Models

The application of mouse models has found increasing favor for the characterization of the cellular and molecular mechanisms of various types of IgE-mediated allergic disease, including asthma to proteins (Kips et al., 2003) and respiratory sensitization to chemicals (Dearman and Kimber, 2001) or to protein detergent enzymes (Robinson et al., 1998; Blaikie and Basketter, 1999). The mouse offers considerable advantages compared with other animal models, particularly with respect to the availability of inbred high-IgE responder strains and various immunological and molecular reagents, including transgenic animals in which particular genes of interest have been overexpressed or deleted. There is increasing interest, therefore, in the characterization of the sensitizing potential of novel food proteins as a function of immune responses provoked in mice.

In common with rat models, some methods incorporate the use of adjuvant. It is claimed that some adjuvants, including, for instance, carrageenan and cholera toxin, selectively promote Th2-type cell activation and IgE antibody responses (Nicklin et al., 1985, 1988; Wilson et al., 1989; Marinaro et al., 1995). This approach has been used with some success to characterize immune responses induced in mice by peanut and CM proteins (Li et al., 1999, 2000). C3H/HeJ strain mice (5 weeks old) were exposed orally (by gavage) to freshly ground whole peanut together with cholera toxin adjuvant. Oral challenge of sensitized mice with peanut extract stimulated anaphylactic reactions, and analysis of serum revealed the presence of murine IgE antibody that recognized the same allergenic epitopes of the major peanut allergen Ara h 2 as IgE from peanut allergic human subjects (Li et al., 2000). In similar experiments with CM proteins, it was necessary to use 3-week-old mice just after weaning, to elicit IgE antibody responses and signs of systemic anaphylaxis (including vascular leakage and elevated serum histamine) (Li et al., 1999). The adjuvant alum has also been utilized to augment IgE antibody responses to proteins, including IgE production to the milk allergen β-lactoglobulin (Adel-Patient et al., 2000) and to ovalbumin (van Halteren et al., 1997). There is no doubt, therefore, that the use of appropriate adjuvants will induce or enhance IgE antibody responses. What is not clear at present, however, is the extent to which the use of different adjuvants will perturb or modify the inherent properties of proteins to induce IgE antibody responses. That is, the use of adjuvant may compromise the ability to discriminate between proteins with respect to allergenic potential, possibly generating false positives and conferring the appearance of sensitizing potential on nonallergens. This issue is currently unresolved, as to date few nonallergenic proteins have been examined using this type of approach. There is some preliminary evidence to suggest that intraperitoneal exposure of mice to protein and carrageenan does indeed result in some loss of selectivity of IgE responses to proteins (Dearman et al., 1999). The use of adjuvant may represent a potential source of false positives in hazard identification.

Another important consideration is the choice of route of exposure. Although it has been argued that in the context of assessing the safety of dietary proteins oral administration is the preferred route of exposure, it has been our experience, and the experience of others, that such a regimen may lack the sensitivity required for effective identification of inherent sensitizing potential. This is probably attributable, at least in part, to the fact that oral exposure may be associated with the development of tolerance (Strobel and Mowat, 1998). Serological responses induced by three proteins of differing sensitizing potential have therefore been compared following systemic (intraperitoneal injection) or gavage administration (Dearman et al., 2001). Mice were exposed to the allergen peanut agglutinin, ovalbumin, or a crude potato protein extract (PPE) (10%) containing acid phosphatase activity, which is assumed to lack significant sensitizing activity (Jeannet-Peter et al., 1999). BALB/c strain mice received a daily gavage dose of 1 mg of test protein for 28 or 42 days. Alternatively, mice received various concentrations of protein (0.2% peanut agglutinin; 2% OVA or 10% PPE) by intraperitoneal injection with the same treatment repeated 7 days later. Serum was collected 14, 28, or 42 days following the start of exposure and analyzed for specific IgG and IgE antibody production by ELISA and homologous PCA, respectively. Gavage administration of peanut agglutinin resulted in vigorous IgG antibody responses, whereas similar treatment of mice with either OVA or PPE resulted in weaker and more variable IgG responses. Marked differences were observed between proteins with respect to IgE antibody. Relatively high titer IgE responses were provoked by gavage administrations of peanut agglutinin, whereas identical exposure to OVA or PPE stimulated no detectable, or only low titer of IgE. A different pattern of responses was recorded following intraperitoneal exposure of mice to the same proteins. All proteins induced IgG responses that were generally of higher titer than those observed after gavage administration. Intraperitoneal exposure to both peanut agglutinin and ovalbumin induced high titer IgE antibody. In contrast, PPE failed to stimulate vigorous IgE antibody responses. Thus, oral exposure of BALB/c strain mice appears to result in a substantial underestimation of the allergenic potential of ovalbumin. It is possible, but as yet unproven, that in the context of stimulation of allergic responses, mouse strains may vary with respect to the relative activity of different proteins and/or the sensitivity of different routes of exposure. Although this might prove to be a productive line of enquiry, the fact remains at present that (in BALB/c strain mice at least) for the purposes of hazard identification and assessment of inherent sensitizing activity, systemic exposure may be more appropriate than oral dosing.

The alternative approach is therefore to examine the quality and vigor of immune responses induced in mice following systemic exposure (in the absence of adjuvant) and to define food proteins as having inherent sensitizing potential if they provoke clear IgE antibody responses. In effect, the strategy is to distinguish between immunogenic proteins (that are able to induce specific IgG antibody responses) and potentially allergenic proteins (that are able to induce both IgG and IgE antibody production). It has been possible using this approach to demonstrate clear differences between proteins with respect to IgE antibody production. Thus, under conditions of exposure where different proteins elicit largely comparable IgG antibody responses, there can be very substantial variations in their ability to induce specific IgE antibody (Dearman et al., 2000, 2001). In these initial experiments, serological responses induced by proteins such as OVA or PPE were examined at a single exposure concentration. In subsequent experiments, dose–response analyses have been performed using proteins that are known food allergens (peanut agglutinin, OVA, and BSA), and materials considered to be lacking in allergenicity (PPE and a purified potato protein, potato agglutinin) (Dearman et al., 2003b). All five proteins were immunogenic in mice, inducing IgG antibody responses at all doses tested, although there was some variation with respect to vigor of IgG responses. Distinct differences were recorded in IgE profiles after exposure to peanut agglutinin, ovalbumin, and BSA compared with those to potato agglutinin and PPE. The potato proteins provoked detectable (low) titer IgE antibody only at the highest concentration tested, whereas the food allergens each stimulated IgE antibody over the complete dose range used. Importantly, differences in IgE antibody production have been observed against a background of equivalent overall immunogenicity (IgG antibody responses). Recently, the reliability and robustness of this method has been assessed in the first phase of an interlaboratory collaborative trial (Dearman et al., 2003a). In each of two laboratories, two independent experiments were performed in which BALB/c strain mice were exposed by intraperitoneal injection to peanut agglutinin, ovalbumin, or potato agglutinin. Serological assessments were conducted in one of the laboratories and it was found that all three proteins induced vigorous IgG antibody responses in every experiment, but only the allergens peanut agglutinin and ovalbumin stimulated IgE antibody production.

17.4 SUMMARY

The evidence summarized above suggests that with continuing effort, it should be possible to develop rodent models that will be of value in identifying proteins with the inherent potential to cause type 1 allergy. Certainly, there has been some success in the use of guinea pig and mouse models for characterization of the allergenic potential of a specific class of proteins (detergent enzymes). Whether these or similar models will prove equally effective for evaluation of the allergenic properties of a wider range of proteins is one focus of research activities. It must be acknowledged that the pathogenesis of food allergy is complex, particularly with regard to inter-individual variations in susceptibility, and it will therefore be difficult to define a method using experimental animals that will predict accurately the likely prevalence, severity, and persistence of food allergy in human populations exposed to a novel protein. Accurate risk assessment with respect to food allergy will represent an important challenge for the future. There are real opportunities, however, for the development of experimental approaches that will allow the first step in the safety assessment process, the identification of intrinsic hazard, which in the context of food allergy is the inherent potential of a protein to cause allergic sensitization.

REFERENCES

Adel-Patient, K., et al., Evaluation of a high IgE-responder mouse model of allergy to bovine beta-lactoglobulin (BLG): development of sandwich immunoassays for total and allergen-specific IgE, IgG1 and IgG2a in BLG-sensitized mice, *J. Immunol. Methods*, 235, 21, 2000.

American Conference of Governmental Industrial Hygienists, *Threshold Limit Values and Biological Exposure Indices*, ACGIH, Cincinnati, OH, 2003.

Atkinson, H.A., and Miller, K., Assessment of Brown Norway rat as a suitable model for the investigation of food allergy, *Toxicology*, 91, 281, 1994.

Atkinson, H.A., et al., Brown Norway rat model of food allergy: effect of plant components on the development of oral sensitization, *Food. Chem. Toxicol.*, 34, 27, 1996.

Blaikie, L., and Basketter, D.A., Experience with a mouse intranasal test for the predictive identification of respiratory sensitization potential of proteins, *Food. Chem. Toxicol.*, 37, 889, 1999.

Blaikie, L., Basketter, D.A., and Morrow, T., Experience with a guinea pig model for the assessment of respiratory allergens, *Hum. Exp. Toxicol.*, 9, 743, 1994.

Burks, A.W., et al., Identification of peanut agglutinin and soybean trypsin inhibitor as minor legume allergens, *Int. Arch. Allergy Immunol.*, 105, 143, 1994.

Clark, E.D., et al., Aspiration vs. intranasal instillation lead to comparable immune responses to Alcalase in BDF1 mice, *Toxicologist*, 54, 584, 2000.

Dearman, R.J., Basketter, D.A., and Kimber, I., Anti-protein antibody isotype distribution: influence of adjuvant, *Immunol. Lett.*, 69, 102, 1999.

Dearman, R.J., and Kimber, I., Cytokine fingerprinting and hazard assessment of chemical respiratory allergy, *J. Appl. Toxicol.*, 21, 153, 2001.

Dearman R.J., et al., Divergent antibody isotype responses induced in mice by systemic exposure to proteins: a comparison of ovalbumin with bovine serum albumin, *Food Chem. Toxicol.*, 38, 351, 2000.

Dearman, R.J., et al., Characterization of antibody responses induced in rodents by exposure to food proteins: influence of route of exposure, *Toxicology*, 167, 217, 2001.

Dearman, R.J., et al., Induction of IgE antibody responses by protein allergens: inter-laboratory comparisons, *Food Chem. Toxicol.*, 41, 1509, 2003a.

Dearman, R.J., et al., Evaluation of protein allergenic potential in mice: dose-response analyses, *Clin. Exp. Allergy*, 33, 1586, 2003b.

Flindt, M.L.H., Pulmonary disease due to inhalation of derivatives of Bacillus subtilis containing proteolytic enzyme, *Lancet*, 1, 1177, 1969.

Griffith-Johnson, D.A., Jin, R., and Karol, M.H., The role of purified IgG1 in pulmonary hypersensitivity responses of the guinea pig, *J. Toxicol. Environ. Health*, 40, 117, 1993.

Groneberg, D.A., et al., Animal models of allergic and inflammatory conjunctivitis, *Allergy*, 58, 1101, 2003.

Hillibrand, J., Thorne, P.T., and Karol, M.H., Experimental sensitization to subtilisin: production of specific antibodies following inhalation exposure of guinea pigs, *Toxicol. Appl. Pharmacol.*, 89, 449, 1987.

Horn, P.A., et al., Detection of allergic inflammation and enzyme specific IgE antibody in the mouse intranasal test, *Toxicologist*, 48, 1-S, 1967, 1999.

Ishida, M., et al., Antigen induced allergic rhinitis model in the guinea pig, *Ann. Allergy*, 72, 240, 1994.

Jeannet-Peter, N., Piletta-Zanin, P.A., and Hauser, C., Facial dermatitis, contact urticaria, rhinoconjunctivitis, and asthma induced by potato, *Am. J. Contact Dermatitis*, 10, 40, 1999.

Kawabata, T., Babcock, L.S., and Horn, P.A., Specific IgE and IgG1 responses to subtilisin Carlsberg (Alcalase) in mice: development of an intratracheal exposure model, *Fundam. Appl. Toxicol.*, 29, 238, 2000.

Kemeny, M., in Proceedings, Conference on Scientific Issues Related to Potential Allergenicity in Transgenic Food Crops, U.S. Food and Drug Administration, Washington, DC (FDA Docket 94N-0053), 1994.

Kimber, I., et al., Assessment of protein allergenicity on the basis of immune reactivity: animal models, *Environ. Health Perspect.*, 111, 1125, 2003.

Kips, J.C., et al., Murine models of asthma, *Eur. Respir. J.*, 22, 374, 2003.

Knippels, L.M., Penninks, A.H., and Houben, G.F., Continued expression of anti-soy protein antibodies in rats bred on a soy protein-free diet for one generation: the importance of dietary control in oral sensitization research, *J. Allergy Clin. Immunol.*, 101, 815, 1998a.

Knippels, L.M., et al., Oral sensitization to food proteins: a Brown Norway rat model, *Clin. Exp. Allergy*, 28, 368, 1998b.

Knippels, L.M., et al., Immune-mediated effects upon oral challenge of ovalbumin-sensitized Brown Norway rats: further characterization of a rat food allergy model, *Toxicol. Appl. Pharmacol*, 156, 161, 1999a.

Knippels, L.M., et al., Humoral and cellular immune responses in different rat strains on oral exposure to ovalbumin, *Food Chem. Toxicol.*, 37, 881, 1999b.

Knippels, L.M., et al., Comparison of antibody responses to hen's egg and cow's milk proteins in orally sensitized rats and food-allergic patients, *Allergy*, 55, 251, 2000.

Li, X-M., et al., A murine model of IgE-mediated cows milk hypersensitivity, *J. Allergy Clin. Immunol.*, 103, 206, 1999.

Li, X-M., et al., A murine model of peanut anaphylaxis: T- and B-cell responses to a major peanut allergen mimic human responses, *J. Allergy Clin. Immunol.*, 106, 150, 2000.

Marinaro, M., et al., Mucosal adjuvant effect of cholera toxin in mice results from induction of T helper 2 (Th2) cells and IL-4, *J. Immunol.*, 155, 4621, 1995.

Markham, R.J.F., and Wilkie, B.N., Influence of detergent on aerosol allergic sensitization with enzymes of Bacillus subtilis, *Int. Arch. Allergy Appl. Immunol.*, 51, 529, 1976.

Miller, K., et al., Allergy to bovine β-lactoglobulin: specificity of immunoglobulin E generated in the Brown Norway rat to tryptic and synthetic peptides, *Clin. Exp. Allergy*, 29, 1696, 1999.

Mosmann, T.R., and Coffman, R.L., Heterogeneity of cytokine secretion patterns and functions of helper T cells, *Adv. Immunol.*, 46, 111, 1989.

Mosmann, T.R., et al., Two types of murine helper T cell clones. I. Definition according to profiles of lymphokine activities and secreted proteins, *J. Immunol.*, 136, 2348, 1986.

Nicklin, S., Atkinson, H.A., and Miller, K., Iota-carrageenan induced reaginic antibody production in the rat. I. Characterization and kinetics of the response, *Int. J. Immunopharmacol.*, 7, 677, 1985.

Nicklin, S., Atkinson, H.A., and Miller K., Adjuvant properties of polysaccharides: effect of iota-carrageenan, pectic acid, pectin, dextran and dextran sulphate on the humoral immune response in the rat, *Food Add. Contam.*, 5, 573, 1988.

Oettgen, H.C., et al., Active anaphylaxis in IgE-deficient mice, *Nature*, 370, 367, 1994.

Oshiba, A., et al., Passive transfer of immediate hypersensitivity and airway hyperresponsiveness by allergen-specific immunoglobulin (Ig)E and IgG1 in mice, *J. Clin. Invest.*, 97, 1398, 1996.

Parris, J.S., Clark, E.D., and Sarlo, K., Association of IgG 1 and IgE antibody responses to bovine albumin and ovalbumin in BDF1 and Balb/c mice, *Toxicologist*, 60, 831, 2001.

Parris, J.S., et al., Allergic inflammation and enzyme specific IgG1 and IgE antibody to various enzymes in the mouse intranasal test, *Toxicologist*, 54, 583, 2000.

Penninks, A.H., and Knippels, L.M., Determination of protein allergenicity: studies in rats, *Toxicol. Lett.*, 120, 171, 2001.

Pepys, J., et al., Allergic reactions of the lungs to enzymes of Bacillus subtilis, *Lancet*, 1, 1181, 1969.

Ratner, B., Jackson, H.C., and Gruehl, H.L., Respiratory anaphylaxis: sensitization, shock, bronchial asthma and death induced in the guinea pig by nasal inhalation of dry horse dander, *Am. J. Dis. Child.*, 34, 23, 1927.

Ritz, H.L., et al., Respiratory and immunological responses of guinea pigs to enzyme containing detergents: a comparison of intratracheal and inhalation modes of exposure, *Fundam. Appl. Toxicol.*, 21, 31, 1993.

Robinson, M.K., et al., Specific antibody responses to subtilisin Carlsberg (Alcalase) in mice: development of an intranasal exposure model, *Fundam. Appl. Toxicol.*, 34, 15, 1996.

Robinson, M.K., et al., Use of the mouse intranasal test to determine the allergenic potency of detergent enzymes: comparison to the guinea pig intratracheal test, *Toxicol. Sci.*, 43, 39, 1998.

Romagnani, S., et al., An update on human Th1 and Th2 cells, *Int. Arch. Allergy Immunol.*, 113, 153, 1997.

Sarlo, K., and Kirchner, D.B., Occupational asthma and allergy in the detergent industry: new developments, *Curr. Opin. Allergy Clin. Immunol.*, 2, 97, 2002.

Sarlo, K., et al., Respiratory allergenicity of detergent enzymes in the guinea pig intratracheal test: association with sensitization of occupationally exposed individuals, *Fundam. Appl. Toxicol.*, 38, 44, 1997a.

Sarlo, K., et al., Proteolytic detergent enzymes enhance the allergic antibody responses of guinea pigs to non-proteolytic detergent enzymes in a mixture: implications for occupational exposure, *J. Allergy Clin. Immunol.*, 100, 480, 1997b.

Sarlo, K., et al., Influence of MHC background on the antibody response to detergent enzymes in the mouse intranasal test, *Toxicol. Sci.*, 58, 299, 2000.

Schweigert, M.K, Mackenzie D.P., and Sarlo K., Occupational allergy and asthma associated with the use of enzymes in the detergent industry: a review of the epidemiology, toxicology and methods of prevention, *Clin. Exp. Allergy*, 30, 1511, 2000.

Strobel, S., and Mowat, A.M., Immune responses to dietary antigens: oral tolerance, *Immunol. Today*, 19, 173, 1998.

Thorne, P.T., et al., Experimental sensitization to subtilisin: production of immediate and late onset pulmonary reactions, *Toxicol. Appl. Pharmacol.*, 86, 112, 1986.

van Halteren, A.G.S., et al., IgE and mast cell response on intestinal allergen exposure: a murine model to study the onset of food allergy, *J. Allergy Clin. Immunol.*, 99, 94, 1997.

Vaz, N.M., Vaz, E.M., and Levine, B.B., Relationship between H–2 genotype and immune responsiveness to low doses of ovalbumin in the mouse, *J. Immunol.*, 104, 572, 1970.

Vaz, N.M., et al., H–2 linked genetic control of immune responsiveness to ovalbumin and ovamucoid, *J. Exp. Med.*, 134, 1335, 1971.

Ward, M.D., Sailstad, D.M., and Selgrade, M.K., Allergic responses to the biopesticide Metarhizium anisopliae in Balb/c mice, *Toxicol. Sci.*, 45, 195, 1998.

Ward, M.D., et al., Allergen-triggered airway hyperresponsiveness and lung pathology in mice sensitized with the biopesticide Metarhizium anisopliae, *Toxicology*, 143, 141, 2000.

Wills-Karp, M., and Chiaramonte, M., Interleukin 13 in asthma, *Curr. Opin. Pulm. Med.*, 9, 21, 2003.

Wilson, A.D., Stokes, C.R., and Bourne, F.J., Adjuvant effect of cholera toxin on the mucosal immune response to soluble proteins: differences between mouse strains and protein antigens, *Scand. J. Immunol.*, 29, 739, 1989.

CHAPTER 18

Hypersensitivity Reactions: Nonrodent Animal Models

Ricki M. Helm

CONTENTS

18.1 INTRODUCTION

The use of rodent models have contributed significantly to our knowledge of cellular, molecular, and functional mechanisms in immunology and immunotoxicology. However, large animal models have unique experimental advantages that contribute to the physiological relevance to humans, and thus can be easily extrapolated to human diseases. Natural disease models, including host–pathogen interactions and hypersensitivity reactions that are observed in swine, cattle, sheep, and dogs, provide unique experimental opportunities for investigations related to developmental immunity, relationships between innate and mucosal immunity, regional immunity, surgical intervention, vaccination strategies, and therapeutic options. Quoting Hein and Griebel (2003), "[I]t is not reasonable to rely solely on rodent models to fill the gaps that exist in our knowledge ... or to believe that experimentation in rodents alone will find cures or treatments for all human and animal diseases."

Traditionally, small animals, such as the guinea pig and rat, have been used as models in toxicologic and immunologic studies for a variety of hypersensitivity reactions (Atherton et al.,

Table 18.1   Considerations for Establishing Hypersensitivity Reactions in Animal Models

1. Genetic predisposition
   a. Inbred strains with high/low IgE responses (mice)
   b. Outbred animals with natural allergy (atopic dog, calves, pigs)
2. Maturation of innate/adoptive immunity
   a. Tolerance vs. immunogenicity
   b. Immediate hypersensitivity (immunoglobulin isotype, e.g., IgE, IgG1, IgG2a)
   c. Primary/secondary immune system development
      i. Th1/Th2/Treg cells
      ii. Cytokine profiles
3. Target organs: physiology of skin, and respiratory and gastrointestinal tract
4. Antigen/allergen presentation to break natural tolerance
   a. Route of administration (intravenous, intraperitoneal, subcutaneous, oral gavage)
   b. Use of adjuvants (known inflammatory agents)
   c. Dose and duration of exposure.
   d. Age (neonatal, infant, adult)
   e. Threshold doses for sensitization and allergic response
5. Biologic responses
   a. Skin tests
   b. Histamine release

2002; Piacentini et al., 1994; van Och et al., 2001; Knippels and Penninks, 2002). However, natural airborne and food allergies in nonrodent animals, such as the atopic dog and swine, are being used for investigations aimed at defining mechanisms of IgE-mediated hypersensitivity (mechanistic studies). Furthermore, these species are used in qualitative approaches to determine the relative allergenicity of novel proteins in genetically modified foods (predictive studies). In each of the animal models, the parameters highlighted in Table 18.1 should be taken into consideration.

Regardless of the animal model, neither the complex process for sensitization nor the allergic response will completely extrapolate to human hypersensitivity. The immunopathogenesis of hypersensitivity reactions may be driven by more than one immunologic mechanism, including immediate hypersensitivity (manifestations mediated by allergen-specific IgE), delayed hypersensitivity (reactions associated with specific T-lymphocytes), and inflammatory reactions (caused by immune complexes) or a combination of these immune responses. Mucosal tissues represent natural barriers and transport systems and are regarded as the primary site for sensitization. Maturation of these tissues or a breakdown in barrier structure of the mucosal tissue could result in the introduction of agents that may lead to the induction, elicitation, or a microenvironment conducive to hypersensitivity.

18.2 MECHANISTIC STUDIES

Food hypersensitivity, as defined in this review, is the ability of an agent to induce specific IgE antibody of sufficient avidity/affinity to cause sensitization, that is, binding of allergen-specific IgE to effector cells (mast cells and basophils). Upon reexposure, cell-bound IgE is cross-linked by the allergen resulting in the release of preformed (histamine) and synthesized leukotrienes and prostaglandins. These are chemical mediators that result in observable manifestations (rashes, urticaria, gastrointestinal/systemic anaphylaxis, and diarrhea) of allergic disease. The clinical manifestations of Type I hypersensitivity reactions include vomiting, diarrhea, and if chronic, weight loss. This type of reaction may cause a potential epithelial mucosal barrier dysfunction leading to increased macromolecule absorption. This can exacerbate an allergic reaction or lead to multiple sensitizations by absorption of bystander antigens/allergens. The mechanism of pathogenesis is complex and is likely due to inadequate digestion, mucosal membrane barrier breakdown, a decreased secretion of secretory IgA, and loss of tolerance by the gut-associated lymphoid tissue.

18.3 PREDICTIVE STUDIES

In the mid-1980s and early 1990s, bioengineered food crops were introduced into the food supply. DNA from any species inserted into plant tissue could introduce novel proteins or alter the protein composition of the food crop. This novel industrial approach raised a number of concerns regarding the safety of the food supply. These concerns were raised in several conference reports and in various workshops held to design appropriate approaches in assessing the potential allergenicity of the introduced protein (Food and Agriculture Organization/World Health Organization, 2001; Helm, 2002a). Animal models, with known IgE responses, continue to be investigated for their potential usefulness in distinguishing between immunogenicity and allergenicity of known food sources and to provide a relative food allergen profile to which novel proteins could be compared.

18.4 THERAPEUTIC OPTIONS

Several animal models, primarily murine, are being investigated to determine both preventive and intervention therapy for the treatment of food allergy. Under consideration are cytokine, recombinant-modified allergen, DNA immunizations, and probiotic therapy regimens. Each of these is currently under study in various research institutions investigating large animal food-allergy models. It is anticipated that these studies will spawn a surge of articles that are expected to appear in the literature in the near future. For the remainder of this chapter, the guinea pig, atopic dog, and neonatal swine animal models are reviewed with respect to immunoregulatory mechanisms, predictive assessment of neoallergens, and therapeutic options for prevention and treatment of allergic responses.

18.5 ANIMAL MODELS

18.5.1 Guinea Pig Model

Because the mechanisms of sensitization remain to be conclusively elucidated, a brief review of guinea pig models related to hypersensitivity responses is presented. The guinea pig allergy model has been in place for over 30 years. The principal use of the guinea pig model has been for hazard identification and risk assessment of potentially sensitizing chemicals and for the prediction of skin allergic responses (contact dermatitis). In the standardized guinea pig maximization test (GPMT), a determination of potency based upon the lowest effective dose derived from a single concentration of intracutaneous or topical induction of a chemical, can be assessed qualitatively for skin responses such as irritant and urticarial allergy (Frankild et al., 1996). Although the guinea pig has also been the focus of studies for asthma and airway hyperreactivity, the focus here is on food allergy and gastrointestinal anaphylaxis.

The allergenicity of cow's milk proteins and milk-based infant formulas have been evaluated by examining altered intestinal permeability, intestinal anaphylaxis, and passive cutaneous anaphylaxis (Kitagawa et al., 1995). Animals were sensitized orally with whole cow's milk or cow's milk protein-based formula (SMA and Enfamil) and challenged with beta-lactoglobulin. Cow's milk and infant formula-fed sensitized animal colon tissue responded with antigen-induced, anaphylactically mediated elevation of the transmural short circuit (net chloride secretion). Bronchospasm was evident in all cow's milk–sensitized animals whereas only infant formula-fed animals that responded with gastrointestinal symptoms developed respiratory symptoms. Cow's milk–based infant formula was regarded as having less sensitizing ability compared to whole cow's milk (Kitagawa et al., 1995). This suggested that the animal model could be effective for determining allergen-sensitizing profiles with intestinal symptoms as an endpoint criterion. Results from guinea pigs sensitized to either conventional cow's milk formula or partially hydrolysate formula also confirmed that the hydrolysate formula had less sensitizing ability than cow's milk (Piacentini et al., 1994).

Terpend et al. (1999) investigated the possible mechanisms of the hygiene hypothesis in which the beneficial effects of lactic acid bacteria on gastrointestinal function appeared as a decrease in symptoms of diarrhea in the allergy-prone guinea pig. Hypotheses investigated included (1) lactic acid bacteria interference in the sensitization process by altering degradation of milk antigens, (2) altering antigen absorption and processing at the intestinal level, and (3) stimulation of the immune system or modification of the intestinal epithelium. Results suggested that the intestinal barrier capacity for food proteins could be reinforced by dehydrated, but not fresh, fermented milk; however, milk and fermented milk were equally efficient in inducing cow's milk allergy.

The close association between mast cells in the gastrointestinal mucosa and nerve involvement following mast cell degranulation as a consequence of antigen challenge, has been investigated in the guinea pig. Increasing evidence of nerve involvement in mast cell degranulation and induction of digestive disturbances indicate that tachykinin receptors are involved in the oral sensitization of beta-lactoglobulin. Observations included intestinal hypermastocytosis, secretory responses, and IgE/IgG titers (Gay et al., 1999). Luis et al. (2003) suggested that signaling between mast cells and the enteric nervous system is one potential mechanism by which food allergen stimulation may lead to intestinal but not gastric anaphylaxis.

In these guinea pig models, caution is warranted with respect to differences in anaphylactic antibody. Two isotypes, IgG1 and IgE, are responsible for anaphylactoid symptoms, but neither the immune nor the physiology of the guinea pig gastrointestinal/enteric nervous system are completely characterized and well-suited for extrapolation to humans.

18.5.2 Atopic Dog Models

Spontaneous allergic disease exists in canines and a description of this is predominantly published in the veterinary literature. Initial reports described primarily hay fever/asthmatic signs and pulmonary immune mechanisms and therapeutic treatment of allergic asthma continue to be a mainstream effort (Witich, 1941; Patterson, 1960). Allergic beagles with high levels of serum IgE and eosinophilia have been bred and sensitized with ragweed by intraperitoneal or intrapulmonary exposure (Redman et al., 2001; Out et al., 2002). Bronchoalveolar lavage and segmental lung challenges revealed (1) increased serum-specific IgE and IgG, (2) higher eosinophil counts and airway responses to histamine, and (3) a statistically significant increase in the percentage of major histocompatiblity complex class II–positive CD4 and CD8 T cells. Continued studies of pulmonary hyperresponsiveness in the allergic dog model will provide useful information on the pathophysiological mechanisms and future treatment regimens that may extrapolate to human pulmonary disease.

The more common manifestation and studied allergic disease in dogs is that of atopic dermatitis as first characterized by Halliwell (1971). These early observations suggested that in view of observed similarities in allergic disease signs and symptoms between humans and dogs, the latter could be used as models for human disease (Butler et al., 1983). Although cutaneous adverse reactions and atopic dermatitis in dogs are often indistinguishable from each other on historical and clinical grounds, allergies to foods are known to induce skin lesions (Reedy et al., 1997). In atopic dogs, the third most common type of allergic disease is food allergies, with flea allergy dermatitis and generalized atopy being the most common. It has been estimated that about 8% of the canine population is affected, regardless of age, sex, or breed, albeit selective breeding can result in high IgE responders (Frick, 1996).

Atopic dermatitis in dogs has many clinical features that are identical to those of the same disorder in humans (Nuttal et al., 2002; Sinke et al., 2002). Convincing evidence appears in the veterinary literature regarding the immunopathogenesis of atopic dermatitis (AD). A relevant number of CD4$^+$ T cells suggests a similar cellular role for T cells in the dog model that may be applied to studies of human atopic dermatitis (Sinke, 1997). Using RT-PCR, mRNA encoding cytokines characteristic of a type 2 cytokine pattern were observed in peripheral blood mononuclear cells (PBMCs) directly *ex vivo* without *in vitro* stimulation (Hayashiya et al., 2002). Levels of IL-4 and IL-5 mRNA expression and eosinophil counts were found to be significantly higher in AD dogs than in controls. Further evidence of similar pathogenic similarities in canine and human atopic dermatitis was identified in the incidence of lesions in the thymus, and in the levels of activation-regulated chemokine (TARC) and the inflammatory cytokines IL-beta, INF-gamma, and TNF-alpha (Maeda et al., 2002).

Research using the canine model of allergic disease has been greatly facilitated by the availability of pure canine IgE of known allergen specificity. This IgE is generated from a mouse × canine heterohybridoma that resulted from fusion of canine B cells from a nematode-infected lymph node (Gebhard et al., 1995). With respect to IgE, canine AD was shown to be associated with more variable numbers of circulating monocytes expressing IgE, and there was no correlation between circulating IgE and cell-bound IgE. This is in contrast to human AD, suggesting high levels of IgE in the dog may be related to greater numbers of circulating B cells committed to IgE production (Jackson et al., 2002).

As in humans, diagnosis of food allergy is dependent on skin testing and food elimination diets followed by re-challenge to confirm food-related disease. However, eosinophilia, ocular, and pulmonary allergic mucosal tissues, including cell-bound IgE in bronchioalveolar lavage fluid, have been suggested as more sensitive and lasting surrogate markers for hypersensitivity in allergic dog models than detection of allergen-specific serum IgE levels (Zemann et al., 2002). While avoidance of the incriminated food is the most successful threatment, other treatment options are being investigated. Canine PBMC stimulated with recombinant IL-13 *in vitro* produced allergen-specific IgE that could be specifically inhibited by recombinant canine IL-13Rα and Fc fragment of canine IgE, which suggests that allergic dogs may be excellent models for research on IgE-mediated disease therapies (Tang et al., 2001).

Sensitization and clinical responses to food allergens in amounts comparable to doses reported to elicit disease in humans were induced in Maltese × beagle dogs as spontaneous food hypersensitivity. This was manifested as atopic dermatitis with enteric signs associated with elevated food-specific IgE (Kimber et al., 2003). Clinical signs of nonseasonal atopic dermatitis appeared in 1- to 3-year-old dogs fed a standard dog chow containing corn, wheat, and soy proteins 8 to 13 days after ingestion of food tablets containing 1 to 2 mg/kg of body weight. Elevated specific IgE was significantly increased by day 20, suggesting that dogs maintained on a hypoallergenic diet could be exploited to identify naturally sensitizing and eliciting food allergens (Kimber et al., 2003).

A colony of inbred, high IgE–producing, atopic spaniel/basenji, with genetic predisposition sensitivity to pollens and foods, has been used to investigate food allergy (Frick, 1996; Ermel et al., 1997). Clinical manifestations included involvement of the skin, gastrointestinal and respiratory tract, central nervous system, and any combination of these symptoms. It has been estimated that approximately 15% of dogs with dermatological signs have concurrent GI signs including vomiting, diarrhea, bloating, and cramping. Newborn puppies were subcutaneously injected with food allergen extract in alum, and followed by routine live attenuated virus vaccinations, after which the animals were maintained on the food allergen extract with bimonthly injections. Allergen-specific IgE presented by 3 to 4 months. Fecal index scoring, skin testing to specific and nonrelated allergens, and repeated mucosal skin testing by gastric endoscopy (gastroscopic food sensitivity testing [GFST]) have been used to document immediate and late-phase reactions. GFST results at 3 to 5 minutes revealed an acute-phase inflammatory response highlighted by congestion and interstitial edema, gastric submucosal periglandular edema, and increased mediator release. At 24 to 48 hours postinjection, a late-phase inflammatory response characterized by gastric submucosal epithelial vacuolar degeneration with a neutrophilic, eosinophilic, and mononuclear cell infiltrate was observed. These combined findings confirm the clinical and immunological presence of food allergy, and will continue to provide valuable data for the characterization of the underlying mechanisms of food allergy.

In more recent studies, thioredoxin-reduced food allergenicity (Buchannan et al., 1997) provided evidence that thioredoxin treatment of milk, beta-lactoglobulin (BLG), and wheat enhanced pepsin- or gastric-simulated digestion and lowered allergenicity. In the wheat studies, in which allergenic strength was assessed by skin testing, thioredoxin treatment led to increased susceptibility to proteolysis by reducing the intramolecular (intrachain) disulfide bonds. This led to unfolding the protein for denaturation suggesting that the thioredoxin system might be useful in the production of hypoallergenic digestible food sources (Buchannan et al., 1997). In a subsequent investigation, skin test comparisons of the allergenicity of major milk proteins (alpha-lactoglobulin, BSA, BLG, and the alpha-, beta- and kappa-caseins), BLG was identified for 80% of the total activity in milk (del Val et al., 1999). Thioredoxin treatment of milk or BLG showed decreased ability to elicit allergic responses in milk-sensitized puppies, as evidenced by both skin testing and clinical symptoms of gastrointestinal food allergy. The expense of the thioredoxin treatment and the clinical relevance of the system are still unresolved questions.

Of interest, two dog models are currently being investigated for determining relative allergenicity of various food allergens. On the basis of allergen concentration, allergic sensitization/responses to peanuts was greater than tree nuts, wheat, soy, and barley, in respective order of eliciting food allergic responses by intradermal skin testing in the spaniel/basenji colony (Teuber et al., 2002). Gastrointestinal symptoms were more pronounced in tree nut–sensitized and orally challenged animals and required pharmacological intervention, whereas challenged peanut-sensitive dogs recovered spontaneously. The observed oral challenge results were suggested to be related to protein concentration during sensitization and oral challenge (Teuber et al., 2002). A similar study using the soft-coated wheaten terrier model is currently under investigation (Vaden et al., 2000). Although positive gastroscopic food-sensitivity testing, dietary challenges, and increased IgE levels are evident in this model, careful evaluation of the diagnosis of food allergy is required to determine if food allergies are the cause or result of enteric disease. Significant limitations in the use of ELISA for the detection of IgG anti-IgE and IgE × IgG immune complexes in the evaluation of IgE-mediated disease in dogs and its potential in treatment regimens have been recorded (Hammerberg et al., 1997). In addition, there is a gap in our knowledge regarding the role of the stem cell factor in mast cell activation and inflammatory mediator release (Hammerberg et al., 2001).

18.5.3 Neonatal Swine Model

The pig has a gastrointestinal physiology comparable to that of humans and an immune response that is more accessible, although less immunologically characterized, than that of the mouse or rat. Published data during the last decade have enabled the production of a substantial number of immunological reagents that can be used in the development of the pig as a useful immunological model (Pescovitz, 1998; Haverson et al., 2001). Additionally, the immune responses to ingested proteins in weaned piglets are well documented (Wilson et al., 1989; Dreau et al., 1994; Li et al., 1991; Bailey et al., 1994). As in humans, the immune response to ingested proteins in piglets suggested that these animals mounted an inappropriate immune response to the antigens, but that these responses were subsequently controlled by classical "oral tolerance" pathways. Notably, the piglets' responses to ingested soybeans were comparable with those produced in response to the injection of soybean together with adjuvant (Bailey et al., 1994). The gastrointestinal-associated adverse immunological responses included emesis, diarrhea, bleeding, and weight loss, all of which are associated with human gastrointestinal food allergy. This literature suggested that neonatal infants and piglets exposed to mucosal antigens were likely to respond inappropriately by generating allergy. Thus, the dietary influences on the development of the mucosal immune system in the neonate are thought to be due to an imbalance of effector and regulatory function, which may result in chronic allergy or disease (Bailey et al., 2001).

Although the immunological and gastrointestinal disturbances in the piglets are considered transient, and the development of the mucosal immune system proceeds normally, Helm et al. (2002b) utilized the neonatal pig to investigate food allergy. Gross physical appearance and histopathology investigations supported a relative food-induced sensitization following an oral food challenge. Oral challenges with the respective sensitizing food resulted in emesis, malaise, tremors, convulsions, minor and major body rashes, physical evidence of respiratory distress and anaphylactic shock. Peanut-sensitized animals challenged with a nonrelevant food source or nonpeanut sensitized animals challenged with peanut did not respond with any symptoms of gastrointestinal food allergy (Helm et al., 2002). Skin tests and food challenges to the sensitizing food were administered on alternate weeks over a 14-week period. Immunologically, antigen-specific IgG was identified by ELISA while heat-treated and nonheat-treated serum was investigated by passive cutaneous anaphylaxis, which clearly suggested an IgE-mediated response (Helm et al., 2002). Histologic evidence of vascular congestion, hemorrhage, and epithelial denudation occurred primarily in the small intestine (Helm et al., 2002). Mucus extrusion and submucosal edema of the stomach was also evident by endoscopic analysis in food-sensitized/oral-challenged animals (Helm et al., 2002).

18.6 ADVANTAGES OF USING LARGE ANIMALS

Humans and large animals have developed as outbred populations with their immune systems being shaped by exposure to similar infectious agents and environmental stimuli. The confirmed clinical and immunologic natural food allergies in large animal models is a significant advantage over smaller animal models. With respect to mechanistic studies of IgE-mediated hypersensitivity, the anatomical and physiologic similarities are important considerations for using either the atopic dog or neonatal swine as food allergy models. The manifestations of gastrointestinal food allergy in these animals are more closely aligned to the human response. Consequently, the underlying immunopathogenic mechanisms and therapeutic intervention strategies will more closely extrapolate to human food gastrointestinal allergy. Repeated endoscopic analysis offers a continuum of intestinal dysfunction not available in smaller rodent models (Vaden et al., 2000).

18.7 DISADVANTAGES OF USING LARGE ANIMALS

The principal disadvantage in the use of large animals is that of the size, husbandry, and cost of maintaining a large animal facility. In addition, several species have a privileged status (e.g., as pets) in some societies, and may not be adopted readily for use as experimental models. Limited research reagents, differences in species and strains, and the availability of knockout strains continue to be problematic despite the increasing development and availability of immunologic reagents applicable to animal studies. Differences in developmental, regional, and mucosal immunity within each animal model must be considered when assessing the biological relevance of a particular animal model. Finally, as in all other animal studies, the major limitation is the difficulties encountered in interspecies extrapolation and application of these data to human disease situations, which requires the use of sound, unbiased knowledge of the animal model.

18.8 CONCLUSIONS

The underlying natural mechanisms of sensitization and the allergic response of gastrointestinal food allergies in large animal models will contribute to the immunopathogenesis of IgE-mediated food allergy. The similarities in immunophysiology and pathological features of allergic diseases between the large animals and humans will provide opportunities for mechanistic studies, novel protein allergenicity prediction, and the application of therapeutic intervention strategies.

REFERENCES

Atherton, K.T, Dearman, R.J., and Kimber, I., Protein allergenicity in mice: a potential approach for hazard identification, *Ann. N. Y. Acad. Sci.*, 964, 163–171, 2002.

Bailey, M., et al., Altered immune response to proteins fed after neonatal exposure of piglets to the antigen, *Int. Arch. Allergy Appl. Immunol.*, 103, 183–187, 1994.

Bailey, M., et al., Regulation of mucosal immune responses in effector sites, *Proc. Nutr. Soc.*, 60, 427–35, 2001.

Buchannan, B.B., et al., Thioredoxin-linked mitigation of allergic responses to wheat, *Proc. Natl. Acad. Sci.*, 94, 5372–5377, 1997.

Butler, J.M., et al., Pruritic dermatitis in asthmatic basenji-greyhound dogs: a model for human atopic dermatitis, *J. Am. Acad. Dermatol.*, 8, 33–38, 1983.

del Val, G., et al., Thioredoxin treatment increases digestibility and lowers allergenicity of milk, *J. Allergy Clin. Immunol.*, 103, 690–697, 1999.

Dreau, D., et al., Local and systemic immune responses to soybean protein ingestion in early-weaned pigs, *J. Anim. Sci.*, 72, 2090–2098, 1994.

Environmental Health Perspectives, Mini-monograph: genetically modified food, *Environ. Health Perspect.*, 111, 1110–1141, 2003.

Ermel, R.W., et al., The atopic dog: a model for food allergy, *Lab. Anim. Sci.*, 47, 40–49, 1997.

Food and Agriculture Organization/World Health Organization, Evaluation of Allergenicity of Genetically Modified Foods, Report of a Joint FAO/WHO Expert Consultation on Allergenicity of Foods Derived from Biotechnology, January 22–25, 2001, Rome, Italy, available at: www.who.int/fsf/GmFood/index.htm.

Frankild, S., Basketter, D.A., and Andersen, K.E., The value and limitations of re-challenge in the guinea pig maximization test, *Contact Dermatitis*, 35, 135–140, 1996.

Frick, O.L., Food allergy in atopic dogs, in *New Horizons in Allergy Immunotherapy*, Sehon, A., Hayglass, K.T., and Kraft, D., Eds., Plenum Press, New York, 1996, pp. 1–7.

Gay, J., et al., Involvement of tachykinin receptors in sensitization to cow's milk proteins in guinea pigs, *Gut*, 44, 497–503, 1999.

Gebhard, D., et al., Canine IgE monoclonal antibody specific for a filarial antigen: production by a canine x murine heterohybridoma using B cells from a clinically affected lymph node, *Immunology*, 85, 429–434, 1995.

Halliwell, R.E., Atopic disease in the dog, *Vet. Rec.*, 89, 209–214, 1971.

Hammerberg, B., et al., Auto IgG anti-IgE and IgG x IgE immune complex presence and effects of ELISA-based quantitation of IgE in canine atopic dermatitis, demodectic acariasis and helminthiasis, *Vet. Immunol. Immunopathol.*, 60, 33–46, 1997.

Hammerberg, B., Olivry, T., and Orton, S.M., Skin mast cell histamine release following stem factor and high-affinity immunoglobulin E receptor cross-linking in dogs with atopic dermatitis, *Vet. Dermatol.*, 12, 339–344, 2001.

Haverson, K., et al., Overview of the third international workshop on swine leukocyte differentiation antigens, *Vet. Immunol. Immunopathol.*, 80, 5–23, 2001.

Hayashiya, S., et al., Expression of T helper 1 and T helper 2 cytokine mRNAs in freshly isolated peripheral blood mononuclear cells from dogs with atopic dermatitis, *J. Vet. Med.*, 49, 27–31, 2002.

Hein, W.R., and Griebel, P.J., A road less traveled: large animal models in immunological research, *Nat. Rev.*, 3, 79–84, 2003.

Helm, R.M., Food allergy animal models. An overview, *Ann. N. Y. Acad. Sci.*, 964, 139–150, 2002a.

Helm, R.M., et al., A neonatal swine model for peanut allergy, *J. Allergy Clin. Immunol.*, 109, 136–142, 2002b.

Jackson, H.A., Orton, S.M., and Hammerberg, B., IgE is present on peripheral blood monocytes and B cells in normal dogs and dogs with atopic dermatitis but there is no correlation with serum IgE concentrations, *Vet. Immunol. Immunopathol.*, 85, 225–232, 2002.

Jeffers, J.G., Meyer, E.K., and Sosis, E.J., Responses of dogs with food allergies to single-ingredient dietary provocation, *J. Am. Vet. Med. Assoc.*, 209, 608–611, 1996.

Kimber, I., et al., Assessment of allergenicity based on immune reactivity: animal models, *Environ. Health Perspect.*, 111, 1125–30, 2003.

Kitagawa, S., et al., Relative allergenicity of cow's milk and cow's milk-based infant formulas in an animal model, *Am. J. Med. Sci.*, 310, 183–187, 1995.

Knippels, L.M., and Penninks, A.H., Assessment of protein allergenicity: studies in Brown Norway rats, *Ann. N. Y. Acad. Sci.*, 964, 151–161, 2002.

Li, D.F., et al., Interrelationship between hypersensitivity to soybean and proteins to growth performance in early-weaned pigs, *J. Anim. Sci.*, 69, 4062–4069, 1991.

Luis, S., et al., Neuroimmune interactions in guinea pig stomach and small intestine, *Am. J. Physiol. Gastrointestinal Liver Physiol.*, 284, G154–164, 2003.

Maeda, S., et al., Lesional expression of thymus and activation-regulated chemokine in canine atopic dermatitis, *Vet. Immunol. Immunopathol.*, 88, 79–87, 2002.

Nuttal, T.J., et al., T-helper 1, T-helper 2 and immunosuppressive cytokines in canine atopic dermatitis, *Vet. Immunol. Immunopathol.*, 87, 379–384, 2002.

Out, T.A., et al., Local T-cell activation after segmental allergen challenge in the lungs of allergic dogs, *Immunology*, 105, 499–508, 2002.

Patterson, R., Investigations of spontaneous hypersensitivities of the dog, *J. Allergy*, 31, 351–363, 1960.

Pescovitz, M.D., Immunology of the pig, in *Handbook of Veterbrate Immunology*, Pastoret, P.-P., Greibel, P., Bazin, H., et al., Eds., Academic Press, London, 1998.

Piacentini, G.L., et al., Ability of a new infant formula prepared from partially hydrolyzed bovine whey to induce anaphylactic sensitization: evaluation in a guinea pig model, *Allergy*, 49, 361–364, 1994.

Redman, T.K., et al., Pulmonary immunity to ragweed in a beagle dog model of allergic asthma, *Exp. Lung Res.*, 27, 433–451, 2001.

Reedy, L.M., Miller, W.H., Willemese, T., *Allergic Skin Diseases in Dogs and Cats*, W.B. Saunders, Philadelphia, 1997.

Sinke, J.D., Rutten, V.P., and Willemse, T., Immune dysregulation in atopic dermatitis, *Vet. Immunol. Immunopathol.*, 87, 351–356, 2002.

Sinke, J.D., Immunophenotyping of skin-infiltrating T-cells subsets in dogs with atopic dermatitis, *Vet. Immunol. Immunopathol.*, 57, 13–23, 1997.

Tang, L., et al., Recombinant canine IL-13 receptor alpha2-Fc fusion protein inhibits canine allergen-specific IgE production in vitro by peripheral blood mononuclear cells from allergic dogs, *Vet. Immunol. Immunopathol.*, 83, 115–122, 2001.

Terpend, K., et al., Intestinal barrier function and cow's milk sensitization in guinea pigs fed milk or fermented milk, *J. Pediatr. Gastroenterol. Nutr.*, 28, 191–198, 1999.

Teuber, S.S., et al., The atopic dog as a model of peanut and tree nut food allergy, *J. Allergy Clin. Immunol.*, 110, 921–927, 2002.

Vaden, S.L., et al., Food hypersensitivity in soft coated wheaten terriers with protein-losing enteropathy or protein-losing nephropathy or both: gastroscopic food sensitivity testing, dietary provocation, and fecal immunoglobulin E, *Eur. J. Vet. Med.*, 14, 60–67, 2000.

van Och, F.M., et al., Comparison of dose-responses of contact allergens using the guinea pig maximization test and the local lymph node assay, *Toxicology*, 167, 207–215, 2001.

Wilson, A.S., Stokes, C.R., and Bourne, F.J., Effect of age on absorption and immune responses to weaning or introduction of novel dietary antigens in pigs, *Res. Vet. Sci.*, 46, 180–186, 1989.

Witich, F.W., Spontaneous allergy (atopy) in the lower animal-seasonal hay fever (fall type) in a dog, *J. Allergy*, 12, 247–251, 1941.

Zemann, B., et al., Allergic pulmonary and ocular tissue responses in the absence of serum IgE antibodies (IgE) in an allergic dog model, *Vet. Immunol. Immunopathol.*, 87, 373–378, 2002.

CHAPTER 19

Recent Developments in Allergic Contact Dermatitis

Ian Kimber and Rebecca J. Dearman

CONTENTS

19.1 INTRODUCTION

Skin sensitization resulting in allergic contact dermatitis is an important and not uncommon occupational health problem. Many hundreds of chemicals are known to have the potential to cause skin sensitization, and allergic contact dermatitis is the most common manifestation of immunotoxicity in humans. The immunology and immunotoxicology of skin sensitization have been considered elsewhere in recent years (Basketter et al., 1999a; Dearman and Kimber, 2003; Grabbe and Schwarz, 1998; Kimber et al., 2002a; Kimber and Dearman, 2003), and the purpose of this chapter is not to provide another general review. Our aim here is rather to first identify briefly recent advances in our understanding of the immunoregulatory mechanisms that might serve to negatively influence the induction and expression of skin sensitization, and second to survey developments in approaches available for hazard identification and characterization.

19.2 RECENT DEVELOPMENTS IN THE IMMUNOBIOLOGY OF ALLERGIC CONTACT DERMATITIS

Allergic contact dermatitis is often considered to be an example of a cell-mediated type IV hypersensitivity reaction. Such a classification is accurate in part, insofar as: (1) reactions are relatively slow to develop (over a period of hours or days, usually requiring 24 hours for clear symptoms to emerge), compared with other forms of allergic hypersensitivity reactions, and (2) reactions are effected by T-lymphocytes. However, there are differences between contact hypersensitivity and delayed-type hypersensitivity to proteins, primarily with respect to the requirement for discrete molecular events.

Before identifying some recent developments in our appreciation of the immunobiological mechanisms that serve to regulate skin sensitization, it is appropriate to provide an overview of the process itself. By definition, skin sensitization is acquired when an (inherently susceptible) subject is exposed topically to a chemical allergen. The level of exposure must be sufficient to provoke a cutaneous immune response of a magnitude necessary to result in immunological priming for the relevant allergen. That is, the immune response must be of sufficient vigor to effect a level of sensitization that will result in the elicitation of a cutaneous hypersensitivity reaction if the subject is exposed subsequently to the same chemical allergen at the same, or a different, skin site. Such elicitation reactions are triggered by the accumulation of allergen-specific memory/effector T-lymphocytes at the site of contact (the challenge site). These cells become activated, and are stimulated to release a variety of factors that stimulate the infiltration of other leukocytes. These cells and their cellular products act in concert to provoke a local inflammatory reaction in the skin that is recognized clinically as allergic contact dermatitis.

There is much interest in the characteristics of the effector T-lymphocytes that are responsible for the elicitation of allergic contact dermatitis. The picture is complicated, but the available information suggests that according to circumstances, a variety of cells may contribute to the initiation and development of cutaneous inflammatory reactions. Nevertheless, the weight of evidence suggests that in many instances the most important effectors are CD8+ T-lymphocytes of a type 1 phenotype. However, this is not a universal rule and in practice other functional subpopulations of T-lymphocytes may contribute also, according to the characteristics of the contact allergen itself and the conditions under which the chemical was encountered by the immune system (Kimber and Dearman, 2002).

There has been considerable attention given to the mechanisms through which cutaneous immune responses are induced by encounter with chemical allergens on the skin. The cells that play a pivotal role in this process, and that orchestrate the processing and presentation of chemical allergens, are Langerhans cell (LC). These cells (part of a wider family of immunologically active dendritic cells [DC]) reside in the epidermis where they are thought to serve as sentinels for the adaptive immune system, surveying the skin for changes in the local microenvironment, and providing information about incursions by foreign antigen. In practice, LC recognize, internalize, and process allergens, probably following the formation of a stable association between host protein and parent chemical, or with a protein-reactive metabolite. Concurrently, a proportion of LC local to the site of antigen exposure, including antigen-bearing cells, become mobilized, and are induced to migrate from the epidermis, via afferent lymphatics, to regional (skin draining) lymph nodes. Here they locate within the paracortical region of the nodes where they come into close contact with T-lymphocytes. During transit to the lymph nodes, LC lose their antigen-processing properties and acquire instead the immunostimulatory phenotype of mature DC that are able effectively to present antigen to responsive T-lymphocytes. Allergen-activated T-lymphocytes are stimulated to divide and differentiate; cell division results in a clonal expansion of allergen-responsive T-lymphocytes that, if of sufficient magnitude, provides the cellular basis for the acquisition of skin sensitization. The activity of LC, their mobilization in response to dermal trauma and antigen exposure, and their directed movement from the skin and subsequent localization within regional lymph nodes, are processes now known to be initiated and regulated by cytokines and chemokines. Our understanding of the pivotal roles played by such molecules during the acquisition of skin sensitization has been reviewed in detail elsewhere (Cumberbatch et al., 2000, 2003a, 2004; Kimber et al., 1998a, 2000; Raju, 2004; Sebastiani et al., 2002), and a similar treatment of the subject is beyond the scope of this chapter. Previously, attention has focused largely on the positive immunostimulatory roles played by cytokines and other factors in the development of skin sensitization and allergic contact dermatitis. However, there is a growing body of evidence for the existence of negative counterregulatory mechanisms that, presumably, provide some measure of homeostasis.

One likely candidate for such activity is interleukin 10 (IL-10), a product of epidermal cells that is upregulated during skin sensitization (Enk and Katz, 1992), and that is known to inhibit skin inflammation (Berg et al., 1995). One series of investigations implicated cutaneous IL-10 as a negative regulator of LC migration during skin sensitization. The evidence indicated that IL-10 is able to inhibit the local production in skin of the proinflammatory cytokines tumor necrosis factor-α (TNF-α) and interleukin 1β (IL-1β) that are required for LC mobilization during skin sensitization (Wang et al., 1999). In addition to affecting epidermal LC function, IL-10 may also be instrumental in the development of T-lymphocytes that are able to negatively regulate contact hypersensitivity responses (Maurer et al., 2003). Two other cytokine family members have been shown to have the potential to inhibit contact hypersensitivity. One of these is the recently described T-lymphocyte cytokine interleukin 21 (IL-21). It was found that pretreatment of DC with this factor reduced their ability to stimulate CD8+ T-lymphocyte proliferative responses *in vitro*, and to induce contact sensitization *in vivo* (Brandt et al., 2003). The second potential regulator is the naturally occurring inhibitor of interleukin 18 (IL-18), known as IL-18 binding protein (IL-18BP). This secreted protein binds with high affinity to IL-18, and prevents the interaction of the cytokine with its natural membrane receptor. Previous studies have suggested that IL-18 plays an important role in one or more aspects of contact sensitization, and therefore it is not entirely unexpected that recent investigations have demonstrated an inhibitory role for IL-18BP (Plitz et al., 2003).

Another molecule that may contribute to the regulation of skin sensitization is lactoferrin (LF), an iron-binding protein that is found in exocrine secretions and also in the skin. It has now been demonstrated, both in humans and in mice, that topical application of LF is able to inhibit contact allergen-induced migration, secondary to an inhibition of the *de novo* production of cutaneous TNF-α (Cumberbatch et al., 2003b; Kimber et al., 2002b). The speculation is that LF provides a molecular mechanism for regulating proinflammatory cytokine responses, and that one consequence of this in the skin is to impose a constraint on the levels and longevity of one or more of the signals that stimulate LC mobilization following skin sensitization.

Finally, there is increasing interest in the possibility that interactions exist between the immune and nervous systems that influence immunological responses, including skin sensitization. One example is provided by recent investigations in which it was revealed that if nerve fibers closely associated with epidermal LC were eliminated, then there was a resultant inhibition of contact sensitization (Beresford et al., 2004).

Taken together, the available data suggest that there are a variety of ways in which immune responses required for the induction and elicitation of contact hypersensitivity can be negatively regulated. The assumption is that such mechanisms play an important physiological role in preventing excessive and undesirable immune responses in the skin.

19.3 RECENT DEVELOPMENTS IN METHODS FOR HAZARD CHARACTERIZATION

19.3.1 Local Lymph Node Assay for Hazard Identification

One of the most significant recent advances in the area of hazard identification of contact sensitizers has been development of the local lymph node assay (LLNA), a test that has now been validated as an alternative to existing guinea pig tests (Gerberick et al., 2000). As part of the validation and acceptance process, LLNA data were scrutinized by a peer review panel appointed by the Interagency Coordinating Committee on the Validation of Alternative Methods, an organization established in the United States by 14 separate federal regulatory agencies to harmonize the development, validation, and acceptance of new toxicological methods. The conclusions reached were that, compared with other predictive methods, the LLNA offers important animal welfare advantages in terms of both reduction and refinement, and is suitable for use as a stand-alone alternative for the purposes of hazard identification (i.e., a method that can be used not only for the identification of skin sensitization hazard, but also for confirmation of the absence of such hazard) (Dean et al., 2001; Haneke et al., 2001; National Institutes of Health, 1999; Sailstad et al., 2001). As a result, the assay has now been adopted formally by several regulatory agencies in the United States. A similar endorsement by the European Centre for the Validation of Alternative Methods followed (Balls and Hellsten, 2000). The LLNA has now been incorporated into Test Guideline No. 429 by the Organization for Economic Cooperation and Development (2002). In parallel, the European Union has prepared a new test guideline for the LLNA (B.42).

In contrast to the guinea pig tests used previously, sensitizing activity is assessed as a function of events occurring in the induction, rather than of the elicitation, phase of sensitization. Specifically, the LLNA measures the ability of topically applied chemicals to induce proliferative responses in lymph nodes draining the site of exposure (Kimber et al., 1994). This is a relevant read-out for the evaluation of skin sensitizing potential as the activation and clonal expansion of allergen-responsive T-lymphocytes is the central event in the acquisition of skin sensitization, with the vigor of lymph node cell proliferative responses being a major factor in determining the extent of sensitization (Kimber and Dearman, 1991).

The standard assay is conducted as follows. Groups of mice (CBA strain) receive topical applications, on the dorsum of both ears, of various concentrations of the test chemical (or of the relevant vehicle control) daily for 3 consecutive days. Five days after the initiation of exposure, mice receive an intravenous injection of tritiated thymdine (^{3}H-TdR). Animals are sacrificed 5 hours later and draining (auricular) lymph nodes excised. These are either pooled for each experimental group (Kimber and Basketter, 1992), or alternatively, are pooled on a per animal basis (Gerberick et al., 1992). Single cell suspensions of LNC are prepared, and the cells washed and suspended in trichloroacetic acid (TCA) for a minimum of 12 hours at 4°C. Precipitates are resuspended in TCA, transferred to a scintillation fluid and the incorporation by LNC of ^{3}H-TdR measured by scintillation counting. Data are recorded as disintegrations per minute (dpm) for each experimental group, or for each animal. There is currently some debate about the need for positive controls for the LLNA, and whether such controls should be incorporated routinely into each experiment, or should be conducted by laboratories at regular intervals (Basketter et al., 2002). If it is necessary to include a positive control, then the most suitable chemical allergen for this purpose is hexyl cinnamic aldehyde (HCA) (Dearman et al., 1998, 2001).

For each concentration of chemical tested in the LLNA, a stimulation index (SI) is derived using as the denominator the value obtained with the concurrent vehicle control. Skin sensitizers are defined as chemicals that at one or more test concentrations, induce a threefold or greater increase in LNC proliferation compared with concurrent vehicle controls (an SI of ≥ 3). This threshold for classification of chemicals as sensitizers was based on initial investigations of both contact allergens and nonsensitizing chemicals, and was confirmed by continued and more extensive experience with a wider range of chemicals. Furthermore, a retrospective analysis of results obtained in the LLNA with over 130 chemicals has now been conducted. The data were subjected to a rigorous mathematical assessment using receiver operator characteristic (ROC) curves in order to derive the threshold value that would discriminate most accurately between sensitizing and non-sensitizing chemicals. The conclusion was that an SI value of 3 provides the most appropriate basis for identifying chemicals as skin sensitizers (Basketter et al., 1999b). Although hazard identification based on an SI value of ≥ 3 is of proven utility, there are circumstances where some flexibility is required. For example, if a chemical causes a dose-related increase in proliferation, but just fails at the highest concentration tested to elicit an SI of 3, it would be unwise to draw the automatic conclusion that the material lacked sensitizing potential. The sensible course of action would be to perform a repeat analysis, using, if possible, higher concentrations of the chemical and/or a different vehicle.

The LLNA has been the subject of both national and international interlaboratory trials (Kimber and Basketter, 1992; Basketter et al., 1991, 1996; Kimber et al., 1991, 1995, 1998b; Loveless et al., 1996; Scholes et al., 1992), and has been compared extensively with guinea pig predictive tests and human contact sensitization data (Basketter and Scholes, 1992; Basketter et al., 1991, 1992, 1993, 1994; Kimber et al., 1990, 1994; Ryan et al., 2000). Collectively, these investigations incorporated analyses of a wide variety of chemical allergens and nonsensitizing chemicals, and demonstrated that the method is robust and that equivalent results are generated in independent laboratories. In addition, the sensitivity, selectivity, and overall accuracy of the LLNA were found to be comparable with, or better than, commonly used guinea pig tests. However, no test method employed for the purposes of toxicological evaluation and safety assessment will provide unequivocal results with 100% accuracy. One issue regarding the performance of the LLNA is the potential for false-positive results. This stems in part from the fact that the nonsensitizing skin irritant sodium lauryl sulfate (SLS) has been found in some instances to elicit weak, but nonetheless positive, responses (Kimber et al., 1994; Loveless et al., 1996; Montelius et al., 1994). However, the majority of nonsensitizing skin irritants fail to cause positive responses in the LLNA (Basketter et al., 1998; Gerberick et al., 1992; Kimber et al., 1991, 1995). SLS may be something of a special case, as there is evidence in mice that topical exposure to this chemical causes the migration of epidermal LC and their accumulation in skin draining lymph nodes, presumably as a result of induced epidermal cytokine expression (Cumberbatch et al., 1993). The fact that some skin irritants, including SLS, are able at high doses to stimulate relatively low-level activity in the LLNA, does not necessarily mean that the accurate identification of potential contact allergens will be prevented. Experience with the assay has shown that there are a number of criteria that can be used to exclude potential false positives (including lack of a structural alert, possession of significant irritant activity, and weakly positive at high doses) that ensure that this is not a problem in practice (Basketter et al., 1998).

19.3.2 Local Lymph Node Assay for Risk Assessment

The identification of potential hazard is an important first step in any toxicological evaluation. For accurate assessment of risks to human health, however, it is necessary to integrate information on the likely conditions and extent of exposure with an understanding of potency. In the context of defining the risk of inducing contact sensitivity, potency is defined as a function of the amount of chemical required for the acquisition of skin sensitization. From experimental studies in humans, it is clear that the important parameter in this respect is the amount of chemical experienced per unit area of skin, rather than the total amount of chemical to which a subject is exposed (Friedmann, 1990). There has therefore been considerable interest recently in defining how the LLNA can be used to measure experimentally the relative potency of contact allergens (Basketter et al., 2001a; Kimber and Basketter, 1997; Kimber et al., 2001, 2003). The acquisition of skin sensitization is dependent on the stimulation of a cutaneous immune response in lymph nodes draining the site of exposure to the inducing chemical allergen, with the extent of sensitization achieved being a function of the clonal expansion of allergen-specific T-lymphocytes. Consequently, the endpoint of the LLNA, the overall vigor of LNC proliferative responses induced following topical exposure to a chemical allergen, is believed to provide a correlate of skin-sensitizing activity (Kimber and Dearman, 1991; Kimber et al., 1999).

The approach taken is to derive mathematically from dose responses in the LLNA an EC3 value (estimated concentration of chemical required to provoke an SI of 3); the minimum amount of chemical that is required to induce a threshold positive response in the assay. It is possible to express EC3 values in terms of total amount of chemical applied, concentration of chemical (either as a percentage or molar value), or amount of chemical applied per unit area of skin. In practice, however, percentage concentrations are used most commonly. In initial studies, three possible mathematical approaches were compared: quadratic regression analysis, Richard's model, and simple linear interpolation. The conclusion reached was that linear interpolation between values on either side of the threefold SI on a LLNA dose–response curve provides the most robust and most convenient method for the routine calculation of EC3 values (Basketter et al., 1999c). This approach can be expressed mathematically as

$$EC3 = c + [(3 - d)/(b - d)] \times (a - c)$$

where (a, b) and (c, d) are the coordinates, respectively, of data points lying immediately above and immediately below the SI value of 3.

It has been demonstrated that EC3 values are very robust parameters of sensitizing activity, both with respect to time within a single laboratory, and between independent laboratories. Using the recommended positive control material, HCA, studies conducted in one laboratory over a period of 10 months revealed that EC3 values were extremely consistent (ranging between 6.9% and 9.6%) (Dearman et al., 1998). A similar level of consistency was observed in experiments conducted with *p*-phenylenediamine (PPD) over a period of 4 months (Warbrick et al., 1999a). In addition, comparisons between laboratories have demonstrated that similar EC3 values are obtained when the same material is analyzed in separate locations (Kimber et al., 1995, 1998b; Loveless et al., 1996; Warbrick et al., 1999a).

There is now considerable experience in the use of EC3 values in characterizing the relative skin-sensitizing potency of groups of related chemicals (Basketter et al., 1997, 1999d, 2001c). However, the most important test is the correlation between skin-sensitizing potency of chemical allergens measured in the LLNA (EC3 values) and what is known of their relative activity in humans. This issue has been addressed by collaborating with experienced clinical dermatologists who provided a view as to the relative skin-sensitizing potency of two series of known human contact allergens. These classifications of relative potency based on clinical judgment were compared with EC3 values derived from LLNA dose responses. For both series of chemicals, a very close correlation was found between clinical assessments of potency and EC3 values, with the chemicals considered to be the most active contact allergens in humans having the lowest EC3 values (relatively small amounts of compound required to induce a threshold positive response) (Basketter et al., 2000; Gerberick et al., 2001).

Chemical allergens differ by several orders of magnitude in terms of their relative potency, and therefore with respect to EC3 values. For example, EC3 values of 0.08% and 35% have been recorded, respectively, for the potent contact allergen 2,4-dinitrochlorobenzene (DNCB), and the considerably weaker allergen ethyleneglycol dimethacrylate (Basketter et al., 2000). The view currently is that classification schemes should be predicated on no less than ten-fold differences in EC3 values, as it will be neither necessary, nor helpful, to measure small (and probably biologically insignificant) differences between chemical allergens in terms of EC3 values. The most recent proposal for a classification scheme for contact allergens based on LLNA data derives from a recent European Centre for Ecotoxicology and Toxicology Task Force charged with the objective of recommending approaches for the measurement of potency and the definition of thresholds for both the induction and elicitation of contact sensitization (Kimber et al., 2003). The recommendation was that four categories would be used, based on percentage concentration EC3 values, and identified by the descriptors extreme, strong, moderate, and weak. The suggestion was that the scheme should distinguish between contact allergens on the basis of ten-fold differences in potency as follows: extreme, EC3 < 0.1; strong, EC3 > 0.1 to < 1; moderate, EC3 > 1 to < 10; and weak, < 10 to > 100.

As described above, EC3 values can be used to measure differences between contact allergens with respect to sensitizing potency. It is important to note, however, that the matrix in which a chemical is encountered at skin surfaces may influence the effectiveness of sensitization. It is clear that the vehicle can have an important influence on sensitization (Cumberbatch et al., 1993; Dearman et al., 1996; Heylings et al., 1996), and this is reflected in derived EC3 values (Basketter et al., 2001b; Lea et al., 1999; Warbrick et al., 1999b; Wright et al., 2001). Although the vehicle matrix in which a chemical is encountered at skin surfaces may impact on potency assessment, the choice of vehicle in the context of hazard assessment is, however, unlikely to affect the accuracy with which skin sensitization hazard can be identified using the LLNA. In addition, it is not yet possible to draw any general conclusions regarding the influence of vehicle matrix on the effectiveness of skin sensitization. Experience to date suggests that the influence of the vehicle matrix may vary significantly with the physicochemical properties and dose of the chemical allergen, and may impact on sensitization through one or more of several independent mechanisms (including skin penetration or epidermal cytokine expression).

19.3.3 Alternative Endpoints for the Local Lymph Node Assay

There has been some interest in the development of alternative endpoints for the LLNA, including those that assess turnover of LNC without the use of radiolabeled isotopes. One such is the measurement of proliferation as a function of the *in vivo* cellular incorporation of the nonradioisotopic analogue of thymidine, bromodeoxyuridine (BrdU), which can be detected using an anti-BrdU antibody (Takeyoshi et al., 2001, 2003). Limited experience of this method suggests that it may have some potential for the identification of contact allergens, although the sensitivity with respect to allergen-induced fold changes in BrdU incorporation is less than that achieved using radiolabeled thymidine. With the reference contact allergen HCA, for example, the SIs achieved at the top dose tested (50%) were 14.4 and 3.6, using *in situ* incorporation of thymidine and BrdU, respectively (Dearman et al., 2001; Takeyoshi et al., 2003).

A novel method for tracking cell division that has been applied recently to the LLNA utilizes the fluorescent label carboxyfluorescein succinimidyl ester (CFSE) (Humphreys et al., 2003). LNC are cultured with the precursor molecule 5,6-carboxyfluorescein diacetate succinimidyl ester (CFSASE), a membrane permeable chemical that is cleaved intracellularly to CFSE, which binds stably to cytosolic proteins and is retained within the cell (Lyons and Parish, 1994).

It had been demonstrated that following cell division, CFSE divided equally between daughter cells in each successive division, with up to ten discrete populations detectable on the basis of decreasing fluorescence intensity (Lyons and Parish, 1994).

An advantage of this method is that not only can proliferating cells be identified in complex cell mixtures, but also the use of two or three color immunofluorescence allows immunophenotyping of the dividing cells using monoclonal antibodies conjugated to other fluorescent dyes. Dose–response studies have been performed with the contact allergens dinitrochlorobenzene (DNCB), HCA, and oxazolone and the nonsensitizing irritant methyl salicylate (MS) using the LLNA exposure protocol with the CFSE endpoint (Humphreys et al., 2003). For DNCB and HCA, the sensitivity of this technique was comparable with that achieved using the conventional LLNA protocol, with significant increases in proliferating cells recorded at 0.05% DNCB and 5% HCA, doses close to EC3 values obtained in the standard LLNA (Loveless et al., 1996). For oxazolone, however, the measurement of cell turnover using CFSE incorporation appeared to be rather less sensitive, with the numbers of proliferating cells returning to background levels (similar to those observed for cells derived from concurrent vehicle-treated control animals) following treatment of mice with 0.01% oxazolone. This concentration of chemical is some orders of magnitude higher than the EC3 value derived using a conventional LLNA (Loveless et al., 1996). Selectivity of this endpoint was determined using the nonsensitizing skin irritant MS. At all concentrations tested, this chemical failed to stimulate increased cell division measured as a function of CFSE incorporation. These preliminary experiments suggest that this method is sufficiently sensitive to identify potent skin-sensitizing chemicals. Further evaluation will be required, however, to confirm the sensitivity and selectivity of this technique using a range of allergens and nonallergens.

An interesting strategy for the provision of a supplementary or alternative endpoint for the LLNA that does not rely on the measurement of cell division is to characterize instead changes in the cellular composition of draining lymph nodes during responses to contact allergens. Although contact hypersensitivity is a T-lymphocyte-mediated reaction, it has been demonstrated that the draining lymph nodes of mice exposed topically to contact allergens display an increased frequency of B-lymphocytes measured as a function of B220+ cells or IgG/IgM+ cells (Gerberick et al., 1997, 1999, 2002; Manetz and Meade, 1999; Sikorski et al., 1996). Of particular interest is the observation that exposure of mice to contact allergens, but not to nonsensitizing skin irritants, causes an increase in B220+ cell numbers. As described previously, certain nonsensitizing skin irritants have been found to elicit low-level responses in conventional LLNA. Although such positive responses do not in practice cause problems with regard to accurate hazard identification (Basketter et al., 1998), there is, nevertheless, some interest in discriminating experimentally between allergens and irritants. A classification model has been proposed in which chemicals that cause a defined increase in the percentage of B220+ cells (such that the ratio of the percentage of B220+ cells in test lymph nodes compared with lymph nodes isolated from vehicle controls is ≥ 1.25) are classified as allergens. Those chemicals that fail to cause an increase in B220+ cells compared with controls, or where the ratio is less than 1.25, are considered to lack sensitizing activity (Gerberick et al., 2002). Comparable results have been generated in independent laboratories, and the proposed model incorporating a cut-off value of 1.25 has an overall accuracy of 93% (Gerberick et al., 2002). To date, however, performance has been judged on the basis of data obtained with a relatively small number of chemicals. Furthermore, there is a relatively small dynamic range for observed increases in B220+ cells. Thus, exposure even to very potent contact allergens such as DNCB provoked only a maximal threefold increase in B220+ cell frequency (Gerberick et al., 2002), whereas when using the conventional LLNA, maximal SIs in excess of 75 have been recorded for potent allergens including DNCB (Loveless et al., 1996). Further evaluation will require a more extensive analysis, with a larger number of chemical allergens and chemical irritants of varying potency.

Other markers of activation have been identified in regional lymph nodes during skin sensitization, including the increased expression and secretion of various chemokines and cytokines. There has been some interest in examining whether chemical-induced changes in the expression of certain cytokines might provide the basis of an alternative readout for the LLNA, with interleukins 2, 6, and 12 (IL-2, IL-6, and IL-12) and interferon γ (IFN-γ) among the cytokines that have been considered (Dearman and Kimber, 1994; Dearman et al., 1993, 1994, 1999; Hariya et al., 1999). However, the measurement of cytokine responses (at least with respect to cytokines studied to date) appears not to provide a realistic alternative to the standard read-out of the LLNA. We have hypothesized that more holistic assessments of draining lymph node activation using transcription profiling to characterize changes in expression of multiple genes may identify more sensitive and selective markers of immune stimulation than cytokine expression (Pennie and Kimber, 2002; Betts et al., 2003). Initial studies have revealed a number of allergen-induced changes in gene expression, including upregulated expression of onzin and guanylate binding protein 2, and downregulation of mRNA levels for glycosylation-dependent cell adhesion molecule 1 (GlyCAM-1), a molecule that is expressed by endothelial cells in the venules of peripheral lymph nodes, and which is involved in the regulation of lymphocyte homing (Betts et al., 2003). To date, however, such changes appear to lack the dynamic range necessary to provide the required levels of sensitivity in the context of an alternative endpoint for the LLNA. Nevertheless, this is an attractive approach, and as the technology continues to evolve and allow analysis of increasing numbers of genes, the search for potential alternative markers continues.

19.4 CONCLUSIONS

There is no doubt that the last two decades have witnessed a substantial increase in our understanding of the cellular and molecular events through which skin sensitization is acquired, and through which contact hypersensitivity reactions are provoked. In parallel with an improved appreciation of the relevant immunobiological mechanisms have come opportunities to design and develop new methods for safety assessment. Much has been achieved, and our ability to identify skin sensitization hazards and to determine accurately likely risks to human health continues to improve. However, important challenges remain, among them the continued exploitation of our understanding of skin sensitization to refine further methods available for effective safety assessment.

REFERENCES

Balls, M., and Hellsten, E., Statement on the validity of the local lymph node assay for skin sensitisation testing, ECVAM Joint Research Centre, European Commission, Ispra, *Alternatives Lab. Anim.*, 28, 366–367, 2000.

Basketter, D.A., Gerberick, G.F., and Kimber, I., Strategies for identifying false positive responses in predictive skin sensitization tests, *Food Chem. Toxicol.*, 36, 327–333, 1998.

Basketter, D.A., Gerberick, G.F., and Kimber, I., Measurement of allergenic potential using the local lymph node assay, *Trends Pharmacol. Sci.*, 22, 264–265, 2001a.

Basketter, D.A., Gerberick, G.F., and Kimber, I., Skin sensitization, vehicle effects and the local lymph node assay, *Food Chem. Toxicol.*, 39, 621–627, 2001b.

Basketter, D.A., and Scholes, E.W., Comparison of the local lymph node assay with the guinea-pig maximization test for detection of a range of contact allergens, *Food Chem. Toxicol.*, 60, 65–69, 1992.

Basketter, D.A., Scholes, E.W., and Kimber, I., The performance of the local lymph node assay with chemicals identified as contact allergens in the human maximization test, *Food Chem. Toxicol.*, 32, 543–547, 1994.

Basketter, D.A., et al., Interlaboratory evaluation of the local lymph node assay with 25 chemicals and comparison with guinea pig test data., *Toxicol. Methods*, 1, 30–43, 1991.

Basketter, D.A., et al., Sulphanilic acid: divergent results in the guinea pig maximization test and the local lymph node assay, *Contact Dermatitis*, 27, 209–213, 1992.

Basketter, D.A., et al., Results with OECD recommended positive control sensitisers in the maximization, Buehler and local lymph node assays, *Food Chem. Toxicol.*, 31, 63–67, 1993.

Basketter, D.A., et al., The local lymph node assay: a viable alternative to currently accepted skin sensitisation tests, *Food Chem. Toxicol.*, 34, 986–997, 1996.

Basketter, D.A., et al., Dinitrohalobenzenes: evaluation of relative skin sensitization potency using the local lymph node assay, *Contact Dermatitis*, 36, 97–100, 1997.

Basketter, D.A., et al., *Toxicology of Contact Dermatitis: Allergy, Irritancy and Urticaria*, Wiley, Chichester, 1999a.

Basketter, D.A., et al., Threshold for classification as a skin sensitizer in the local lymph node assay: a statistical evaluation, *Food Chem. Toxicol.*, 37, 1167–1174, 1999b.

Basketter, D.A., et al., A comparison of statistical approaches to the derivation of EC3 values from local lymph node assay dose responses, *J. Appl. Toxicol.*, 19, 261–266, 1999c.

Basketter, D.A., et al., Skin sensitization risk assessment: a comparative evaluation of 3 isothiazolinone biocides, *Contact Dermatitis*, 40, 150–154, 1999d.

Basketter, D.A., et al., Use of the local lymph node assay for the estimation of relative contact allergenic potency, *Contact Dermatitis*, 42, 344–348, 2000.

Basketter, D.A., et al., Human potency predictions for aldehydes using the local lymph node assay, *Contact Dermatitis*, 45, 89–94, 2001c.

Basketter, D.A., et al., Local lymph node assay — validation and use in practice, *Food Chem. Toxicol.*, 40, 593–598, 2002.

Beresford, L., et al., Nerve fibres are required to evoke a contact sensitivity response in mice, *Immunology*, 111, 118–125, 2004.

Berg, D.J., et al., Interleukin 10 but not interleukin 4 is a natural suppressant of cutaneous inflammatory responses, *J. Exp. Med.* 182, 99–198, 1995.

Betts, C.J., et al., Assessment of glycosylation dependent cell adhesion molecule 1 (GlyCAM–1) as a correlate of allergen-stimulated lymph node activation, *Toxicology*, 185, 103–117, 2003.

Brandt, K., et al., Interleukin-21 inhibits dendritic cell-mediated T cell activation and induction of contact hypersensitivity in vivo, *J. Invest. Dermatol.*, 121, 1379–1382, 2003.

Cumberbatch, M., et al., Influence of sodium lauryl sulphate on 2, 4-dinitrochlorobenzene-induced lymph node activation, *Toxicology*, 77, 181–191, 1993.

Cumberbatch., M., et al., Langerhans cell migration, *Clin. Exp. Dermatol.*, 25, 413–418, 2000.

Cumberbatch, M., et al., Epidermal Langerhans cell migration and sensitisation to chemical allergens, *APMIS*, 111, 797–804, 2003a.

Cumberbatch, M., et al., IL-1β-induced Langerhans cell migration and TNF-α production in human skin: regulation by lactoferrin, *Clin. Exp. Immunol.*, 132, 352–359, 2003b.

Cumberbatch, M., et al., Langerhans cell migration and the induction phase of skin sensitization, in *Immune Mechanisms in Allergic Contact Dermatitis*, Cavani, A., and Girolomoni, G., Eds., Landes Bioscience, Georgetown, TX, 2004, pp. 29–43.

Dean, J.H., et al., ICCVAM evaluation of the murine local lymph node assay. II. Conclusions and recommendations of an independent scientific peer review panel, *Regul. Toxicol. Pharmacol.*, 34, 258–273, 2001.

Dearman, R.J., and Kimber, I., Cytokine production and the local lymph node assay, in *In Vitro Skin Toxicology*, Rougier, A., Goldberg, A.M., and Maibach, H.I., Eds., Mary Ann Liebert, New York, 1994, pp. 367–372.

Dearman, R.J., and Kimber, I., Factors influencing the induction phase of skin sensitization, *Am. J. Contact Dermatitis*, 14, 188–194, 2003.

Dearman, R.J., et al., Interleukin 6 (IL-6) production by lymph node cells: an alternative endpoint for the murine local lymph node assay, *Toxicol. Methods*, 4, 268–278, 1993.

Dearman, R.J., et al., The local lymph node assay: an interlaboratory evaluation of interleukin 6 (IL-6) production by draining lymph node cells, *J. Appl. Toxicol.*, 14, 287–291, 1994.

Dearman, R.J., et al., Influence of dibutylphthalate on dermal sensitization to fluorescein isothiocyanate, *Fundam. Appl. Toxicol.*, 33, 24–30, 1996.

Dearman, R.J., et al., Temporal stability of local lymph node assay responses to hexyl cinnamic aldehyde, *J. Appl. Toxicol.*, 18, 281–284, 1998.

Dearman, R.J., et al., Cytokine endpoints for the local lymph node assay: consideration of interferon-γ and interleukin 12, *J. Appl. Toxicol.*, 19, 149–155, 1999.

Dearman, R.J., et al., The suitability of hexyl cinnamic aldehyde as a calibrant for the murine local lymph node assay, *Contact Dermatitis*, 44, 357–361, 2001.

Enk, A.H., and Katz, S.I., Identification and induction of keratinocyte-derived IL-10, *J. Immunol.*, 149, 92–95, 1992.

Friedmann, P.S., The immunology of allergic contact dermatitis: the DNCB story, *Adv. Dermatol.*, 5, 175–196, 1990.

Gerberick, G.F., et al., Examination of the local lymph node assay for use in contact sensitization risk assessment, *Fundam. Appl. Toxicol.*, 19, 438–445, 1992.

Gerberick, G.F., et al., Selective modulation of T cell memory markers CD62L and CD44 on murine lymph node cells following allergen and irritant treatment, *Toxicol. Appl. Pharmacol.*, 146, 1–10, 1997.

Gerberick, G.F., et al., Selective modulation of B-cell activation markers CD86 and I-A^k on murine draining lymph node cells following allergen or irritant treatment, *Toxicol. Appl. Pharmacol.*, 159, 142–151, 1999.

Gerberick, G.F., et al., Local lymph node assay: validation assessment for regulatory purposes, *Am. J. Contact Dermatitis*, 11, 3–18, 2000.

Gerberick, G.F., et al., Contact allergenic potency: correlation of human and local lymph node assay data, *Am. J. Contact Dermatitis*, 12, 156–161, 2001.

Gerberick, G.F., et al., Use of a B cell marker (B220) to discriminate between allergens and irritants in the local lymph node assay, *Toxicol. Sci.*, 68, 420–428, 2002.

Grabbe, S., and Schwarz, T., Immunoregulatory mechanisms involved in the elicitation of allergic contact hypersensitivity, *Immunol. Today*, 19, 37–44, 1998.

Haneke, K.E., et al., ICCVAM evaluation of the murine local lymph node assay. III. Data analyses completed by the National Toxicology Program Interagency Center for the Evaluation of Alternative Toxicological Methods, *Regul. Toxicol. Pharmacol.*, 34, 274–286, 2001.

Hariya, T., Hatao, M., and Ichikawa, H., Development of a non-radioactive endpoint in a modified local lymph node assay, *Food Chem. Toxicol.*, 37, 87–93, 1999.

Heylings, J.R., et al., Sensitization to 2, 4-dinitrochlorobenzene: influence of vehicle on absorption and lymph node activation, *Toxicology*, 109, 57–65, 1996.

Humphreys, N.E., Dearman, R.J., and Kimber, I., Assessment of cumulative allergen-activated lymph node cell proliferation using flow cytometry, *Toxicol. Sci.*, 73, 80–89, 2003.

Kimber, I., and Basketter, D.A., The murine local lymph node assay: a commentary on collaborative trials and new directions, *Food Chem. Toxicol.*, 30, 165–169, 1992.

Kimber, I., and Basketter, D.A., Contact sensitization: a new approach to risk assessment, *Hum. Ecol. Risk Assessment*, 3, 385–395, 1997.

Kimber, I., and Dearman, R.J., Investigation of lymph node cell proliferation as a possible immunological correlate of contact sensitising potential, *Food Chem. Toxicol.*, 29, 125–129, 1991.

Kimber, I., and Dearman, R.J., Allergic contact dermatitis: the cellular effectors, *Contact Dermatitis*, 46, 1–5, 2002.

Kimber, I., and Dearman, R.J., What makes a chemical an allergen, *Ann. Allergy Asthma Immunol.*, 90, 28–31, 2003.

Kimber, I., Gerberick, G.F., and Basketter, D.A., Thresholds in contact sensitization: theoretical and practical applications, *Food Chem. Toxicol.*, 37, 553–560, 1999.

Kimber, I., Hilton, J., and Botham, P.A., Identification of contact allergens using the murine local lymph node assay: comparisons with the Buehler occluded patch test in guinea pigs, *J. Appl. Toxicol.*, 10, 173–180, 1990.

Kimber, I., et al., The murine local lymph node assay: results of an interlaboratory trial, *Toxicol. Lett.*, 55, 203–213, 1991.

Kimber, I., et al., The local lymph node assay: developments and applications, *Toxicology*, 93, 13–31, 1994.

Kimber, I., et al., An international evaluation of the murine local lymph node assay and comparison of modified procedures, *Toxicology*, 103, 63–73, 1995.

Kimber, I., et al., Langerhans cells and chemical allergy, *Curr. Opin. Immunol.*, 10, 614–619, 1998a.

Kimber, I., et al., Assessment of the skin sensitizing potential of topical medicaments using the local lymph node assay: an inter-laboratory evaluation, *J. Toxicol. Environ. Health*, 53, 563–579, 1998b.

Kimber, I., et al., Cytokines and chemokines in the initiation and regulation of epidermal Langerhans cell mobilization, *Br. J. Dermatol.*, 142, 401–412, 2000.

Kimber, I., et al., Skin sensitization testing in potency and risk assessment, *Toxicol. Sci.*, 59, 198–208, 2001.

Kimber, I., et al., Allergic contact dermatitis, *Int. Immunopharmacol.*, 2, 201–211, 2002a.

Kimber, I., et al., Lactoferrin: influences on Langerhans cells, epidermal cytokines, and cutaneous inflammation, *Biochem. Cell Biol.*, 80, 103–107, 2002b.

Kimber, I., et al., Classification of contact allergens according to potency: proposals, *Food Chem. Toxicol.*, 41, 1799–809, 2003.

Lea, L.J., et al., The impact of vehicle on assessment of relative skin sensitization potency of 1, 4-dihydroquinone in the local lymph node assay, *Am. J. Contact Derm.*, 10, 231–218, 1999.

Loveless, S.E., et al., Further evaluation of the local lymph node assay in the final phase of an international collaborative trial, *Toxicology*, 108, 141–152, 1996.

Lyons, A. B., and Parish, C. R., Determination of lymphocyte division by flow cytometry, *J. Immunol. Methods*, 171, 131–137, 1994.

Manetz, T.S., and Meade, J.B., Development of a flow cytometry assay for the identification and differentiation of chemicals with the potential to elicit irritation, IgE-mediated, or T cell-mediated hypersensitivity responses, *Toxicol. Sci.*, 48, 206–217, 1999.

Maurer, M., et al., Critical role of IL-10 in the induction of low zone tolerance to contact allergens, *J. Clin. Invest.*, 112, 432–439, 2003.

Montelius, J., et al., Experience with the murine local lymph node assay: inability to discriminate between allergens and irritants, *Acta Dermatol. Venereol.*, 74, 22–27, 1994.

National Institutes of Health. The Murine Local Lymph Node Assay: A Test Method for Assessing the Allergic Contact Dermatitis Potential of Chemicals/Compounds, NIH, Bethesda, MD, 1999 (99-4494).

Organization for Economic Cooperation and Development, Guidelines for Testing Chemicals, Guideline 429: Skin Sensitization, Local Lymph Node Assay, OECD, Paris, 2002.

Pennie, W.D., and Kimber, I., Toxicogenomics: transcript profiling and potential application to chemical allergy, *Toxicol. In Vitro*, 16, 319–326, 2002.

Plitz, T., et al., IL-18 binding protein protects against contact hypersensitivity, *J. Immunol.*, 171, 1164–1171, 2003.

Raju, R., Tracking the "general": tagging skin-derived dendritic cells, *Trends Biotechnol.*, 22, 58–59, 2004.

Ryan, C.A., et al., Activity of human contact allergens in the murine local lymph node assay, *Contact Dermatitis*, 43, 95–102, 2000.

Sailstad, D.M., et al., ICCVAM evaluation of the murine local lymph node assay. I. The ICCVAM review process, *Regul., Toxicol. Pharmacol.*, 34, 249–257, 2001.

Scholes, E.W., et al., The local lymph node assay: results of a final inter-laboratory validation under field conditions, *J. Appl. Toxicol.*, 12, 217–222, 1992.

Sebastiani, S., et al., The role of chemokines in allergic contact dermatitis, *Arch. Derm. Res.*, 293, 552–559, 2002.

Sikorski, E.E., et al., Phenotypic analysis of lymphocyte subpopulations in lymph nodes draining the ear following exposure to contact allergens and irritants, *Fundam. Appl. Toxicol.*, 34, 25–35, 1996.

Takeyoshi, M., et al., Development of a non-radio isotopic endpoint of murine local lymph node assay based on 5-bromo–2′-deoxyuridine (BrdU) incorporation, *Toxicol. Lett.*, 119, 203–208, 2001.

Takeyoshi, M., et al., Assessment of statistical analysis in non-radio isotopic local lymph node assay (non-RI-LLNA) with α-hexylcinnamic aldehyde as an example, *Toxicology*, 191, 259–263, 2003.

Wang, B., et al., Enhanced epidermal Langerhans cell migration in IL-10 knockout mice, *J. Immunol.*, 162, 277–283, 1999.

Warbrick, E.V., et al., Local lymph node responses to paraphenylene diamine: intra and inter-laboratory studies, *J. Appl. Toxicol.*, 19, 255–260, 1999a.

Warbrick, E.V., et al., Influence of application vehicle on skin sensitization to methylchloroisothiazoli-none/methylisothiazolinone: an analysis using the local lymph node assay, *Contact Dermatitis*, 41, 325–329, 1999b.

Wright, Z.M., et al., Vehicle effects on skin sensitizing potency of four chemicals: assessment using the local lymph node assay, *Int. J. Cosmet. Sci.*, 23, 75–83, 2001.

CHAPTER 20

Predictive Methods Specific for Human Immune Response

Fiona A. Harding, David L. Wong, and Donald P. Naki

CONTENTS

0-415-30854-2/05/$0.00+$1.50

20.1 INTRODUCTION

Proteins are capable of inducing immune responses. The migration to human proteins as therapeutics has reduced the overall rate of immune responses to protein therapeutics (for instance, the anti-antibody responses to xenogenic, chimerized, and wholly human monoclonal antibodies) (Chamberlain and Mire-Sluis, 2003). However, immune responses to human sequence–derived proteins are still commonly detected (Antonelli et al., 1998; Basser et al., 2002; Brand et al., 1993; Casadevall et al., 2002; Castelli et al., 2000; Cook et al., 2001; Li et al., 2001; Scagnolari et al., 2002; Wadhwa et al., 1996, 1999).

The most common form of immune response measured to protein therapeutics is an antigen-specific antibody response. Cytotoxic responses to gene therapy vectors have been demonstrated, but most immune responses to adeno-associated virus (AAV) and other vectors have been antibody mediated (Croyle et al., 2001; Sun et al., 2003). Often, the response is characterized as a binding antibody response that does not interfere with the activity of the drug. Binding antibodies may, however, interfere with the pharmacokinetics of drug clearance and localization. For some protein drugs, binding antibodies disappear over time (Brady et al., 1997; Eng et al., 2001; Rup, 2003), a situation reminiscent of allergen-specific IgE production in young children (Bottcher et al., 2002; Julge et al., 2001). However, some allergen-exposed children and some protein therapeutic-exposed patients will develop type I IgE-mediated hypersensitivity reactions. Immediate hypersensitivity reactions can develop into life-threatening anaphylactic reactions.

The most potentially troubling sequelae of the use of human-derived protein therapeutics are accidental establishment of autoimmunity. When a neutralizing antibody response is induced against a drug that is also a functionally nonredundant human protein, the result can be the creation of a *de facto* knockout of the protein in the recipient. Such a neutralizing response has been detected with erythropoietin (Casadevall et al., 2002; Castelli et al., 2000) and thrombopoietin (Basser et al., 2002; Li et al., 2001). Therefore, it is important to develop methods to reduce the risk of developing deleterious immune responses to administered proteins. This chapter describes methods currently used to detect and modify problematic regions of proteins.

It is unclear why wholly human proteins are capable of initiating immune responses in human subjects. Allelotype differences in human antibody constructs are one consideration, as are the potential for allelic differences in protein drug candidates. There are many other factors to consider in assessing the overall immunogenicity of proteins, such as the route of administration, dose, presence or absence of aggregates in the product, immune status of the patient, co-medication, and presence of adjuvant activity from concomitant infections (Braun et al., 1997; Brimnes et al., 2003; Sun et al., 2003). In addition, preexisting antibody responses to some cytokines have been detected in serum samples prior to exposure to the proteins, and in human immunoglobulin preparations (Rosenberg, 2003; Wadhwa et al., 1999, 2000). The presence of such antibodies in the serum strongly suggests that an antigen-specific CD4[+] T-helper cell–mediated response has occurred at some point in natural history of an individual, and may in fact indicate that clearance of rare, tightly regulated proteins such as cytokines is mediated by specific antibody responses.

The induction of neutralizing antibody responses, IgE-mediated responses, and high titers of IgG binding antibodies is thought to be largely dependent on CD4[+] T-helper cell–mediated regulation. Effector CD[+] T-helper cells interact with antigen-specific B cells to induce isotype class switching and the upregulation of antibody production, and differentiate into effector cells after a productive antigen-specific interaction with an antigen-presenting cell (APC). T cells recognize small fragments of protein antigens that are excised from the sequence of the protein during antigen processing. Peptides are presented on the surface of APCs for recognition of T cells in conjunction with major histocompatability molecules (MHC in mice, HLA in humans). Since CD4[+] T-cell helper epitopes are necessary to initiate antigen-specific helper T-cell responses, the identification of putative T-cell peptide epitopes in proteins is an indicator of the potential to induce proliferative responses to the protein.

In this chapter, *in vitro* and *in silico* methods to identify human-specific peptide epitope responses are described and discussed. The reason for this emphasis is that while the considerations listed above, including the presence of adjuvants and aggregates in the protein preparation are critically important, a high-quality antigen-specific immune response cannot occur in the absence of T-cell priming (Guery et al., 1995; Langenkamp et al., 2002; Lanzavecchia and Sallusto, 2000). Methods to detect T-cell epitopes include *in silico* computation methods based on peptide-binding algorithms, *in vitro* methods based on the isolation of presented peptides from the surface of APCs, and the functional testing of peptides on human cells. Considerable discussion of the computational methods is included because such methods are actively being developed at present, and will likely be of primary importance in the future. Therefore, a basic understanding of the derivation of the most common methods is warranted. The potential application of epitope identification to reduce the immunogenicity of proteins by modification of T-cell epitopes is discussed. This chapter also addresses the prediction of relative immunogenic potential of proteins in mouse models, and discusses the merits of a recently developed human cell–based assay to identify T-cell epitopes.

20.2 EPITOPE DETERMINATION USING COMPUTATIONAL METHODS

The computational methods considered here do not attempt to predict epitopes *per se*, but rather the ability of peptides to bind to specific HLA molecules. This is a necessary and critical, but insufficient, step in the accurate prediction of epitopes. A complete computational method for accurately predicting epitopes would require consideration of the other major events in the pathway, such as the processing of proteins to peptides and the recognition of the MHC II–bound peptide by the T-cell. The primary application of these methods is to reduce the number of peptides that must be screened for T-cell recognition in traditional assays, while missing as few positives as possible. False positives are therefore better tolerated than false negatives.

In this section, we provide a brief overview of some of the major computational algorithms for predicting binding by peptides to HLA class II molecules (Table 20.1).

Table 20.1   Selected Ligand-Binding Prediction Methods

Method Name	Algorithm	Developers	Publicly Available	MHC Type
	Motif	Sette	N	
SYFPEITHI	Simple Matrix	Rammansee	Y[a]	MHC I, II
RankPep	Position-Specific Scoring Matrix	Reche, Reinherz	Y[b]	MHC I(82), II(69)
ProPred/TEPITOPE	Binding Matrix	Sturniolo, Hammer	Y[c]	MHC II
EpiMatrix	Matrix	EpiVax[d]	N	?
PERUN	Artificial Neural Network	Brusic	N	MHC II
SVMHC	Support Vector Machine	Dönnes, Elofsson	Y[e]	MHC I
(HMM)	Hidden Markov Model	Mamitsuka	N	MHC I, II
PREDEP	Threading	Altuvia, Yael[f]	N[g]	MHC I
MHC Thread	Threading	Brooks	Y[h]	MHC II
(Threading)	Threading	BioVation[i]	N	MHC II

[a] Available at http://syfpeithi.bmi-heidelberg.com/scripts/MHCServer.dll/home.htm.
[b] Available at www.mifoundation.org/Tools/rankpep.html.
[c] Available at www.imtech.res.in/raghava/propred/.
[d] See http://epivax.com/epimatrix.html.
[e] Available at www.sbc.su.se/svmhc/.
[f] Paper co-written by authors from Hebrew University (Israel) and Epimmune Inc.
[g] Available at http://bioinfo.md.huji.ac.il/marg/Teppred/mhc-bind/Info.
[h] Available at www.csd.abdn.ac.uk/~gjlk/MHC-Thread/.
[i] Footnote 8 at www.biovation.co.uk/Brooks states that BioVation uses a version of their algorithm.

20.2.1 Motif Methods

The earliest and simplest epitope prediction methods were based on motifs. These methods (Rothbard and Taylor, 1988; Sette et al., 1989) characterize the specificity of an HLA class II molecule
by enumerating the amino acids required (and optionally, prohibited) at various "anchor" positions
of the HLA binding cleft. A peptide that has any of the required amino acids at each of the required
positions is predicted to bind with sufficient affinity to be a candidate epitope. These methods say
nothing about the relative strength of the association between an anchor position and its allowed
amino acids — each allowed amino acid is considered equally probable, which runs counter to
observation and is one of the obvious weaknesses of motif-based methods. Typically, motifs only
consider the stronger peptide anchor positions and ignore positions that contribute less strongly to
binding; hence, important information on weaker, but significant, binding interactions is not used.
Only about a third of the peptides chosen by motif-based methods actually bind (Gulukota et al.,
1997). The low performance of motif methods does not mean that sequence patterns of binders
and nonbinders are indistinguishable, but only that simple motifs are inadequate to represent these
patterns.

Both motif- and matrix-based methods implicitly assume that the ability of an amino acid to
occupy a position within the epitope is independent of the other amino acids in the peptide sequence.
This has become known as the independent binding of sidechains assumption (Parker et al., 1994).

20.2.2 Matrix Methods

Matrix-based methods assign a coefficient to each amino acid at each ligand residue binding position
of the MHC binding cleft, resulting typically in a 9×20 matrix of coefficients. Each column
represents a position in the binding cleft, and each row one of the 20 amino acids. To obtain a
prediction of the binding ability of a peptide for a particular MHC class II molecule, the peptide
is aligned with the residue-binding positions of the HLA class II molecule and the coefficients for
the peptide's amino acids for each corresponding position are summed, obtaining the values by
looking them up in the matrix. Values above a particular threshold indicate binders.

One of the earliest and simplest matrix approaches is the SYFPEITHI algorithm (Rammensee
et al., 1999). In this method, matrix values are derived from arbitrarily chosen coefficients based
on observed prevalence of amino acids at each position. In general, the method performs relatively
poorly, suffering from a high rate of false negatives, with a reliability of only 50% for prediction
of MHC class II peptide binders.

Borrowing from bioinformatics methods for detecting sequence homology Reche et al. (2002)
used position-specific scoring matrices (PSSMs) to predict MHC binders. Here, each matrix entry
is derived from the frequency of the observed amino acid at the specified column. The number of
observed sequences contributing to the frequencies is usually too small to produce reliable scores,
so they are often supplemented with information on the frequencies that amino acids are observed
to substitute for each other in nature; these statistics are captured in a substitution matrix such as
PAM or BLOSUM (Gribskov et al., 1987; Henikoff and Henikoff, 1996). The method was validated
by determining whether empirically determined Kb- and Db-restricted epitopes (37 and 34 members, respectively) were among the top-ranking peptides when all overlapping nonamer sequences
from their parent proteins were scored using the relevant PSSMs. Approximately 80% of the known
epitopes were found in the top 2% of scored peptides. Although the authors focused on MHC class I
ligands, they acknowledge that prediction for class II alleles is a significantly more difficult task.

The TEPITOPE algorithm is based on matrices comprised of experimentally derived binding
coefficients (Sturniolo et al., 1999). MHC molecules were found to contain a limited number of
sequence variations of each of the nine so-called "pocket positions" of the binding cleft (each MHC
pocket position accommodates a residue of the ligand), and that MHC alleles were simply different

combinations of these polymorphic pockets. Each of these pocket variants exhibited affinities for the 20 amino acids that could be characterized quantitatively as a "pocket profile." Once a pocket profile had been characterized, it could be reused in any allele that contained it, as long as these pocket profiles were independent. Therefore, a limited number of pocket profiles need be characterized to construct a much larger number of "virtual matrices," each of which was one combination of nine pocket profiles. The HLA polymorphism issue is effectively addressed by this method since there is no need to explicitly characterize each separate HLA allele.

To validate their method, a set of known binders and nonbinders from a random synthetic library were combined, and it was determined whether their algorithm could separate them computationally. Up to 80% of the binders could be predicted using a threshold of 1 to 3% of top-scoring peptides. The method was also tested on natural ligands from the SYFPEITHI database. More than half of all natural ligands could be predicted using a 1 to 3% threshold setting and more than 75% with a 1 to 6% threshold setting.

20.2.3 Machine Learning Approaches

The machine learning approaches Artificial Neural Networks (ANNs) and Hidden Markov Models (HMMs) discussed here attempt to classify peptides into broad categories, such as binders/nonbinders, or nonbinders and high, medium, and low binders. One of the advantages of these approaches is that they are able to use data on nonbinders as well as binders. In addition, these methods do not rely on the assumption of independent binding (IBS); they can implicitly handle dependencies between neighboring amino acids. Finally, they can extract multiple sequence patterns separately from a data set, whereas motifs and matrix methods extract a single mixed pattern (Mamitsuka, 1998).

Machine-learning methods must be used with care because when used improperly (e.g., employing too many hidden node layers), they can overfit the data, that is, they can model the noise in the data rather than solely the process that generated it. Additionally, they are sensitive to the data used to train them, and will find it difficult or impossible to predict valid binder sequences whose patterns were absent from the training data.

20.2.3.1 Artificial Neural Networks

ANNs are networks consisting of interconnecting nodes, loosely similar to human neurons. Like biological neurons, the nodes can be viewed as computational units. Each node receives inputs from other nodes and processes them to obtain an output.

The networks are constructed with architectures or topologies that are optimized for particular problems. The so-called multilayered perceptron (MLP) is often used for classification problems. In the MLP architecture, nodes are arranged in layers, with an input layer that receives important parameter values of the system under study, one or more hidden layers (not observable from the inputs or the outputs), and finally, an output layer. Information flows from the input nodes to the hidden nodes to the output nodes in a unidirectional manner. Each node receives a value from one or more nodes in the previous layer (except for the input nodes). These values are weighted, summed, and a simple nonlinear function is applied to obtain a scalar output.

The ANN is "supervised" when a set of output values is provided for each corresponding set of input values. A "training" procedure is then performed that adjusts the weights between the nodes to minimize the error between the input and output values. Once the ANN is trained, test data are used to determine how well the ANN generalizes to nontraining data. Finally, the ANN is used to make predictions for new input data.

A method called PERUN that uses an MLP neural net for predicting peptide binders to the MHC class II molecule HLA-DR4(B1*0401) has been developed (Brusic et al., 1998). The input to the MLP was a set of nonbinding peptides along with a set of peptides for which there were binding data to the HLA molecule, aligned with respect to the nine pocket positions of the HLA class II–binding cleft. The alignments were created using matrices that were "evolved" using an evolutionary algorithm.

The method was evaluated using binary (binder/nonbinder) as well as four (nonbinding, low-, moderate-, and high-affinity binding) classifications. For the binary classification, 80% of nonbinders were classified correctly, whereas 50, 70, and 80% of low-, moderate-, and high-affinity binders, respectively, were correctly classified as binders.

Both nonbinders and high-affinity binders were well classified compared to low- and moderate-affinity binders. The authors speculate this was due both to the smaller number of intermediate binders in the training set, and the arbitrarily chosen boundaries for the low- and moderate-binding classes. The results were compared to those from the TEPITOPE matrix and a motif-based method. PERUN was found to be comparable to the TEPITOPE, quantitative binding matrix method, and better than the motif method.

20.2.3.2 Hidden Markov Models

A Hidden Markov Model (HMM) can be visualized as a set of nodes connected by edges. The nodes represent states, and the edges represent transitions between the states. Each state contains a set of symbols and their associated probabilities, and each edge contains a value describing the probability of transitioning between the two states it connects.

For biological sequence applications of HMMs, each main state represents a position within a consensus sequence, and the symbol generation probabilities describe the probability of observing each amino acid (the amino acids are the symbols) at that position. In addition to the main states, there are usually states representing deletions and insertions relative to the consensus, along with transition probabilities between these states and the main consensus states; thus HMMs can deal effectively with sequence-length heterogeneity. Once the model has been parameterized (i.e., the symbol-generation and transition probabilities determined) using a set of training data, it can be used to score any sequence for its fit against the model.

A novel HMM for classifying peptides as low, medium, high, or nonbinders to various class I (HLA-A2) and class II (HLA-DR1 and HLA-DR4) molecules was developed (Mamitsuka, 1998). The training data for HLA-DR1 consisted of 440 binding peptides obtained from the MHCPEP database, of which 29 were also known to activate T cells. For HLA-DR4, 560 peptides were used for training, of which 40 were also known to activate T cells. The ability of the HMM to correctly classify peptides into binders/nonbinders was compared with an ANN that was identical to one trained with the same data sets (Gulukota et al., 1997). The accuracy of the HMM was 5 to 10% higher than the ANN, which was previously the best-performing algorithm.

20.2.4 Structure Based Methods

Although threading was conceived to tackle the inverse folding problem (determining the fold a given protein sequence adopts), it has recently been used to predict binding of peptides to HLA molecules (Altuvia et al., 1995) (www.csd.abdn.ac.uk/~gjlk/MHC-Thread/). We are not aware of threading literature for MHC class II binding; however, literature does exist for MHC class I binding, which for the sake of illustration is discussed below.

The idea behind threading is conceptually simple, although the implementation is complex. Threading approaches exploit the fact that amino acids have distinct preferences for occupying different structural environments. Quantitative properties associated with these preferences have been used to produce scoring functions, which reflect the degree to which an amino acid occupies a preferred environment. Further, known protein structures can be represented in a way that makes explicit the structural environment at each position. The sequence can then be spatially aligned (or "threaded") onto the structure in a way that optimizes the score function. For the current application, peptide/HLA alignments that produce good scores are predicted to be binders. The success of threading approaches depends on the descriptors used to characterize structural environments, as well as the scoring functions employed. Both environment descriptors and score functions currently used limit the success of the threading approach for the purpose of predicting the three-dimensional fold of proteins (Salzberg et al., 1998). A limitation for the current application is the obvious need for a solved structure for each different MHC molecule.

A threading approach in which the sequence of the test peptide was threaded through the backbone fold of a peptide in a solved MHC/peptide co-crystal has been described to rank peptides as binders to the MHC class I molecule HLA-A2 (Altuvia et al., 1995; Schueler-Furman et al., 2000). In an initial study, the authors focused on crystal structures of four viral peptides (Altuvia et al., 1995). Each of the four peptides was threaded through its own template as well as the template of the other three peptides. The authors then ranked all nonamer peptides within each of the four parent protein sequences, and found that for three of the proteins, the peptide from the crystal structure was ranked high (at the first through sixth percentiles), but in the fourth, the peptide from the crystal structure was ranked at the 30th through 50th percentiles. The authors ultimately concluded that the lack of adequate representation of hydrophilic interactions in the pairwise potential table led to poor results for class I molecules with hydrophilic pockets.

In a subsequent study, a modified pairwise contact potential table was used that better accounted for hydrophilic interactions, as well as an improved criterion for identifying the MHC and peptide positions that were in contact (Schueler-Furman et al., 2000). The algorithm improved, but for many alleles performance was mediocre.

20.2.5 Conclusions

A rigorous, comprehensive comparison of the various computational methods for predicting peptide binding to MHC class II molecules is currently not practical due to several factors, most notably the lack of publicly available implementations of many of the algorithms, and the limited number of HLA allelotypes supported by some of the methods that do have public implementations available. The situation is somewhat better for MHC class I molecules, and motif, matrix, ANN, and HMM-based methods have been compared (Yu et al., 2002). These studies conclude that when data are scarce, binding motifs are a useful alternative to random guessing, but as more data on peptide binders become available, binding matrices and HMMs become more useful predictors. ANNs and HMMs appear to be methods of choice for MHC alleles with more than 100 known peptide binders.

When these methods are used appropriately, they produce a set of candidate epitopes peptides that are capable of binding with sufficiently high affinity to a particular HLA class II molecule to qualify as an epitope. If the accuracy of the method in question is high, then the list could be considered a superset of the actual epitopes, and protein processing and T-cell recognition would act as additional filters on this set.

Perhaps the greatest value of the computational methods is their ability to scan large sets of proteins, even entire genomes, for proteins with candidate epitopes — a task that is prohibitively labor intensive for traditional assay-based epitope identification. Once a small number of candidate antigenic proteins are identified, "wet" methods can be used for a more rigorous identification of epitopes. When used in this manner, the computational and wet methods are complementary. On the other hand, when the allergenic protein is known, it is usually advisable to identify epitopes by the more accurate, validated approach of screening all overlapping peptides in a functional *in vitro* assay.

20.3 EPITOPE DETERMINATION USING *IN VITRO* METHODS

In addition to the computational methods discussed above, other *in vitro* methods also exist to separate functional epitope peptides from the nonfunctional. In addition, the selection of T-cell epitopes by APCs includes a consideration of the proteolysis of antigenic proteins. A number of current *in vitro* techniques that rely on testing cell samples for a read-out that indicates whether the epitope peptide is capable of inducing responses in T cells are available to test the functionality of the epitope peptides. To determine that a particular peptide is presented from the processed protein, either memory responses must be analyzed, or an analysis of cell surface presented epitopes must be performed. All of these methods are discussed in this section.

20.3.1 Identification of Presented Peptides by Mass Spectral Analysis

Antigen-specific T cells do not recognize intact protein antigens. Protein antigens must be processed by APCs into small peptide fragments, and peptide fragments must be presented by MHC molecules on the surface of APCs in order for T cells to be capable of recognition (Fremont et al., 1996; Germain, 1981). Presentation of peptide fragments by MHC class I and II molecules is usually discrete (Watts and Powis, 1999), arising from separate intracellular pathways and compartments for peptide processing and MHC interaction. Endogenously expressed proteins including tumor and viral proteins are processed within the cytosol by the proteosome complex (Lehner, 2003; Van Kaer, 2002). Peptides of the appropriate size (eight to ten amino acids) are transported by the TAP proteins to the endoplasmic reticulum where they bind to nascent MHC class I molecules (Van Kaer et al., 1992). Exogenous proteins are endocytosed by APCs, resulting in the delivery of proteins into endosomes and lysosomes. A battery of endo- and exo-peptidases cleave proteins, which results in the release of peptide fragments from the protein. The subcellular vesicles containing degraded exogenous protein fuse with vesicles containing nascent class II molecules, and peptide binding occurs under the control of the editor molecule HLA-DM (Busch et al., 2000). Peptide-loaded class I and II molecules are subsequently transported to the cell surface.

The enzymatic events that release class I and II binding peptides from proteins have been extensively studied, but much is still unknown. For example, it is difficult to predict which peptide fragments will be excised from a given protein due to a number of confounding factors such as (1) the comparatively nonspecific proteolytic cleavage motifs demonstrated by most enzymes implicated in protein processing, (2) the presence of disulfide bonds, and (3) the overall tertiary stability of the molecule (Chen et al., 1999; Li et al., 2002; So et al., 2001; Watts, 2001). It has also been difficult to study the modification of sequences in order to affect the efficiency of presentation of particular epitope sequences from polypeptides, with some notable exceptions (Schneider et al., 2000). In fact, the study of processed peptides isolated from the surface of APC has been applied to the understanding of polypeptide cleavage patterns (Lippolis et al., 2002; Phelps et al., 1998).

As presentation of peptide fragments excised from antigens is a necessary prelude to the induction of a T-cell dependent immune response, an understanding of exactly what peptides the APCs present is of interest. The most logical method for identifying presented peptides is to isolate the MHC molecules from the surface of APCs and to determine the sequence of their associated peptides. This is a daunting task, however, for a number of reasons. MHC molecules on the surface of cells primarily present self-derived peptides (Chicz et al., 1993; Dongre et al., 2001), with antigen-derived peptides forming a minority of all the MHC/peptide complexes presented. In addition, class II-presented peptides are represented as sets of related peptides (Lippolis et al., 2002; Phelps et al., 1998; Vignali et al., 1993). This is likely due to the open-ended nature of the class II peptide binding cleft that allows for overhanging residues. This is in contrast to class I–presented peptides where the ends of the peptides must fit precisely within the closed binding pocket. Each peptide within the nested set of class II peptides represents a fraction of the total pool. For class I peptides, a single peptide sequence represents majority of the pool presented. Finally, it is difficult to show antigen-specific presentation of exogenous antigen. Most analyses of class II-presented peptides have relied on cancer antigen presentation (Kao et al., 2001; Pieper et al., 1999). Some solutions to the problem of exogenous protein presentation have entailed the development of antigen delivery systems that up-regulate internalization of exogenous proteins by cross-linking them to lectins (Peakman et al., 1999), while others have relied on cross-presentation to the class II compartment via virus infection (Ovsyannikova et al., 2000, 2003) or transfection (Nepom et al., 2001).

20.3.1.1 Basics of Mass Spectrometry

The method of choice for identifying peptide epitopes isolated from the surface of APCs is mass spectrometry (Chapman, 1996; Grayson, 2002). It is defined as the measurement of ion mass to charge ratios (m/z) usually by direct amplification of ion signals, which are produced either by removing an electron from the molecule or adding a proton (H atom) to form a positively charged cation, or conversely by adding an electron or removing a proton to produce an anion.

The current mass spectrometric (MS) techniques use two major ionization methods: electrospray ionization (ESI) and matrix-assisted laser desorption ionization (MALDI). In ESI/MS, a flow of sample solution is pumped through a narrow-bore metal capillary held with a few kilovolts potential. Due to charging of the liquid, it sprays from the capillary orifice as a mist of very fine, charged droplets, which are evaporated with a drying gas and explode to form a number of much smaller droplets. The ions, which are field desorbed from these droplets, are then sampled through a system of small orifices into the vacuum system for mass analysis. In MALDI/MS, a matrix, mixed with the sample is deposited as a small spot on a sample plate. After the solvent evaporates, the sample and the matrix form a crystal. A laser beam with a wavelength absorbed readily by the matrix is used to vaporize and ionize the sample. The resulting ions, immediately accelerated to a constant energy, enter into a time-of-flight mass spectrometry (TOF/MS) analyzer.

In a TOF/MS analyzer, ions charged in the source by either the ESI or MALDI method are accelerated into the mass analyzer, which is simply a field-free tube. Since all the ions accelerated by the same voltage have the same energy, their velocity becomes a function of their mass. The lighter ions pass through the tube more rapidly than the heavier ions. A high-speed multichannel plate (MCP) detects the arrival of the ions of different mass at different times and creates a mass-to-charge (M/Z) spectrum.

The main advantage of ESI/MS is that large molecules become multiply charged, primarily by the addition of multiple protons. Therefore, a series of peaks appears in the spectrum, representing a distribution of such charges. Each peak represents an independent determination of the molecular weight of the analyte. The MALDI-TOF technique has been used primarily to ionize massive biological molecules that often weigh more than 100,000 daltons. Unlike ESI/MS, which can fragment delicate molecules, MALDI is considered a nondestructive or "soft" ionization technique.

20.3.1.2 Identification of Peptide Epitopes by Mass Spectrometry

The development of a combination of mass spectrometric-based techniques (immunoaffinity isolation, MS peptide mapping, and protein database analysis) has enabled the rapid identification of class I and II epitope peptides from cell lysate samples (Dongre et al., 2001). Many variations on the method presented below are currently in use (Tomlinson and Chicz, 2003).

The selection of the MHC/peptide complex is achieved by passing the antigen-presenting cell lysate through a column with immobilized MHC-specific monoclonal antibodies, followed by removal of supernatant material containing all unbound proteins and peptides. The MHC-associated peptides, which are dissociated from the immunoaffinity column, can be analyzed by either MALDI-TOF or NanoESI-MS/MS mass spectrometric techniques. All of the peptide masses along with their corresponding MS/MS data identified by mass spectrometry can be used to perform an automated protein database search, combined with a search for a common "epitope motif." Matching the raw MS/MS data to the computer program generated theoretical MS/MS patterns from any random peptide sequence in the database, allows unequivocal identification of the epitope peptides.

Recently, we have initiated a new strategy to make the identification of processed peptides in the cytosol and MHC-bound peptides less ambiguous. This method is based on the combined use of specific stable isotope labeling, sensitive NanoESI-LC/MS/MS for peptide sequencing, and a proprietary peptide doublet-searching program. The software screens raw MS data and identifies antigen-related peptides based on their expected doublet masses. Trypsin-digested peptide doublets from an N14/N15 labeled protein mixture mixed into a cell cytosol sample were correctly detected at the femtomolar level, not only by our peptide doublet detection software (IP# WO 02/16952 A2), but also by protein database search programs using the N14-peptide MS/MS fragmentation data. This novel method is expected to speed MS-based epitope mapping and permit another level of confidence in the identification of processed peptides.

20.3.1.3 Applications

The obvious relevance of presented peptide identification has supported the development of these techniques in the face of daunting challenges. The most often cited application of these methods is the identification of immunodominant, processed and presented epitope peptide candidates for vaccine development (Urban et al., 1997). We have used these methods to verify the presentation of peptides identified using synthetic peptides in the I-mune assay and to modify processing sites in an attempt to modify antigen processing. Recently, identified peptides from cytochrome P450 1B1, a pan-cancer antigen, and specific for the HLA-A2 class I molecules, were incorporated into a cancer therapeutic treatment (Maecker et al., 2003).

For this method to be widely applicable in both the creation of vaccines and engineering of hypoimmunogenic proteins, it needs to become easy and robust. In addition, data on antigen presentation by a wide variety of HLA molecules need to be collected, a step currently precluded due to the effort involved. Finally, differential antigen presentation by subsets of APCs should be considered, which adds another level of complexity to this already complex field (Crowe et al., 2003).

20.3.2 Identification of Functional Epitope Determinants by Cell-Based Analysis

The identification of T-cell epitope determinants capable of inducing a functional response usually relies on the detection of cytokine secretion by activated cells (ELISA, ELISPOT, or intracellular cytokine-staining methods) or by the measurement of cellular proliferation. A recently developed method, tetramer binding analysis, uses flow cytometric analysis to detect the presence of T cells bearing a TCR capable of specifically interacting with a peptide-MHC complex. Detection of T cells with peptide/MHC-specific TCRs by flow cytometry can be combined with intracellular cytokine detection to indicate the presence of both antigen-specific and functional cells. Each of these methods are discussed below.

20.3.2.1 Tetramer Binding Analysis

After a productive immune response to a pathogen or other immunogens, it is of interest to identify the peptide specificities that were useful in eradicating the disease. MHC/peptide complexes that could be used as reagents in flow cytometry were initially developed for immediate use in tracking productive immune responses, as well as for describing functionally immunodominant T-cell responses *in vivo*.

MHC molecules have a low affinity for TCR (van der Merwe and Davis, 2003), and are thus unsuitable as reagents for the detection of antigen-specific T cells. To overcome this barrier, MHC molecules with modifications to their carboxy-terminal sequences that allowed the tetramerization of the molecules using biotin as a linker were created (Altman et al., 1996). Tetrameric MHC molecules refolded in the presence of peptide for the correct tertiary conformation to coalesce are labeled with fluorescent tags for flow cytometric detection. Because the peptide/MHC molecule is now in a tetrameric format, low avidity binding to the TCR is overcome. Concomitant studies of T-cell phenotype and functions such as cytokine secretion are possible (Xu and Screaton, 2002). Tetrameric MHC molecules are now available for both MHC class I and II molecules of both humans and mice, including HLA-DQ [Cameron et al., 2002; Doherty and Christensen, 2000; Kwok et al., 2000; McMichael and Kelleher, 1999, Immunomics (www.immunomics.com), and Becton Dickinson (www.bdbiosciences.com)].

T-cell epitope mapping of both class I and II epitopes followed the establishment of tetramers for the analysis of antigen-specific T cells (Kern et al., 1998; Novak et al., 2001). For this application, peripheral blood cells from HLA-typed, disease-positive donors must be used in order to use the correct tetramer, and to ensure enrichment of antigen-specific T cells to a level that can be accurately detected using flow cytometry. Amplification of antigen-specific human cells is recommended if the frequency of specific cells is less than 1:1000 (Novak et al., 2001). This suggests that detection of antigen-specific T cells in an unexposed individual will not be possible. To increase the overall efficiency of the method, peptides are sometimes tested as pools, or computer algorithms used to limit the overall number of peptides for testing (Kwok et al., 2001). If pools of peptides are used, subsequent analyses to isolate the active peptide from the mix are essential. The use of computer algorithms to limit the number of peptides used for epitope mapping is compelling, but caveats associated with the selection of epitopes (see Section 20.1) have to be considered.

The variety of MHC tetramers available for epitope mapping is limited. In the human system, HLA class I constructs are commercially available for A*0101, A*0201, A*0301, A*1101, A*2402, B*0702, B*0801, B*1501, B*2705, B*3501, and B*5701 from Immunomics, as are HLA class II DRB1*0401, DRB1*0101, and DRB1*1501 constructs. HLA-DQ2 tetramers have been published (Kwok et al., 2000). While the number of tetrameric MHC molecular constructs is increasing, the current specificities do not represent the full complexity of human HLA alleles. To assess epitope responses predicted to be relevant for the subset of the human population represented by the available tetramers, a discouraging number of flow cytometric analyses must be performed, which can be limited by automated flow cytometry systems. Even with automated systems in place to facilitate screening of large numbers of samples, there are serious shortcomings in the available repertoire of tetramers. Notably, HLA-DP tetramers are currently unavailable, even though HLA-DP4 is the most commonly expressed human HLA molecule (Castelli et al., 2002). Epitope analysis by tetramer binding analysis with concomitant phenotypic and functional analysis offers a powerful method to identify immunodominant individual-specific memory responses. However, its utility for identifying epitopes for the creation of hypoimmunogenic proteins is limited due to technical issues.

20.3.2.2 ELISPOT Analysis

The ELISPOT method for the quantitation of peptide-specific T cells is based on the ELISA assay. Both assays allow cytokines to bind to an immobilized anticytokine antibody, and then detect binding using another enzyme-coupled anticytokine antibody. The difference is that in an ELISA, detection of cytokines typically occurs in the fluid from an activated cell culture. In the ELISPOT method, antigen-activated T cells, plated directly onto the antibody-coated support, leave a footprint wherever a cytokine-secreting T-cell settles. Each spot represents a cytokine-secreting cell. The size of the spot indicates the productivity of the particular cell. The ELISPOT method is applicable to the detection of both class I and II epitopes (Anthony and Lehmann, 2003; Geginat et al., 2001; Wertheimer et al., 2003), and has been shown to be capable of high specificity and sensitivity when appropriate reagents are selected (Schmittel et al., 1997). The method is most robust when applied to cell populations containing recall responses (Helms et al., 2000). The ELISPOT method is highly sensitive, which allows direct detection and quantitation of antigen-specific responses *ex vivo* (Keilholz, et al., 2002). When circulating precursor frequencies are very low, *ex vivo* stimulation of PBMC is necessary to detect antigen-specific responses (Asai et al., 2000).

T-cell epitope mapping using the ELISPOT method has been applied to the identification of disease-specific peptide epitope responses (Anthony and Lehmann, 2003). The assay has been modified for high throughput in order to fully map large proteins, which has led to some interesting and unexpected results. A comparison of results for predicted HLA class I binding peptides with results from all possible 9-mer peptides tested in an ELISPOT assay using hepatitis C-infected individuals has shown that the two methods do not return the same results for HLA specificity of responses (Anthony et al., 2002). This result reveals the possibility for error when relying on computer algorithms to subset peptide epitopes for subsequent functional testing. In contrast, the ELISPOT assay is ideal for accurate identification of disease-specific antigen epitopes, which is crucial for the design of cancer and viral vaccines (Novitsky et al., 2002), as well as for the potential identification of specific epitope identification in adverse responses to protein therapeutics.

20.3.2.3 Proliferation Responses

Proliferative response to peptides is the classical method to identify epitopes in a protein of interest (Brett et al., 1991; Bungy et al., 1994; Reece et al., 1993, 1994; Walden, 1996). Cells from exposed individuals, or immunized mice, are cultured with peptides derived from the protein of interest, and cellular proliferation is measured usually by tritiated thymidine incorporation. Cells are sometimes re-stimulated *in vitro* to create an antigen-specific cell line, or further subcloned to create a panel of antigen-specific T-cell clones. All of these methods have been used successfully to identify class I- and II-restricted T-cell epitopes. However, the proliferative method is not as sensitive or reproducible as the methods mentioned above (Bercovici et al., 2003; Tesfa et al., 2003). It is worth remembering that these methods (tetramer binding analysis and ELISPOT) are designed for the detection of effector cells, which have been previously activated and expanded. Signaling requirements for cytokine secretion and cytotoxicity can be different than the requirements to induce proliferation (Carter et al., 1998; Crowe et al., 1998; Lanzavecchia and Sallusto, 2002). It is therefore not surprising that a good correlation was not seen between effector function and proliferation.

20.3.2.4 I-mune® Assay

Cell-based assays for epitope mapping of T-cell responses in proteins of interest are an improvement over purely predictive methods because such assays measure a correlate to immune responsivity. All the methods described (tetramer analysis with or without concomitant functional analysis, ELISPOT, and proliferation methods) have largely relied on access to exposed individuals. The reliance on a pre-primed cell source is a severe limitation to the study of immune responses to novel proteins and to the study of immune response induction to self-proteins. To overcome this limitation, we developed a functional assay (I-mune assay) to detect the first response of a previously unactivated cell population (Stickler et al., 2000). The assumption was made that a community donor PBMC pool would most likely not be exposed to industrial antigens at any significant frequency, and that humans would be either immunologically tolerant or "ignorant" to the majority of human proteins currently in use as therapeutics (i.e., human IFN-beta, human erythropoietin). The assay was designed to optimize the chance of inducing and detecting proliferative responses by CD+ "helper" T cells to synthetic peptides. In this assay, CD+ T cells are enriched from the donor blood sources such that a single culture represents ten cultures of unfractionated PBMC. Cultured autologous dendritic cells differentiate *in vitro* and are used as APCs. Proliferation is detected by assessing tritiated thymidine on day 6 of the culture. The use of dendritic cells to activate primary cell responses *in vitro* has been described by others (Mehta-Damani et al., 1995; Schlienger et al., 2000), and is a critical component of the epitope mapping technique using unexposed donors (Austyn et al., 1988; Steinman, 1991; Young and Steinman, 1990). The assay is also effective if the human donor population has been preexposed to a protein antigen, such as staphylokinase (Stickler et al., 2004a). Another critical component of this assay system is the identification of epitopes on a population basis. Responses from a large donor set (typically 50 to 100 donors) are summarized in order to determine the localization of peptide epitope determinants that induce proliferative responses in a statistically significant subpopulation of the donors. HLA types are also determined, and associations between the expression of particular HLA alleles and responses to peptides are determined. The selection of epitope responses on a population basis is critical for the design of effective vaccines (Novitsky et al., 2002), and is also critical for the design of commercial proteins with a reduced immunogenicity that will be encountered by a general population.

Human proteins such as interferon-beta (IFN-beta) have been demonstrated to possess CD4+ T-cell epitopes by the I-mune assay (Stickler, et al., 2004b). Human IFN-beta is known to induce neutralizing antibody responses in relapsing/remitting multiple sclerosis patients (Antonelli et al., 1998). This result suggests that for at least some human proteins, the potential for an immune response is present, and if the protein is administered in an immunogenic context, an immune response is likely to occur.

20.4 COMPARATIVE IMMUNOGENICITY ASSESSMENTS

The final goal of the identification of T-helper cell epitopes within the context of this review is to be able to modify epitope regions such that the result is a variant protein with a significantly reduced capacity to induce immune responses in human recipients. A potential mechanism to alter immunogenicity could be to change the amino acid sequence of the protein such that antigen processing does not result in releasing intact epitope peptides. This would involve identifying and altering the major cleavage sites within the protein such that cleavage no longer occurs. Such alterations would also obviate peptide-MHC binding. However, the complexities of protein processing for most proteins have yet to be thoroughly resolved. Changing T-cell receptor contacts such that the T-cell interaction no longer occurs is also a potential method. However, given the enormous complexity of T-cell receptor specificities, the latter method is least likely to be successful on a population basis.

Modification of the amino acid sequence of proteins to reduce or eliminate MHC binding is thus the most attractive method to reduce overall immunogenicity of a protein. The complexity of T-cell receptors is not an issue here, nor is an understanding of protein processing. However, the problem with modifying sequences to interfere with MHC binding is that while this may be a straightforward approach for any one particular MHC allele, changes that affect binding to one MHC allele might in fact create epitope sequences for a different MHC allele. Therefore, an assessment of any changes, made to an epitope sequence have to be tested in a second round of analysis to be certain that a new epitope has not been created. The new sequence must be checked for "fit" in a set of alleles that would represent majority of the general population and should be checked in all possible binding frames. This caution applies to both predictive algorithm-based and function-based methods.

20.4.1 Quantitative Assessment of Immunogenicity from Human Cell Data

A method to predict the potential immunogenicity of novel proteins in humans would be an extremely valuable tool in the development of commercial proteins. The identification of T-cell epitopes in proteins of interest is a good first step, but the identification of presentable peptides does not provide information on the magnitude of the potential response, nor the population penetration of responses. The I-mune assay, developed to identify CD4+ T-cell peptide epitopes within proteins using community donors as the cell source, requires testing a large set of donors in order to select immunodominant epitopes on a population-wide basis. The data therefore represent the robustness of peptide responses on a human population basis, which includes the testing of a wide set of HLA alleles, present at their usual frequency within the population. The magnitude and frequency of peptide responses thus could be interpreted as the relative immunogenicity of the test protein. With this in mind, a statistical method to calculate the activity within an I-mune data set was developed and compared to data in the literature for the relative immunogenicity of a set of industrial proteins. This method based on a difference from linearity measure, results in a value between 0 and 2.0. The higher the value, the more activity is present within the data set. Using this method, the structure index values of four known industrial allergens, their immunogenic rankings in animal models and occupational exposure sensitization rates showed a reasonable correlation. The calculation also correctly defined known immunogenic proteins as "immunogenic" and putative nonimmunogenic proteins (such as human beta-2 microglobulin) as "nonimmunogenic." An analysis of the structure index values of known immunogenic versus putatively nonimmunogenic proteins provides a cut-off value for immunogenic proteins (Stickler et al., 2004a).

Using the structure index value as a guide, one could select protein homologues that display comparatively reduced values for further commercial development. This would be useful where a number of proteins with the same function could all fulfill the same application. In addition, epitope regions of proteins with relatively high values could be reduced to background "*in silico*," and whether epitope modifications reduce the structure index value to below the cut-off value could be determined. If a protein has a number of regions that would need modifications in order to reduce the structure index value, such a protein would not be suitable for further development.

20.4.2 *In Vivo* Testing

While the selection of modified epitope variants that significantly reduce MHC binding and/or functional response is a good start, the demonstration that an immune response to the intact protein molecule is reduced is highly desired. *In vivo* testing of modified whole protein variants is more problematic than selection of variant peptides. There are a number of *in vivo* methods for comparative testing of proteins, most involving the use of animals as a model system. For comparative testing of the relative allergenicity of bacterial proteins, both the guinea pig intratracheal test (GPIT) and mouse intranasal test (MINT) have been shown to produce results that correlate well with occupational sensitization rates (Robinson et al., 1998; Sarlo et al., 1997). However, while MINT and GPIT data correspond to human exposure data, testing human-epitope modified variants in these systems may not reflect differences that could be seen in humans (Birmingham et al., 2002; Bussiere, 2003). Differences in antigen presentation by mouse and guinea pig MHC molecules as compared to humans are the likely explanation. In some cases, inbred strains of mice do recognize the same epitopes as defined by human-derived data. However, careful screening of strains is necessary, which limits the use of this method. The use of HLA class II allele transgenic mice might be able to overcome this problem (Chen et al., 2003; Chen et al., 2002; Ito et al., 1996; Sonderstrup et al., 1999). Human transgenes encoding class II genes are transferred into mice genetically modified to reduce or eliminate murine class II expression. HLA transgenic mice have been used to epitope map known allergens and proteins implicated in autoimmunity (Black et al., 2002; Chapoval and David, 2003; Das et al., 2000; Papouchado et al., 2000; Raju et al., 2002). An additional confounding principle for the use of mouse models to predict immunogenicity is that humans are largely tolerant of highly expressed human proteins while animals are not (Dalum et al., 1996; Goodnow, 1992; Guery et al., 1995; Ramsdell and Fowlkes, 1992). Animals made transgenic with relevant human proteins are becoming available as test models for inducing immune responses to tolerizing proteins, but this method is unwieldy to apply generally. In addition, HLA allelic representation is an issue for these models.

20.4.3 The Proof is in the Pudding

No matter which *in vitro* or *in vivo* method shows that a modified protein now displays reduced immunogenicity, the final proof is in its demonstration in humans. There have been no published reports thus far on the treatment of human patients with successful CD4+ T-cell epitope–modified proteins. A report on the T-cell epitope modification of staphylokinase showed that the response to a specific T-cell epitope could be reduced, but the overall immune response to the protein remained intact (Warmerdam et al., 2002). The change from mouse toward more fully human therapeutic antibody constructs has coincided with a general reduction in the percent of anti-antibody responses (Chamberlain and Mire-Sluis, 2003), suggesting that a reduction in T-cell priming reduces overall immunogenicity. While the use of human sequence-derived proteins is not the final answer, it suggests that a reduction in sequences recognized as foreign by the adaptive immune system results in reduced immune responses. This observation continues to support the development of CD+ T-cell epitope-modified proteins. Since a number of companies now use this approach, it is anticipated that its utility will be validated in clinical trials within the next few years.

20.5 FUTURE PERSPECTIVES

Most computational methods for predicting T-cell epitopes focus on MHC/peptide binding. The accuracy of such methods to predict the presence of functional epitopes within proteins will increase as they develop and incorporate methods for modeling other important aspects of the cognate T-cell antigen-presenting cell interaction. Examples are the effect of flanking residues on MHC class II/peptide recognition by the T-cell receptor (Carson et al., 1997; Godkin et al., 1998, 2001; Moudgil et al., 1998; Vignali and Strominger, 1994) and the enzymatic processing of proteins (Watts and Powis, 1999). In addition, peptide presentation by different subsets of antigen-presenting cells (Crowe et al., 2003), and under different environmental stimuli such as the presence of type I and type II interferons (Heystek et al., 2003; Luft et al., 1998; Montoya et al., 2002), must also be carefully characterized. Any diversity in the processing and presentation of proteins must be incorporated into developing computational methods.

The generation of such an integrative algorithm that uses both motif and flanking residue data from HLA-specific, cell-surface presented peptides that have been confirmed with functional data, should allow the eventual establishment of a predictive model for both class I and II T-cell epitopes. Such a model would allow the discovery of functional epitopes utilizing human cell-derived data without the need for *in vitro* and *in vivo* validation. A validated, specific, and sensitive model will also be useful for predicting residue modifications for modulation of immunity, both in the construction of efficient vaccine constructs and in the replacement of immune response-initiating regions of therapeutic proteins. Finally, functional data from both previously sensitized and primary responses should be considered, and separate validations performed, since functional differences in peptide epitope recognition between naive and memory T cells for protein therapeutics and other consumer end-use proteins are seen in the human population.

As the current predictive models improve, they will become more useful for the *a priori* selection of regions of proteins of interest that need to be functionally tested *in vitro*. While computational methods are currently used (Schirle et al., 2001), their accuracy is not impressive, and accounts of missed epitopes suggest that significant epitopes are not predicted (Anthony et al., 2002; Warmerdam et al., 2002). The first critical step to improve current predictive methods will be to collect data for a larger set of MHC alleles, and incorporate epitope peptide sequence data from more processed and presented proteins.

REFERENCES

Altman, J.D., et al., Phenotypic analysis of antigen-specific T lymphocytes, *Science*, 274, 94–96, 1996.

Altuvia, Y., et al., Ranking potential binding peptides to MHC molecules by a computational threading approach, *J. Mol. Biol.*, 249, 244–250, 1995.

Anthony, D.D., and Lehmann, P.V., T-cell epitope mapping using the ELISPOT approach, *Methods*, 29, 260–269, 2003.

Anthony, D.D., et al., Comprehensive determinant mapping of the hepatitis C-specific CD8 cell repertoire reveals unpredicted immune hierarchy, *Clin. Immunol.*, 103, 264–276, 2002.

Antonelli, G., et al., Development of neutralizing antibodies in patients with relapsing-remitting multiple sclerosis treated with IFN-beta1a, *J. Interferon Cytokine Res.*, 18, 345–350, 1998.

Asai, T., et al., Evaluation of the modified ELISPOT assay for gamma interferon production in cancer patients receiving antitumor vaccines, *Clin. Diagn. Lab. Immunol.*, 7, 145–154, 2000.

Austyn, J.M., et al., Clustering with dendritic cells precedes and is essential for T-cell proliferation in a mitogenesis model, *Immunology*, 63, 691–696, 1988.

Basser, R.L., et al., Development of pancytopenia with neutralizing antibodies to thrombopoietin after multi-cycle chemotherapy supported by megakaryocyte growth and development factor, *Blood*, 99, 2599–2602, 2002.

Bercovici, N., et al., Multiparameter precursor analysis of T-cell responses to antigen, *J. Immunol. Methods*, 276, 5–17, 2003.

Birmingham, N., et al., Relative immunogenicity of commonly allergenic foods versus rarely allergenic and nonallergenic foods in mice, *J. Food Protein*, 65, 1988–1991, 2002.

Black, K.E., et al., HLA-DQ determines the response to exogenous wheat proteins: a model of gluten sensitivity in transgenic knockout mice, *J. Immunol.*, 169, 5595–5600, 2002.

Bottcher, M.F., et al., Immune responses to birch in young children during their first 7 years of life, *Clin. Exp. Allergy*, 32, 1690–1698, 2002.

Brady, R.O., et al., Management of neutralizing antibody to Ceredase in a patient with type 3 Gaucher disease, *Pediatrics*, 100, E11, 1997.

Brand, C.M., et al., Antibodies developing against a single recombinant interferon protein may neutralize many other interferon-alpha subtypes, *J. Interferon Res.*, 13, 121–125, 1993.

Braun, A., et al., Protein aggregates seem to play a key role among the parameters influencing the antigenicity of interferon alpha (IFN-alpha) in normal and transgenic mice, *Pharm. Res.*, 14, 1472–1478, 1997.

Brett, S.J., et al., Human T cell recognition of influenza A nucleoprotein. Specificity and genetic restriction of immunodominant T helper cell epitopes, *J. Immunol.*, 147, 984–991, 1991.

Brimnes, M.K., et al., Influenza virus-induced dendritic cell maturation is associated with the induction of strong T cell immunity to a coadministered, normally nonimmunogenic protein, *J. Exp. Med.*, 198, 133–144, 2003.

Brusic, V., et al., Prediction of MHC class II-binding peptides using an evolutionary algorithm and artificial neural network, *Bioinformatics*, 14, 121–130, 1998.

Bungy, G., et al., Mapping of T cell epitopes of the major fraction of rye grass using peripheral blood mononuclear cells from atopic and non-atopics. II. Isoallergen clone 5A of Lolium perenne group I (Lol p I), *Eur. J. Immunol.*, 24, 2098–2103, 1994.

Busch, R., et al., Accessory molecules for MHC class II peptide loading, *Curr. Opin. Immunol.*, 12, 99–106, 2000.

Bussiere, J.L., Animal models as indicators of immunogenicity of therapeutic proteins in humans, *Dev. Biol. (Basel)*, 112, 135–139, 2003.

Cameron, T.O., et al., Labeling antigen-specific CD4+ T cells with class II MHC oligomers, *J. Immunol. Methods*, 268, 51–69, 2002.

Carson, R.T., et al., T cell receptor recognition of MHC class II–bound peptide flanking residues enhances immunogenicity and results in altered TCR V region usage, *Immunity*, 7, 387–399, 1997.

Carter, L.L., et al., Regulation of T cell subsets from naive to memory, *J. Immunother.*, 21, 181–187, 1998.

Casadevall, N., et al., Pure red-cell aplasia and antierythropoietin antibodies in patients treated with recombinant erythropoietin, *N. Engl. J. Med.*, 346, 469–475, 2002.

Castelli, F.A., et al., HLA-DP4, the most frequent HLA II molecule, defines a new supertype of peptide-binding specificity, *J. Immunol.*, 169, 6928–6934, 2002.

Castelli, G., et al., Detection of anti-erythropoietin antibodies in haemodialysis patients treated with recombinant human-erythropoietin, *Pharmacol. Res.*, 41, 313–318, 2000.

Chamberlain, P., and Mire-Sluis, A.R., An overview of scientific and regulatory issues for the immunogenicity of biological products, *Dev. Biol. (Basel)*, 112, 3–11, 2003.

Chapman, J.R., ed., *Protein and Peptide Analysis by Mass Spectrometry*, vol. 61, Humana Press, Totowa, NJ, 1996.

Chapoval, S.P., and David, C.S., Identification of antigenic epitopes on human allergens: studies with HLA transgenic mice, *Environ. Health Perspect.*, 111, 245–250, 2003.

Chen, D., et al., Characterization of HLA DR3/DQ2 transgenic mice: a potential humanized animal model for autoimmune disease studies, *Eur. J. Immunol.*, 33, 172–182, 2003.

Chen, W., et al., Modification of cysteine residues in vitro and in vivo affects the immunogenicity and antigenicity of major histocompatibility complex class I–restricted viral determinants, *J. Exp. Med.*, 189, 1757–1764, 1999.

Chen, Z., et al., A 320-kilobase artificial chromosome encoding the human HLA DR3-DQ2 MHC haplotype confers HLA restriction in transgenic mice, *J. Immunol.*, 168, 3050–3056, 2002.

Chicz, R.M., et al., Specificity and promiscuity among naturally processed peptides bound to HLA-DR alleles, *J. Exp. Med.*, 178, 27–47, 1993.

Cook, S.D., et al., Serum IFN neutralizing antibodies and neopterin levels in a cross- section of MS patients, *Neurology*, 57, 1080–1084, 2001.

Crowe, P.D., et al., Differential signaling and hierarchical response thresholds induced by an immunodominant peptide of myelin basic protein and an altered peptide ligand in human T cells, *Hum. Immunol.*, 59, 679–689, 1998.

Crowe, S.R., et al., Differential antigen presentation regulates the changing patterns of CD8+ T cell immunodominance in primary and secondary influenza virus infections, *J. Exp. Med.*, 198, 399–410, 2003.

Croyle, M.A., et al., "Stealth" adenoviruses blunt cell-mediated and humoral immune responses against the virus and allow for significant gene expression upon readministration in the lung, *J. Virol.*, 75, 4792–4801, 2001.

Dalum, I., et al., Breaking of B cell tolerance toward a highly conserved self protein, *J. Immunol.*, 157, 4796–4804, 1996.

Das, P., et al., HLA transgenic mice as models of human autoimmune diseases, *Rev. Immunogenet.*, 2, 105–114, 2000.

Doherty, P.C., and Christensen, J.P., Accessing complexity: the dynamics of virus-specific T cell responses, *Ann. Rev. Immunol.*, 18, 561–592, 2000.

Dongre, A.R., et al., In vivo MHC class II presentation of cytosolic proteins revealed by rapid automated tandem mass spectrometry and functional analyses, *Eur. J. Immunol.*, 31, 1485–1494, 2001.

Eng, C.M., et al., Safety and efficacy of recombinant human alpha-galactosidase A-replacement therapy in Fabry's disease, *N. Engl. J. Med.*, 345, 9–16, 2001.

Fremont, D.H., et al., Structures of an MHC class II molecule with covalently bound single peptides, *Science*, 272, 1001–1004, 1996.

Geginat, G., et al., A novel approach of direct ex vivo epitope mapping identifies dominant and subdominant CD4 and CD8 T cell epitopes from Listeria monocytogenes, *J. Immunol.*, 166, 1877–1884, 2001.

Germain, R.N., Accessory cell stimulation of T cell proliferation requires active antigen processing, Ia-restricted antigen presentation, and a separate nonspecific 2nd signal, *J. Immunol.*, 127, 1964–1966, 1981.

Godkin, A.J., et al., Use of complete eluted peptide sequence data from HLA-DR and -DQ molecules to predict T cell epitopes, and the influence of the nonbinding terminal regions of ligands in epitope selection, *J. Immunol.*, 161, 850–858, 1998.

Godkin, A.J., et al., Naturally processed HLA class II peptides reveal highly conserved immunogenic flanking region sequence preferences that reflect antigen processing rather than peptide-MHC interactions, *J. Immunol.*, 166, 6720–6727, 2001.

Goodnow, C.C., Transgenic mice and analysis of B-cell tolerance, *Annu. Rev. Immunol.*, 10, 489–518, 1992.

Grayson, M.A., Ed., *Measuring Mass: From Positive Rays to Proteins*, Chemical Heritage Foundation, Philadelphia, 2002.

Gribskov, M., et al., Profile analysis: detection of distantly related proteins, *Proc. Natl. Acad. Sci. U.S.A.*, 84, 4355–4358, 1987.

Guery, J.C., et al., Constitutive presentation of dominant epitopes from endogenous naturally processed self-beta 2-microglobulin to class II-restricted T cells leads to self-tolerance, *J. Immunol.*, 154, 545–554, 1995.

Gulukota, K., et al., Two complementary methods for predicting peptides binding major histocompatibility complex molecules, *J. Mol. Biol.*, 267, 1258–1267, 1997.

Helms, T., et al., Direct visualization of cytokine-producing recall antigen-specific CD4 memory T cells in healthy individuals and HIV patients, *J. Immunol.*, 164, 3723–3732, 2000.

Henikoff, J.G., and Henikoff, S., Using substitution probabilities to improve position-specific scoring matrices, *Comput. Appl. Biosci.*, 12, 135–143, 1996.

Heystek, H.C., et al., Type I IFNs differentially modulate IL-12p70 production by human dendritic cells depending on the maturation status of the cells and counteract IFN-gamma-mediated signaling, *Clin. Immunol.*, 107, 170–177, 2003.

Ito, K., et al., HLA-DR4-IE chimeric class II transgenic, murine class II-deficient mice are susceptible to experimental allergic encephalomyelitis, *J. Exp. Med.*, 183, 2635–2644., 1996.

Julge, K., et al., Development of allergy and IgE antibodies during the first five years of life in Estonian children, *Clin. Exp. Allergy*, 31, 1854–1861, 2001.

Kao, H., et al., A new strategy for tumor antigen discovery based on in vitro priming of naive T cells with dendritic cells, *Clin. Cancer Res.*, 7, 773s–780s, 2001.

Keilholz, U., et al., Immunologic monitoring of cancer vaccine therapy: results of a workshop sponsored by the Society for Biological Therapy, *J. Immunother.*, 25, 97–138, 2002.

Kern, F., et al., T-cell epitope mapping by flow cytometry, *Nat. Med.*, 4, 975–978, 1998.

Kwok, W.W., et al., HLA-DQ tetramers identify epitope-specific T cells in peripheral blood of herpes simplex virus type 2-infected individuals: direct detection of immunodominant antigen-responsive cells, *J. Immunol.*, 164, 4244–4249, 2000.

Kwok, W.W., et al., Rapid epitope identification from complex class-II-restricted T-cell antigens, *Trends Immunol.*, 22, 583–588, 2001.

Langenkamp, A., et al., T cell priming by dendritic cells: thresholds for proliferation, differentiation and death and intraclonal functional diversification, *Eur. J. Immunol.*, 32, 2046–2054, 2002.

Lanzavecchia, A., and Sallusto, F., Dynamics of T lymphocyte responses: intermediates, effectors, and memory cells, *Science*, 290, 92–97, 2000.

Lanzavecchia, A., and Sallusto, F., Progressive differentiation and selection of the fittest in the immune response, *Nat. Rev. Immunol.*, 2, 982–987, 2002.

Lehner, P.J., The calculus of immunity: quantitating antigen processing, *Immunity*, 18, 315–317, 2003.

Li, J., et al., Thrombocytopenia caused by the development of antibodies to thrombopoietin, *Blood*, 98, 3241–3248, 2001.

Li, P., et al., Role of disulfide bonds in regulating antigen processing and epitope selection, *J. Immunol.*, 169, 2444–2450, 2002.

Lippolis, J.D., et al., Analysis of MHC class II antigen processing by quantitation of peptides that constitute nested sets, *J. Immunol.*, 169, 5089–5097, 2002.

Luft, T., et al., Type I IFNs enhance the terminal differentiation of dendritic cells, *J. Immunol.*, 161, 1947–1953, 1998.

Maecker, B., et al., The shared tumor-associated antigen cytochrome P450 1B1 is recognized by specific cytotoxic T cells, *Blood*, 102, 3287–94, 2003.

Mamitsuka, H., Predicting peptides that bind to MHC molecules using supervised learning of hidden Markov models, *Proteins*, 33, 460–474, 1998.

McMichael, A.J., and Kelleher, A., The arrival of HLA class II tetramers, *J. Clin. Invest.*, 104, 1669–1670, 1999.

Mehta-Damani, A., et al., Generation of antigen-specific CD4+ T cell lines from naive precursors, *Eur. J. Immunol.*, 25, 1206–1211., 1995.

Montoya, M., et al., Type I interferons produced by dendritic cells promote their phenotypic and functional activation, *Blood*, 99, 3263–3271, 2002.

Moudgil, K.D., et al., Modulation of the immunogenicity of antigenic determinants by their flanking residues, *Immunol. Today*, 19, 217–220, 1998.

Nepom, G.T., et al., Identification and modulation of a naturally processed T cell epitope from the diabetes-associated autoantigen human glutamic acid decarboxylase 65 (hGAD65), *Proc. Natl. Acad. Sci. U.S.A.*, 98, 1763–1768, 2001.

Novak, E.J., et al., Tetramer-guided epitope mapping: rapid identification and characterization of immunodominant CD4+ T cell epitopes from complex antigens, *J. Immunol.*, 166, 6665–6670, 2001.

Novitsky, V., et al., Magnitude and frequency of cytotoxic T-lymphocyte responses: identification of immunodominant regions of human immunodeficiency virus type 1 subtype C, *J. Virol.*, 76, 10155–10168, 2002.

Ovsyannikova, I.G., et al., Isolation and rapid identification of an abundant self-peptide from class II HLA-DRB1*0401 alleles induced by measles vaccine virus infection, *J. Immunol. Methods*, 246, 1–12, 2000.

Ovsyannikova, I.G., et al., Naturally processed measles virus peptide eluted from class II HLA-DRB1*03 recognized by T lymphocytes from human blood, *Virology*, 312, 495–506, 2003.

Papouchado, B.G., et al., HLA-DQ/human CD4-restricted immune response to cockroach allergens in transgenic mice, *Tissue Antigens*, 55, 303–311, 2000.

Parker, K.C., et al., Scheme for ranking potential HLA-A2 binding peptides based on independent binding of individual peptide side-chains, *J. Immunol.*, 152, 163–175, 1994.

Peakman, M., et al., Naturally processed and presented epitopes of the islet cell autoantigen IA-2 eluted from HLA-DR4, *J. Clin. Invest.*, 104, 1449–1457, 1999.

Phelps, R.G., et al., Presentation of the Goodpasture autoantigen to CD4 T cells is influenced more by processing constraints than by HLA class II peptide binding preferences, *J. Biol. Chem.*, 273, 11440–11447, 1998.

Pieper, R., et al., Biochemical identification of a mutated human melanoma antigen recognized by CD4(+) T cells, *J. Exp. Med.*, 189, 757–766, 1999.

Raju, R., et al., Cryptic determinants and promiscuous sequences on human acetylcholine receptor: HLA-dependent dichotomy in T-cell function, *Hum. Immunol.*, 63, 237–247, 2002.

Rammensee, H., et al., SYFPEITHI: database for MHC ligands and peptide motifs, *Immunogenetics*, 50, 213–219, 1999.

Ramsdell, F., and Fowlkes, B.J., Maintenance of in vivo tolerance by persistence of antigen, *Science*, 257, 1130–1134, 1992.

Reche, P.A., et al., Prediction of MHC class I binding peptides using profile motifs, *Hum. Immunol.*, 63, 701–709, 2002.

Reece, J.C., et al., Mapping the major human T helper epitopes of tetanus toxin: the emerging picture, *J. Immunol.*, 151, 6175–6184, 1993.

Reece, J.C., et al., Scanning for T helper epitopes with human PBMC using pools of short synthetic peptides, *J. Immunol. Methods*, 172, 241–254, 1994.

Robinson, M.K., et al., Use of the mouse intranasal test (MINT) to determine the allergenic potency of detergent enzymes: comparison to the guinea pig intratracheal (GPIT) test, *Toxicol. Sci.*, 43, 39–46, 1998.

Rosenberg, A.S., Immunogenicity of biological therapeutics: a hierarchy of concerns, *Dev. Biol. (Basel)*, 112, 15–21, 2003.

Rothbard, J.B., and Taylor, W.R., A sequence pattern common to T cell epitopes, *EMBO J.*, 7, 93–100, 1988.

Rup, B., Immunogenicity and immune tolerance coagulation Factors VIII and IX, *Dev. Biol. (Basel)*, 112, 55–59, 2003.

Salzberg, S.L., et al., Eds., *Computational Methods in Molecular Biology*, Amsterdam: Elsevier, 1998.

Sarlo, K., et al., Respiratory allergenicity of detergent enzymes in the guinea pig intratracheal test: association with sensitization of occupationally exposed individuals, *Fundam. Appl. Toxicol.*, 39, 44–52, 1997.

Scagnolari, C., et al., Neutralizing and binding antibodies to IFN-beta: relative frequency in relapsing-remitting multiple sclerosis patients treated with different IFN-beta preparations, *J. Interferon Cytokine Res.*, 22, 207–213, 2002.

Schirle, M., et al., Combining computer algorithms with experimental approaches permits the rapid and accurate identification of T cell epitopes from defined antigens, *J. Immunol. Methods*, 257, 1–16, 2001.

Schlienger, K., et al., Efficient priming of protein antigen-specific human CD4+ T cells by monocyte-derived dendritic cells, *Blood*, 96, 3490–3498, 2000.

Schmittel, A., et al., Evaluation of the interferon-gamma ELISPOT-assay for quantification of peptide specific T lymphocytes from peripheral blood, *J. Immunol. Methods*, 210, 167–174, 1997.

Schneider, S.C., et al., Cutting edge: introduction of an endopeptidase cleavage motif into a determinant flanking region of hen egg lysozyme results in enhanced T cell determinant display, *J. Immunol.*, 165, 20–23, 2000.

Schueler-Furman, O., et al., Structure-based prediction of binding peptides to MHC class I molecules: application to a broad range of MHC alleles, *Protein Sci.*, 9, 1838–1846, 2000.

Sette, A., et al., Prediction of major histocompatibility complex binding regions of protein antigens by sequence pattern analysis, *Proc. Natl. Acad. Sci. U.S.A.*, 86, 3296–3300, 1989.

So, T., et al., Contribution of conformational stability of hen lysozyme to induction of type 2 T-helper immune responses, *Immunology*, 104, 259–268, 2001.

Sonderstrup, G., et al., HLA class II transgenic mice: models of the human CD4+ T-cell immune response, *Immunol. Rev.*, 172, 335–343, 1999.

Steinman, R.M., The dendritic cell system and its role in immunogenicity, *Ann. Rev. Immunol.*, 9, 271–296, 1991.

Stickler, M.M., et al., CD4+ T-cell epitope determination using unexposed human donor peripheral blood mononuclear cells, *J. Immunother.*, 23, 654–660, 2000.

Stickler, M., et al., The HLA-DR2 haplotype is associated with an increased proliferative response to the immunodominant CD4+ T-cell epitope in human interferon-beta, *Genes Immun.*, 5, 1–7, 2004b.

Stickler, M., et al., An in vitro human cell-based assay to rank the relative immunogenicity of proteins, *Toxicol. Sci.*, 77, 280–289, 2004a.

Sturniolo, T., et al., Generation of tissue-specific and promiscuous HLA ligand databases using DNA microarrays and virtual HLA class II matrices, *Nat. Biotechnol.*, 17, 555–561, 1999.

Sun, J.Y., et al., Immune responses to adeno-associated virus and its recombinant vectors, *Gene Ther.*, 10, 964–976, 2003.

Tesfa, L., et al., Comparison of proliferation and rapid cytokine induction assays for flow cytometric T-cell epitope mapping, *Cytometry*, 52A, 36–45, 2003.

Tomlinson, A.J., and Chicz, R.M., Microcapillary liquid chromatography/tandem mass spectrometry using alkaline pH mobile phases and positive ion detection, *Rapid Commn. Mass Spectrom.*, 17, 909–916, 2003.

Urban, R.G., et al., The discovery and use of HLA-associated epitopes as drugs, *Crit. Rev. Immunol.*, 17, 387–397, 1997.

van der Merwe, P.A., and Davis, S.J., Molecular interactions mediating T cell antigen recognition, *Ann. Rev. Immunol.*, 21, 659–684, 2003.

Van Kaer, L., Major histocompatibility complex class I-restricted antigen processing and presentation, *Tissue Antigens*, 60, 1–9, 2002.

Van Kaer, L., et al., TAP1 mutant mice are deficient in antigen presentation, surface class I molecules, and CD4-8+ T cells, *Cell*, 71, 1205–1214, 1992.

Vignali, D.A., and Strominger, J.L., Amino acid residues that flank core peptide epitopes and the extracellular domains of CD4 modulate differential signaling through the T cell receptor, *J. Exp. Med.*, 179, 1945–1956, 1994.

Vignali, D.A., et al., Minute quantities of a single immunodominant foreign epitope are presented as large nested sets by major histocompatibility complex class II molecules, *Eur. J. Immunol.*, 23, 1602–1607, 1993.

Wadhwa, M., et al., Production of neutralizing granulocyte-macrophage colony-stimulating factor (GM-CSF) antibodies in carcinoma patients following GM-CSF combination therapy, *Clin. Exp. Immunol.*, 104, 351–358, 1996.

Wadhwa, M., et al., Immunogenicity of granulocyte-macrophage colony-stimulating factor (GM-CSF) products in patients undergoing combination therapy with GM-CSF, *Clin. Cancer Res.*, 5, 1353–1361, 1999.

Wadhwa, M., et al., Neutralizing antibodies to granulocyte-macrophage colony-stimulating factor, interleukin-1alpha and interferon-alpha but not other cytokines in human immunoglobulin preparations, *Immunology*, 99, 113–123, 2000.

Walden, P., T-cell epitope determination, *Curr. Opin. Immunol.*, 8, 68–74, 1996.

Warmerdam, P.A., et al., Elimination of a human T-cell region in staphylokinase by T-cell screening and computer modeling, *Thromb. Haemost.*, 87, 666–673, 2002.

Watts, C., Antigen processing in the endocytic compartment, *Curr. Opin. Immunol.*, 13, 26–31, 2001.

Watts, C., and Powis, S., Pathways of antigen processing and presentation, *Rev. Immunogenet.*, 1, 60–74, 1999.

Wertheimer, A.M., et al., Novel CD4+ and CD8+ T-cell determinants within the NS3 protein in subjects with spontaneously resolved HCV infection, *Hepatology*, 37, 577–589, 2003.

Xu, X.N., and Screaton, G.R., MHC/peptide tetramer-based studies of T cell function, *J. Immunol. Methods*, 268, 21–28, 2002.

Young, J.W., and Steinman, R.M., Dendritic cells stimulate primary human cytolytic lymphocyte responses in the absence of CD4+ helper T cells, *J. Exp. Med.*, 171, 1315–1332, 1990.

Yu, K., et al., Methods for prediction of peptide binding to MHC molecules: a comparative study, *Mol. Med.*, 8, 137–148, 2002.

SECTION V

Autoimmunity

CHAPTER 21

Chemical-Induced Animal Models of Human Autoimmunity

Pierluigi E. Bigazzi

CONTENTS

21.1 INTRODUCTION

The term "autoimmune disease" is commonly used to define disorders in which damage of various organs and tissues is mediated by immune ("autoimmune") responses, characterized by the production of antibodies ("autoantibodies") and/or effector T-cell reactions against normal components of the body or "autoantigens" (Bigazzi, 1997). A total of about 40 human diseases can be defined as autoimmune, and even though some of them are rare, the general prevalence of autoimmunity is quite high, approximating 5% of populations in the Western world (Parikh-Patel, 1999). According to some estimates, 1 in 31 Americans is likely to develop an autoimmune disease during life. In addition, the incidence of some autoimmune disorders such as thyroiditis and type 1 diabetes mellitus has recently increased, in part as the outcome of better diagnosis but possibly caused by actual growth. Therefore, autoimmune disease is a significant cause of morbidity and mortality.

Unfortunately, the causes of autoimmunity are still unknown. Viral, bacterial, and parasitic infections are frequently associated with autoimmune responses and/or disease; therefore, they have

Table 21.1   Chemical-Induced Animal Models of Human Autoimmune Disease

Xenobiotic	Animal Species (and inbred strain)	Autoimmune Responses[a]	Human Autoimmune Disease
Pristane	Mice [BALB/c, SJL/J]	ANA	SLE
Procainamide hydroxylamine	Mice [(C57BL6xDBA/2)F1]	ANA	Drug-related lupus
Mercury	Mice [C57B1/6, BALB/c, AKR]		
Bleomycin	Mice [BALB/c]	Increased TFG-β	Systemic sclerosis (fibrosis)
Mercury	Mice [SJL/J, other H2^S strains]	Autoantibodies to fibrillarin	Systemic sclerosis (subset)
Pristane, squalene	Mice [BALB/c, CBA, DBA] Rats [LEW]	Rheumatoid factor, autoantibodies to collagen type II	Rheumatoid arthritis
Streptozotocin	Mice [C57BL/6]	Autoantibodies to islet cell antigens	Type I diabetes mellitus
Iodine, methylcholanthrene, trypan blue	Mice [NOD], chickens [OS], rats [BUF, BB]	Autoantibodies to thyroglobulin	Autoimmune thyroiditis
Mercury, gold	Rats [BN, MAXX, DZB]	Autoantibodies to laminin	Nephrotic syndrome
6-Bromohexanoate-BSA	New Zealand white rabbits	AMA	Primary biliary cirrhosis
TNBSS, dextran	Mice [BALB/c, SJL/J]	Increased cytokines (IFN-γ, IL-4, etc.)	Inflammatory bowel disease

[a] ANA = autoantibodies to nuclear antigens; AMA = autoantibodies to mitochondrial antigens.

been suggested as possible triggers of autoimmunity (Gibofsky et al., 2000; McMurray, 2000; Naides, 2000). Xenobiotics may be alternative or additional etiologic agents of autoimmunity. They are environmental foreign substances of synthetic, natural, or biological origin, and comprise industrial chemicals, drugs, and cytokines. They may cause suppression or stimulation of immune responses, possibly leading to increased susceptibility to infection, development of neoplasia, allergic reactions, and autoimmunity (Bigazzi, 1999, 2000; Powell et al., 1999).

Investigations of the etiology of autoimmune diseases in humans are very difficult, because these conditions develop in a slow fashion and are usually diagnosed only after the occurrence of functional disturbances. On the other hand, such studies are possible in animal models. Experimental animals develop autoimmune disease after immunization with autoantigens or other manipulations. Animal autoimmunity can also develop spontaneously, that is, without human intervention. Some of the "spontaneous" animal models have characteristics similar to human organ- or tissue-specific disorders, whereas others resemble systemic disease. As is the case for human autoimmunity, we ignore the etiology of "spontaneous" animal autoimmune disease. However, recent evidence points to the causative role of microorganisms in some animal models (Greiner et al., 2001; Zipris, 2003). Alternatively, the importance of xenobiotics in the induction of autoimmunity has been demonstrated in other animal models. These studies have shown the connection between administration of the chemicals, genetic factors, kinetics of autoimmune responses, and eventual development of disease (see previous reviews in Bigazzi, 1999; Parikh-Patel, 1999; Powell et al., 1999; Bigazzi, 2000).

Due to editorial guidelines and space limitations, the present chapter is not meant to be comprehensive, but provides a critical review of the most useful chemical-induced animal models of autoimmunity. I have divided them in groups, on the basis of their similarities to human autoimmune diseases (Table 21.1). For the sake of brevity, I have focused on recent advances and preferentially quoted recently published reviews and only a few original papers.

21.2 SYSTEMIC LUPUS ERYTHEMATOSUS

Systemic lupus erythematosus (SLE) is a systemic autoimmune disease involving various organs and tissues (Lahita, 2000). Its clinical manifestations may range from mild to severe, and are due to an inflammatory process caused by the deposition of immune complexes. There are many "spontaneous" murine models of SLE, from the traditional NZW/NZB to MRL, BXB, and other strains of mice. Recent studies have shown that the autoimmune syndromes occurring in these mouse strains can be accelerated by the administration of certain chemicals and in particular mercury (Pollard et al., 1999, 2001). In this case, the xenobiotic does not induce autoimmunity *de novo*, but facilitates its development in animals genetically prone to autoimmune disease. This interesting observation contrasts with the fact that there are very few animal models of SLE induced by xenobiotics. It is often claimed that mercury and gold can induce an SLE-like syndrome in animals that under normal conditions do not develop autoimmune responses. However, neither the autoantibodies produced (antifibrillarin in H2s mice, antilaminin in BN rats) nor the severity of immune complex–mediated damage resembles SLE.

In my opinion the most interesting model with close similarities to SLE is obtained by a single intraperitoneal injection of pristane (2,6,10,14-tetramethylpentadecane) in inbred mice (Shaheen, 1999). Pristane is found in high concentration in mineral oil, a substance extensively used in food preparation and therefore ingested by humans on a daily basis. Intestinal absorption of mineral oil carries it to the portal lymph nodes, with possible formation of lipogranulomas. Inhalation of mineral oil induces similar lesions in the lung. Pristane-injected BALB/c mice develop autoanti-bodies to nRNP/Sm, Su, and ssDNA, whereas similarly treated SJL/J mice produce antiribosomal P autoantibodies. These autoimmune responses result in an immune complex–mediated glomeru-lonephritis in both strains of mice. Pristane-induced lupus is associated with marked hypergamma-globulinemia that is influenced by microbial stimulation. This animal model with considerable similarities to human SLE can provide interesting explanations of the various molecular and cellular pathways involved in autoimmunity induction. For example, recent experiments have shown that CD1d may have an immunoregulatory role in the development of this model (Yang, 2003).

21.3 DRUG-RELATED LUPUS

The administration of certain drugs results in a syndrome defined as "drug-related lupus" (also called "drug-induced lupus") (Hess, 1999). As compared to SLE this syndrome is characterized by the absence of autoantibodies to double-stranded DNA, scarce or absent renal involvement, and generally milder clinical features. It usually regresses when the offending chemical is withdrawn. Its prevalence is still unknown, but according to some estimates it may affect approximately 15,000 to 30,000 patients in the United States alone (Mongey and Hess, 1999; Vaile and Russell, 2000). Animal models induced by the administration of isoniazid, hydralazine, procainamide, propylthio-uracil, certain aminoacids, or oleic acid anilide have been previously reviewed (Rubin, 1999). More recently, a murine model of drug-induced lupus has been obtained by the intrathymic injection of procainamide-hydroxylamine (a reactive metabolite of procainamide) (Rubin and Kretz-Rommel, 1999). Of particular interest is also an SLE model induced in DBA/2 mice injected with CD4[+] T cells that had been previously treated *in vitro* with 5-azacytidine (5-azaC) (Richardson, 2003). Studies of this model have confirmed the observation that failures to maintain DNA methylation result in autoimmune responses. Drugs such as procainamide or hydralazine provide an example: They cause defective maintenance of DNA methylation and may induce drug-related lupus. Finally, mice from some inbred strains may produce autoantibodies against nuclear antigens (histone, chromatin) after administration of mercury. This is a different response from that observed in H2s mice, which mostly produce autoantibodies to fibrillarin, a nucleolar antigen (see Section 21.4). Surprisingly, even strains that are not usually considered good mercury responders (e.g., C57BL/6, BALB/c, AKR) have been reported to produce autoantibodies that react only with nuclear antigens (Hu et al., 1999; Pollard, 1999, 2001). Similarly to patients with drug-induced lupus, these mice have scarce or absent immune deposits in their kidneys.

To conclude this section, I include two animal models that have only limited similarities to drug-related lupus: mice with a toxic oil syndrome and rats with penicillamine-induced autoimmunity. The human toxic oil syndrome (TOS) is a multisystemic disease that occurred in Spain during 1981 and affected about 20,000 people, 400 of whom died (Koller et al., 2002). Some of these patients exhibited an autoimmune-like syndrome with immunologic and clinical features resembling (but not identical to) those of scleroderma, SLE, or Sjögren's syndrome. It is mentioned here because it is one notable example of xenobiotic-induced human autoimmune disease. It also demonstrates that autoimmunity caused by chemicals does not necessarily repeat all aspects of "idiopathic" autoimmunity. Animal models of TOS are quite scarce; however, one may have recently been obtained in MRL/lpr mice, which showed accelerated systemic autoimmune disease after treatment with oils containing a mixture of diesters of phenilamine-propane-diol (Koller et al., 2002).

Interest in another animal model has been stimulated by the clinical findings observed in approximately 50 to 70% of patients treated with D-penicillamine because of rheumatoid arthritis, cystinuria, or Wilson's disease. These patients develop a variety of autoimmune disorders that range from a lupus-like syndrome to pemphigus, myasthenia gravis, or membranous nephropathy. The mechanisms leading to these effects are poorly understood and have been investigated in BN rats after prolonged oral administration of D-penicillamine. Weight loss, dermatitis, and production of antinuclear antibodies were observed in approximately 73% of D-penicillamine–treated BN rats (Sayeh and Uetrecht, 2001). Similarities to the autoimmune syndrome induced by mercury in rats of the same strain (see Section 21.8) suggested that the process might be mediated by Th2 cytokines. However, more recent investigations indicate that both Th1 and Th2 cytokines are involved in this animal model of penicillamine-induced autoimmunity (Sayeh and Uetrecht, 2001). It is of considerable therapeutic interest that this syndrome is suppressed by the administration of misoprostol (a prostaglandin-E analog that inhibits Th1 cytokines) and aminoguanidine (an inhibitor of iNOS) (Sayeh and Uetrecht, 2001; Seguin et al., 2003). On the other hand, nonselective cyclooxygenase inhibitors seem to increase the incidence of the disease (Seguin et al., 2003).

21.4 SYSTEMIC SCLEROSIS

Systemic sclerosis (SSc), also called scleroderma, is a clinically heterogeneous autoimmune disease characterized by variable involvement of the skin and internal organs as well as production of autoantibodies to nuclear and nucleolar antigens (Smith, 2000; White, 2001). The various specificities of these autoantibodies are useful in the identification of disease subsets, degree of specific organ involvement, and ultimate outcome. Clinical and serologic heterogeneity have suggested that SSc might be a family of diseases with different etiologies and pathogenetic processes, a suggestion that still lacks proof. Thus, it may be important to explore the cellular and molecular pathways underlying the various subsets of SSc. In particular, some SSc patients at risk of developing internal organ involvement are characterized by the production of autoantibodies to fibrillarin (a nucleolar antigen also defined as U3-RNP) (Tormey, 2001; Mahler et al., 2003). A high frequency of isolated pulmonary hypertension, myositis, and renal disease occurs at a young age in these patients, mostly of African-American and Afro-Caribbean origin.

Several environmental factors (e.g., silica) and drugs have been associated with the development of SSc and scleroderma-like disorders; however, there is no definite proof of a role of xenobiotics in SSc. Its pathogenesis is also unknown, even though numerous fibrogenic cytokines are involved in the fibrotic process. As previously mentioned, our present ignorance justifies the use of animal models to gain some insight in the etiology and pathogenesis of SSc. Considering the clinical and immunologic heterogeneity of this condition, it is not surprising that animal models similar to the various subsets of SSc are not available. On the other hand, some aspects of the disorder are reproduced in "spontaneous" models, such as tight skin mice and UCD200 chickens. With a few notable exceptions, studies of animal models of SSc have focused on the fibrotic process. For example, repeated local treatment with bleomycin, an antitumor antibiotic, induces dermal sclerosis in Balb/c mice after 4 weeks (Yamamoto, 2001). The histopathology observed in these mice mimicks that of human SSc, with thickened collagen bundles and deposits of homogeneous material with cellular infiltrates. Cytokines such as TGF-β are extremely important for the development of sclerosis in bleomycin-treated mice. This is an interesting finding that may suggest appropriate therapeutic interventions. However, there is also a need to explore the initial events leading to autoimmunity in SSc.

Mercury-treated mice of certain inbred strains may provide such an opportunity. Exposure to mercury results in a murine model that can allow the study of both environmental factors and cell contact-dependent lymphocyte stimulatory pathways in SSc. Mercuric chloride administered to mice from H-2s susceptible strains (e.g., A.SW, B10.S, SJL/J) induces the production of autoantibodies reacting with fibrillarin, a 34-kDa nucleolar U3 RNP protein. These autoantibodies recognize the same epitopes detected by autoantibodies from some patients with SSc (see above) (Yang, 2001). The antinucleolar autoimmune responses of $H2^s$ mice persist for 10 to 12 weeks or longer. These mice also show involvement of internal organs, since their kidneys usually contain granular immune deposits, mostly localized in the glomerular mesangium and the walls of interlobular arterioles and arteries. Similar deposits are found in the wall of splenic arteries and arterioles as well as intramyocardial arteries. The autoimmune effects of mercury in mice are in large part determined by the MHC, but background non-MHC genes are also involved. Initially, it was thought that murine autoimmunity caused by mercury was due to hyperactivity of Th2 cells secreting IL-4 associated with decreased Th1 cytokine secretion. However, the hypothesis of an imbalance between Th1 and Th2 cells in this chemically induced model of autoimmunity has not been confirmed (reviewed in Bigazzi, 2000). Studies of gene-targeted mice and observations from other models of autoimmunity (murine lupus, experimental autoimmune myasthenia gravis, etc.) suggest that early production of IFN-γ is essential for the initiation of autoantibody-mediated autoimmune disease.

21.5 RHEUMATOID ARTHRITIS

Rheumatoid arthritis (RA) is a systemic autoimmune syndrome, characterized by an inflammatory response in the synovium leading to the formation of tissue that is invasive and destructive of joints (Weyand and Goronzy, 2000). Autoimmune responses to a variety of autoantigens are thought to play a major pathogenic role in RA, but its cause and pathogenesis are still unknown and are currently being investigated in various animal models.

Autoimmune arthritis with similarities to RA has been induced in experimental animals by the injection of various xenobiotics (pristane, squalene, complete Freund's adjuvant). The intraperitoneal injections of pristane in certain susceptible strains of mice (BALB/c, CBA, DBA) results in an inflammatory joint disease defined as pristane-induced arthritis (PIA) (Lu and Holmdahl, 1999). Preferential involvement of ankle and wrist joints develops between 100 and 200 days after the initial pristane injection. A variety of autoantibodies, including rheumatoid factor, autoantibodies to collagen type II, and antibodies to stress proteins are detected in the serum of affected mice. Polymorphonuclear cell infiltration, synoviocyte hyperplasia, cartilage erosion, and formation of pannus characterize the histopathology. Contact with environmental microorganisms is necessary for the induction of PIA, which does not develop in specific pathogen-free mice maintained in isolator cages. CD4+ T cells specific for mycobacterial heat shock protein were found to protect against PIA. Intranasal administration of type IX collagen also inhibits rat PIA (Carlson et al., 2000). Finally, the subcutaneous injection of pristane or squalene (an endogenous cholesterol precursor) in DA rats results in an autoimmune arthritis, with characteristics similar to those of human RA (Lu and Holmdahl, 1999; Carlson et al., 2000).

21.6 TYPE 1 DIABETES MELLITUS

Type 1 or insulin-dependent diabetes mellitus (IDDM) is a disorder characterized by hyperglycemia, polyuria, polydipsia, and weight loss in children and young adults (Greiner et al., 2001). It is responsible for approximately 10% of all cases of diabetes mellitus, and its pathogenesis is likely autoimmune. Concordance in identical twins is approximately 50%, suggesting the causative role of environmental factors. There are several well-investigated animal models of type 1 diabetes, but currently all research directed at prediction and prevention (i.e., "translational" research) involves the NOD mouse and the BB rat (Greiner et al., 2001). A variety of environmental conditions, including diet and viral infections, can modulate the onset of diabetes in these animals (Greiner et al., 2001). Protein derived from cow's milk is one of the factors that can influence incidence and/or onset of human type 1 diabetes. However, its relevance remains controversial, and recent research interest is mostly focused on viral causes of these animal models.

At the same time, less attention seems to be given to an older model of type 1 diabetes induced by the administration of low doses of streptozotocin (SZ), a naturally occurring compound isolated from *Streptomyces achromogenes*. Repeated injections of this chemical into mice or rats of certain inbred strains induce insulitis and hyperglycemia (Pieters and Albers, 1999; Mensah-Brown et al., 2002; Muller et al., 2002; Nierkens, 2002; Nicoletti, 2003). The insulitis is a T-cell–dependent process, likely mediated by IFN-γ even though other cytokines (IL-2, IL-4, TNF-α) are also found within the pancreatic lesions. In addition, SZ-treated animals may produce autoantibodies to islet cell antigens. SZ-induced diabetes is favored by a concurrent viral infection or by the administration of adjuvants (e.g., CFA). Recent studies have shown that the popliteal lymph node response to SZ is under type 1, MHC class–I restricted CD8$^+$ T-cell control (Choquet-Kastylevsky et al., 2000). In conclusion, the SZ-induced animal model of autoimmune diabetes can still contribute useful information. An excellent example has been provided by studies of transgenic mice expressing activated Notch3 on thymocytes and T cells (Anastasi, et al., 2003). These animals do not develop autoimmune diabetes after SZ administration, a failure associated with an increase of CD4$^+$CD25$^+$ T-regulatory cells.

21.7 AUTOIMMUNE THYROIDITIS

Autoimmune thyroiditis (also known as "Hashimoto's disease") is the most common form of thyroiditis (Herold and Sarne, 2000). It is characterized by enlargement of the thyroid gland and variable changes in thyroid function that may eventually result in hypothyroidism. Circulating autoantibodies to thyroid peroxidase and thyroglobulin are usually detected in affected subjects. Thyroid damage is likely mediated by T-lymphocytes, but etiology and pathogenesis of the disorder are still unclear in spite of numerous studies in patients and various animal models.

Iodine, possibly because of its effects on thyroglobulin epitopes, can induce autoimmune thyroid disease in genetically susceptible animals (rats, mice, and chickens) (Ruwhof and Drexhage, 2001; Lam-Tse et al., 2002; Rose et al., 2002). Approximately 20 to 26% of inbred BUF rats develop "spontaneous" autoimmune thyroiditis (SAT), characterized by inflammatory infiltration of the thyroid and production of autoantibodies to thyroglobulin (Bigazzi, 1998). The administration of iodine (and/or neonatal thymectomy) increases the percentage of affected animals. DP BB rats are another strain affected by iodine. These animals spontaneously develop autoimmune type 1 diabetes (Greiner et al., 2001). With age, some sublines of DP BB rats produce autoantibodies to thyroglobulin, and are affected by autoimmune thyroiditis. Various experiments have shown that the onset of thyroiditis is accelerated in DP rats exposed to high levels of iodine in their drinking water.

NOD-H-2h4 mice can also be affected by iodine (Rasooly et al., 1996). Animals from this strain, obtained from a cross of NOD and B10.A(4R), spontaneously develop autoimmune thyroiditis but not IDDM. When treated with iodine in their drinking water, approximately 54% of female and 70% of male mice developed autoimmune thyroiditis at a time when only 5% of untreated controls showed thyroid lesions. Finally, approximately one-fourth of chickens of the Cornell C strain (CS) receiving excess dietary iodine produced antibodies to thyroglobulin, T3, and T4 that persisted throughout the course of the experiment (Sundick, 1990). The administration of iodized water also resulted in lymphocytic thyroiditis, with a dose-related response. In contrast, four other strains of chickens did not develop autoimmunity after exposure to iodine. CS chickens on a high iodide diet likely synthesize a different thyroglobulin molecule with many iodide atoms that stimulates an autoantibody response to highly iodinated epitopes.

Other xenobiotics (3-methylcholanthrene, trypan blue) induce autoimmune thyroiditis in BUF rats (Bigazzi, 1993). Early studies did not reveal major differences between SAT and chemical-induced thyroiditis. However, rats treated with trypan blue or 3-methylcholanthrene were later found to have a very mild disease when compared to neonatally thymectomized animals (Cohen and Weetman, 1987). Both SAT and chemical-induced autoimmune thyroiditis of BUF rats seem to have disappeared in recent years, possibly because of improved breeding and housing conditions resulting in microorganism-free animals (Bigazzi, 1998). Interestingly, mercury stimulates the production of autoantibodies to thyroglobulin in BB-DP rats, but does not increase the incidence of thyroiditis (Kosuda et al., 1997). These effects are somewhat similar to those observed after lithium administration. Lithium does not induce autoimmune thyroiditis *de novo*, but favors autoimmune responses to thyroglobulin. Rats of the AUG strain immunized with rat thyroglobulin in CFA and exposed to lithium show an increased production of autoantibodies to thyroglobulin when the drug was administered during the initial stages. These autoantibodies decreased if lithium was administered at later stages when the thyroiditis was spontaneously resolving. Lithium had no effects on the degree of lymphocytic infiltration of the thyroid or on serum TSH. It did not induce antithyroglobulin production in a group of naive unimmunized rats, suggesting that the drug may have an immunomodulatory effect, but cannot induce autoimmune thyroid disease.

21.8 NEPHROTIC SYNDROME

The nephrotic syndrome is characterized by increased glomerular permeability with massive leakage of albumin ("proteinuria," more than 3.5 g/day), hypoalbuminemia, and edema (Robson and Walport, 2000). These functional changes can be associated with a spectrum of morphological alterations of renal glomeruli; however, membranous, nephropathy is the most common kidney disease responsible for the occurrence of the nephrotic syndrome in adults. Membranous nephropathy is a distinct noninflammatory histopathologic entity, characterized by thickening of glomerular capillary walls. The disease has a slowly progressive course, sometimes with spontaneous remissions; however, approximately half of the patients develop renal insufficiency or end-stage disease. Its cause is unknown in most patients (idiopathic membranous nephropathy, which accounts for 62 to 86% of all membranous nephropathies). In a minority of subjects (approximately 17%), the disease is related to various antigens or environmental agents (secondary membranous nephropathies associated with infectious diseases, neoplasias, gold, mercury, and therapeutic and illicit drugs). The likely pathogenesis of membranous glomerulonephropathy is immunologically mediated glomerular damage, possibly through the *in situ* formation of immune complexes and/or binding of antibodies to epitopes of the renal glomerular basement membrane (GBM). However, a putative antigen has been identified in only a few cases of human membranous nephropathy, and both mechanisms of tissue injury and factors affecting the progression of disease are still not completely understood. Experimental animal studies of immunologically mediated membranous nephropathy have been performed by active immunization with renal antigens in complete Freund's adjuvant or by passive transfer of antibodies or lymphocytes obtained from animals previously immunized with renal antigens (Robson and Walport, 2000). These studies have provided important information but, possibly because of the immunization protocols and the use of complete Freund's adjuvant, they have biased the renal disease to an inflammatory pattern that is usually not observed in human membranous glomerulonephropathy. The only studies obviating this bias are based on the experimental induction of autoimmune responses to renal autoantigens as a result of the exposure to xenobiotics such as mercury and gold.

The repeated administration of $HgCl_2$ to rats of some inbred strains results in both autoimmune responses and disease (reviewed in Bigazzi, 2000; Bigazzi et al., 2003). Mercury-treated Brown Norway (BN), MAXX, and Dorus Zadel Black (DZB) rats produce high levels of autoantibodies to laminin 1. Low levels of autoantibodies to type IV collagen, heparan sulfate proteoglycan, entactin, and thyroglobulin have been detected in the circulation of BN rats after exposure to mercury, but they do not seem to correlate with proteinuria. Autoantibodies to laminin 1 are first detectable in the circulation after 6 to 7 days, reach peak titers and incidence by day 14 to 15, and then decrease within 20 to 30 days. Other effects of mercury include splenomegaly, lymph node hyperplasia, and thymic atrophy. Renal immune deposits are observed after mercury treatment with an initial stage showing linear deposits of IgG, containing autoantibodies to laminin 1, at the level of both GBM and tubular basement membrane (TBM). A second stage shows granular deposits of IgG both in GBM and TBM, again containing autoantibodies to laminin 1. The histopathology of the kidneys consists mostly in a membranous glomerulonephropathy without a major inflammatory component.

Damage to other tissues was not described in early studies of mercury-treated BN rats, but later reports have described a graft-versus-host-like pathology with cutaneous, intestinal, hepatic, salivary gland, and joint inflammatory lesions (reviewed in Bigazzi, 2000; Bigazzi et al., 2003). Our own experience did not agree with those reports; therefore, we recently examined paraffin-embedded H&E-stained as well as plastic-embedded, toluidine-blue–stained thin sections of tissues from a large number of mercury-treated and control BN rats. In this retrospective study, we did not detect any histopathologic alterations in the skin and intestine of these animals. Gross examination of these tissues had previously shown no lesions. Thus, two primary target organs for cell-mediated immune damage during graft-versus-host disease were completely normal. Instead, we did find signs of acute and/or chronic inflammation in other tissues from control and experimental rats. Areas of chronic inflammation were noted in the thyroids of control (6%) and mercury-treated (4%) BN rats. Signs of acute and chronic inflammation were detected in the pancreas (27% in both experimental and controls), liver (60% in control and 27% experimental) and lungs (100% of control and 84% mercury-injected BN rats). The differences between our results and those published by other authors are likely due to synergistic effects of mercury, the stress of anesthesia, and repeated injections of mercury or distilled water in rats that were not specific-pathogen free. Our finding of acute and chronic inflammatory changes in the lungs and other tissues of control as well as mercury-treated rats confirms this possibility.

21.9 PRIMARY BILIARY CIRRHOSIS

Primary biliary cirrhosis (PBC) is a chronic inflammatory disease of the liver, characterized by obliteration of small intrahepatic bile ducts and portal inflammation resulting in fibrosis and eventually liver cirrhosis (Parikh-Patel, 1999; Wesierska-Gadek et al., 2000). It is primarily a disease of middle-aged women, with production of autoantibodies to mitochondrial antigens and in particular the E2 component of the 2-oxo acid dehydrogenase pathway (PDC-E2). It is most common in Western countries, and its incidence may be increasing. Elegant structural analysis studies have recently suggested that xenobiotics metabolized in the liver may modify PDC-E2 and initiate autoimmune responses to this autoantigen (Long et al., 2001). This hypothesis has been pursued in an experimental animal model: rabbits immunized with 6-bromohexanoate (a xenobiotic organic compound) coupled to BSA developed antimitochondrial autoantibodies, the hallmark of human PBC (Leung et al., 2003). These autoantibodies react with PDC-E2 and other principal autoantigens of PBC. They are true autoantibodies, since they recognize rabbit mitochondrial proteins by immunoblotting and give a typical antimitochondrial staining pattern on rabbit epithelial cells. This model is unique since mice from five different strains did not produce antimitochondrial antibodies after similar immunizations. Additional studies are necessary to ascertain whether liver histopathology lesions similar to human PBC occur in experimental animals with autoimmune responses to mitochondria.

21.10 INFLAMMATORY BOWEL DISEASE

Inflammatory bowel disease (IBD) comprises two forms of chronic intestinal inflammation, Crohn's disease and ulcerative colitis, which differ in clinical presentation, course, diagnostic criteria, prognosis, and response to treatment (Wesierska-Gadek et al., 2000). Crohn's disease is a chronic relapsing disorder that can affect the entire gastrointestinal tract, and is characterized by focal, symmetric, transmural, and occasionally granulomatous inflammation. A chronic inflammatory process confined to the mucosa and superficial submucosa of the colon and rectum characterizes ulcerative colitis. Cause and pathogenesis of IBD are unknown, even though inflammatory cytokines are considered to have a main pathogenic role, possibly with different expression in its two major forms. Recently, mucosal inflammation resembling IBD has been observed in laboratory animals after a variety of experimental manipulations (including genetic engineering) (Strober et al., 2002). Some of the available animal models are caused by the administration of certain xenobiotics. Irrespective of the cause, the eventual outcome is an inflammatory process driven by an excessive production of IL-12, IFN-γ and TNF-α (so-called Th1 cytokines), or IL-4 and IL-5 (so-called Th2 cytokines).

An experimental colitis that has histologic resemblance to human ulcerative colitis is induced by skin presensitization with oxazolone (4-ethoxymethylene-2-phenyl-2-oxazoline-5-one) followed by intrarectal injection with the same compound. Oxazolone colitis (OC) is mediated by NK T cells, since mice depleted of these cells or lacking NK T-cell functions do not develop the disease (Heller et al., 2002). IL-13, produced by NK T-lymphocytes is a significant factor in the pathogenesis of OC, and its neutralization by specific antibodies inhibits the disease. Another model of chronic colitis is experimentally induced by the intrarectal injection of the hapten 2,4,6-trinitrobenzene sulfonic acid (TNBS) in BALB/c and SJL/J mice (Neurath, 1995; Dohi et al., 2001). The disorder induced in SJL/J mice has the characteristics of a Th1 model. IFN-γ-producing T cells are involved in the lesions, and treatment with antibodies against IL-12 inhibits the disease. On the other hand, the disease of BALB/c mice has Th2 characteristics. An older model of xenobiotic-induced IBD is obtained by the administration of dextran sulfate sodium (DSS) in drinking water (Strober et al., 2002; Tsuchiya, 2003). The T-cell response in this condition varies, with an initial Th1 response followed by a mixed Th1/Th2 stage. The severity of DSS-induced colitis may be regulated by cytokines secreted by $\gamma\delta$-T cells. The impact of these various animal models on therapeutic trials for human IBD has been considerable and has led to the development of agents (monoclonal antibodies, soluble receptors) with great curative potential.

21.11 CONCLUSIONS AND FUTURE NEEDS

This brief review of chemical-induced animal models of human autoimmunity has attempted to summarize some of the most recent experimental work in this area. I will now try to derive a few conclusions and suggestions for future studies. First, various pitfalls should be avoided by investigators of animal models (Bigazzi, 1998). Investigators using mice often disregard evidence obtained in rats or other animal species. Progress from experimentation with mice is undeniable; however, differences between humans and experimental animals must be kept in mind. Autoantibodies to mitochondria have been induced in rabbits but not mice by immunization with xenobiotics, possibly because of species differences in the metabolism and toxicity of xenobiotics (Leung et al., 2003). The diabetes of the NOD mouse has been cured by numerous procedures that have no effects on human diabetes (Greiner et al., 2001). Oral tolerance inhibited experimental autoimmune encephalitis in mice and rats, but did not improve multiple sclerosis in human therapeutic trials. Another pitfall that can lead to inaccurate and misleading conclusions is the frequent division between investigators utilizing models of systemic autoimmunity and those studying organ-specific autoimmune disease. Finally, the tendency to disregard older evidence in favor of more attractive new experimentation and an exaggerated propensity to follow novel trends (so-called "paradigms") can also have a negative influence. As an example, the Th1/Th2 dichotomy may be useful as a reductionist model, but does not seem to adequately explain the mechanisms involved in chemical-induced autoimmunity or idiopathic autoimmunity.

Progress will be easier if we recognize that most animal models (either spontaneous or experimentally induced) differ in some aspects from human autoimmunity (Bigazzi, 1998; Greiner et al., 2001). The pathophysiology created by experimental immunization with autoantigens mixed with adjuvant is obviously artificial. Transgenic and knockout mice are not completely natural because of genetic manipulations. Spontaneous animal models of autoimmunity show differences based on species and inbred strain. Similarly, chemical-induced animal models of autoimmunity seldom resemble human disease. Thus, translational research (aimed at preventive and therapeutic measures applicable to humans) requires great caution and careful evaluation of advantages and disadvantages of each animal model (Greiner et al., 2001).

The available evidence briefly reviewed in this chapter shows that xenobiotics can induce autoimmune responses and disease in experimental animals. The pathogenesis of tissue damage in chemical-induced animal models of autoimmunity varies according to model and species. Autoantibodies to nuclear antigens are responsible for the renal damage of the pristane murine model. Interestingly, autoantibodies against fibrillarin induced by mercury in H2s mice cause immune deposits at the mesangial level, but do not seem to have a strong pathogenic potential. On the other hand, autoantibodies against laminin induced by mercury treatment in BN, MAXX, and DZB rats have a strong correlation with proteinuria. Effector T cells are variously involved in STZ-induced diabetes and IBD models.

Chemicals may cause autoimmune responses and disease by a variety of mechanisms. They can alter the structure of autoantigens and/or the function and viability of immune cells through their effects on protein thiols, calcium channels, and other non-epitope–dependent mechanisms (Pollard, 1997; Lutz et al., 1998; Manome et al., 1999; Audesirk et al., 2000; Pheng et al., 2000; Araragi et al., 2003). The importance of xenobiotics in autoimmune disease has also been strengthened by recent evidence of the interactions between innate and adaptive immunity (Shi et al., 2001). Xenobiotics can affect dendritic cells, macrophages, NK cells, and other components of the innate immune system. In addition, studies of chemical-induced animal models have shown that both microorganisms and xenobiotics may be necessary to induce autoimmune responses and disease in genetically predisposed individuals. Their combined activity usually induces more intense dysregulation of the cytokine network, powerful stimulation of the appropriate subsets of CD4+ and CD8+ T-lymphocytes, proliferation of B-lymphocytes, and overt autoimmune disease. The effects of the administration of CFA and IFA provide good examples of the combined role of chemicals and microorganisms in the induction of autoimmunity.

Finally, various ecogenetic and immunogenetic factors are involved in xenobiotic-induced autoimmune responses. The relevant ecogenetic factors likely control the metabolism of xenobiotics, susceptibility to bacteria and viruses, cell cycle, apoptosis, and inflammation. Immunogenetic elements regulate innate and adaptive immune responses leading to autoimmunity. Recent progress in genotyping and proteomics methods will hopefully improve our understanding of the factors involved (Maas et al., 2002; Morahan and Morel, 2002; Landi et al., 2003). These procedures should be preferentially applied to chemical-induced animal models to provide guidelines for human studies.

How can we define the best xenobiotic-induced animal models of autoimmune disease? At present, pristane-induced SLE, mercury-induced autoimmunity, and some IBD and PBC models are the most useful since they help us understand the possible etiology of human autoimmune disease. However, an analysis of the published literature suggests that the majority of immunologists agree that "the virus is the most plausible etiologic agent of autoimmune disease" (Morse et al., 1999). In contrast, relatively scarce attention seems to be given to the autoimmune effects of xenobiotics. The viral hypothesis has been investigated for a long time, but with a few exceptions there is still no solid proof that viruses or other microorganisms are the initial causative agent of autoimmunity. On the other hand, we have reliable evidence in favor of chemical-induced human autoimmune disease, as shown by the numerous patients affected by the eosinophilia-myalgia syndrome and the toxic oil epidemic.

The current focus on a viral etiology of autoimmune disease may be due to the historical development of immunology as an offshoot of microbiology and the intellectual connections between the two areas. Thus, adaptive immune responses to epitopes of microorganisms and possible molecular mimicry with autoantigens may appear most meaningful for the development of autoimmunity, whereas similar interactions with chemicals are thought less likely. Another reason for the rather scarce scientific interest in xenobiotic-induced autoimmunity may be the present legal climate, exemplified by the well-known silicone gel implant story. Several studies reported the association of connective tissue diseases with silicone gel implants (Espinoza and Cuellar, 2000). These conditions, often atypical and not easily diagnosed, resulted in a number of legal suits and the decision by the U.S. Food and Drug Administration to restrict the use of silicone gel implants. Even though epidemiological studies have not confirmed well-defined pathologic consequences of these implants in humans, it is still questionable whether the suspected autoimmune effects of silicone have been investigated to complete scientific satisfaction by a series of well-controlled and unbiased experiments. Granted, specific immune responses to silicone are highly unlikely (Saxon, 2000). On the other hand, short-term experimental animal studies have provided contrasting results (White, 1998; Naim et al., 2000). Therefore, it is surprising that long-term studies of primates and autoimmune-susceptible experimental animals are not currently in progress. Over time, leaking silicone gel implants can generate reactive inflammatory responses. Assuming that the "danger" theory has any validity (Vance, 2000; Gallucci and Matzinger, 2001), inflammation might stimulate autoimmunity in genetically susceptible subjects. Such a possibility should be investigated by unbiased scientists who have not been involved in the controversy as paid consultants of plaintiffs or industry. Instead, this scientific issue seems to have been settled in the courts based on expert testimony showing the present lack of irrefutable evidence. Under these circumstances, it is not surprising that possible autoimmune effects of drugs and other environmental chemicals are ignored or minimized by a large part of the immunology community.

The task of investigating chemical-induced autoimmunity may very well belong to immuno-toxicologists. Because of their intellectual derivation from toxicology, they are more likely to understand the problems and search for the appropriate solutions. Immunotoxicology is a relatively recent discipline that combines aspects of immunology and toxicology. Its initial efforts were directed to investigations of possible immunosuppressive effects of xenobiotics, but in recent years it has addressed both chemical-induced allergy and autoimmunity. This trend must be encouraged and graduate programs in toxicology should have a strong immunotoxicology component, providing solid information on immunopathology. There are only a few older textbooks of immunotoxicology, but the current volume provides a complete and detailed overview of the progress reached by this discipline. What is now needed is an introductory basic text, the updated English-language equivalent of an excellent small volume published in Lyon several years ago by Jacques Descotes (1992). That book examined the clinical consequences of immunotoxicity and the preclinical detection of this phenomenon, and included some of the legislation available at that time.

In conclusion, chemical-induced animal models have provided very useful information on the etiology and pathogenesis of human autoimmune disease. We still need to identify key target autoantigens and self-peptides involved in the initiation of autoimmune responses, quantitate and localize autoaggressive T cells, define possible regulatory cells, and clarify the various genetic (ecogenetic and immunogenetic) factors involved in xenobiotic-induced animal models. These studies will likely provide interesting explanations of many aspects of human autoimmunity that remain unresolved.

ACKNOWLEDGMENTS

My research on animal models of xenobiotic-induced autoimmune disease was supported by the U.S. Public Health Service (grant ES03230).

REFERENCES

Anastasi, E., et al., Expression of activated Notch3 in transgenic mice enhances generation of T regulatory cells and protects against experimental autoimmune diabetes, *J. Immunol.*, 171, 4504–4511, 2003.

Araragi, S., et al., Mercuric chloride induces apoptosis via a mitochondrial-dependent pathway in human leukemia cells, *Toxicology*, 184, 1–9, 2003.

Audesirk, G., et al., Calcium channels: critical targets of toxicants and diseases, *Environ. Health Perspect.*, 108, 1215–1218, 2000.

Bigazzi, P.E., Autoimmunity in Hashimoto's disease, in *The Molecular Pathology of Autoimmune Disease*, Bona, C.A., Siminovitch, K., Theofilopoulos, A.N., et al., Eds., Harwood Academic Publishers, Chur, Switzerland, 1993, pp. 493–510.

Bigazzi, P.E., Autoimmunity caused by xenobiotics, *Toxicology*, 1997, 119, 1–21.

Bigazzi, P.E., Animal models of autoimmunity: spontaneous and induced, in *The Autoimmune Diseases*, Rose, N.R., and Mackay, I., Eds., Academic Press, San Diego, CA, 1998, pp. 211–244.

Bigazzi, P.E., Metals and kidney autoimmunity, *Environ. Health Perspect.*, 107, 753–765, 1999.

Bigazzi, P.E., Autoimmunity induced by metals, in *Textbook of the Autoimmune Diseases*, Lahita, R.G., Chiorazzi, N., and Reeves, W.H., Eds., Lippincott Williams & Wilkins, Philadelphia, 2000, pp. 753–782.

Bigazzi, P.E., et al., Lack of graft-versus-host-like pathology in mercury-induced autoimmunity of Brown Norway rats, *Clin. Immunol.*, 109, 229–237, 2003.

Carlson, B.C., et al., The endogenous adjuvant squalene can induce a chronic T-cell-mediated arthritis in rats, *Am. J. Pathol.*, 156, 2057–2065, 2000.

Choquet-Kastylevsky, G. et al., The popliteal lymph node response to streptozotocin is under type 1, MHC class-I restricted, CD8+ T-cell control, *Toxicology*, 146, 73–82, 2000.

Cohen, S.B., and Weetman, A.P., Characterization of different types of experimental autoimmune thyroiditis in the Buffalo strain rat, *Clin. Exp. Immunol.*, 69, 25–32, 1987.

Descotes, J., Ed., *Introduction à l'immuno-toxicologie*, Editions Alexandre Lacassagne, Lyon, 1992.

Dohi, T., et al., Elimination of colonic patches with lymphotoxin β receptor-Ig prevents Th2 cell-type colitis, *J. Immunol.*, 167, 2781–2790, 2001.

Espinoza, L.R., and Cuellar, M.L., Diseases associated with silicone implants, in *Textbook of the Autoimmune Diseases*, Lahita, R.G., Chiorazzi, N., and Reeves, W.H., Eds., Lippincott Williams & Wilkins, Philadelphia, 2000, pp. 719–725.

Gallucci, S., and Matzinger, P., Danger signals: SOS to the immune system, *Curr. Opin. Immunol.*, 13, 114–119, 2001.

Gibofsky, A., Visvanathan, K., Kerwar, S., et al., Rheumatic fever, in *Textbook of the Autoimmune Diseases*, Lahita, R.G., Chiorazzi, N., and Reeves, W.H., Eds., Lippincott Williams & Wilkins, Philadelphia, 2000, pp. 679–692.

Greiner, D.L., Rossini, A.A., and Mordes, J.P., Translating data from animal models into methods for preventing human autoimmune diabetes mellitus: caveat emptor and primum non nocere, *Clin. Immunol.*, 100, 134–143, 2001.

Heller, F., et al., Oxazolone colitis, a Th2 colitis model resembling ulcerative colitis, is mediated by IL-13-producing NK-T cells, *Immunity*, 17, 629–638, 2002.

Herold, K.C., and Sarne, D.H., Autoimmune endocrine disorders in *Textbook of the Autoimmune Diseases*, Lahita, R.G., Chiorazzi, N., and Reeves, W.H., Eds., Lippincott Williams & Wilkins, Philadelphia, 2000, pp. 377–406.

Hess, E.V., Are there environmental forms of systemic autoimmune diseases? *Environ. Health Perspect.*, 107, 709–711, 1999.

Hu, H., Möller, G., and Abedi-Valugerdi, M., Mechanism of mercury-induced autoimmunity: both T helper 1- and T helper 2-type responses are involved, *Immunology*, 96, 348–357, 1999.

Koller, L.D., et al., Immunoglobulin and autoantibody responses in MRL/lpr mice treated with "toxic oils," *Toxicology*, 178, 119–133, 2002.

Kosuda, L.L., Greiner, D.L., and Bigazzi, P.E., Effects of HgCl2 on the expression of autoimmune responses and disease in diabetes-prone (DP) BB rats, *Autoimmunity*, 26, 173–187, 1997.

Lahita, R.G., Systemic lupus erythematosus, in *Textbook of the Autoimmune Diseases*, Lahita, R.G., Chiorazzi, N., and Reeves, W.H., Eds., Lippincott Williams & Wilkins, Philadelphia, 2000, pp. 537–547.

Lam-Tse, W.K., Lernmark, A., and Drexhage, H.A., Animal models of endocrine/organ-specific autoimmune diseases: do they really help us to understand human autoimmunity? *Springer Semin. Immunopathol.*, 24, 297–321, 2002.

Landi, S., et al., Evaluation of a microarray for genotyping polymorphisms related to xenobiotic metabolism and DNA repair, *Biol. Techniques*, 35, 816–827, 2003.

Leung, P.S.C., et al., Immunization with a xenobiotic 6-bromohexanoate bovine serum albumin conjugate induces antimitochondrial antibodies, *J. Immunol.*, 170, 5326–5332, 2003.

Long, S.A., et al., Immunoreactivity of organic mimeotopes of the E2 component of pyruvate dehydrogenase: connecting xenobiotics with primary biliary cirrhosis, *J. Immunol.*, 167, 2956–2963, 2001.

Lu, S., and Holmdahl, R., Different therapeutic and bystander effects by intranasal administration of homologous type II and type IX collagens on the collagen-induced arthritis and pristane-induced arthritis in rats, *Clin. Immunol.*, 90, 119–127, 1999.

Lutz, C.T., Browne, G., and Petzold, C.R., Methylcholanthrene causes increased thymocyte apoptosis, *Toxicology*, 128, 151–167, 1998.

Maas, K., et al., Molecular portrait of human autoimmune disease, *J. Immunol.*, 169, 5–9, 2002.

Mahler, M., Blüthner, M., and Pollard, K.M., Advances in B-cell epitope analysis of autoantigens in connective tissue diseases, *Clin. Immunol.*, 107, 65–79, 2003.

Manome, H., Aiba, S., and Tagami, H., Simple chemicals can induce maturation and apoptosis of dendritic cells, *Immunology*, 98, 481–490, 1999.

McMurray, R.W., Hepatitis C–associated autoimmunity, in *Textbook of the Autoimmune Diseases*, Lahita, R.G., Chiorazzi, N., and Reeves, W.H., Eds., Lippincott Williams & Wilkins, Philadelphia, 2000, pp. 669–678.

Mensah-Brown, E.D. et al., Downregulation of apoptosis in the target tissue prevents low-dose streptozotocin-induced autoimmune diabetes, *Mol. Immunol.*, 38, 941–946, 2002.

Mongey, A.-B., and Hess, E.V., Drug and environmental lupus: clinical manifestations and differences, in *Systemic Lupus Erythematosus*, Lahita, R.D., Ed., Academic Press, San Diego, 1999, pp. 929–943.

Morahan, G., and Morel, L., Genetics of autoimmune disease in humans and in animal models, *Curr. Opinion Immunol.*, 14, 803–811, 2002.

Morse, S.S., Sakaguchi, N., and Sakaguchi, S., Virus and autoimmunity: induction of autoimmune disease in mice by mouse T lymphotropic virus (MTLV) destroying CD4+ T cells, *J. Immunol.*, 162, 5309–5316, 1999.

Muller, A., et al., Differential regulation of Th1-type and Th2-type cytokine profiles in pancreatic islets of C57BL/6 and BALB/c mice by multiple low doses of streptozotocin, *Immunobiology*, 205, 35–50, 2002.

Naides, S.J., Postinfectious autoimmunity, in *Textbook of the Autoimmune Diseases*, Lahita, R.G., Chiorazzi, N., and Reeves, W.H., Eds., Lippincott Williams & Wilkins, Philadelphia, 2000, pp. 693–708.

Naim, J.O. et al., Induction of hypergammaglobulinemia and macrophage activation by silicone gels and oils in female A.SW mice, *Clin. Diagn. Lab. Immunol.*, 7, 366–370, 2000.

Neurath, M. et al., Antibodies to interleukin 12 abrogate established experimental colitis in mice, *J. Exp. Med.*, 182, 1281–1290, 1995.

Nicoletti, F. et al., Essential pathogenetic role of endogenous IL-18 in murine diabetes induced by multiple low doses of streptozotocin. Prevention of hyperglycemia and insulitis by a recombinant IL-18-binding protein: Fc construct, *Eur. J. Immunol.*, 33, 2278–2286, 2003.

Nierkens, S. et al., Selective requirement for CD40-CD154 in drug-induced type 1 versus type 2 responses to trinitrophenyl-ovalbumin, *J. Immunol.*, 168, 3747–3754, 2002.

Parikh-Patel, A. et al., The geoepidemiology of primary biliary cirrhosis: contrasts and comparisons with the spectrum of autoimmune diseases, *Clin. Immunol.*, 91, 206–218, 1999.

Pheng, S.-R., Chakrabarti, S., and Lamanotagne, L., Dose-dependent apoptosis induced by low concentrations of methylmercury in murine splenic Fas+ T cell subsets, *Toxicology*, 149, 115–128, 2000.

Pieters, R., and Albers, R., Screening tests for autoimmune-related immunotoxicity, *Environ. Health Perspect.*, 107, 673–677, 1999.

Pollard, K.M. et al., The autoimmunity-inducing xenobiotic mercury interacts with the autoantigen fibrillarin and modifies its molecular and antigenic properties, *J. Immunol.*, 158, 3521–3528, 1997.

Pollard, K.M. et al., Xenobiotic acceleration of idiopathic systemic autoimmunity in lupus-prone BXSB mice, *Environ. Health Perspect.*, 109, 27–33, 2001.

Pollard, K.M. et al., Lupus-prone mice as models to study xenobiotic-induced acceleration of systemic autoimmunity, *Environ. Health Perspect.*, 107, 729–735, 1999.

Powell, J.J., Van de Water, J., and Gershwin, M.E., Evidence for the role of environmental agents in the initiation or progression of autoimmune conditions, *Environ. Health Perspect.*, 107, 667–672, 1999.

Rasooly, L., Burek, C.L., and Rose, N.R., Iodine-induced autoimmune thyroiditis in NOD-H–2h4 mice, *Clin. Immunol. Immunopathol.*, 81, 287–292, 1996.

Richardson, B., DNA methylation and autoimmune disease, *Clin. Immunol.*, 109, 72–79, 2003.

Robson, M.G., and Walport, M.J., Renal autoimmunity, in *Textbook of the Autoimmune Diseases*, Lahita, R.G., Chiorazzi, N., and Reeves, W.H., Eds., Lippincott Williams & Wilkins, Philadelphia, 2000, pp. 323–349.

Rose, N.R., Bonita, R., and Burek, C.L., Iodine: an environmental trigger of thyroiditis, *Autoimmun. Rev.*, 1, 97–103, 2002.

Rubin, R.L., Etiology and mechanisms of drug-induced lupus, *Curr. Opin. Rheumatol.*, 11, 357–360, 1999.

Rubin, R.L., and Kretz-Rommel, A., Initiation of autoimmunity by a reactive metabolite of a lupus-inducing drug in the thymus, *Environ. Health Perspect.*, 107, 803–806, 1999.

Ruwhof, C., and Drexhage, H.A., Iodine and thyroid autoimmune disease in animal models, *Thyroid*, 11, 427–436, 2001.

Saxon, A., The antigen that wasn't — silicone, *Clin. Immunol.*, 95, 171–172, 2000.

Sayeh, E., and Uetrecht, J.P., Factors that modify penicillamine-induced autoimmunity in Brown Norway rats: failure of the Th1/Th2 paradigm, *Toxicology*, 163, 195–211, 2001.

Seguin, B., Teranishi, M., and Uetrecht, J.P., Modulation of D-penicillamine-induced autoimmunity in the Brown Norway rat using pharmacological agents that interfere with arachidonic acid metabolism or synthesis of inducible nitric oxide synthase, *Toxicology*, 190, 267–278, 2003.

Shaheen, V.M. et al., Immunopathogenesis of environmentally induced lupus in mice, *Environ. Health Perspect.*, 107, 723–727, 1999.

Shi, F.-D., Ljunggren, H.-G., and Sarvetnick, N., Innate immunity and autoimmunity: from self-protection to self-destruction, *Trends Immunol.*, 22, 97–101, 2001.

Smith, M.D., Scleroderma, in *Textbook of the Autoimmune Diseases*, Lahita, R.G., Chiorazzi, N., and Reeves, W.H., Eds., Lippincott Williams & Wilkins, Philadelphia, 2000, pp. 557–567.

Strober, W., Fuss, I.J., and Blumberg, R.S., The immunology of mucosal models of inflammation, *Annu. Rev. Immunol.*, 20, 495–549, 2002.

Sundick, R.S., Iodine in autoimmune thyroiditis, in *Organ-Specific Autoimmunity*, Bigazzi, P.E., Wick, G., and Wicher, K., Eds., Marcel Dekker, New York, 1990, pp. 213–228.

Tormey, V.J. et al., Anti-fibrillarin antibodies in systemic sclerosis, *Rheumatology*, 40, 1157–1162, 2001.

Tsuchiya, T. et al., Role of γδT cells in the inflammatory response of experimental colitis mice, *J. Immunol.*, 171, 5507–5513, 2003.

Vaile, J.H., and Russell, A.S., Drug-induced autoimmunity, in *Textbook of the Autoimmune Diseases*, Lahita, R.G., Chiorazzi, N., and Reeves, W.H., Eds., Lippincott Williams & Wilkins, Philadelphia, 2000, pp. 783–798.

Vance, R.E., A Copernical revolution? Doubts about the danger theory, *J. Immunol.*, 165, 1725–1728, 2000.

Wesierska-Gadek, J., Reinisch, W., and Penner, E., Autoimmunity of the gastrointestinal tract, in *Textbook of the Autoimmune Diseases*, Lahita, R.G., Chiorazzi, N., and Reeves, W.H., Eds., Lippincott Williams & Wilkins, Philadelphia, 2000, pp. 229–272.

Weyand, C.M., and Goronzy, J.J., Rheumatoid arthritis, in *Textbook of the Autoimmune Diseases*, Lahita, R.G., Chiorazzi, N., and Reeves, W.H., Eds., Lippincott Williams & Wilkins, Philadelphia, 2000, pp. 573–594.

White, B., Systemic sclerosis, in *Clinical Immunology: Principles and Practice*, Rich, R.R., Fleisher, T.A., Shearer, W.T., et al., Eds., Mosby, London, 2001, pp. 64.1–64.10.

White, K.L., Silicone gel and animal models of autoimmune disease, *Clin. Immunol. Immunopathol.*, 87, 205–206, 1998.

Yamamoto, T., Animal model of sclerotic skin induced by bleomycin: a clue to the pathogenesis of and therapy for scleroderma? *Clin. Immunol.*, 102, 209–216, 2001.

Yang, J.-M. et al., Fibrillarin and other snoRNP proteins are targets of autoantibodies in xenobiotic-induced autoimmunity, *Clin. Immunol.*, 101, 38–50, 2001.

Yang, J.-Q. et al., Immunoregulatory role of CD1d in the hydrocarbon oil-induced model of lupus nephritis, *J. Immunol.*, 171, 2142–2153, 2003.

Zipris, D. et al., Infections that induce autoimmune diabetes in BBDR rats modulate CD4+CD25+ T cell populations, *J. Immunol.*, 170, 3592–3602, 2003.

Emerging Issues in Immunotoxicology

CHAPTER 22

Transgenic Rodent Models in Immunotoxicology

Robert V. House

CONTENTS

22.1 INTRODUCTION

The discipline of immunotoxicology has made phenomenal advances within the past decade, evolving from the early conceptual work by Vos (1977) to the development of a relatively standardized panel of functional assays that allow identification of test materials with immunotoxic potential and investigation of broad mechanisms of action (the so-called "tier" testing approach) (reviewed in Luster et al., 1994; Vos and van Loveren, 1998), to the incorporation of immunotoxicology as a safety determinant for human pharmaceuticals (Hastings, 2002; Putman et al., 2003). Over the course of the past three decades, advances in biological technology have facilitated techniques and approaches in immunotoxicology that provide greater understanding in mechanisms, thereby enhancing the usefulness (and, hopefully, the predictivity) of assays (Dean et al., 2001).

Of course, it may not always make technical or economic sense to incorporate the newest technology simply because it is available. Likewise, powerful technologies can provide multiple answers to as yet unasked questions. A good example of this former case is the evaluation of cytokines. Cytokines are an almost intuitive target of immunomodulators due to their central role in the regulation of all phases of the immune response, as well as their multiple roles in other organ systems such as the nervous system. Indeed, much effort is currently underway to therapeutically target cytokines to control multiple disease conditions. However, although cytokines are straight-forward to evaluate (House, 1999), the full promise of this methodology remains unrealized. The same is true for many other recent advances such as immunomics (Maecker et al., 2001).

One approach that may provide greater insight into immunotoxicology is the use of transgenic animals. Although the possible use of transgenic animals in immunotoxicology was previously reviewed (Løvik, 1997), the pace of technology in this field has advanced greatly and much new information has been gained. In the current review, a broader scope of possibilities will be explored. More importantly, however, the appropriate use of this technology will be to serve as a tool to help understand specific mechanisms of immunotoxicity, rather than for its indiscriminate use to generate answers in search of a question. To this end, it may be of benefit to propose a new paradigm of immunotoxicology that incorporates much of the knowledge gained in basic and clinical immu-nology since the initial description of the tiered testing approach.

22.2 PARADIGMS OF IMMUNOTOXICOLOGY

To date, the most commonly used model of immunotoxicology is sometimes referred to as the *continuum of immune response* (Figure 22.1). In the continuum model, the immune response is visualized as a Poisson distribution, assuming that the immune response of a population falls within a narrow range of values. By perturbing this response negatively, a condition of immunosuppression results. Immunosuppression is a diminished functional capability of one or several of the various components of the immune system. These perturbations, which can be either structural, functional, or a combination thereof, result in a graded decrease in the ability of the host to control infection or neoplasia (a condition I refer to as *homeostasis of identity*).

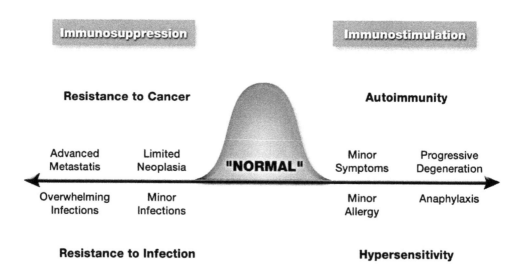

Figure 22.1   Continuum of immune response.

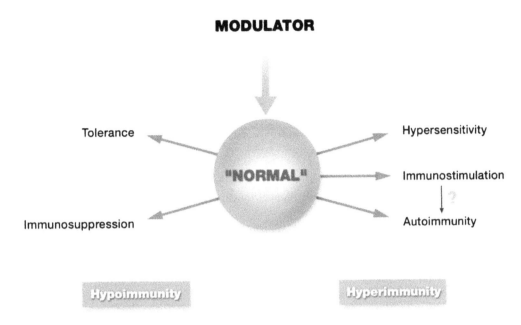

Figure 22.2 Alternative model of immunotoxicology.

One problem with the traditional model is that a condition of increased immune function is seen as invariably leading to autoimmunity or hypersensitivity. This conclusion is not always supported by experimental or empirical evidence. Another problem with the model is its assumption that the immune response is linear. The role of hypersensitivity in this continuum is even more uncertain because these types of reactions, although undesirable from the standpoint of human health, are certainly within the range of normal human immune function.

Assumptions built into the model limit the practicality of applying the results of animal studies to human risk assessment. For example, the traditional model does not account for the multiplicity of effector and control mechanisms, but rather simplistically groups multiple interrelated cellular networks together as the "immune system." In addition, the model does not account for interaction of the immune network with other organ systems, particularly the nervous and endocrine systems. Finally, the traditional model does not account for environmental and individual heterogeneity. For these reasons, it is obvious that the two-dimensional traditional model of immunotoxicology does not adequately explain the complexity of immune system modulation resulting from exogenous influences. What is required is a new way of thinking about this interaction.

A proposed alternative model is illustrated in Figure 22.2. In this model, the immune response may be imagined as a flexible "balloon," capable of expanding (corresponding to an increased level of activity following microbial challenge), shrinking (the resolution of an infection and a return to resting levels of activity), and being "squeezed" in different directions when acted on by an outside agent (termed a "modulator.")

The nature of the modulator and its interaction with the immune system determine the consequence of this interaction. The modulator may be an antigen, an external agent such as an environmental chemical, a physical agent such as ultraviolet light, or a physiological action in a related organ system, such as psychogenic stress. As a consequence of this modulator's action, a condition of *hypoimmunity* may be induced in which immune function is decreased either generally or specifically. Depending on the nature of the modulator, a condition of *hyperimmunity* may also result. Hyperimmunity may take the form of hypersensitivity, immunostimulation, or autoimmunity. To extend this analogy, hypoimmunity would limit the degree to which the balloon could expand in response to an immune stimulus, whereas hyperimmunity would cause the balloon to expand beyond its normal limits.

When immunotoxicology is therefore perceived as multidimensional immunomodulation, it is clear that our greatest insights into mechanisms will come from models that can address many different parameters, preferably simultaneously. Moreover, these models should address physiological alterations at the cellular, and preferably the molecular, level. Transgenic animal models show promise in both these respects.

22.3 TRANSGENIC TECHNOLOGY

The term "transgenic," in its simplest definition, refers to the transfer of genetic elements from one organism to another. By implication, an organism in which genetic constitution has been altered is a mutant, although this term should not necessarily carry its usual negative connotation. For the purposes of this review, we will consider transgenic animals only in the context of genetically engineered models to evaluate gene function, and to be used as experimental models. More specifically, we will examine the utility of models in which genetic information is inserted into the germ-line of a complex vertebrate model of immunomodulation, in this case the laboratory mouse and, less commonly, the laboratory rat.

To best understand the advantages that transgenic technology can provide to immunotoxicology, it is necessary to understand the basics of the technique. Several excellent recent reviews are available (Hofker and Breuer, 1998; Müller, 1999; Gao et al., 1999; Babinet, 2000; Court et al., 2002). In the following sections, we review the elementary background; the interested reader is referred to the abovementioned reviews for a more thorough coverage of this technology, including specific methodologies.

22.3.1 Nonhomologous Recombination (Random Integration) Transgenesis

Early efforts at creating transgenic mice employed microinjection of cloned DNA elements into the pronuclei of fertilized eggs, resulting in the stable integration of the DNA into the mouse genome. The modified zygote was then implanted into pseudo-pregnant dams; offspring carrying the transgene are known as transgenic founders. Due to the random nature of the gene insertion, these founder animals expressed varied phenotypes. Subsequent breeding to homozygosity for expression of transgene results in the establishment of transgenic lines. A distinct disadvantage of this approach is the random integration of the transgene. Although inclusion of an appropriate promoter may facilitate expression of the gene in an inappropriate location, certain genetic sequences near the insertion were found to interfere with the gene's regulation (this phenomenon is known as *position effect*). This interference makes it difficult, if not impossible, to control copy number and function of the inserted gene. Although this approach is still used, it lacks the control necessary for immunotoxicology model development, and therefore will not be described further.

22.3.2 Homologous Recombination Transgenesis

The ability to specifically target genetic transfer was the result of the development of embryonic stem (ES) cell technology in the late 1980s. ES cells derived from the inner cell mass of preimplantation blastocysts are totipotent and, if injected into a mouse embryo at an early stage of development, are capable of differentiating into both somatic and germ-line tissues when reintroduced into blastocysts. Mouse ES cells can be grown in cell culture, and when maintained under certain culture conditions remain in an undifferentiated state indefinitely (Müller, 1999).

The genetic background of the ES cells is of practical concern. To date, most ES cells used in transgenesis are of the 129 mouse strain, although ES from other strains such as C57BL/6 and BALB/c are also available. The selection of strain is important because strain characteristics of the "parent" (such as genetic flanking sequences) may have an effect on the phenotype of the transgenic

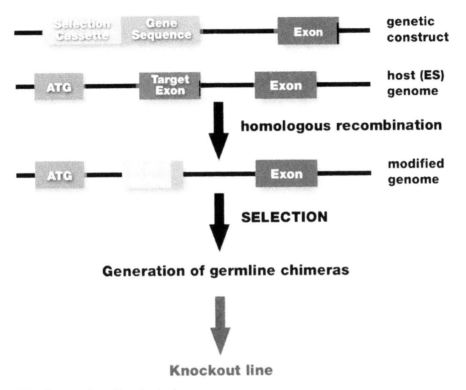

Figure 22.3 Construction of knockout mice.

offspring (Ledermann, 2000). Suggestions for appropriate selection criteria and breeding strategies have been published (Wolfer et al., 2002).

22.3.2.1 Knockout Mice

Homologous recombination in which the function of a native gene is disrupted is referred to as a *knockout*. In this system, a genetic construct is prepared that contains sequences of the target gene as well as a selection cassette inserted into an exon, disrupting the unit of transcription ("null allele"). Following introduction into ES cells, selection is performed (based on the selection cassette), and surviving ES clones are screened for homologous recombinants. Following production of founder mice from these germ-line chimeras, mice carrying the null allele are obtained (Figure 22.3).

Although the process for building a knockout animal is straightforward, there are a number of technical difficulties still inherent in the process. For example, problems with design or preparation of the genetic construct may result in an incomplete knockout in which mutant forms of the knocked-out protein may still be secreted ("leaky mutation"). In addition, replacement of the native gene with a construct may disrupt adjacent coding regions or ablate other genes; this greatly complicates the ability to assess the role of the knocked-out gene (Müller, 1999).

22.3.2.2 Knockin Mice

Homologous recombination in which the code for a regulatory element or a different gene (allelic substitution) is inserted is referred to as a *knockin*. In this system, a genetic construct is prepared that contains the gene of interest (foreign gene, mutated cDNA, oncogene, etc.), and a selection cassette in frame in an exon being targeted. Because the new gene is placed under the control of *cis*-acting regulatory elements, the expression of the new gene is tightly controlled (Figure 22.4).

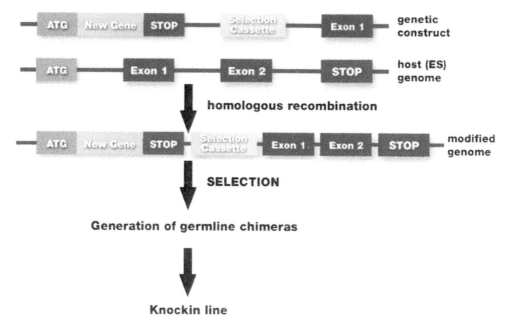

Figure 22.4 Construction of knockin mice.

22.3.3 Conditional Mutation

A concept that has advanced transgenic technology enormously is that of *conditional mutation*, also referred to as *conditional gene targeting*. It was discovered that deletion of certain genes resulted in embryo lethality. Therefore, it is desirable to have available a technique whereby the transgene would not be expressed until triggered by an exogenous influence. Several systems have been described (Gao et al., 1999); one representative of conditional expression is the *Cre/loxP* system (Figure 22.5). *Cre/loxP* is a site-specific recombination system that allows the construction of stage- or tissue-specific gene targeting. The *Cre* protein, derived from bacteriophage P1, is a recombinase catalyzing site-specific DNA recombination between 34 base pair repeats (termed *loxP* sites). In this technique, a transgene is created containing at least two *loxP* sites. This gene is added to ES cells, resulting in homologous recombination, with the target gene residing between two *loxP* sites (referred to as a "floxed" gene). Founder animals expressing the floxed gene are then mated with a second line of transgenic animals that contain a fusion gene expressing the *Cre* recombinase (these animals are produced using oocyte injection or standard knockin techniques in which the *Cre* molecule is under the control of an appropriate promoter). Depending on the tissue in which the *Cre* recombinase is expressed, the double-transgenic offspring will exhibit tissue-specific transgene excision (Rajewsky et al., 1996).

22.4 SPECIFIC TRANSGENIC MODELS WITH POTENTIAL UTILITY FOR IMMUNOTOXICOLOGY

22.4.1 Cytokines and Immunomodulation

Cytokines are small molecule (protein and peptide) regulators of the immune response that are critical for all aspects of adaptive immunity, and serve as important mediators between adaptive and immune immunity. Due to this central role, they are natural targets for investigational immunotoxicology. (On

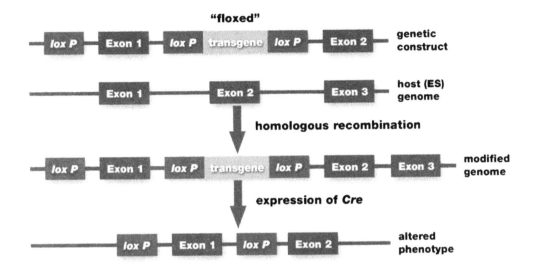

Figure 22.5 Conditional gene targeting.

the other hand, their multiple overlapping roles in various processes, as well as the redundant function of many cytokines, such as IL-1 and IL-6 or IL-4 and IL-13, makes them less desirable as a screening tool.

The use of transgenic models in evaluating the role of cytokines in evaluating different forms of immunotoxicology is covered in the following sections. However, one of the most common uses of cytokines in immunotoxicology to date has been in the evaluation of immunomodulation, particularly immunosuppression. Cytokines play a central role in many naturally occurring immunodeficiency conditions (Leonard, 2001), as well as in immunodeficiencies resulting from exposure to test materials.

A number of cytokine gene transgenic mice have been described; examples may be found in various Internet databases, as well as in selected reviews (Ryffel, 1995; Slattery et al., 2000; Moser et al., 2001; Power, 2003).

22.4.2 Host Resistance

Although a wide variety of *in vitro* and *in vivo* assays are available to evaluate immunotoxicology/immunomodulation, they are all by necessity at least one step removed from reality. Although alterations in immune parameters detected by many of the immunotoxicology assays have shown some degree of correlation with known immunotoxicants, this limited approach can never adequately address the redundancy and synergism of the *in vivo* immune response. For this reason, host resistance assays that directly assess an animal's ability to resist an infection or neoplasia remain an important "gold standard" for immunotoxicology assessment (Luster et al., 1992). Over the history of immunotoxicology, a wide variety of animal models have been developed including viral, bacterial, protozoa, and fungal (Luster et al., 1994; Burleson et al., 1995; Burleson, 2000).

Paradoxically, given the general acknowledgment of the gold standard assigned to host resistance assays, they are performed relatively infrequently in comparison to other evaluations such as functional tests or even histopathological analysis of lymphoid tissue. This is due in great part to the labor-intensive nature of these assays, as well as the difficulty in reproducibly performing these studies.

Given the perceived disadvantages to performing host resistance assays, it may be difficult to understand the potential role of transgenic technology in furthering research in this area. Indeed,

Table 22.1   Host Resistance Mechanisms of Commonly Used
Immunotoxicology Models Investigated Using
Transgenic Rodents

Organism	Type	Reference
Listeria monocytogenes	Bacterium	Lecuit and Cossart (2002)
Trichinella spiralis	Nematode parasite	Dent et al. (1997)
Influenza	Virus	Caton et al. (1998)
Toxoplasma gondii	Protozoan parasite	Yap and Sher (2002)
B16F10	Murine melanoma	Fuzii and Travassos (2002)

the transgenic technology's major contribution to host resistance assays is through a better under-standing of host defense mechanisms. It is axiomatic that for a host resistance assay to be maximally instructive, it is necessary to understand the mechanisms whereby elimination of the challenge is effected. For example, resistance to the transplantable B16F10 melanoma, frequently used in immunotoxicology assessment, has been shown to correlate with natural killer (NK) cell activity (Wilson et al., 2001). Unfortunately, the exact mechanism(s) are not known for all of the host resistance assays currently in use. Transgenic models may provide a greater understanding of these mechanisms, leading to greater value in the data. Table 22.1 provides some examples of commonly used immunotoxicology host resistance models, as well as transgenic systems that have been used to elucidate the mechanisms of host resistance.

22.4.3 Evaluating Immunotoxicology of Recreational Adjuncts (Abused Drugs)

It has long been established that immunomodulation may be caused not just by inadvertent exposure to environmental and industrial chemicals (to which we might choose to limit or prevent our exposure), or by carefully controlled administration of pharmaceutical agents to treat or prevent disease. Rather, humans readily use a variety of natural and synthetic pharmacologic agents to recreationally modify their brain chemistry. Whether this takes the form of socially acceptable agents such as nicotine and ethanol, ethical but misapplied therapeutics such as opiates and amphetamines, or the forthrightly illegal such as lysergic acid diethylamide (LSD) and phencycli-dine (PCP), many of these agents are known to modulate the immune response; this modulation is always an undesired consequence of the agent's use (House et al., 1997a; Friedman et al., 2003).

A number of studies have been conducted to evaluate abused drugs for immunomodulatory potential. To date, most such investigations have been conducted using the standard tier-testing paradigm; this approach, although broadly instructive, is inadequate for understanding the mech-anisms of immunomodulation produced by these agents, which (1) may not produce their effects quickly, and (2) may operate via complex receptor mechanisms (House et al., 1997b; McCarthy et al., 2001).

Perhaps the most widely used (and abused) recreational drug is ethanol. In addition to its multiple deleterious effects on health, ethanol is a known immunosuppressant (Szabo, 1999; Friedman et al., 2003). Although immunomodulation by a number of other abused drugs is known to be receptor dependent, as yet no receptor for ethanol has been described. Therefore, understanding this mechanism requires novel systems; the transgenic mouse may represent just such a system.

Transgenic mouse models have been used to some effect in evaluating the effect of ethanol on immune function. For example, Schodde et al. (1996) used a transgenic mouse in which the majority of the CD4-bearing T cells had T-cell receptors reacting with ovalbumin (OVA). This construct produces mice that "naturally" react with a specific antigen (in this case OVA), but would require antigen priming to develop an immune response to other alloantigens. The results of this study demonstrated that ethanol feeding had no effect on development of a vigorous delayed hypersen-sitivity response to OVA. However, these animals exhibited a profound inhibition of cell-mediated immunity when immunized with sheep red blood cells. This discrepancy demonstrated that immunomodulation by ethanol occurs independently of antigen priming.

Likewise, Colombo et al. (2001a) used the Oncomouse (a transgenic mouse carrying an activated v-Ha-ras oncogene under the control of the mouse mammary tumor virus promoter) to evaluate cytokine modulation by alcohol, cocaine, or morphine. Oncomouse thymocytes released less sIL-2R than FVB thymocytes. Oncomouse thymocytes and splenocytes responded in a nearly opposite fashion to their FVB counterparts, demonstrating that the v-Ha-ras oncogene plays a role in defining the host immune response. Although the practical implications of this finding remain to be determined, these results illustrate the high degree of specificity provided by this model that would not be achievable in nonmutant mice.

Another popular recreational drug is cocaine, which has also been shown to be an immunomodulator (Pellegrino and Bayer, 1998; Friedman et al., 2003). The immunomodulation produced by cocaine is intriguing, and can be manifested as either immunostimulation or immunosuppression. This feature makes understanding the effects of cocaine especially difficult using standard tier-testing methodology, which is ill suited to evaluate such "mixed" effects. Although the use of transgenics may be only slightly more instructive, some studies have been performed. For example, Chen et al. (1996) and Colombo et al. (2001b) used the Oncomouse model to evaluate the effect of cocaine on cytokines, similar to studies described above for ethanol. These results suggest that the product of the v-Ha-ras oncogene appears to play a role in drug-related immunomodulation, although the precise mechanism whereby this is achieved has not yet been clarified.

Another important group of abused drugs (albeit ones with legitimate therapeutic use) is the opiates. Heavy drug abuse has long been associated with alterations in host defense against infectious diseases, and its ability to alter specific immune functions has been demonstrated experimentally (House et al., 1997a). Numerous studies have demonstrated that the immunomodulatory activity of opiates is due largely to their interaction with opioid receptors, principally those of the mu (μ) class, but also of the delta (δ) and, to a lesser extent, the kappa (κ) classes (McCarthy et al., 2001).

Although early work in establishing this link used standard immunology assays and techniques, the high degree of specificity provided by transgenic technology is well suited to evaluate mechanisms of immunomodulation. For example, Roy et al. (1998) demonstrated in mu opioid receptor (MOR) knockout that morphine-induced modulation of certain immune functions (e.g., macrophage phagocytosis and secretion of tumor necrosis factor) was eliminated, whereas other alterations (e.g., reduction of splenic and thymic cell number and mitogen-induced proliferation were unaffected). This interesting result may indicate that at least some of morphine's effects may be mediated by either delta or kappa opioid receptors. The work of Roy et al. was confirmed and expanded upon by Gaveriaux-Ruff et al. (1998), who demonstrated other immunomodulatory effects of morphine that were eliminated in MOR knockout mice, including lymphocyte subset ratios and NK cell activity. This group also demonstrated that ablation of the MOR itself was not associated with any immunomodulation, verifying that immunomodulation results from processes subsequent to MOR binding to opiates.

Although the mu opioid receptors have been extensively studied in relation to immunomodulation, less work has been performed with the kappa opioid system. Gaveriaux-Ruff et al. (2003) have investigated the role of kappa-opioid receptor (KOR) activity in immunity *in vivo* by studying immune responses in KOR knockout mice. These mice displayed a slight reduction in thymus cellularity and the $CD4^+:CD8^+$ cell ratio, along with a slight increase in immature CD4/CD8 lymphocytes. When immunized with keyhole limpet hemocyanin, KOR knockout mice produced significantly higher levels of antigen-specific total Ig, IgM, IgG1, and IgG2a antibodies. A similar enhancement in function was not observed in mu-opioid receptor and delta opioid–receptor knockout animals, suggesting that KOR may play a role in endogenous immunomodulation. These results demonstrate that much work remains to be done in evaluating the role of opioid receptors in the immune system; however, they also clearly demonstrate that transgenic animals can play a crucial role in understanding how this system may be modulated, whether artificially or inadvertently.

Although the delta opioid receptors have been demonstrated pharmacologically to have immuno-modulatory potential (House et al., 1996), there have been as yet no reported studies using transgenic animals to investigate this activity.

22.4.4 Autoimmunity

A central dogma of immunology has long been that the primary role of the immune system is to differentiate "self" from "nonself." During ontology, the immune system educates itself to effect this distinction, and at a certain point mounts a defense against anything foreign. This model, however, is complicated by the condition known as autoimmunity, in which the immune system recognizes self as foreign. To a limited degree, autoimmunity appears to be part of the normal immune process, possibly as a means of regulating an active immune response. However, when autoimmunity becomes pathological, damaging tissues or organ systems, it becomes autoimmune disease.

Using the self/nonself paradigm, numerous hypotheses have been advanced to account for autoimmunity. These hypotheses include molecular mimicry (similarity between self antigens and the inadvertent modification of self proteins as by modification following covalent binding of a chemical to self proteins), the breakdown of regulatory mechanisms that normally prevent recognition of self (e.g., suppression of autoreactivity, regulation of apoptosis, induction of anergy), or polyclonal lymphocyte activation, as by exposure to bacterial endotoxins or superantigens. This latter possibility connects immunostimulation with autoimmunity as per the original paradigm, although this connection is by no means consistent. A more recent model of autoimmunity invokes the concept of "danger." This concept presumes that the immune system is unconcerned with notions of self and nonself; rather, the goal is to recognize danger signals (e.g., tissue damage, microbial components), and it is within this context that reactions consistent with autoimmunity become manifest (Matzinger, 1998).

The etiology of autoimmunity and autoimmune disease is incompletely understood, but is generally recognized as having a multifactorial origin including viral infection (Barnaba, 1996), paraneoplastic (tumor autoantigens), and environmental factors, among others (Lernmark, 1997). Some examples of chemical (i.e., environmental) agents associated with autoimmunity are listed in Table 22.2. Regardless of the ultimate etiology, it is well established that propensity toward autoimmune disease depends largely on genetic predisposition. In this respect, it is similar to many types of hypersensitivity, in which certain individuals exhibit a greater predisposition toward reactivity. It differs, however, in that predicting the potential of materials to induce reactivity is not as straightforward. In fact, predicting the potential of drugs to induce autoimmunity is currently one of the most pressing needs in immunotoxicology assessment. Representative examples of autoimmune conditions, and the use of transgenic models to understand them, are given below.

Table 22.2 Autoimmune Conditions Associated with Exposure to Chemicals

Autoimmune Condition	Agents Associated with Condition	Studies with Transgenics
Systemic lupus erythematosus	Hydralazine, heavy metals	Ravirajan and Isenberg (2002); Higuchi et al. (2002)
Immune-complex glomerulonephritis	Penicillamin, chlorpromazine, anticonvulsants, solvents, procainamide	Kikuchi et al. (2002); Zhang et al. (1995)
Hemolytic anemia	Methyldopa, penicillin, mefenamic acid, diphenylhydantoin, interferons	Nisitani et al. (1997); Sakiyama et al. (1999)
Thrombocytopenia	Acelazolamide, chlorothiazide, gold salts, p-amionsalicylic acid, quinidine	McKenzie (2002); Reilly and McKenzie (2002)
Scleroderma-like disease	Vinyl chloride, silica, L-tryptophan	Saito et al. (1999, 2002)

22.4.4.1 Diabetes

Type 1 (insulin-dependent) diabetes is a multifactorial autoimmune disease characterized by destruction of the insulin-producing pancreatic islet cells. The disease is polygenic, although the major locus of susceptibility in both humans and rodents appears to be the MHC complex. The etiology of this disease remains unknown, although it may involve an environmental or infectious agent trigger. Many transgenic models of autoimmune diabetes have been described including the BDC2.5 mouse that carries a transgene encoding a T-cell antigen receptor specific for pancreatic antigen. These animals develop a massive destruction of the islet cells (Ji et al., 1999).

Another interesting animal model has been described by Ludewig et al. (2000) for evaluating the role of viral infection in the induction of autoimmune disease. In this model, dendritic cell-mediated antigen presentation appeared to be responsible for the induction of autoimmune responses against microbial "neoself" antigens. This intriguing possibility will require further investigation to fully elucidate the role of microbial infection in diseases such as diabetes.

A number of other transgenic models of autoimmune diabetes have been described. The interested reader is directed to several recent reviews (Wong et al., 1999; Suri and Katz, 1999; Das et al., 2000; Adorini et al., 2002).

22.4.4.2 Rheumatoid Arthritis

Rheumatoid arthritis (RA) is a chronic inflammatory disease that results in destruction of bone and joints, and is increasing in incidence. The etiology of this autoimmune disease is complex and only partially understood. However, several transgenic mouse models are available such as the KRN mouse (Kyburz and Corr, 2003). This mouse is transgenic for a T-cell receptor specific for an epitope of bovine Rnase; these animals are bred on a nonobese diabetic mouse line, and develop a severe arthritis resembling human RA. The mechanisms of this disease involve an interaction between T cells and B cells that is dependent on the MHC class II molecule I-A(g7). In addition, the development of disease requires B-cell activation by antigen; an additional CD40–CD40 ligand interaction results in rise to the production of autoantibodies specific for glucose-6-phosphate isomerase. This model also appears to implicate the involvement of the innate immune system, complement components, Fc receptors, and neutrophils. Thus, many of the features of human RA are present in this system.

Another useful model for studying RA is the transgenic mouse that over-expresses human TNF-alpha, originally described by Keffer et al. (1991) and recently reviewed (Li and Schwarz, 2003). This mouse develops an erosive polyarthritis with many characteristics observed in RA, and has provided much of the evidence for the central role that this factor plays in the cytokine cascade associated with RA. A variety of other transgenic models for arthritis are available.

Recently, Sonderstrup (2003) reported the development of a humanized mouse model of RA that shows some promise in understanding the etiology of this disease, although this model has yet to be fully characterized.

22.4.5 Hypersensitivity

Hypersensitivity is a condition of reactivity to an antigen that exceeds what would be considered "normal" for that particular antigen; in almost all cases, the result of this reactivity is deleterious rather than protective. Although the particular cellular and molecular mechanisms involved in these responses are appropriate to an antigenic challenge, the vigor of the reactions results in consequences that can range from minor to acutely pathological to lethal. Evaluation of the specific mechanisms involved in the induction and maintenance of hypersensitivity, as well as the ability to screen test materials for their potential to induce these reactions, is one of the most active areas in immunotoxicology at present (Hastings, 2002; Kimber and Dearman, 2002). Although a number

Table 22.3   Investigation of Cytokines in Hypersensitivity Using Transgenic Mice

Cytokine	Condition Studied	Reference
Interleukin-5	Immunity to aeroallergens	Dent et al. (1997)
Interleukin-9	Allergic asthma	Soussi-Gounni et al. (2001)
Interleukin-10	Allergic dermatitis	Laouini et al. (2003)
Interleukin-13	Respiratory allergy	Hershey (2003); Elias et al. (2003)

of models are under development for prediction of materials that might provoke hypersensitivity, details of the mechanisms involved still require elucidation. This is of particular importance in that the incidence of allergy (a type of hypersensitivity) is increasing. The reasons for this increase are unknown, although a number of hypotheses have been advanced, including a lack of microbial challenge due to better living conditions (the so-called "hygiene hypothesis") (Prescott, 2003), exposure to environmental chemicals (Levetin and Van de Water, 2001), and even a transmissible agent such as a virus (Hussain and Smith, 2003).

The mechanisms of hypersensitivity are multiple and highly complex, and are beyond the scope of this review. Nevertheless, it is increasingly obvious that regulation (or perhaps dysregulation) of cytokines is pivotal. In particular, understanding the differential expression and modulation of type 1 (Th1) and type 2 (Th2) cytokines provides great insight into the pathophysiology of this condition, in addition to the other exemplar of hyperimmunity, autoimmunity (Singh et al., 1999; Moser et al., 2001). As previously described, transgenic technology provides an ideal tool for evaluating the role of cytokines in pathological conditions, including hypersensitivity (Burns and Gaspari, 1996; Traidl et al., 2000; Moser et al., 2001; Wang et al., 2003). Some representative cytokine investigations of hypersensitivity mechanisms are listed in Table 22.3.

22.4.6 Humanized Mice for Assessment of Human Biologies

A particularly appealing use of transgenic technology for immunomodulation and immunotoxicology is the creation of "humanized" mice, that is, mice whose immune systems have been replaced with a human immune system. One approach for humanized mice is the SCID-hu model. This model is based on severe-combined immunodeficient (SCID) mice, which lack a functional murine immune system due to a lack of functional T- and B-lymphocytes. These animals are given a donor human immune system consisting of human fetal liver and human fetal thymus. These implanted tissues colonize the animal, begin essentially normal human hematopoiesis, and subsequently mature into functional immunocytes. (Since the SCID lacks a functioning immune system, the human implants are not rejected.) The SCID-hu model has been used extensively for research into transplantation and HIV research, where normal laboratory animal models are inappropriate due to species specificity (McCune, 1996; Boehncke, 1999). Although seemingly representing an ideal model for evaluation of immunomodulators (either therapeutic or unintended), there are practical limitations on the widespread use of this animal. For example, construction of these animals is laborious and time consuming, the animals are prohibitively expensive due to the necessity of maintaining SCID mice under gnotobiotic conditions prior to their immune reconstitution, and the availability of human tissues for implantation is potentially problematic due to the limited supply and possible ethical considerations regarding the use of fetal human tissue.

Given the limitations of using SCID-hu mice, a more feasible alternative would be the construction of a "selectively human" mouse, in which only certain specific portions of the immune system have been humanized. Transgenic animals, within certain constraints, are well suited to this purpose.

One of the most complete reported uses of humanized mice for assessment of immunomodulation was a series of studies evaluating the activity of keliximab, a human-cynomolgus chimeric monoclonal antibody with specificity for the human CD4 molecule. These studies employed the

HuCD4/Tg transgenic mouse, a murine CD4 knockout/human CD4 knockin model first described by Killeen et al. (1993). These mice were found to be a good model for evaluating the pharmaco-dynamics of keliximab, which produced a reduction in circulating CD4 levels; this study therefore demonstrated the utility of the model (Sharma et al., 2000). In a subsequent study (Bugelski et al., 2000), these researchers used the humanized mouse to evaluate preclinical safety of keliximab; studies included chronic toxicity, reproductive toxicology, genotoxicity, and immunotoxicology. The only effects observed were those directly related to the pharmacologic activity of keliximab. Finally, using models of host resistance to infectious agents and tumors, keliximab was found to exert a generalized immunomodulatory activity rather than simply acting as an immunosuppressant by depleting circulating CD4 cells. This was evidenced by a differential effect on type 1 and type 2 cytokine expression in splenocytes stimulated *ex vivo*; keliximab caused an up-regulation of inter-leukin-2 (IL-2) and gamma interferon, followed by transient down-regulation of IL-4 and IL-10 (Herzyk et al., 2001, 2002). This later work confirmed the findings of Podolin et al. (2000), who demonstrated multiple types of immunomodulation using a contact hypersensitivity model.

The value of such a system for evaluating the efficacy of immunomodulatory pharmaceuticals cannot be overstated. Due to the species specificity of certain molecules, the only options available may be (1) the parallel development of a homologous murine molecule studied in mice (which will subsequently have to be bridged to human or primate studies); or (2) the exclusive use of primates in safety and efficacy studies. The disadvantages of the former approach include substantially higher development costs and extended timelines (even with parallel tracking), as well as persistent uncertainties in extrapolating the results of murine studies to human safety and efficacy. The disadvantages of the latter approach are extremely high development costs due to the use of primates, the uncertainties of extrapolating nonhuman primate results to humans, and the specter of having to use higher primate species such as baboons or chimpanzees, which are in exceedingly short supply and carry ethical concerns (e.g., inability to perform post-life evaluations).

The usefulness of humanized mice as models for evaluating the unintended immunotoxicology of test materials is less certain, although still promising. For example, humanized transgenics are increasingly being used to limit uncertainties related to species-specific metabolism of xenobiotics (Elmquist and Miller, 2001; Törnell and Snaith, 2002; Gonzalez, 2003). It is conceivable that a multiply transgenic mouse could be developed that would exhibit human-like metabolism, as well as multiple human genes such as human CD4 and MHC molecules (Laub et al., 2000). Such a model would help allay concerns about species specificity as well as relevant metabolism, resulting in an improved screening model. At present, the economics of constructing such an animal is probably prohibitive, since nonmutant models are acceptable to the regulatory agencies and there is little incentive to develop such a model.

22.4.7 Transgenic Rats

The mouse has historically been the workhorse model for immunotoxicology. However, in recent years there has been a move to switch, as least partially, to the rat as a model; this is because almost all standard toxicology testing is performed in this rodent, rather than the mouse. Thus, toxicological findings in the rat following exposure to test materials can be used to assist in understanding toxicity to the immune system in a more comprehensive manner. In addition, immunotoxicology testing in rats obviates the need for additional studies in the mouse, and eliminates concerns about species-related differences in response to test materials. Therefore, it may be desirable to eventually have transgenic rat models as a supplement to immunotoxicology testing.

Although mouse models represent the greatest bulk of transgenic rodents, a number of transgenic rat models are also being developed. Progress in this arena has been impeded by the failure to develop embryonic stem cell cultures analogous to the mouse system. Several methods have been devised to counter this obstacle including gene transduction into male germ-cell lines (Hamra et al., 2002), and transgenically supplying small interfering RNA (siRNA) as a means of gene silencing.

The latter approach suggests that transgenic RNA interference could function as an alternative method of gene silencing by applying homologous recombination to ES cells, and may be successful even in species where ES cell lines have not been established (Hasuwa et al., 2002).

Up to the present, transgenic rats have been used primarily in the study of cardiovascular function and disease, rather than for immunological investigations. This has changed in recent years, as an increasing number of models became available. For example, the rat is becoming an extremely valuable model for understanding the mechanisms of autoimmunity. Breban (1998) reported that rats transgenic for HLA-B27 and human beta 2-microglobulin develop inflammatory conditions similar to human spondylarthropathies. Sugaya et al. (2002) and Higuchi et al. (2003) have described rats transgenic for the env-pX gene of HTLV-I that develop a constellation of collagen vascular diseases and a chronic destructive arthritis similar to rheumatoid arthritis in humans. These models may hold promise for mechanistic understanding of autoimmunity and autoimmune disease; unfortunately, at present it is impossible to predict whether the transgenic rat models can be successfully adapted to develop predictive models of autoimmunity.

Another area of research using transgenic rats that may eventually be useful for immunotoxicology include development of models for a greater understanding of microbial pathogenesis and, consequently, the development of better host defense models (Warner et al., 1996). The dearth of models currently available makes this utility uncertain at present.

One exciting possibility is the creation of a humanized rat, similar to the humanized mouse models described in detail above. A step in this direction has been taken by Keppler et al. (2002), who developed a rat transgenic for human CD4 and human CCR5 (chemokine receptor 5) as a model permissive for HIV-1 infection. Cells from these animals were capable of being infected by HIV, demonstrating functionality similar to human cells. Although these investigators did not describe immune function in these rats, it may be presumed that such tests could be performed. The ability to construct animals with multiple transgenes could eventually lead to the development of humanized rats, with all the attendant benefits described above for humanized mice.

22.5 SUMMARY

Transgenic animal technology has the potential to greatly enhance the field of immunotoxicology, particularly in understanding mechanisms of immunomodulation at a highly detailed level. In this chapter, we reviewed only a few of the many possible uses of this technology, with some uses of immediate potential applicability and others more speculative at present. We also examined a new paradigm of immunotoxicology that expands the historical definition of this field; this expanded definition suggests many possible avenues for exploring immune dysregulation such as autoimmunity and hypersensitivity. Although the use of these animals *specifically* for immunotoxicology research is becoming more common (e.g., Mitchell and Lawrence, 2003; Page et al., 2003), it is anticipated that this powerful tool will greatly accelerate work in this field.

REFERENCES

Adorini, L., Gregori, S., and Harrison, L.C., Understanding autoimmune diabetes: insights from mouse models, *Trends Mol. Med.*, 8, 31, 2002.

Babinet, C., Transgenic mice: an irreplaceable tool for the study of mammalian development and biology, *J. Am. Soc. Neprhol.*, 11, S88, 2000.

Barnaba, V., Viruses, hidden self-epitopes and autoimmunity, *Immunol. Rev.*, 152, 47, 1996.

Boehncke, W.H., The SCID-hu xenogeneic transplantation model: complex but telling, *Arch. Dermatol. Res.*, 291, 367, 1999.

Breban, M., HLA-B27 transgenic rats model, *Ann. Med. Interne (Paris)*, 149, 139, 1998.

Bugelski, P.J., et al., Preclinical development of keliximab, a primatized anti-CD4 monoclonal antibody, in human CD4 transgenic mice: characterization of the model and safety studies, *Hum. Exp. Toxicol.*, 19, 230, 2000.

Burleson, G., Dean, J., and Munson, A., Eds., *Methods in Immunotoxicology*, Wiley-Liss, New York, 1995.

Burleson, G.R., Models of respiratory immunotoxicology and host resistance, *Immunopharmacology*, 48, 315, 2000.

Burns, R.P. Jr., and Gaspari, A.A., The use of transgenic mouse models to investigate the immune mechanisms of allergic contact dermatitis: an area of emerging opportunities, *Am. J. Contact Dermatitis*, 7, 120, 1996.

Caton, A.J., Cerasoli, D.M., and Shih, F.F., Immune recognition of influenza hemagglutinin as a viral and a neo-self-antigen, *Immunol. Res.*, 17, 23, 1998.

Chen, G., et al., Effect of ethanol and cocaine treatment of the immune system of v-Ha-ras-transgenic mice, *Int. J. Immunopharmacol.*, 18, 251, 1996.

Colombo, L.L., et al., In vitro response of v-Ha-ras transgenic mouse lymphocytes after in vivo treatment with alcohol, *Immunopharmacol. Immunotoxicol.*, 23, 597, 2001a.

Colombo, L.L., et al., In vitro response of v-Ha-ras transgenic mouse lymphocytes after in vivo treatment with cocaine, *Immunopharmacol. Immunotoxicol.*, 23, 607, 2001b.

Court, D.L., Sawitzke, J.A., and Thomason, L.C., Genetic engineering using homologous recombination, *Annu. Rev. Genet.*, 36, 361, 2002.

Das, P., Abraham, R., and David, C., HLA transgenic mice as models of human autoimmune diseases, *Rev. Immunogenet.*, 2, 105, 2000.

Dean, J.H., House, R.V., and Luster, M.I., Immunotoxicology: effects of, and response to, drugs and chemicals, in *Principles and Methods of Toxicology*, 4th ed., Hayes, A.W., Ed., Taylor & Francis, London, 2001, pp. 1415–1450.

Dent, L.A., et al., Immune responses of IL-5 transgenic mice to parasites and aeroallergens, *Mem. Inst. Oswaldo Cruz*, 92, 45, 1997.

Elias, J.A., et al., Transgenic modeling of interleukin–13 in the lung, *Chest*, 339S, 2003.

Elmquist, W.F., and Miller, D.W., The use of transgenic mice in pharmacokinetic and pharmacodynamic studies, *J. Pharm. Sci.*, 90, 422, 2001.

Friedman, H., Newton, C., and Klein, T.W., Microbial infections, immunomodulation, and drugs of abuse, *Clin. Microbiol. Rev.*, 16, 209, 2003.

Fuzii, H.T., and Travassos, L.R., Transient resistance to B16F10 melanoma growth and metastasis in CD43-/- mice, *Melanoma Res.*, 12, 9, 2002.

Gao, X., Kemper, A., and Popko, B., Advanced transgenic and gene-targeting approaches, *Neurochem. Res.*, 24, 1181, 1999.

Gaveriaux-Ruff, C., et al., Abolition of morphine-immunosuppression in mice lacking the mu-opioid receptor gene, *Proc. Natl. Acad. Sci. U.S.A..*, 95, 6326, 1998.

Gaveriaux-Ruff, C., et al., Enhanced humoral response in kappa-opioid receptor knockout mice, *J. Neuroimmunol.*, 134, 72, 2003.

Gonzalez, F.J., Role of gene knockout and transgenic mice in the study of xenobiotic metabolism, *Drug Metab. Rev.*, 35, 319, 2003.

Hamra, F.K., et al., Production of transgenic rats by lentiviral transduction of male germ-line stem cells, *Proc. Natl. Acad. Sci. U.S.A..*, 99, 14931, 2002.

Hastings, K.L., Implications of the new FDA/CDER immunotoxicology guidance for drugs, *Int. Immunopharmacol.*, 2, 1613, 2002.

Hasuwa, H., et al., Small interfering RNA and gene silencing in transgenic mice and rats, *FEBS Lett.*, 532, 227, 2002.

Hershey, G.K., IL-13 receptors and signaling pathways: an evolving web, *J. Allergy Clin. Immunol.*, 111, 677, 2003.

Herzyk, D.J., et al., Immunomodulatory effects of anti-CD4 antibody in host resistance against infections and tumors in human CD4 transgenic mice, *Infect. Immun.*, 69, 1032, 2001.

Herzyk, D.J., et al., Practical aspects of including functional end points in developmental toxicity studies. Case study: immune function in HuCD4 transgenic mice exposed to anti-CD4 MAb in utero, *Hum. Exp. Toxicol.*, 21, 507, 2002.

Higuchi, M., et al., Functional alteration of peripheral CD25(+)CD4(+) immunoregulatory T cells in a transgenic rat model of autoimmune diseases, *J. Autoimmun.*, 20, 43, 2003.

Higuchi, T., et al., Cutting edge: ectopic expression of CD40 ligand on B cells induces lupus-like autoimmune disease, *J. Immunol.*, 168, 9, 2002.

Hofker, M.H., and Breuer, M., Generation of transgenic mice, *Methods Mol. Biol.*, 110, 63, 1998.

House, R.V., Thomas, P.T., and Bhargava, H.N., A comparative study of immunomodulation produced by in vitro exposure to delta opioid receptor antagonist peptides, *Peptides*, 17, 75, 1996.

House, R.V., Thomas, P.T., and Bhargava, H.N., Immunotoxicology of opioids, inhalants, and other drugs of abuse, *NIDA Res. Monogr.*, 173, 175, 1997a.

House, R.V., Thomas, P.T., and Bhargava, H.N., In vitro exposure to peptidic delta opioid receptor antagonists results in limited immunosuppression, *Neuropeptides*, 31, 89, 1997b.

House, R.V., Theory and practice of cytokine assessment in immunotoxicology, *Methods*, 19, 17, 1999.

Hussain, I., and Smith, J., Evidence for the transmissibility of atopy: hypothesis, *Chest*, 124, 1968, 2003.

Ji, H., et al., Different modes of pathogenesis in T-cell-dependent autoimmunity: clues from two TCR transgenic systems, *Immunol. Rev.*, 169, 139, 1999.

Keppler, O.T., et al., Progress toward a human CD4/CCR5 transgenic rat model for de novo infection by human immunodeficiency virus type 1, *J. Exp. Med.*, 195, 719, 2002.

Kikuchi, S., et al., A transgenic mouse model of autoimmune glomerulonephritis and necrotizing arteritis associated with cryoglobulinemia, *J. Immunol.*, 169, 4644, 2002.

Killeen, N., Sawada, S., and Littman, D.R., Regulated expression of human CD4 rescues helper T cell development in mice lacking expression of endogenous CD4, *EMBO J.*, 12, 1547, 1993.

Kimber I., and Dearman, R.J., Immune responses: adverse versus non-adverse effects, *Toxicol. Pathol.*, 30, 54, 2002.

Kyburz, D., and Corr, M., The KRN mouse model of inflammatory arthritis, *Springer Semin. Immunopathol.*, 25, 79, 2003.

Laouini, D., et al., IL-10 is critical for Th2 responses in a murine model of allergic dermatitis, *J. Clin. Invest.*, 112, 1058, 2003.

Laub, R., et al., A multiple transgenic mouse model with a partially humanized activation pathway for helper T cell responses, *J. Immunol. Methods.*, 246, 37, 2000.

Lecuit, M., and Cossart, P., Genetically-modified-animal models for human infections: the Listeria paradigm, *Trends Mol. Med.*, 8, 537, 2002.

Ledermann, B., Embryonic stem cells and gene targeting, *Exp. Physiol.*, 85, 603, 2000.

Leonard, W.J., Cytokines and immunodeficiency diseases, *Nat. Rev. Immunol.*, 1, 200, 2001.

Lernmark, A., Immune surveillance: paraneoplastic or environmental triggers of autoimmunity, *Crit. Rev. Immunol.*, 17, 437, 1997.

Levetin, E., and Van de Water, P., Environmental contributions to allergic disease, *Curr. Allergy Asthma Rep.*, 1, 506, 2001.

Li, P., and Schwarz, E.M., The TNF-alpha transgenic mouse model of inflammatory arthritis, *Springer Semin. Immunopathol.*, 25, 19, 2003.

Løvik, M., Mutant and transgenic mice in immunotoxicology: an introduction, *Toxicology*, 199, 65, 1997.

Ludewig, B., Zinkernagel, R.M., and Hengartner, H., Transgenic animal models for virus-induced autoimmune diseases, *Exp. Physiol.*, 85, 653, 2000.

Luster, M.I., et al., Qualitative and quantitative experimental models to aid in risk assessment for immunotoxicology, *Toxicol. Lett.*, 64–65, 71, 1992.

Luster, M.I., et al., Use of animal studies in risk assessment for immunotoxicology, *Toxicology*, 192, 229, 1994.

Maecker, B., et al., Linking genomics to immunotherapy by reverse immunology — "immunomics" in the new millennium, *Curr. Mol. Med.*, 1, 609, 2001.

Matzinger, P., An innate sense of danger, *Semin. Immunol.*, 10, 399, 1998.

McCarthy, L., et al., Opioids, opioid receptors, and the immune response, *Drug Alcohol Dependency*, 62, 111, 2001.

McCune, J.M., Development and applications of the SCID-hu mouse model, *Semin. Immunol.*, 8, 187, 1996.

McKenzie, S.E., Humanized mouse models of FcR clearance in immune platelet disorders, *Blood Rev.*, 16, 3, 2002.

Mitchell, K.A., and Lawrence, B.P., T cell receptor transgenic mice provide novel insights into understanding cellular targets of TCDD: suppression of antibody production, but not the response of CD8(+) T cells, during infection with influenza virus, *Toxicol. Appl. Pharmacol.*, 192, 275, 2003.

Moser, R., Quesniaux, V., and Ryffel, B., Use of transgenic animals to investigate drug hypersensitivity, *Toxicology*, 158, 75, 2001.

Müller, U., Ten years of gene targeting: targeted mouse mutants, from vector design to phenotype analysis, *Mech. Dev.*, 82, 3, 1999.

Nisitani, S., Murakami, M., and Honjo, T., Anti-red blood cell immunoglobulin transgenic mice. An experimental model of autoimmune hemolytic anemia, *Ann. N. Y. Acad. Sci.*, 815, 246, 1997.

Page, T.J., O'Brien, S., Holston, K., MacWilliams, P.S., Jefcoate, C.R., and Czuprynski, C.J., 7, 12-Dimethylbenz[a]anthracene-induced bone marrow toxicity is p53-dependent, *Toxicol. Sci.*, 74, 85, 2003.

Pellegrino, T., and Bayer, B.M., In vivo effects of cocaine on immune cell function, *J. Neuroimmunol.*, 83, 139, 1998.

Podolin, P.L., et al., Inhibition of contact sensitivity in human CD4+ transgenic mice by human CD4-specific monoclonal antibodies: CD4+ T-cell depletion is not required, *Immunology*, 9, 287, 2000.

Power, C.A., Knock out models to dissect chemokine receptor function in vivo, *J. Immunol. Methods*, 273, 73, 2003.

Prescott, S.L., Allergy: the price we pay for cleaner living? *Ann. Allergy Asthma Immunol.*, 90, 64, 2003.

Putman, E., van der Laan, J.W., and van Loveren, H., Assessing immunotoxicity: guidelines, *Fundam. Clin. Pharmacol.*, 17, 615, 2003.

Rajewsky, K., et al., Conditional gene targeting, *J. Clin. Invest.*, 98, 600, 1996.

Ravirajan, C.T., and Isenberg, D.A., Transgenic models of tolerance and autoimmunity: with special reference to systemic lupus erythematosus, *Lupus*, 11, 843, 2002.

Reilly, M.P., and McKenzie, S.E., Insights from mouse models of heparin-induced thrombocytopenia and thrombosis, *Curr. Opin. Hematol.*, 9, 395, 2002.

Roy, S., Barke, R.A., and Loh, H.H., MU-opioid receptor-knockout mice: role of mu-opioid receptor in morphine mediated immune functions, *Brain Res. Mol. Brain Res.*, 61, 190, 1998.

Ryffel, B., Cytokine knockout mice: possible application in toxicological research, *Toxicology*, 105, 69, 1995.

Saito, S., Kasturi, K., and Bona, C., Genetic and immunologic features associated with scleroderma-like syndrome of TSK mice, *Curr. Rheumatol. Rep.*, 1, 34, 1999.

Saito, S., et al., Induction of skin fibrosis in mice expressing a mutated fibrillin-1 gene, *Mol. Med.*, 6, 825, 2000.

Sakiyama, T., et al., Requirement of IL-5 for induction of autoimmune hemolytic anemia in anti-red blood cell autoantibody transgenic mice, *Int. Immunol.*, 11, 995, 1999.

Schodde, H., et al., Ethanol ingestion inhibits cell-mediated immune responses of unprimed T-cell receptor transgenic mice, *Alcohol Clin. Exp. Res.*, 20, 890, 1996.

Sharma, A., et al., Comparative pharmacodynamics of keliximab and clenoliximab in transgenic mice bearing human CD4, *J. Pharmacol. Exp. Ther.*, 293, 33, 2000.

Singh, V.K., Mehrotra, S., and Agarwal, S.S., The paradigm of Th1 and Th2 cytokines: its relevance to autoimmunity and allergy, *Immunol Res.*, 20, 147, 1999.

Slattery, D.M., Gerard, N., and Gerard, C., Gene targeting of chemokines and their receptors, *Springer Semin. Immunopathol.*, 22, 417, 2000.

Sonderstrup, G., Development of humanized mice as a model of inflammatory arthritis, *Springer Semin. Immunopathol.*, 25, 35, 2003.

Soussi-Gounni, A., Kontolemos, M., and Hamid, Q., Role of IL-9 in the pathophysiology of allergic diseases, *J. Allergy Clin. Immunol.*, 107, 575, 2001.

Sugaya, T., et al., Clonotypic analysis of T cells accumulating at arthritic lesions in HTLV-I env-pX transgenic rats, *Exp. Mol. Pathol.*, 72, 56, 2002.

Suri, A., and Katz, J.D., Dissecting the role of CD4+ T cells in autoimmune diabetes through the use of TCR transgenic mice, *Immunol. Rev.*, 169, 55, 1999.

Szabo, G., Consequences of alcohol consumption on host defence, *Alcohol Alcoholism*, 34, 830, 1999.

Törnell, J., and Snaith, M., Transgenic systems in drug discovery: from target identification to humanized mice, *Drug Discovery Today*, 7, 461, 2002.

Traidl, C., et al., New insights into the pathomechanisms of contact dermatitis by the use of transgenic mouse models, *Skin Pharmacol. Appl. Skin Physiol.*, 13, 300, 2000.

Vos, J.G., Immune suppression as related to toxicology, *CRC Crit. Rev. Toxicol.*, 5, 67, 1977.

Vos, J.G., and van Loveren, H., Experimental studies on immunosuppression: how do they predict for man? *Toxicology*, 129, 13, 1998.

Wang, B., et al., Cytokine knockouts in contact hypersensitivity research, *Cytokine Growth Factor Rev.*, 14, 381, 2003.

Warner, T.F., et al., Human HLA-B27 gene enhances susceptibility of rats to oral infection by Listeria monocytogenes, *Am. J. Pathol.*, 149, 1737, 1996.

Wilson, S.D., et al., Correlation of suppressed natural killer cell activity with altered host resistance models in B6C3F1 mice, *Toxicol. Appl. Pharmacol.*, 177, 208, 2001.

Wolfer, D.P., Crusio, W.E., and Lipp, H.P., Knockout mice: simple solutions to the problems of genetic background and flanking genes, *Trends Neurosci.*, 25, 336, 2002.

Wong, F.S., Dittel, B.N., and Janeway, C.A., Jr., Transgenes and knockout mutations in animal models of type 1 diabetes and multiple sclerosis, *Immunol. Rev.*, 169, 93, 1999.

Yap, G.S. and Sher, A., The use of germ line-mutated mice in understanding host-pathogen interactions, *Cell. Microbiol.*, 4, 627, 2002.

Zhang, L., et al., An immunological renal disease in transgenic mice that overexpress Fli-1, a member of the ets family of transcription factor genes, *Mol. Cell Biol.*, 15, 6961, 1995.

CHAPTER **23**

Applications of Genomics in Immunotoxicology

Sami Haddad, Michele D'Elia, Jacques Bernier, Alain Fournier, and Daniel G. Cyr

CONTENTS

23.1 INTRODUCTION

The term "genomics" has come to encompass a large number of disciplines including gene sequencing, analyses of gene expression at the level of the transcriptome, genetic variability, proteomics, and metabolomics. The wide array of information generated from the "-omics" has also resulted in the development of complex analyses of these data, and consequently, the development of bioinformatics. Toxicologists quickly realized the immense potential of these emerging sciences for understanding complex biochemical pathways leading to pathological conditions. Hence, toxicogenomics, which refers to the effects of toxic chemicals on genome analysis, has become a flourishing area of toxicology, and represents an excellent field of research for the application of genomics technologies.

While we consider this science as novel, the term *genomics* itself was coined in 1920 by Winkler to describe the complete set of chromosomes (McKusick, 1997). The sequencing of the human, mouse, and rat genomes, as well as those of the yeast and other microorganisms such as plants and lower vertebrates, represents a unique opportunity to use the tools developed in these megaprojects to understand the regulation of large clusters of genes that regulate cellular signaling pathways necessary for cellular function. Perhaps more importantly, such tools permit the examination, at the gene level, of how these pathways can be modulated by chemicals and pathological conditions. The new tools have been commercialized and thus have become widely available to most research laboratories. As such, the application of these technologies, and how they might be modified to address complex biological problems, remains in its infancy and its potential for toxicology is particularly promising. In this chapter, we review aspects of functional gene expression and proteomics and how they may be applied to the area of immunotoxicology.

23.2 FLOW OF GENETIC INFORMATION

23.2.1 Genomics

The genome is composed of nucleotides whose sequence represents the complete set of genes of an organism. This sequence contains the entire genetic information necessary for a living organism. In 2001, the first draft of the human genome was published. The sequence of this genome was thought to consist of approximately 3×10^9 nucleotides. The number of genes that comprise the human genome is controversial, with estimates varying from 23,000 to 150,000 (Lander et al., 2001; Venter et al., 2001), and accurate determination of the number is not expected for several years. Within the human genome, there exist many variations in the sequence of DNA from one individual to another, primarily as single nucleotide polymorphisms (SNP). Although there have been 3×10^6 SNP discovered so far, it is estimated that over 11×10^6 exist in the human genome (Joos et al., 2003). At the moment, the genome has been sequenced for only two other mammalian species: the mouse and the rat (http://public.bcm.tmc.edu/pa/rgsc-genome.htm), although the sequencing of other species is underway.

23.2.2 Functional Genomics

The level of gene expression depends on several factors, including cell type, tissue type, age of the organism, developmental stage, and environmental factors. The expression of mRNA transcripts constitutes the transcriptome. While genomics focuses on the DNA, functional genomics concentrates on the transcriptome (i.e., the level of a transcript or specific mRNA present in the cell). The primary goal of functional genomics is to identify and assign gene expression patterns to tissues or cells at normal and abnormal physiological states. The clustering of genes into specific biochemical signaling pathways enables the analyses of cellular processes and how these processes may be activated, or deactivated, under toxic stress.

23.3 TECHNOLOGICAL ADVANCES FOR ANALYZING
FUNCTIONAL GENOMICS

23.3.1 DNA Arrays

Many have been developed to study functional genomics, including differential display polymerase chain reaction (PCR), SAGE (serial analysis of gene expression), and subtractive hybridization. In recent years, the development of DNA macro- and micro-arrays has enabled the simultaneous analyses of hundreds and/or thousands of genes, respectively. These analyses are based on well-characterized hybridization techniques that molecular biologists have perfected over the past 30 years.

For many years now, molecular biologists were restricted to low-throughput techniques to analyze gene expression. Microarray techniques have come to complement existing low-throughput molecular biology techniques by offering large-scale high-throughput capabilities. The concept of the microarray techniques is relatively simple (Figure 23.1). RNA is harvested from a cell type or tissue and labeled to generate a target (i.e., the free nucleic acid whose identity or abundance is being detected). These targets are then hybridized to the fastened DNA sequences (i.e., probes) corresponding to specific genes. The cDNA probes are fixed directly onto a solid matrix or, in the case of oligonucleotides, can be synthesized directly onto the matrix. The hybridization between target and probes is based on the base-pairing capability of DNA. The abundance of RNA for a specific gene in the sample is then reflected by the amount of bound-labeled target on a specific spot of the array, which is captured digitally. There are basically two types of array platforms that are commonly used: macroarrays and microarrays. Fluid arrays (or suspension bead arrays) are another array format that is currently in development (Willis et al., 2003).

23.3.2 Macroarrays

Macroarrays, also called filter-based arrays, differ from microarrays in their probe density and the use of ^{33}P-labeled targets. Probes (cDNA clones, PCR amplicons, or oligonucleotides) are deposited onto charged nylon filters or membranes. The porous nature of these nylon filters does not allow for miniaturization. These arrays typically contain between 200 and 8000 genes. A limitation in using radiolabeled targets in these arrays is that only one sample at a time can be hybridized. Unsaturated phosphorimage screens or radioautography with x-ray films can then be used to quantify the levels of specific transcripts. This type of array platform has several advantages: (1) they are simple to use, particularly for labs with experience in Northern or Southern blotting; (2) they do not require specialized equipment; (3) they are reusable several times and therefore relatively inexpensive (Arfin et al., 2000); and (4) the detection of radioactive targets is more sensitive than fluorochromes, thereby allowing for linear measurements over a four- to five-log range.

23.3.3 Microarrays

In contrast to macroarrays, microarray probes are affixed to a glass matrix and contain a high density of genes (i.e., more than 10,000 genes). The use of glass as a matrix has several advantages. First, DNA probes can be covalently attached to the matrix. Second, because glass is nonporous, the hybridization volumes can be kept low, enhancing the base-pairing kinetics between the probe and target. Third, glass is a durable material that is resistant to high temperatures and washes of high ionic strength, allowing washes that greatly reduce background levels. Finally, glass contributes little to background noise because of its low inherent fluorescence.

For microarray analyses, fluorescent dyes (Table 23.1) are used instead of radioactivity. Therefore, by using different fluorescent dyes, many samples can be hybridized simultaneously to the

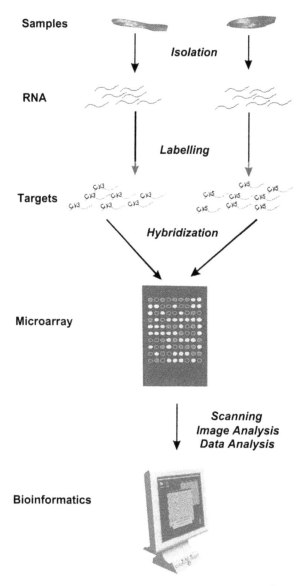

Figure 23.1   Steps in a microarray experiment. RNA is extracted and isolated from the biological samples of interest. Targets are prepared from RNA by incorporating labeling dyes during reverse trancription. Targets are then hybridized to the probes on the microarray. Finally, the target-probe complex levels are detected through a scanner and converted to data that are later analyzed with the use of bioinformatics.

same glass slide array. This has the advantage of comparing control and experimental samples under identical hybridization conditions with the same array. As a result, analyses can be done using two different samples and, in some experiments, researchers have used up to four different samples with specific fluorochromes that can be excited by lasers at different wavelengths and distinguished from one another (Lindroos et al., 2002; Lovmar et al., 2003). This has been particularly useful for studying single-nucleotide polymorphisms. While the use of different fluorochromes offers a major advantage for microarrays as compared to macroarrays, certain fluorochromes more easily incorporate into certain transcripts than others. Consequently, dye-swap experiments must always be performed to correct for differences in preferential incorporation of one dye over

Table 23.1 List of Commonly Used Fluorescent Dyes

Dye	Wavelengths Excitation	Emission	Laser Color
Alexa 488	495	520	Green-blue
FITC (fluorescein isothiocyanate)	490	520	Green-blue
Fluor X	494	520	Green-blue
FAM	495	535	Green-blue
Alexa 532	525	550	Green
Alexa 546	555	570	Green
Alexa 555	555	565	Green
Cy3	550	570	Green
Cy3.5	581	596	Green
JOE	525	557	Green
TAMRA	543	570	Green
TMR (tetramethyl rhodamine)	537	566	Green
Alexa 594	590	615	Yellow
ROX	594	614	Yellow
Texas red	595	615	Yellow
Alexa 647	650	670	Red
BODIPY	630	650	Red
Cy5	649	670	Red
Cy5.5	675	694	Red

another. This is accomplished by changing the given dye between control and experimental samples of hybridizations.

A growing industry has developed and commercialized microarrays designed for use either for specific animal model species or with target specific pathologies or processes (e.g., cancer, ADME, toxicology, metabolism, SNPs, etc.). It should be noted, however, that the quality of DNA arrays can vary from one distributor to another, as does the sensitivity of the arrays. It is therefore important that preliminary experiments be done to establish whether a specific array will provide reproducible results both within a given batch and between batches in the case of long-term experiments.

In toxicology, the simplest studies involve a reference group of animals (or cells) compared to a group that has been exposed to a toxicant. To generate added insight regarding which genes are directly linked to the toxicological insult, more groups may be added in order to generate dose responses and different timepoints. Apart from the biological aspects of the experimental design, statistical considerations also play a key role in the experimental design. Aspects to consider in this process include the various factors to be tested, sample size per factor (i.e., biological replicates), and the number of replicates per sample (i.e., technical replicates). In addition, when carrying out two-color array experiments, the design of the sample pairing (i.e., two samples per slide), is an additional consideration (Figure 23.2). Several resources for statistical design issues for microarray experiments are available in the literature (Churchill, 2002; Dobbin et al., 2003; Pan et al., 2002; Simon and Dobbin, 2003; Yang and Speed, 2002). As toxicologists have documented over the years, biological replicates are particularly important, as gene expression between individuals and between cells can vary quite a bit. Consequently, it is generally necessary to use between three and five replicates in order to generate reliable data, particularly for targeting specific genes whose expression is changing. Confirmation of changes is also necessary when doing microarray analyses. Changes to specific genes should be confirmed using an alternative method such as real-time PCR.

23.4 BIOINFORMATICS

Numerous software programs have been developed for the various levels of analyses that must be carried out (a comprehensive list can be found at http://ihome.cuhk.edu.hk/~b400559/arraysoft.html).

Reference Design

Standard

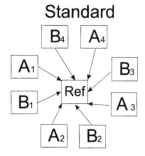

Balanced BlockDesign

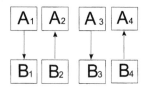

Loop Design

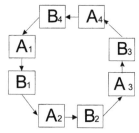

Figure 23.2 Three basic dual-color microarray experimental designs for comparing samples A and B, available in four replicates. The samples at beginning of arrows are labeled in green (Cy3), and those at the end are labeled in red (Cy5). In the reference design, samples are cohybridized with a reference sample. In the balanced block design, the samples are hybridized against each other, and dyes are inversed between each replicate. In the loop design, samples are hybridized between each other in order to form a loop pattern. The latter design requires two aliquots of each sample to form the loop.

Some of these programs are highly specialized and deal with only one aspect of the array analysis process, while others cover almost the entire analysis. As one would expect, price and user-friendliness vary enormously. It is therefore important to consider your needs, competence level in informatics, and budget in the acquisition of microarray software. The levels of microarray analysis involving bioinformatics are briefly described in this section.

23.4.1 Image Acquisition and Analysis

When fluorescent dyes are used to label the targets for hybridization with the DNA array, a scanner must be used to excite the fluorophores and detect the light emitted. What these instruments basically do is to convert the captured signals emitted from the labeled targets on the arrays to images consisting of pixels of 256,000 shades of gray. The resolution of these images can vary between

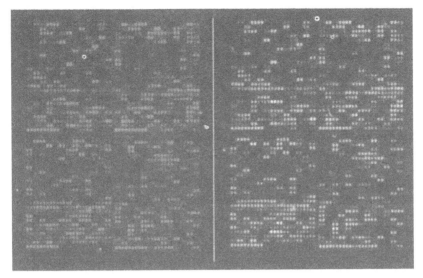

Figure 23.3   (See color insert following page 236) Two images of the same microarray are generated in a two-color microarray experiment. The left image is captured following excitation of the Cy5-labeled targets by a red laser. The right image is captured following excitation of the Cy3-labeled targets by a green laser. The ratio of intensities of a spot in a particular coordinate in both images is indicative of the differential expression of the gene that the spot represents.

1 to 100 μm/pixel depending on the imager or scanner. Evidently, when using two color arrays, two different images are generated (Figure 23.3).

Once the images are generated, software is used to analyze them. The software converts the intensity of each gray pixel into a numerical value and determines the median and mean intensity of each spot and its associated background. These software programs usually give the user the possibility to visualize the spots on the array image and view the data that are linked to it. This allows the user to apply quality control to the data. Based on quality criteria, spots can be manually or automatically filtered in order to eliminate them from further analysis. It is particularly important to identify irregularities in the spots, such as shape, background level, presence of dust particles, and degree of saturation.

The data must then be normalized in order to minimize experimental and systematic variations in gene expression levels. Experimental variations can result from differences in amounts of (1) starting biological sample, (2) amounts of RNA isolated, and (3) cDNA generated. Well-known sources of systematic variation are biases associated with the use of different dyes in two-color experiments. The dye bias originates from several factors such as varying physical properties of the dyes, dye incorporation efficiency, hybridization efficiencies, and signal measurement sensitivity at the data collection step. Several normalization protocols have been proposed to correct for these sources of variability. The most commonly used methods of normalization are express gene expression levels relative to levels of housekeeping genes, normalization by using reference RNA, normalization by global scaling, and dye swap.

The expression levels of housekeeping genes are thought to be invariable under different experimental conditions. The method basically consists of normalizing array data with the values of the housekeeping genes of the same array, in order to render comparisons between slides/chips possible. One must be careful when using this method, because some genes considered to be housekeeping genes have been altered by the experimental treatment (Lee et al., 2002; Suzuki et al., 2000).

Normalizing to a reference RNA consists of adding a given amount of RNA standard from another organism to each total RNA sample. Based on the reference RNA, each hybridized sample can then be compared within and across arrays. This allows for corrections for variance in mRNA isolation, labeling, and hybridization efficiencies. The downfall of this method is that variance can be introduced by errors in the measurement of the standard, and in their application to each of the arrays.

Normalization by global scaling consists of expressing the intensity value for each gene on an array divided by the sum of all intensities on that same array, bringing the value of the total signal to 1. This is based on the assumption that the total amount of mRNA on the slides is kept constant under different experiment conditions. When small or specialized arrays are used this assumption may no longer hold. A commonly used normalization method for two-color arrays is the locally weighted, linear regression Lowess analysis (Cleveland, 1979). This method has been proposed as a normalization process in order to correct for artifacts caused by nonlinear rates of dye incorporation, as well as inconsistencies in the relative fluorescence intensity between some red and green dyes (Chen et al., 1997; Yang et al., 2002). In the absence of bias, the values of the $\log_2$ ratio on the MA plot would be expected to scatter evenly around the zero values along the $\log_2$ intensities (Figure 23.4). The Lowess method detects systematic deviations in the plot, and corrects them by carrying out a local weighted linear regression of the data and subtracting the best-fit $\log_2$ ratio from the measured ratio of each data point. Dye swapping is used for correcting for dye biases in two-color experiments. This consists in making a technical replicate for each array-containing chip in which a dye inversion is done during the labeling. This method effectively normalizes for labeling and hybridization efficiencies.

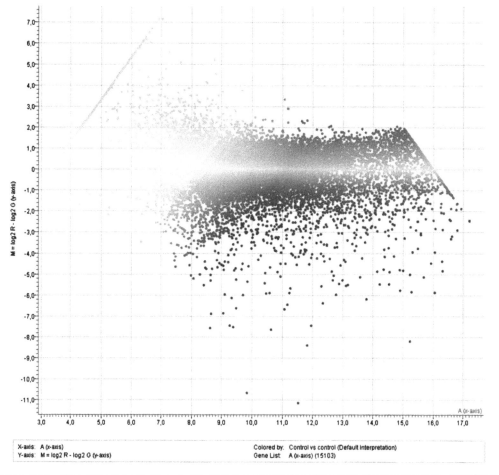

Figure 23.4   (See color insert) M vs. A plot. M represents the log of the ratio of intensities ($\log_2$ Red/Green), whereas A is the log of the root of the product of intensities ($\log_2 \sqrt{Red^*Green}$).

23.4.2 Data Analysis

23.4.2.1 Differential Gene Expression

Once the data have been normalized, the investigator can attempt to identify genes that are differentially expressed, groups of genes that are coregulated, or genome-wide patterns of differential gene expression. Determination of which genes are differentially expressed by the experimental condition can be done by different methods. A simple method is to declare "significant" the change in expression of genes for which the x-fold change is greater than an arbitrary ratio (usually twofold ratio). Although this fold-change method has been used extensively, it lacks biological significance, and variability is not considered. False negatives are more likely to occur in highly expressed genes, whereas false positives are more likely to occur in poorly expressed genes. Although the use of this method is still reported in publications, it is now generally agreed that statistical methods are much more valuable (Chen et al., 1997).

The advent of microarray data has challenged statisticians to arrive at a standardized approach for assessing statistical significance. This is perhaps most evident in toxicology, where responses can vary significantly between individuals. The challenge for statisticians, who are used to dealing with large sample sizes and relatively few parameters, is to deal with a small sample size and extremely large numbers of different parameters. A simple statistical test used to address this is the Student's t-test. The problem with a t-test is that because of the large number of genes that are measured on a microarray, the likelihood of finding false positives is high. For example, if 10,000 genes are measured on an array, and genes with p values of 0.05 are selected, 500 of these genes are likely to be false positives. Several corrections have been suggested for using t-tests in microarray experiments, including other assays such as the Bonferonni, Sid·k, Holm, Benjamini-Hochberg false discovery rate test, the Westfall and Young test, and bootstrapping (Draghici, 2002). Variants to the two-sample t-statistical method have been proposed for use for microarrays (i.e., the regression model approach [Pan, 2002], the mixture modeling approach [Pan, 2002], the SAM t-test [Tusher et al., 2001], and B-statistic [Lönnstedt and Speed, 2002]). When more than two conditions are analyzed in a microarray experiment, analysis of variance should be used for determining statistical significance (Kerr et al., 2000).

23.4.2.2 Pattern Recognition

A necessary goal for toxicogenomics is to develop databases in which signature profiles of gene expression can be assigned to a given chemical or mixture of chemicals. As this information becomes part of large public databases, it is hoped that it will be possible to associate gene expression profiles to those that have been identified from controlled laboratory experiments or well-characterized exposures. This will enable scientists to predict specific chemical exposures based on gene expression profiles or to obtain information on the mechanisms of transcriptional regulatory networks (Alon et al., 1999; Tavazoie et al., 1999; Xing and Karp, 2001).

The numerous pattern recognition methods that have been developed (Quackenbush, 2001; Valafar, 2002) can be divided into two groups: supervised and unsupervised. In this context, supervision refers to the use of existing biological information on the genes or predetermined clusters of genes to guide the clustering of new genes. The supervised methods use statistical techniques known as linear and quadratic discriminant analysis and decision trees, neural networks, or support vector machines. The unsupervised methods, often referred to as clustering methods, are the most commonly used methods in microarray analysis. Among the most commonly used clustering algorithms for microarray data are hierarchical clustering approaches, K-means clustering approaches, principal component analysis (PCA), and self-organizing maps (SOM) (Baldi and Hatfield, 2002; Quackenbush, 2001; Valafar, 2002).

Hierarchical clustering is considered to be relatively simple and can be easily visualized as dendrograms. Hierarchical clustering algorithms are examples of either agglomerative or the divisive approaches. Agglomerative methods basically consist of iteratively grouping pairs based on similarities from individual genes until only one group remains (Quackenbush, 2001). In the created dendrograms, each node represents a cluster of similar objects. Hierarchical clustering has been used extensively in microarray studies (Eisen et al., 1998; Hamadeh et al., 2002; Wen et al., 1998). In contrast, divisive methods start from one cluster in which the most dissimilar elements are separated to form a new cluster; this is done iteratively until each gene falls into a separate cluster (Smolkin and Ghosh, 2003).

Unlike hierarchical clustering, K-means clustering produces no dendrograms. Instead, objects (i.e., genes) are initially randomly separated into a fixed number (k) of clusters and the distance within and between each cluster is then calculated. The objects are then shuffled from one cluster to another iteratively until the distance within and between each cluster is optimal (i.e., low intracluster and high intercluster distance) (Quackenbush, 2001). A disadvantage of this method is that the K-value is not known in advance. This can be circumvented by using PCA to generate a K-value (Quackenbush, 2001).

Karhunen-Loéve expansion, which is also known as singular value decomposition or PCA in statistics, is a widely used mathematical technique that reduces the effective dimensionality of gene-expression space without significant loss of information in order to find patterns in the data (Raychaudhuri et al., 2000). This complex technique basically iteratively views the data from various perspectives, that is, different variables until it discovers the angle that best separates the data into different groupings (Quackenbush, 2001). Although the precise boundaries between clusters are often difficult to define, this technique has proven to be powerful in conjunction with clustering techniques that require specifications of the number of clusters to use (e.g., K-means clustering and SOM).

The SOM clustering approach is based on neural networks, and was first proposed by Kohonen (1990). The SOM assigns genes to a series of partitions on the basis of the similarity of their expression vectors to reference vectors that are defined for each partition. The process of defining the reference vector is what differentiates this method from K-means clustering (Quackenbush, 2001). Much like K-means clustering, the number of partitions are first assigned by the user; therefore, this method is well complemented with PCA when biological information is lacking. This method has been used to uncover patterns in gene expression (Golub et al., 1999; Tamayo et al., 1999; Toronen et al., 1999).

23.4.3 Microarray Information Necessary for Data Storage

When conducting microarray experiments, huge amounts of data are generated, resulting in new challenges. One of these challenges is how to standardize the information associated with any given experiment in order to eventually have these data stored in a database so that they can be shared, used, and compared among laboratories. In order to be able to track and exchange the data and their related information efficiently, some common standards have been adopted by the scientific community. Projects to this effect have been pursued by the Microarray Gene Expression Data (MGED) Society (i.e., MIAME, MAGE, and its ontologies). The MIAME project focuses on the issue of the minimum information about any given microarray experiment that is required in order to interpret and compare the results of an experiment with those of others (www.mged.org/Workgroups/MIAME/miame.html). MIAME compliancy has become a requirement for the publication of data in many peer-reviewed scientific journals. The MAGE project is oriented toward the establishment of a data exchange format (MAGE-ML) and object model (MAGE-OM) for microarray experiments (www.mged.org/Workgroups/MAGE/mage.html). The Ontologies project focuses on the development of standard vocabulary for describing microarray experiments, and in particular, biological material (biomaterial) annotations (http://mged.sourceforge.net/ontologies/index.php).

Another project that has been created is the Transformation project, whose goal is to provide recommendations on the methods of microarray data transformation and normalization. Together these initiatives should allow for the development of databases and exchange of standardized data information among scientists.

As indicated previously, the human genome appears to contain between 30,000 and 150,000 genes that may be translated into anywhere from 100,000 to 1 million different proteins because of alternative splicing and post-translational modifications, such as phosphorylation, glycosylation, or myristylation, to name a few (LoPachin et al., 2003; Sanchez, 2002; Werner, 2004). The ensemble of the proteins derived from the genome was identified in the 1990s as the "proteome" (Figeys, 2003; Wilkins et al., 1996). The proteome is particularly complex because proteins play key roles in nearly all aspects of cell functions. Moreover, in any given cell, they change continuously, as genes are turned on and off in response to their environment. The therapeutic or diagnostic potential resulting from the characterization of the proteome, as well as its capacity to lead to the understanding of molecular mechanisms related to physiology and pathologies, favored the development of new strategies to study protein features and functions. From this situation emerged "proteomics," a research field aiming at obtaining an integrated view of normal and abnormal cellular processes (Blackstock and Weir, 1999). Clearly, the proteome constituents are the targets of a plethora of compounds, including of course toxic agents. Therefore, investigating changes in the proteome upon exposure to toxicants can provide valuable data about the related biochemical mechanisms and cellular effects in a given physiological system.

23.5 PROTEOMICS

Early proteomic studies relied essentially on the combination of two techniques: two-dimensional gel electrophoresis and mass spectrometry. However, over the past few years, the field has evolved substantially by incorporating a broad range of technologies that have given rise to three subdomains known as "expression" or "profiling" proteomics, "functional" proteomics, and "structural" proteomics (Figeys, 2003; LoPachin et al., 2003). Expression (profiling or mapping) proteomics aims to identify specific targets and markers present in a biological sample or proteins that are differentially expressed between samples. Profiling the proteins of a complex mixture can be obtained by separating the components, followed by their digestion into peptides, or by hydrolyzing first the proteins and subsequently, isolating the resulting peptides. Several techniques based on various physicochemical properties such as size, solubility, and charge can be used for protein or peptide enrichment (affinity chromatography, isoelectric focusing, extraction, SDS-PAGE, etc.). Sometimes, digested samples can even be analyzed by mass spectrometry without further purification procedures. However, the most standard procedure favors, as a first step, the separation of proteins using two-dimensional (2D) gel electrophoresis, a technology that has the ability to resolve complex mixtures of proteins in a single gel. Thus, a protein sample is applied on a gel and, in the first dimension, the constituents are separated by isoelectric focusing within an appropriate pH range. The second dimension of separation consists of gel electrophoresis to separate the proteins according to their molecular weight. Resolved proteins can subsequently be visualized by numerous staining methods, but care must be taken in order to select a procedure compatible with the next steps of the analyses, that is, in-gel digestion and mass spectrometry (MS).

In various biological studies, the effects are measured in comparison to control conditions. Consequently, mapping proteomics carried out for toxicological studies will usually compare differences in the protein 2D patterns of complex mixtures obtained in the presence or absence of a presumed toxic state. Hence, image analysis performed by straightforward visualization or computer-assisted techniques identifies substantial changes observed between control and toxic exposure conditions. Specific protein spots are then excised from the gel and destained before being digested with proteases (frequently trypsin) in order to generate a set of peptides whose molecular masses usually range between 500 and 3000 daltons. Thereafter, the peptide combination is analyzed by MS and, depending on the MS technology applied, peptide mass fingerprinting or peptide sequencing is obtained. This MS information is used to explore protein sequence databases in order to compare homologies with known proteins, and thereby identify the protein(s) exhibiting a significant modulation of its expression under a given toxic or pathological state. For peptide mass fingerprinting, the predominant MS method is matrix-assisted laser desorption ionization (MALDI) time of flight, while the electrospray ionization (ES) procedure, paired with tandem mass analyzers (MS/MS), is used for peptide sequence analysis (LoPachin et al., 2003; Naylor and Kumar, 2003). As described by its acronym, MALDI spectrometry is carried out using an aromatic matrix able to absorb light from a laser source. Upon laser pulses, the matrix and the peptide analytes are ionized and the corresponding masses are calculated from their "time of flight" in a high electric field, after calibration with peptides of known molecular weight. On the other hand, ES-MS/MS also provides information about the mass of peptide compounds. However, the instrument is designed in such a way that it is possible to first isolate a single peptide ion species, and second, to fragment this entity after collision of the ion with an inert gas. The fragmentation corresponds to the loss of amino acid fragments. Therefore, mass analysis of the derived new fragments, in comparison to the mass of the original ion, leads to the elucidation of the peptide sequence.

Functional and structural proteomics were subsequently developed with the application of complementary analytical techniques such as the surface plasmon resonance (SPR) technology, the printed protein–peptide array microchip analysis, x-ray crystallography, and nuclear magnetic resonance (NMR) spectroscopy (Figeys, 2003; Sem et al., 2003; Zieziulewicz et al., 2003). These fields provide a detailed knowledge of protein roles, interactions with other protein or peptide ligands, cellular location, and key structural features. For instance, protein–peptide microarrays and the SPR technology can facilitate the detection of a physiologically relevant protein or peptide complex, as well as the identification of nonnatural ligands, a central observation in toxicological studies. Moreover, with the latter technique, characterization of dynamic processes can be achieved, since binding data are measured in real time. In parallel, structural proteomics attempt to determine the crucial geometry of peptide and proteins by means of powerful technologies such as x-ray diffraction and NMR spectroscopies. The molecular arrangement of thousand of proteins have already been determined by crystallography. For instance, it was used to explore the proteins involved in cell motility (Pollard and Borisy, 2003), and to explain the mode of action of anticancer drugs such as Taxol (Jimenez-Barbero et al., 2002). Similarly, a large number of peptide and protein structures were described in the literature following their NMR analysis. In fact, NMR plays a prominent role in "metabonomics" (Nicholson et al., 2002; Sem et al., 2003), a growing research field associated with metabolism-related proteomics. Interestingly, NMR has also applied a scheme from the traditional 2D protein gels. This method referred to as 2D electrophoretic NMR allows for the separation of proteins in an electric field, but in solution rather than a gel (He et al., 1998; Sem et al., 2003). This strategy reveals an enormous potential as a tool when studying multiple targets in parallel.

23.6 APPLICATIONS FOR IMMUNOTOXICOLOGY

23.6.1 Applications of Genomics in Immunotoxicology

Traditional research has focused on understanding how one or a few genes at a time can be linked to a certain response caused by a chemical. Recently, new and improved methods in molecular biology have been developed to identify mechanisms responsible for immunotoxicologic responses. Genomic approaches have allowed investigators to view the entire pattern of expression of numerous genes associated with different biological processes at any specific time. In fact, application of gene expression profiling of the immune system has revealed that lymphocyte differentiation and activation are accompanied by changes of hundreds of genes in parallel (Staudt and Brown, 2000). Knowing that the immune response is a complex mechanism that requires the participation of various cell types, genomic analyses could precisely identify which immune cell population is affected at a specific time interval after the exposure to a chemical by sorting the gene expression profile of the immune response. Unfortunately, very few immunotoxicologic studies have used either the genomic or the proteomic approach to investigate the effect of a specific chemical on the immune system response. Here, we report some studies that have focused on evaluating gene expression profile of specific immune response parameters such as cytokines or cytokine receptors mRNA expression when elicited by a chemical. We also describe various immune response processes that could be good candidates to microarray-based gene expression analyses, and which could potentially serve as biomarkers of sensitivity to chemical compounds.

23.6.2 Cytokine Analysis in Immunotoxicology

Among the most useful tools in immunotoxicology is the assessment of cytokines, which are responsible for regulating a variety of processes including immunity, inflammation, apoptosis, and hematopoiesis (House, 2001). Cytokines are signaling molecules that play a central role during the initiation of the immune response. In innate immunity, the effector cytokines are produced primarily by mononuclear phagocytes, and are therefore often called monokines. These molecules elicit neutrophil-rich inflammatory reactions that serve to contain and, when possible, eradicate microbial infections. When the immune response is specific for an antigen, most of the cytokines are produced by activated T-lymphocytes. Such molecules are commonly known as lymphokines. T-cells produce various cytokines that serve primarily to regulate the growth and differentiation of various lymphocyte populations and thus play an important role in the activation phase of T-cell dependent immune responses. Other cytokines derived from different T-cells function to activate and regulate inflammatory processes by acting on various immune system effectors such as mononuclear phagocytes, neutrophils, and eosinophils. In addition, both lymphocytes and mononuclear cells produce colony-stimulating factors (CSFs), which stimulate the growth and differentiation of immature leukocytes in bone marrow. More importantly, cells that do not belong to the immune system, such as endothelial cells, produce cytokines.

Assessing cytokine production can be achieved at two different levels, that is, by measuring either mRNA or protein levels. Protein levels represent the biologically active cytokines, and are therefore a better indicator of cytokine production than the measurement of mRNA levels. Protein levels can be assayed by ELISA or by flow cytometric analysis using specific fluorescent-conjugated monoclonal antibodies. However, it is not always possible to measure the amount of cytokines produced due to the limited sensitivity of the method, and the fact that cytokines are not always stored intracellularly. Indeed, measurement of a detectable amount of cytokine production ex-vivo is often performed when sensitized cells are exposed to an additional nonspecific stimulant such as lectins. In this case, cytokine mRNA measurement appears to be an appropriate endpoint for evaluating the effects of chemical exposure.

A large number of cytokines and their receptors have been cloned, thus facilitating the exploitation of molecular biology in immunotoxicology. Vandebriel et al. (1998b) have compiled a list containing the sequence data of cDNA clones from mouse, rat, and human available for cytokine and cytokine receptors.

Sensitizers such as drugs, chemicals, and protein allergens trigger immune responses that are regulated by cytokines. The particular profile of cytokine production may provide important information regarding the nature of an immunotoxic response, such as hypersensitivity. Immune reactions are often dominated by the type of cytokines produced. These specific immune responses are mediated by CD4+ cells. Two subsets of these cells exist and are referred to as T-helper 1 (Th1) and T-helper 2 (Th2). The biochemical pathways that regulate cellular differentiation of CD4+ cells are determined by the specific cytokines that stimulate the cells at the time of antigen recognition.

Moreover, once either T-helper cell subset develops, it suppresses the differentiation of the other subset.

Th1-like responses are characterized by the production of pro-inflammatory cytokines, such as interferon γ (IFN-γ). This cytokine stimulates the microbicidial activity of phagocytes. Therefore, the principal effector function of Th1 T cells is to promote the phagocyte defense response against pathogens. Th1 cells also secrete IL-2, which functions as an autocrine growth factor by stimulating cellular proliferation, differentiation, and activity. This permits the development of CD8+ T cells and promotes their cytolytic activity.

Th2 cells produce IL-4 and IL-5. These cells are mediators of allergic reactions and defense against parasites. In addition, Th2 cells also produce both IL-10 and IL-13, which antagonize the action of IFN-γ and suppress macrophage activation. Excessive or uncontrolled Th2 development is associated with deficient cell-mediated immunity to infections with intracellular microbes.

Several reports have documented the elicitation of the type of T-helper response following exposure to strong sensitizers (contact or respiratory sensitisers). Ulrich et al. (2001) have completed an exhaustive study in which they have analyzed cytokine profile expression in murine contact allergy following sensitization (induction) and elicitation (challenge) with various chemicals (dinitro-chlorobenzene, dinitrofluorobenzene, oxazolone, glutaraldehyde, formaldehyde, trimellitic anhy-dride). They have demonstrated that co-expression of Th1 and Th2 cytokines is an important feature of murine contact allergy in responsive mice, and that chemicals differ in their potency to induce the expression of these cytokines. Gaspard et al. (1999) have demonstrated that quantifying the mRNA expression of cytokines represents a good way to predict and diagnose xenobiotic-induced hypersensitivity reactions. Because antigen-presenting cells (APC) such as dendritic cells play a key role during the sensitization to xenobiotics and activation of specific T-lymphocytes, they characterized IL-1β mRNA levels in dendritic cells derived from human blood cord and treated them with dinitrochlorobenzene. Using this approach they were able to detect and quantify IL-4 and IFN-γ mRNA expression in lymph nodes of mice treated with a penicillin-protein conjugate, and to establish a ratio that could predict a shift between Th1 and Th2 immune response.

Other reports have demonstrated that several chemicals have deleterious effects on cytokine production. Tributyltin oxide (TBTO) exerts selective immunotoxicity in the rat. This was shown by in vitro exposure studies to TBTO in a T-cell activation model which resulted in a dose-dependent inhibition in the mRNA levels of the IL-2 receptor (IL-2R), serine protease, and increased expression of IFN-γ (Vandebriel et al., 1998a). Moreover, decreased mRNA levels of IL-2 and IL-2R correlated with thymic atrophy, which is observed following exposure to TBTO. This suggests that one of the early events in TBTO immunotoxicity may be the down-regulation of IL-2R expression, and that this may be involved in the sequence of events which leads to impairment of thymocyte maturation. The same group has also reported that exposure of spleen cells to other chemicals, such as hexachlorobenzene and benzo-a-pyrene, produced an increase in IL-2R, IL-2, and IFN-γ mRNA levels. These results indicate that the response of the cells to chemical exposure is chemical-specific with respect to the expression of cytokines and their receptors.

23.6.3 Genomic-Scale Analysis of Gene Expression in Immune System

Although a complete repertoire is not yet available, gene expression profiles have provided new insights in the processes of immune response. Lymphochip, a DNA microarray developed by the National Institutes of Health, is a specialized human cDNA microarray that contains many genes implicated in immune function (Alizadeh et al., 1999). The lymphochip microarray is composed of more than 17,000 cDNA clones derived from three sources, including genes that are known to be implicated in cell proliferation, apoptosis, or oncogenesis.

Systemic exploration of gene expression profiles during human lymphocyte development and activation, such as the biochemical differentiation pathways between Th1 and Th2, was character-ized by Granucci et al. (2001). Chemical compounds are known to affect immunoglobulin production by B cells, causing hypersensitivity and allergic reactions. B-cell maturation occurs in germinal centers where these cells interact with activated T cells and proliferate. Genes related to various stages of development were identified by Ollila and Vihinen (2002). Microarray analysis of gene expression in antigen-stimulated naive B cells demonstrated that 59 genes were significantly induced or repressed after 1 hour of stimulation, whereas the expression of more than 300 genes was altered after 6 hours (Glynne et al., 2000).

Immune cells that circulate throughout the body act as sentinels of disease. The sensitive and diverse repertoire of receptors and signal transduction pathways that these cells use to monitor and respond to alterations at any given site in the body may induce specific patterns of altered gene expression that could reflect the nature of the toxicity. In this case, a large-scale analysis of peripheral blood mononuclear cells using a microarray based on lymphocyte activation could reveal a distinct pattern of changes associated with chemical-specific exposure. This information would enable the development of new biomarkers allowing for the identification of the family of chemicals causing the toxic effect. It is difficult to assess immunotoxicity in humans. Biro et al. (2002) reported that the exposure to various chemicals such as styrene, benzene, polycyclic aromatic hydrocarbons, and mixed solvents can impact lymphocyte subpopulations. By studying surface antigens, they were able to conclude that these chemicals influence lymphocyte activation and caused a change in the incidence of a specific CD25$^+$ T-regulatory cell population. It is not difficult to imagine a wider range of clinical settings in which peripheral blood gene expression profiles could be useful in patient management.

Several studies using microarray analysis have focused on evaluating the response of many immunological cell types exposed to various chemical or biological compounds. Molecules such as fluoroquinolones have immunomodulatory properties and interfere with the cytokine production. T cells are a key element in effective specific immunity, recognizing MHC-antigen peptide complexes on the surface of antigen-presenting cells and translating these signals into cytotoxic effector T cells or antibody-producing B cells. Zhang et al. (2002) used cDNA array techniques to investigate the expression profile of 514 immunologically relevant genes in naive and tumor-specific activated murine T-cell populations. The list of significantly altered genes includes numerous cytokines and their receptors (e.g., IL-2Ralpha, IL-6Ralpha, IL-7Ralpha, IL-16, IL-17, TGF-beta); chemokines and chemokine receptors (e.g., RANTES, CCR7, CXCR4); surface proteins (e.g., integrins-alphaL and -beta7, L-selectin, CD6, CD45); cytoplasmic signaling intermediates (e.g., GATA-3, SMAD4, JAK-1); and an array of other molecules. Several genes are associated with T-cell self-regulation and migration. These data contribute to our understanding of the generalized processes that accompany T-cell activation.

Eriksson et al. (2003) characterized the expression profile of the superinduced mRNA levels in activated human lymphocytes incubated with ciprofloxacin. They noted that several gene transcripts were modulated. These were distributed between major signaling pathways including interleukins, signal-transduction molecules, adhesion molecules, tumor necrosis factor, and transforming growth factor-beta superfamilies, cell-cycle regulators, and apotosis-related molecules. In another study, Kang et al. (2003) used gene expression profiles to identify novel biomarkers in peripheral blood lymphocytes (PBLs) exposed to ionizing radiation. They found that only four genes were induced, including the TRAIL receptor 2, FHL2, cyclin G, and cyclin protein gene. These results suggest that relative expression of these genes may provide a reliable biomarker to assess radiation exposure.

Cadmium (Cd2$^+$) is a ubiquitous toxic metal with well-established apoptotic and genotoxic effects. Tsangaris et al. (2002) studied the effects of Cd2$^+$ on the gene expression profiles implicated in apoptosis and cytotoxicity in a T-cell line. They showed that exposure to Cd2$^+$ produces time- and dose-dependent molecular signaling cascades, inducing disturbances in different subcellular compartments, and resulting in permanently altered cellular differentiation and function.

Integrating proteomic analyses to gene profiling results in a much more powerful tool and can reveal post-translational modulation of many genes. Using this approach, Grolleau et al. (2002) demonstrated that rapamycin affected protein translation in Jurkat T cells. First, they compared the levels of polysome-bound and total RNA using oligonucleotide microarrays containing 6300 genes. They then determined the level of protein synthesis using 2D polyacrylamide gel electrophoresis. Their results indicated that the translation of 6% of the total number of genes was repressed by rapamycin, and 7% were unaffected. Furthermore, they were able to assess that 16 genes were modulated by the exposure to rapamycin.

Dendritic cells (DC) are essential to the initiation of an immune response due to their ability to take up and process antigens, translocate to lymph nodes, and present processed antigens to naive T cells. Their phenotypic and functional characteristics are intimately linked to their stage of maturation. The specific biochemical pathways and genes whose expression mediates differentiation of progenitors to DCs and their maturation are largely undefined. Richards et al. (2002) used two approaches — DNA microarrays and proteomics — to analyze the expression profile of human CD14+ blood monocytes and their derived DCs. Approximately 4% of genes or proteins expressed were regulated during DC differentiation. These genes are involved in antigen presentation, cell adhesion, lipid metabolism, and signaling pathways. Moreover, they observed that in contrast to DC differentiation, very few genes were modified at the transcript level during DC maturation, as determined by microarray experiments. The use of proteomics is therefore necessary for full comprehension of the DC maturation process. Many chemokines, chemokine receptors, and other G-protein–coupled receptors are implicated in the various aspects of DC biology. Maturation of dendritic cells from monocytes is an important but poorly understood mechanism. Skelton et al. (2003) compared through microarray analysis the expression levels of chemokines, their cognate receptors, and selected G-protein–coupled receptors in human monocytes and *in vitro* derived immature and mature DC. They found a differential expression of the GPR105 receptor on immature and mature DC, suggesting a role for this receptor in DC activation.

23.7 RESEARCH NEEDS

Over the past several decades, toxicologists have developed a variety of tools to assess and predict the potential toxicity of a variety of chemicals. These tools include numerous specific biomarkers, and animal and cellular models, as well as many mathematical approaches for assessing risk. The development of new tools that are emerging for the fields of genomics and proteomics identifies a need for reassessing and integrating our toxicological approaches with those of these emerging fields. A particularly critical gap of information is how we can relate the toxicokinetic modeling approach to toxicant receptor occupancy and changes in gene expression. The development of new models to address this should allow for stronger predictability in risk assessment and predicting pathological outcomes of exposure to toxicants. Furthermore, in immunotoxicology, as for other fields of toxicology, the use of gene expression and protein profiling should allow for a more rapid screening and identification of mechanism of toxicity. This approach, however, is likely to require integrating genomics, proteomics, and metabonomics in studies in order to fully understand the metabolism and toxicity of different compounds and their mixtures. While these emerging fields offer a wide range of possibilities, their cost and analytical complexity still limit the types of studies that address many current needs. However, as the competition for tool development increases, it is likely that costs will decrease, thereby permitting many gaps to be filled for toxicological analyses.

23.8 SUMMARY

Toxicogenomics and proteomics are becoming valuable tools for understanding the activation of processes implicated in immunotoxic response. The long-term development of databases with characteristic chemical signature expression profiles and the combination of different approaches such as gene expression profiling with proteomics offers a new understanding for assessing and predicting the immunotoxicity of environmental toxicants and pharmaceuticals. Furthermore, as we link these analyses to well-characterized toxicological approaches and endpoints, we can expect that this information will allow for a stronger and more thorough assessment of the risks associated with chemical exposure.

ACKNOWLEDGMENT

The authors would like to thank Mary Gregory (INRS-Institut Armand Frappier) for her assistance.

REFERENCES

Alizadeh, A., et al., The lymphochip: a specialized cDNA microarray for the genomic-scale analysis of gene expression in normal and malignant lymphocytes, *Cold Spring Harbor Symp. Quant. Biol.*, 64, 71–78, 1999.

Alon, U., et al., Broad patterns of gene expression revealed by clustering analysis of tumor and normal colon tissues probed by oligonucleotide arrays, *Proc. Natl. Acad. Sci. U.S.A.*, 96, 6745–50, 1999.

Arfin, S.M., et al., Global gene expression profiling in Escherichia coli K12. The effects of integration host factor, *J. Biol. Chem.*, 275, 29672–84, 2000.

Baldi, P., and Hatfield, G.W., *DNA Microarrays and Gene Expression*, Cambridge University Press, Cambridge, 2002.

Biro, A., et al., Lymphocyte phenotype analysis and chromosome aberration frequency of workers occupationally exposed to styrene, benzene, polycyclic aromatic hydrocarbons or mixed solvents, *Immunol. Lett.*, 81, 133–140, 2002.

Blackstock, W.P., and Weir, M.P., Proteomics: quantitative and physical mapping of cellular proteins, *Trends Biotechnol.*, 17, 121–127, 1999.

Chen, Y., Dougherty, E.R., and Bittner, M.L., Ratio-based decisions and the quantitative analysis of cDNA microarray images, *J. Biomed. Optics*, 2, 364–374, 1997.

Churchill, G.A., Fundamentals of experimental design for cDNA microarrays, *Nat. Genet.*, 32 (Suppl.), 490–495, 2002.

Cleveland, W.S., Robust locally weighted regression and smoothing scatterplots, *J. Am. Stat. Assoc.*, 74, 829–836, 1979.

Dobbin, K., Shih, J.H., and Simon, R., Questions and answers on design of dual-label microarrays for identifying differentially expressed genes, *J. Natl. Cancer Inst.*, 95, 1362–1369, 2003.

Draghici, S., Statistical intelligence: effective analysis of high-density microarray data, *Drug Discovery Today*, 7, S55–63, 2002.

Eisen, M.B., et al., Cluster analysis and display of genome-wide expression patterns, *Proc. Natl. Acad. Sci. U.S.A.*, 95, 14863–14868, 1998.

Eriksson, E., Forsgren, A., and Riesbeck, K., Several gene programs are induced in ciprofloxacin-treated human lymphocytes as revealed by microarray analysis, *J. Leukocyte Biol.*, 74, 456–463, 2003.

Figeys, D., Proteomics in 2002: a year of technical development and wide-ranging applications, *Anal. Chem.*, 75, 2891–2905, 2003.

Gaspard, I., et al., Quantitation of cytokine mRNA expression as an endpoint for prediction and diagnosis of xenobiotic-induced hypersensitivity reactions, *Methods*, 19, 64–70, 1999.

Glynne, R., et al., B-lymphocyte quiescence, tolerance and activation as viewed by global gene expression profiling on microarrays, *Immunol. Rev.*, 176, 216–246, 2000.

Golub, T.R., et al., Molecular classification of cancer: class discovery and class prediction by gene expression monitoring, *Science*, 286, 531–537, 1999.

Granucci, F., et al., Gene expression profiling in immune cells using microarray, *Int. Arch. Allergy Immunol.*, 126, 257–266, 2001.

Grolleau, A., et al., Global and specific translational control by rapamycin in T cells uncovered by microarrays and proteomics, *J. Biol. Chem.*, 277, 22175–22184, 2002.

Hamadeh, H.K., et al., Methapyrilene toxicity: anchorage of pathologic observations to gene expression alterations, *Toxicol. Pathol.*, 30, 470–482, 2002.

He, Q., Liu, Y., and Nixon, T., High-field electrophoretic NMR of protein mixtures in solution, *J. Am. Chem. Soc.*, 120, 1341, 1998.

House, R.V., Cytokine measurement techniques for assessing hypersensitivity, *Toxicology*, 158, 51–58, 2001.

Jimenez-Barbero, J., Amat-Guerri, F., and Snyder, J.P., The solid state, solution and tubulin-bound conformations of agents that promote microtubule stabilization, *Curr. Med. Chem. Anti-Cancer Agents*, 2, 91–122, 2002.

Joos, L., Eryuksel, E., and Brutsche, M.H., Functional genomics and gene microarrays — the use in research and clinical medicine, *Swiss Med. Wkly.*, 133, 31–38, 2003.

Kang, C.M., et al., Possible biomarkers for ionizing radiation exposure in human peripheral blood lymphocytes, *Radiat. Res.*, 159, 312–9, 2003.

Kerr, M.K., Martin, M., and Churchill, G.A., Analysis of variance for gene expression microarray data, *J. Comput. Biol.*, 7, 819–837, 2000.

Lander, E.S., et al., Initial sequencing and analysis of the human genome, *Nature*, 409, 860–921, 2001.

Lee, P.D., et al., Control genes and variability: absence of ubiquitous reference transcripts in diverse mammalian expression studies, *Genome Res.*, 12, 292–297, 2002.

Lindroos, K., et al., Multiplex SNP genotyping in pooled DNA samples by a four-colour microarray system, *Nucleic Acids Res.*, 30, e70, 2002.

Lönnstedt, I., and Speed, T., Replicated microarray data, *Statistica Sinica*, 12, 31–46, 2002.

LoPachin, R.M., et al., Application of proteomics to the study of molecular mechanisms in neurotoxicology, *Neurotoxicology*, 24, 761–775, 2003.

Lovmar, L., et al., Quantitative evaluation by minisequencing and microarrays reveals accurate multiplexed SNP genotyping of whole genome amplified DNA, *Nucleic Acids Res.*, 31, 29, 2003.

McKusick, V.A., Genomics: structural and functional studies of genomes, *Genomics*, 45, 244–249, 1997.

Naylor, S., and Kumar, R., Emerging role of mass spectrometry in structural and functional proteomics, *Adv. Protein Chem.*, 65, 217–248, 2003.

Nicholson, J.K., et al., Metabonomics: a platform for studying drug toxicity and gene function, *Nat. Rev. Drug Discovery*, 1, 153–61, 2002.

Ollila, J., and Vihinen, M., Microarray analysis of B-cell stimulation, *Vitam. Horm.*, 64, 77–99, 2002.

Pan, W., A comparative review of statistical methods for discovering differentially expressed genes in replicated microarray experiments, *Bioinformatics*, 18, 546–554, 2002.

Pan, W., Lin, J., and Le, C.T., How many replicates of arrays are required to detect gene expression changes in microarray experiments? A mixture model approach, *Genome Biol.*, 3, 22, 2002.

Pollard, T.D., and Borisy, G.G., Cellular motility driven by assembly and disassembly of actin filaments, *Cell*, 112, 453–465, 2003.

Quackenbush, J., Computational analysis of microarray data, *Nat. Rev. Genet.*, 2, 418–427, 2001.

Raychaudhuri, S., Stuart, J.M., and Altman, R.B., Principal components analysis to summarize microarray experiments: application to sporulation time series, *Pac. Symp. Biocomput.*, 455–466, 2000.

Richards, J., et al., Integrated genomic and proteomic analysis of signaling pathways in dendritic cell differentiation and maturation, *Ann. N. Y. Acad. Sci.*, 975, 91–100, 2002.

Sanchez, M., The proteome puzzle, *Biotechnol. Focus*, 5, 16–18, 2002.

Sem, D., Villar, H., and Kelly, M., NMR on Target, *Mod. Drug Discovery*, 6, 26–31, 2003.

Simon, R.M., and Dobbin, K., Experimental design of DNA microarray experiments, *Biotechniques*, (Suppl.), 16–21, 2003.

Skelton, L., et al., Human immature monocyte-derived dendritic cells express the G protein-coupled receptor GPR105 (KIAA0001, P2Y14) and increase intracellular calcium in response to its agonist, uridine diphosphoglucose, *J. Immunol.*, 171, 1941–1949, 2003.

Smolkin, M., and Ghosh, D., Cluster stability scores for microarray data in cancer studies, *BMC Bioinformatics*, 4, 36, 2003.

Staudt, L.M., and Brown, P.O., Genomic views of the immune system, *Annu. Rev. Immunol.*, 18, 829–859, 2000.

Suzuki, T., Higgins, P.J., and Crawford, D.R., Control selection for RNA quantitation, *Biotechniques*, 29, 332–337, 2000.

Tamayo, P., et al., Interpreting patterns of gene expression with self-organizing maps: methods and application to hematopoietic differentiation, *Proc. Natl. Acad. Sci. U.S.A.*, 96, 2907–2912, 1999.

Tavazoie, S., et al., Systematic determination of genetic network architecture, *Nat. Genet.*, 22, 281–285, 1999.

Toronen, P., et al., Analysis of gene expression data using self-organizing maps, *FEBS Lett.*, 451, 142–146, 1999.

Tsangaris, G., Botsonis, A., Politis, I., and Tzortzatou-Stathopoulou, F., Evaluation of cadmium-induced transcriptome alterations by three color cDNA labeling microarray analysis on a T-cell line, *Toxicology*, 178, 135–160, 2002.

Tusher, V.G., Tibshirani, R., and Chu, G., Significance analysis of microarrays applied to the ionizing radiation response, *Proc. Natl. Acad. Sci. U.S.A.*, 98, 5116–5121, 2001.

Ulrich, P., et al., Cytokine expression profiles during murine contact allergy: T helper 2 cytokines are expressed irrespective of the type of contact allergen, *Arch. Toxicol.*, 75, 470–479, 2001.

Valafar, F., Pattern recognition techniques in microarray data analysis: a survey, *Ann. N. Y. Acad. Sci.*, 980, 41–64, 2002.

Vandebriel, R.J., et al., Effects of in vivo exposure to bis(tri-n-butyltin)oxide, hexachlorobenzene, and benzo(a)pyrene on cytokine (receptor) mRNA levels in cultured rat splenocytes and on IL-2 receptor protein levels, *Toxicol. Appl. Pharmacol.*, 148, 126–136, 1998a.

Vandebriel, R.J., Van Loveren, H., and Meredith, C., Altered cytokine (receptor) mRNA expression as a tool in immunotoxicology, *Toxicology*, 130, 43–67, 1998b.

Venter, J.C., et al., The sequence of the human genome, *Science*, 291, 1304–1351, 2001.

Wen, X., et al., Large-scale temporal gene expression mapping of central nervous system development, *Proc. Natl. Acad. Sci. U.S.A.*, 95, 334–339, 1998.

Werner, T., Proteomics and regulomics: the yin and yang of functional genomics, *Mass Spectrom. Rev.*, 23, 25–33, 2004.

Wilkins, M.R., et al., From proteins to proteomes: large scale protein identification by two-dimensional electrophoresis and amino acid analysis, *Biotechnology (N. Y.)*, 14, 61–65, 1996.

Willis, E., Allauzen, S., and Vlasenko, S., The emergence of suspension bead arrays, *BioRadiations*, 111, 30–35, 2003.

Xing, E.P., and Karp, R.M., CLIFF: clustering of high-dimensional microarray data via iterative feature filtering using normalized cuts, *Bioinformatics*, 17(Suppl. 1), S306–315, 2001.

Yang, Y.H., et al., Normalization for cDNA microarray data: a robust composite method addressing single and multiple slide systematic variation, *Nucleic Acids Res.*, 30, e15, 2002.

Yang, Y.H., and Speed, T., Design issues for cDNA microarray experiments, *Nat. Rev. Genet.*, 3, 579–588, 2002.

Zhang, X., et al., DNA microarray analysis of the gene expression profiles of naive versus activated tumor-specific T cells, *Life Sci.*, 71, 3005–3017, 2002.

Zieziulewicz, T.J., et al., Shrinking the biologic world-nanobiotechnologies for toxicology, *Toxicol. Sci.*, 74, 235–244, 2003.

The Role of Immunotoxicology in Risk Assessment and the Regulatory Process

Biomarkers of Immunotoxic Effects in the Context of Environmental Risk Assessment

Stéphane Pillet and Jean-Marc Nicolas

CONTENTS

24.1 INTRODUCTION

Human activity results in the constant release of natural and synthetic substances into the environment. A large number of these environmental contaminants are persistent (e.g., heavy metals, polyhalogenated aromatic hydrocarbons, polybrominated compounds, etc.), and past emissions will have an effect on ecosystems for years and even decades to come. Furthermore, it has been established that these persistent toxic substances (PTS) can be found in remote environments with no direct anthropogenic impact, and it is highly probable that anthropogenic contaminants are ubiquitous on the planet. This situation has driven environmental scientists and managers to develop models that would permit the quantification of risk based on measurable endpoints. In this chapter, we describe and discuss biomarkers that have the potential of detecting environmentally relevant chemical-induced immunomodulation.

24.2 CONCEPTS IN ENVIRONMENTAL RISK ASSESSMENT

Depledge and Fossi (1994) define environmental risk assessment (ERA) as the "procedure by which the likely or actual adverse effects of pollutants and other anthropogenic activities on ecosystems and their components are estimated with a known degree of certainty using scientific methodologies." The ultimate objective of ERA is to predict the nature and scope of the effects resulting from anthropogenic activities to exposed organisms in order to determine if risk-reducing action should be taken. The principles of ERA are almost *universal* in that most government agencies and environmental scientists and managers apply extremely similar frameworks (Figure 24.1).

The complexity of the toxic actions of environmental contaminants and their derivatives prohibits risk assessment based solely on environmental levels of chemicals. In the case of low levels of contaminants, deleterious effects on populations are often difficult to detect since these effects tend to manifest themselves only after long periods of exposure. When the effect becomes obvious, the destruction process may have gone beyond the point where it can be reversed by remedial actions or risk reduction. This has lead the scientific community to seek out early-warming signals of toxic effects in exposed organisms, or biomarkers.

24.3 DEFINITION OF BIOMARKERS

Biomarkers are the forensic tools that will provide the clues to evaluate the likelihood of risk to the environment. Biomarkers are defined by the U.S. National Academy of Sciences as "a xenobiotically-induced variation in cellular or biochemical components or processes, structures, or functions that are measurable in a biological system or samples" (National Research Council, 1989). Biomarkers provide an integrated measure of exposure to biologically available contaminants over time and reflect the combined effects/results of simultaneous and/or consecutive exposures to several contaminants. However, it is often impossible to attribute the observed adverse effects to any particular chemical. Furthermore, some of the biomarkers routinely used for assessing the health

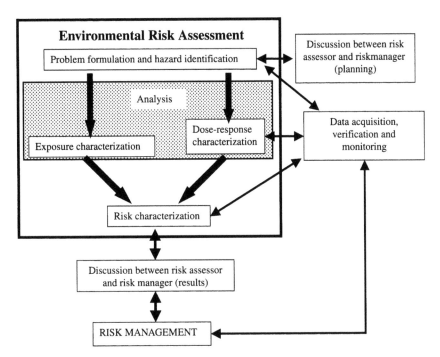

Figure 24.1 Conceptual framework for environmental risk assessment and risk management. (From U.S. Environmental Protection Agency, Proposed guidelines for ecological risk assessment, *Federal Register*, 61, 47552–47631, 1996.)

of an organism can be substantially affected by confounding factors. Most biological functions are influenced by abiotic and biotic factors. All of these factors vary in both time and space. Most of the biochemical and physiological biomarkers studied to date may depend on temperature, season, animal age, sex, nutritional status, reproductive stage, and genotype. Moreover, most of these factors are strongly interdependent.

However, there are several advantages to applying biomarkers (Handy et al., 2003):

1. Biomarker responses may indicate the presence of a biologically available contaminant, rather than a biologically inert form of contamination.
2. Using a set of a biomarkers may reveal the presence of contaminants that were not suspected initially.
3. Biomarker responses often persist long after a transient exposure to a contaminant that has by then been degraded and is no longer detectable. Thus, biomarkers may detect intermittent pollution events that routine chemical monitoring may miss.
4. Biomarker analyses are, in many cases, much easier to perform and are considerably less expensive than a wide range of chemical analyses.

According to the National Research Council (1987) and World Health Organization (1993), biomarkers can be subdivided into three classes: exposure, effect, and susceptibility. Biomarkers of exposure cover the detection and measurement of an exogenous substance or its metabolites or the product of an interaction between a xenobiotic agent and some target molecule or cell that is measured in a compartment within an organism. Fundamental properties of biomarkers of exposure rely mainly on the following two founding principles of the dose–response relationship:

- Contaminant concentration at/in the cell or tissue is clearly the cause of the biomarker response.
- The contaminant concentration in cells/tissues can be correlated with changes in environmental concentrations of the contaminant (albeit the "bioavailable" fraction).

Biomarkers of effect include measurable biochemical, physiological, or other alterations within cells, tissues, or body fluids of an organism that can be identified as associated with an established or possible health impairment or disease. According to Handy et al. (2003), biomarkers of effect must:

- Measure deleterious effects of pollution on biological functions, perhaps against long-term changes in other stresses (e.g., nutrition, disease, or general environmental quality).
- Relate to a body system or cell type that accumulates injury over long time scales, or at least has some form of biochemical memory to store the response that is associated with some critical biological function.
- Show negligible temporal variability to confounding factors or be easily normalized for such factors.
- Remain effective, even when an organism redistributes the body burden, or chemically/spatially alters cellular storage of accumulated contaminants over time to improve storage efficiency.

Biomarkers of effect can be used to document either preclinical alterations or adverse health effects due to external exposure and absorption of a chemical. This would include molecular and cellular modifications that can be linked to pathologies, lesions, reduction of reproductive potential, and population or community disruptions.

Biomarkers of susceptibility indicate the inherent or acquired ability of an organism to respond to the challenge of exposure to a specific contaminant, including genetic factors and changes in receptors that alter the susceptibility of an organism to that exposure.

All effects of toxicants have their origins in chemical processes at a molecular level. If the impact of the toxicant at any level (biochemical, cytological, and physiological) is great enough to exceed the "compensatory" responses at that level, then its effects are passed on to successively higher levels of biological organization (Figure 24.2). It is often difficult to identify the early effects of contamination. The most visible biomarkers are generally those that have the slowest response to the contaminants. In the evaluation of toxicological risk, it is therefore important to include biomarkers from several organizational levels (i.e., from the molecular to the organism's level).

Van der Oost et al. (2003) proposed several criteria for the selection of biomarkers:

1. The assay to quantify the biomarker should be reliable, relatively cheap, and easy to perform.
2. The biomarker response should be sensitive to pollutant exposure and/or effects in order to serve as an early warning parameter.
3. Baseline data of the biomarker should be well defined in order to distinguish between natural variability (noise) and contaminant-induced stress (signal).
4. The impacts of confounding factors to the biomarker response should be well established.
5. The underlying mechanism of the relationships between biomarker response and pollutant exposure (dosage and time) should be established.
6. The toxicological significance of the biomarker, such as the relationships between its response and the (long-term) impact to the organism, should be established.

In addition to these criteria, it has been suggested that biomarkers should preferentially be noninvasive or nondestructive to allow or facilitate environmental monitoring of pollution effects in protected or endangered species.

The nature of the contamination, anticipated effects, and choice of endpoints will condition the selection of biomarkers to be used. Conversely, the choice of biomarkers will determine which questions will be addressed, and will therefore influence the outcome of the ERA. The use, implementation, and appropriate interpretation of biomarkers is dependent on the knowledge of the characteristics of a specific marker. Inappropriate use or misinterpretation of biomarker responses can occur when this fundamental research is either unknown, ignored, or misunderstood.

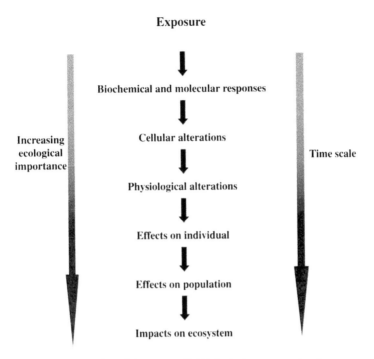

Figure 24.2 Schematic representation of the sequential biological responses to stress.

The immune, endocrine, and nervous systems play a central role for the health of an organism. Therefore, assessing the impacts of contaminants on these systems is extremely relevant in the context of toxicological risk assessment.

24.4 THE IMMUNE SYSTEM

The immune system has evolved to protect the host from potentially pathogenic agents including microorganisms (viruses and bacteria), parasites, and fungi; to eliminate neoplastic cells; and to reject non-self components (Roitt et al., 1993). The immune response is comprised of a range of biological processes involved in recognition and elimination of non-self items. The immune response can be separated into two parts with distinct characteristics: the nonspecific immune response, and the specific immune response which involves cell-mediated immunity (CMI) and humoral-mediated immunity (HMI). For a more comprehensive review of the immune system, see Chapter 1.

24.5 IMMUNOTOXIC EFFECTS OF CHEMICAL SUBSTANCES

A number of chemicals or their metabolites have been demonstrated to affect immune functions and alter immune response. According to recommendations of the International Seminar on the Immunological System, "A chemical substance should be considered immunotoxic when undesired events of the chemical are (i) a direct and/or indirect action of the xenobiotic (and/or its biotransformation products) on the immune system; or (ii) an immunologically based host response to the compounds and/or its metabolite(s), or host antigens are modified by the compound or its metabolite(s)" (Krzystyniak et al., 1995).

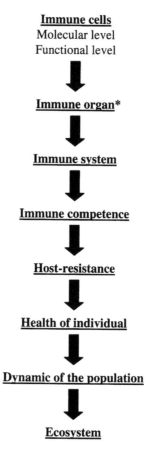

Figure 24.3 Schematic representation of the sequential biological responses to stress applied to the immune system. *One of the features of the immune system is that an important proportion of immune cells function outside of any organized structure such as the immune organs.

Immunotoxicology is a relatively new discipline that studies the toxic effects of chemicals on the immune system, and consequences on the immune response and the health of an organism. The immunotoxicity of many compounds has been demonstrated since the late 1960s and early 1970s. Major classes of contaminants such as heavy metals or polychlorinated biphenyls (PCBs) have undoubted immunotoxic effects. However, most of the studies documenting these effects have been conducted mainly on mice and rats (reviewed by Bernier et al., 1995; Krzystyniak et al., 1995; Tryphonas, 1995, 2001).

Impairments of the immune response can lead to increased susceptibility to infections, cancers, and autoimmune diseases. Therefore, in view of the central role of the immune system for the health of organisms, it is essential to consider the immunotoxic potential of contaminants in the evaluation process of their toxic effects on the environment. The general hierarchical structure approach of biomarkers presented previously can be applied to biomarkers of the immune system (i.e., immuno-markers) (Figure 24.3).

24.6 IMMUNOMARKERS OF EFFECTS

The range of chemicals known to disrupt the immune system is wide and their number is continually increasing as the immunotoxicity of new molecules and mixtures is tested. Several critical endpoints of the immune response have been demonstrated to be sensitive to chemical exposures, and have

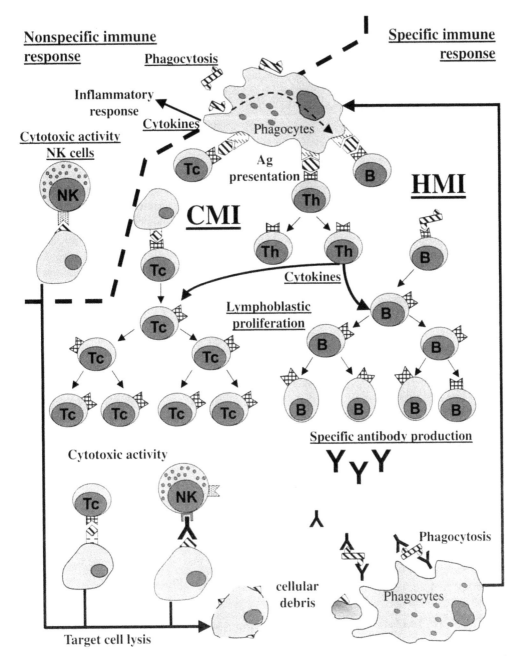

Figure 24.4 General structure diagram of the immune response with special emphasis on immunomarker-related endpoints. Cell-mediated immunity (CMI) involves lymphoblastic proliferation of cytotoxic T cells (Tc) and helper T cells (Th). Humoral-mediated immunity (HMI) is based on the lymphoblastic proliferation of differentiated B cells (B), which produce an antigen-specific antibody.

therefore been proposed for use as immunomarkers of effects (Luster et al., 1988, 1992, 1993). Immunomarkers cover several biological scales, and encompass markers from the macroscopic scale through to the cellular scale. The general structure of the immune response with special emphasis on the most commonly used immunomarkers is presented in Figure 24.4. Methods to assess these immunomarkers have been extensively described in several manuals (Lefkovits, 1997; Brousseau et al., 1999).

24.6.1   Weight, Organ:Body Weight Ratio, and Cellularity of Immune Organs

Considered as a rough appreciation of immunotoxic effects, alterations of thymus and spleen weight or organ:body weight ratio, associated with modifications of the organ cellularity (i.e., the number of cells per organ) can give some insight on the impact of contaminants at the organ level. These markers can also be valuable indicators of effects on immune functions. However, they are not extremely sensitive, and an absence of effect on these markers does not mean an absence of effect on the immune response.

24.6.2   Hematology

Immune cells are highly heterogeneous. Major peripheral blood immune cell populations (granulocytes, monocytes, lymphocytes) have been identified based on morphological and functional features. More sophisticated techniques such as dye retention and more recently phenotyping (discussed below) have subsequently identified subpopulations of these cell groups.

24.6.3   Nonspecific Immune Response Immunomarkers

24.6.3.1   Phagocytic Activity of Macrophages and Other Phagocytes

Phagocytosis represents the first line of defense of the immune system against invading agents, especially bacteria (Van Oss, 1986; Verhoef, 1992). The phagocytosis assay measures the ability of phagocytes to ingest stained foreign material (bacteria or beads).

24.6.3.2   Oxydative Burst

The generation of reactive oxygen species (ROS), or oxydative burst, is one of the major pathways of degradation of the material engulfed by phagocytes (Roitt et al., 1993). The oxydative burst assay measures the production of ROS as an estimation of the ability of phagocytes to destroy phagocytosed material.

24.6.3.3   Cytotoxic Activity of Natural Killer Cells

Natural killer (NK) cells play a role in tumor resistance, host immunity to viral infection, and regulation of lymphoid and other hematopoietic cell populations. NK cells are present in peripheral blood and the spleen (Trinchieri, 1992; Roitt et al., 1993). The NK cell activity assay measures the ability of NK cells to kill tumor cells.

24.6.4   Specific Immune Response Immunomarkers

24.6.4.1   Delayed-Type Hypersensitivity

Delayed-type hypersensitivity (DTH) is used for the evaluation of *in vivo* CMI (Figure 24.4). Generally, DTH reactions are developed against all protein antigens that are introduced at low concentrations in the skin, where specific antigen-presenting cells will take them up and migrate to the draining lymph nodes to activate specific T cells. This *in vivo* response therefore involves the cooperation of several elements of the specific immune response, and reflects the efficiency of interactions between these elements in order to build a specific immune response. One of the advantages of *in vivo* assays is that they encompass indirect effects that are present in the whole animal which are excluded from *in vitro* assays.

24.6.4.2 Lymphoblastic Proliferation

In order to protect the host against pathogens, immunocompetent cells require continued proliferation and differentiation. One way to evaluate the proliferation potential of leukocytes is to test their lymphoproliferative response to mitogens. Phytohemagglutinin (PHA) and concanavalin A (Con-A) are mitogens known to activate the proliferation of the lymphocyte subpopulation involved in CMI (i.e., the T cells), while lipopolysaccharides (LPS) induce proliferation of lymphocyte subpopulations involved in HMI (i.e., the B cells), and pokeweed mitogens (PWN) activate DNA synthesis in both T and B cells.

24.6.4.3 Mixed Lymphocyte Reaction

Recognition is a fundamental phenomenon of the immune response. Histocompatibility antigens present on the surface of cells allows organisms to distinguish between self and non-self. Mixed lymphocyte reaction (MLR) occurs when lymphoid cells from genetically distinct organisms of the same species are in contact. This results in stimulation of lymphoid cells of each donor, which then start to proliferate. The MLR assay measures this ability of lymphoid cells to proliferate when stimulated by the contact of cells from genetically distinct organisms of the same species. The "stimulating cells" are pretreated to prevent their proliferation while their surface antigens remain unchanged. This allows for the assessment of the proliferation of the "responding cells."

24.6.4.4 Humoral-Mediated Immunity

Humoral-mediated immunity (HMI) is based on the production of specific proteins — antibodies (or immunoglobulines [Ig]) — by secreting cells differentiated from the B cells. Antibody production results from the complex sequence of events involved in antigen processing and presentation, which lead to activation of B cells that differentiate into antibody-secreting cells. This requires interactions and cooperation between several cell populations, including T cells, phagocytes, and B cells. A deficit in any of the links of this chain of events has the potential to impair the humoral response. The humoral response can be evaluated through the number of antibody-secreting cells or the level of specific antibody secreted. Detection and quantification of specific antibody production relies on antigen-specific recognition properties. The most prevalent methods for specific antibody quantification are radioimmunoassay (RIA) and enzyme-linked immunosorbent assay (ELISA). The number of antibody-secreting cells can be assessed using plaque-forming cells (PFC) or enzyme-linked immuno-spot (ELISpot) assays.

24.6.4.5 Phenotyping

The level of expression of several specific cell-surface proteins (i.e., cluster of differentiation [CD]) reflects the state of differentiation or activation of specific immune cells. These proteins can be identified using monoclonal antibody and flow cytometry techniques that lead to the determination of a cellular phenotype. Phenotyping can provide access to absolute number and relative proportion of the various immune cell subpopulations in tissues or organs. Phenotyping is a powerful tool. However, its application in the case of feral species has been often hindered because of the low interspecies homology between CD that can lead to a poor cross-reactivity with available antibodies (generally developed for humans or rodents).

24.6.5 Challenge Against Infectious Agents

The challenge of an organism's immune system by an infectious agent represents the ultimate assay to assess the immunocompetence of a host. This assay has the advantage of being extremely representative. However, its implementation is often difficult due to logistical and ethical constraints. This is particularly true in the case of wildlife species.

Functional immunocompetence is under the control of many compensatory mechanisms, which reflects the necessity for an organism to maintain an efficient immune response. Consequently, the disruption of one immune function may not necessarily alter the immunocompetence and resistance of the host. Because of the redundancy of the mechanisms involved, it is difficult to predict whether and to what extent functional immune changes induced by stress, including chemical exposure, are likely to induce alterations of host resistance that can lead to increased susceptibility to infections or cancers. In order to establish which tests are most predictive of the adverse outcomes, and with the purpose of clarifying the relationship between immune test changes and clinical disease, Luster et al. (1988, 1992, 1993) investigated the ability of individual and pairwise immune tests to predict an alteration of host resistance in mice. The highest positive association was found with the T-cell dependent antibody PFC assay (78% of concordance) or phenotyping (83% of concordance). The combination of one of these two tests with almost any other test markedly increased the ability to predict an immunotoxic outcome, and some combinations provided greater than 90% concordance. These studies demonstrated a good correlation between changes in the immune tests and altered host resistance in that there were no instances when host resistance was altered without affecting the immune test(s). However, in many instances, immune changes may be observed in the absence of detectable changes in host resistance which illustrates the presence of the compensatory mechanisms.

24.6.6 New Potential Immunomarkers: Emergence of Genomic and Proteomic Techniques

Recent developments in our knowledge of the cellular and molecular patterns that underpin processes such as cell activation, differentiation, maturation, proliferation, and apoptosis have identified critical proteins. Chemicals that could act on these endpoints could therefore alter the whole cellular process, leading to serious consequences at higher levels of organization (tissues, organs). These proteins can thus be considered as interesting potential biomarkers or immunomarkers. In this regard, cytokines (e.g., interleukin) are particularly interesting to consider. Cytokines are the most important "messengers" within the immune system. These proteins play a central role in the immune response by promoting the activation of specific and nonspecific immune effector mechanisms and tissue repair. There are a large variety of cytokines, each of which has a specific function (Feldmann, 1992; Roitt et al., 1993). Alteration of their synthesis has been linked to the functional impairment of the effector cells that could have consequences on the entire immune response (Voss et al., 1994). Cytokine levels can be measured in body fluid (serum), but genomic and proteomic techniques allow us to determine the rate of gene expression and protein synthesis in cells. Cytokines mediate their effects through binding to high-affinity receptors on the cell surface. The modulation of the expression of these receptors can also have important consequences on cell functions. Chemicals have been demonstrated to alter both cytokines and cytokine receptor synthesis (Theocharis et al., 1991; Kimber, 1994; Kwon et al., 2002; Seth et al., 2003). Several other proteins involved in essential processes (e.g., cell proliferation and apoptosis) such as caspases have been shown to be modulated by contaminants including heavy metals and dioxin-like compounds (Tian et al., 2002; Pulido and Parrish, 2003), and could be considered as potential biomarkers in the future.

Metallothioneins (MT) are small proteins with unique properties that make them a regulator of essential cation homeostasis, as well as an important modulator of heavy metal toxicity (Moffatt and Denizeau, 1997; Miles et al., 2000). MT are mostly produced in liver and kidneys where they are traditionally considered as a biomarker of heavy metal exposure (Viarengo et al., 1999; Das et al., 2000). However, MT have been found to be induced by heavy metals in immune cells of humans and rodents (Yamada and Koizumi, 1991; Yurkow and Makhijani, 1998), and more recently of pinnipeds (Pillet et al., 2002). Moreover, MT have been demonstrated to be involved in many cellular pathways regulating the immune response by direct or indirect interactions (Lynes et al.,

1990, 1993; Leibbrandt et al., 1994; Borghesi et al., 1996; Crowthers et al., 2000). Although the precise role and mechanisms remain to be elucidated, MT are involved in processes associated with resting and activated immune cells. Hence, a modulation in the level of MT in these cells may have important consequences for the immune response.

The example of MT illustrates the limitations of distinctions among biomarkers established in the past. These proteins, which were initially used as a biomarker of exposure, can also be considered immunomarkers of effect. This change is likely going to become more widespread, as we are gaining access to tools that enable us to measure the costs and consequences of alterations of cellular processes, including detoxification processes. It is also interesting to note that MT can be used as biomarkers of susceptibility, since it has been shown that the sensitivity of an organism to heavy metal toxicity is a function of the capacity of the cells to synthesize MT (Klaassen et al., 1999; Habeebu et al., 2000a, 2000b).

Genomic and proteomic technologies allow us to rapidly measure the expression of hundreds of genes and detect the presence of as many proteins. The use of these techniques in toxicology will greatly increase the number of potential biomarkers by determining the effects of contaminant exposure on the expression of genes (toxicogenomic) and the synthesis of proteins (toxicoproteomic). These techniques are currently beginning to be applied to the field of immunotoxicology.

As previously mentioned, immunocompetence is under the control of many compensatory mechanisms, and disruption of one parameter may not necessarily alter immune function. As studies have been undertaken to characterize the relationship between alterations of functional immunomarkers and a decrease in host resistance (Luster et al., 1988, 1992, 1993), studies identifying the molecular mechanisms that underpin these functions contribute to the identification of new cellular immunomarkers, as well as to the characterization of the relationship between these two levels of organization. However, as it is not known whether there is a set level of immunosuppression that can be universally tolerated, any consistent suppression of a measurable immune function may represent a potential risk to an organism. Logistic and standard regression modeling have indicated a linear relationship between many of the immune tests and host resistance using one tumor and two infectious disease models. While more work is needed, these studies suggest that there is not a set level of immune suppression that can be universally tolerated and, therefore, considered to be without risk. Hence, it is reasonable in hazard identification to consider any statistically significant and consistent alteration of an immune function as a potential risk for the organism (Selgrade, 1999).

24.6.7 Confounding Factors

As previously mentioned, confounding factors can significantly affect the outcome of biomarker data analysis. Genetics, age, gender, nutritional and hormonal status (particularly the timing of the reproductive cycle for females), infections, and stress have all been demonstrated to act on various immune functions in a specific way, with specific consequences on the immune response. These factors can also affect the sensitivity of an organism to immunotoxic effects of toxicants. Confounding factors are a significant limiting factor for the use of immunomarkers in ERA. When possible, steps should be taken to minimize the influence of these factors.

24.6.7.1 *Genetics*

The difference of sensitivity to a toxicant between strains illustrates the influence of genetics. In some cases, the origin of these differences has been identified. For example, the sensitivity of different strains of mice to 2,3,7,8-tetrachlorodibenzo-p-dioxin (TCDD)–induced toxicity including immunotoxicity is related to differences in expression of the aryl hydrocarbon (Ah) receptor. The Ah receptor is a cytosolic receptor that plays an important role in the molecular and cellular mechanisms of TCDD and TCDD-like compounds' cytotoxicity (Vecchi et al., 1983; Kerkvliet et al., 1990).

24.6.7.2 Gender and Age

Several studies have recently demonstrated that immunotoxic effects of several chemicals can be influenced by age and gender. For most of the chemicals tested, exposures during perinatal periods resulted in transitory and persistent immunotoxic effects at doses much lower than those that induced similar effects when exposure occurred in adulthood. Moreover, some of these immunotoxic effects appear to be clearly gender specific (Miller et al., 1998; Holladay and Smialowicz, 2000; Bunn et al., 2001a, 2001b; Rooney et al., 2003). Although the regulation of immune response by sex steroids has been widely described (Grossman, 1989; Schuurs and Verheul, 1990), the mechanisms involved in gender differences in immunotoxic effects have not been investigated. In a recent study, Pillet et al. (in preparation) demonstrated a significant modulation of the immunotoxic effects of cadmium (Cd) by a physiological level of estradiol in female rats.

24.6.7.3 Nutritional Status

Diet and nutritional status can affect the immune response (Chandra, 1997, 2002). Handy et al. (2002) demonstrated that the immunotoxic effects of chronic oral exposure to diazinon, an organophosphorous pesticide, were dependent on the diet. Chen et al. (2004) demonstrated that maternal dietary protein intake can modulate the immunotoxic effects of lead (Pb) exposure during early development. These studies highlight the integration of nutritional status and immunotoxicity.

24.6.7.4 Interactions between Toxic Pathways of Chemicals

Certain contaminants have an impact on the endocrine system, and are therefore named endocrine disrupters (ED). Most of these ED compounds also have demonstrated immunotoxic effects. Considering the strong interaction between the endocrine and immune systems, it is likely that some immunotoxic effects can result from effects on the endocrine system (Fournier et al., 2000). Moreover, it has been suggested that gender differences of immunotoxic effects observed following perinatal exposure to chemicals with no ED potential might be linked to sex hormones (Bunn et al., 2001a, 2001b). Therefore, the disruption of hormonal balance resulting from ED exposure could modulate the immunotoxic effects of other chemicals with no ED potential. Immune function is also under close coordination by neural input. Accordingly, immune competence is vulnerable to changes induced by environmental pollutants that act on the nervous system (Navarro et al., 2001). Some of these contaminants, such as organophosphorous pesticides, may also have ED potential (Galloway and Handy, 2003). This illustrates the complex pattern of direct and indirect pathways that can lead to immunotoxic effects.

These are only the major confounding factors for which an impact on contaminant-induced immunotoxicity has been demonstrated. A number of other biotic and abiotic factors have the potential to affect immunotoxicity. Changes in metabolic rates in response to seasonal and/or geographical variations, for example, can affect the uptake, metabolism, and accumulation of contaminants and, consequently, could influence the impacts on the immune response. Moreover, most confounding factors are strongly interdependent, increasing the complexity of estimating the relative contribution of each factor to the observed modulation of the immunomarkers.

24.7 IMMUNOMARKERS OF EFFECTS IN WILDLIFE SPECIES

General knowledge regarding the immune system of wildlife species is very limited. Hence, the impact of a contaminant on the immune system of these species is difficult to assess, and is often based on similarities drawn from human and rodent models. The following explanations for this can be put forth:

- Technical limitations
  - Existing methods cannot be successfully adapted to assess immune functions of the targeted species.
  - Reagents are not available to assess immune functions.
- Immunomarkers are not available
  - Lack of knowledge about the immune system of the targeted species.
  - Primitive immune system offering fewer possible markers.
- Logistic limitations
  - High cost of sampling and monitoring (limited access, no facilities).
  - Legal constraint (endangered/protected species).

However, the evolutionary conservation of fundamental methods of recognition and response of immune cells provides a basis for the adaptation of major immunomarker assays developed for humans and rodents to other species, and several methods to assess immune functions have been successfully adapted to wildlife species. The use of these immunomarkers has provided evidence of the immunotoxic potential of numerous environmental contaminants to feral species. However, a large proportion of wildlife studies are laboratory or semi-field based. Therefore, despite the evidence of immunotoxic effects generated under experimental conditions, there are few convincing studies of exposed populations in which alteration in immune functions can be conclusively linked to contaminant exposure in their environment. The difficulty in establishing a link between exposure to a contaminant and alterations of immunomarkers in the wild results primarily from the influence of multiple interactive confounding factors. These are often compounded with logistical difficulties inherent to most field studies. Usually, the complexity of interactions precludes the establishment of a clear causal relationship outside of catastrophic events. Furthermore, additional stresses such as capture, handling, and in some cases captivity, can modulate immune responses either directly or indirectly via the neuroendocrine system (Luebke et al., 1997).

24.7.1 Contaminant-Induced Immunodeficiency in Marine Mammals: Illustration of Complexity of Risk Evaluation in Wildlife Species

The work undertaken to assess the impact of xenobiotics on the immune system of marine mammals is a good illustration of the overall approach and of the problems encountered in order to evaluate the risk of immunotoxicity in wildlife species. As top predators, marine mammals exercise strong control over the dynamics of the marine ecosystem. Alterations of the population dynamic of these key species could therefore have major consequences on the equilibrium of the whole ecosystem. In addition, top predators may be particularly vulnerable groups, as their position at the top of the marine food web causes them to bioaccumulate high levels of PTS from their prey organisms (Das et al., 2003; O'Hara and Becker, 2003; O'Hara et al., 2003; Reijnders and Simmonds, 2003; Law et al., 2003). Moreover, based on the relative persistence of some organochlorine congeners in their tissues, cetaceans seem to have a lower capacity for detoxification than terrestrial mammals (Boon et al., 1992; Norstrom et al., 1992; Tanabe et al., 1994).

Some of these contaminants have clear potential immunotoxic effects, albeit demonstrated in humans and rodents. Furthermore, the severity of several outbreaks of morbillivirus and other diseases that resulted in large-scale mortalities of marine mammals inhabiting polluted areas has led to speculations of immunosuppression associated with environmental pollution causing a reduction in host resistance. This could have contributed to amplification of the mortality associated with epizootic episodes. These speculations were supported by several studies reporting higher concentrations of PTS in animals that died from morbillivirus infection, as compared with free-ranging animals or animals that died from physical trauma (Hall et al., 1992; Aguilar and Borrell, 1994; Kuiken et al., 1994; Olsson et al., 1994). Contaminant-induced immunodeficiency has also been put forward to explain the high prevalence of tumors and infections by opportunistic bacteria found in beluga whales (*Delphinapterus leucas*) from the St. Lawrence River, Canada (Martineau et al.,

Table 24.1   Immune Function Assays Successfully Applied to Marine Mammals

Immune Functions	Species	References
Phagocytic activity of peripheral blood granulocytes	Beluga	De Guise et al. (1995a)
	Harbor seal; grey seal	Pillet et al. (2000)
Cytotoxic activity of peripheral blood natural killer cells	Beluga	De Guise et al. (1997)
	Harbor seal	Ross et al. (1996b)
Delayed-type hypersensitivity	Harbor seal	Ross et al. (1995)
Mixed lymphocyte reactions	Harbor seal	De Swart et al. (1995)
Production-specific antibody following immunization	Harbor seal	Ross et al. (1995)
Lymphoblastic proliferation	Beluga	De Guise et al. (1996b)
	Harbor seal	De Swart et al. (1993)
	Bottlenose dolphin	Lahvis et al. (1995)
IL-2–induced lymphocyte proliferation	Harbor seal	De Swart et al. (1993)
Phenotyping: few antibodies have been developed against cluster of differentiation–like proteins	10 cetacean species	De Guise et al. (2002)
	Killer whale (*Orcinus orca*)	De Guise et al. (2004)
IL-2 production by lymphocytes	Beluga; grey seal	St-Laurent et al. (1999)

Note: IL, interleukin.

1994; De Guise et al., 1995b). Higher levels of PCBs and (bioavailable) mercury (Hg) in tissues were also associated to prevalence of infections in harbor porpoises (*Phocoena phocoena*) (Jepson et al., 1999; Siebert et al., 1999; Bennett et al., 2001). The infectious disease reported in these studies (severe parasitic pneumonia often associated with secondary bacterial infections, as well bacterial septicemia) could indicate immunosuppressive status. These studies also highlight the existing complex interrelationships between contaminant tissue concentrations, age, nutritional status, and diseases. As a result, several confounding factors have to be taken into account so that immunotoxic effects can be clearly defined.

In order to better understand the relationship between PTS exposures and immunodeficiency, studies were undertaken in which the use of immunomarkers was applied to marine mammals (Table 24.1). These studies provided useful insights into the potential of immunomarkers to assess immunotoxicity in marine mammals, and have contributed significantly to our knowledge of immunology of marine mammals.

A captive feeding study demonstrated impairments of several immune functions in harbor seals (*Phoca vitulina*) fed with "contaminated" herrings from the Baltic Sea as compared to seals fed with "less contaminated" herrings from the Atlantic Ocean. The estimated daily intakes of 2,3,7,8-TCDD toxic equivalents were ten times higher for seals fed herrings from the Baltic Sea. During the feeding experiment, the Baltic group of seals displayed higher peripheral blood leukocyte counts associated with an increase of peripheral blood neutrophils, diminished NK cell activity, diminished lymphoproliferative response of T cells, diminished DTH, diminished MLR, and diminished antigen-specific lymphocyte proliferation after immunization, associated in some cases with alteration of specific antibody production (summarized in Ross et al., 1996a; Van Loveren et al., 2000; Vos et al., 2003). These effects indicated a global alteration of the immune response.

The logistics as well as technical and ethical constraints associated with this type of study make them difficult to implement. Therefore, *in vitro* exposures represent an alternate and/or complementary approach. Moreover, *in vitro* exposures allow us to discriminate between direct and indirect effects on immune cells. *In vitro* exposures also allow the investigation of mechanisms of toxicity at cellular and molecular levels. De Guise et al. (1996a, 1998) reported reduced lymphoblastic proliferation of beluga whale lymphocytes exposed *in vitro* to heavy metals, PCB congeners, or DDT metabolites. *In vitro* exposure to Aroclor 1254 or methyl-mercury induced strong genotoxic effects that could partially explain immunotoxic impacts of these compounds in peripheral blood lymphocytes in a bottlenose dolphin (*Tursiops truncatus*) (Betti and Nigro, 1996; Taddei et al., 2001). *In vitro* exposure to heavy metals revealed a high heterogeneity in MT induction among the major peripheral blood leukocytes subpopulations (i.e., granulocytes, lymphocytes, and monocytes) from grey seals (*Halichoerus grypus*) (Pillet et al., 2002). *In vitro* exposures also permit for the evaluation of the complex effects of mixtures on immune cells (De Guise et al., 1998). Furthermore, *in vitro* experiments help to assess factors such as gender effects or interspecies differences (Betti and Nigro, 1996; Taddei et al., 2001; Pillet et al., 2000, 2002).

Studies investigating critical molecular and cellular endpoints such as MT or cytokines illustrate a new field of investigation in marine mammal immunology and immunotoxicology (St-Laurent et al., 1999; St-Laurent and Archambault, 2000; Denis and Archambault, 2001; Pillet et al., 2002).

Although often very difficult to implement, field studies are a critical step in the evaluation of the environmental effects of contaminants. They are an essential complement to laboratory studies, as they allow for the validation of laboratory results in a complex environment where numerous factors may alter immunotoxic effects.

Hematological and biochemical parameters of serum, including Ig levels, seem to be affected by exposure to contaminants in some populations of marine mammals. However, geographical, species-specific, as well as age and gender differences compound the potential effects of contaminants (Bernhoft et al., 2000; Beckmen et al., 2003; Nyman et al., 2003). Beckmen et al. (2003) reported that a lesser proportion of northern fur seal pups (*Callorhinus ursinus*) from young dams (probably firstborn pups) were able to mount an efficient humoral immune response after vaccination as compared to old dams' pups. These effects were attributed to a higher exposure to the lipophilic organic contaminants through maternal milk for the firstborn pups (Beckmen et al., 1999, 2003). Lavhis et al. (1995) reported that Con-A–stimulated proliferation of bottlenose dolphin peripheral blood lymphocytes was negatively correlated with levels of tetrachlorinated to octachlorinated biphenyl in blood; as well as with levels of 1,1,1-trichloro-2,2-bis(*p*-chlorophenyl)ethane (p,p′ DDT) and metabolites.

The case of marine mammals illustrates the necessity of a hierarchical multistep approach for the assessment of immunological risk in wildlife species. Following the standard ERA framework based on problem formulation and hazard identification (Figure 24.1), studies have been undertaken to characterize exposure, and to attempt to link exposure to potential immunotoxic effects in order to characterize the risk. These studies allowed for the implementation of well-known immunomarkers, development of new ones, and identification of confounding factors. It is clearly extremely difficult to establish a causal link between an exposure to contaminants and an impairment to the immune system; however, the body of evidence accumulated from these studies has significantly improved our ability to evaluate the immunological risk associated with contaminant exposure. We believe that the critical issue of assessing immunotoxicity in wildlife species should be addressed by combining markers of immunotoxicity from several levels of biological organization, which are summarized in the next section.

24.7.2 Tier Approach for Immunotoxicity Studies in Wildlife Species

1. Effects of *in vitro* exposures on immune cells. This type of research provides an opportunity to test several doses and chemicals on cells from the same organism. This enables us to assess direct effects on immune cells and determine a dose–response curve. It also allows for easy comparisons between species.

2. *In vitro* assessment of immunomarkers with cells from organisms exposed *in vivo*. In this step, effects on the whole organism are assessed including indirect effects that could have consequences on the immune function. This involves laboratory *in vivo* exposure under controlled conditions in order to investigate the specific effects of different conditions of exposure, and allows comparison between environmentally exposed animals sampled at different sites, usually in reference to a control site.

3. Measurement of the ability of the organism to build an immune response against new antigens (vaccination, antibody production, etc.) following *in vivo* laboratory exposures or in environmentally exposed animals.

4. Determine the ability of environmentally exposed organisms to counter infections at large by challenging organisms exposed in the laboratory or sampled at different sites in the environment with bacteria, viruses, parasites, or tumor inducers in a closed laboratory system.
5. Epidemiological studies tracking pathology that could indicate effects on the immune system, such as high incidence of tumors, pathologies of immune tissues, infections by opportunistic bacteria, and so forth.

24.8 RESEARCH NEEDS

Although efforts have been made to increase our understanding of the immune system of wildlife species, there are serious deficiencies in our knowledge that limit our ability to assess immunotoxic effects.

In the context of ERA, tremendous efforts have to be made to determine how responses of early warning immunomarkers (i.e., at the molecular or cellular level) translate at higher biological levels of organization. We have to better characterize the link between response of immunomarkers, alteration of host resistance, and emergence of infections. This is undoubtedly one of the major challenges of wildlife immunotoxicology that needs to be addressed in order to provide the scientific rationale for immunotoxicity risk assessment.

Furthermore, we need to identify confounding factors and how they modulate the response of the immunomarkers under environmental conditions.

Finally, more work has to be done to establish how the immune system interacts with other critical biological systems such as the endocrine and nervous systems, how toxic effects on these three systems are connected, and how this information can be integrated in the context of ERA.

24.9 SUMMARY

ERA relies on the response of biomarkers to characterize the risk of chemical exposure. As the defense mechanism of a host against potential pathogens, the immune system should be a critical endpoint in the assessment of the risk associated with environmental contamination. Hazard identification, exposure, and dose–response characterization are aspects of ERA that can certainly be addressed through the use of immunomarkers. However, the complexity and influence of a wide variety of confounding factors constitute important obstacles to the integration of immunomarkers into a risk assessment strategy. Nevertheless, certain chemical-induced changes on the immune response can have major repercussions on the health of an organism, and through this can potentially affect populations. In the case of key ecological species, this could arguably alter the balance of whole ecosystems. It is therefore essential in our view to develop and implement a framework that would address the limitations of immunomarkers in risk assessment.

Each immunomarker can only assess a part of the complex network that is the immune system, providing us with a window of observation onto an aspect of the immune response. When looking through this window, it is important to keep in mind the significance of the marker, that is, what immune function it is related to and how it integrates into the overall immune response, as well as the potential consequences of a toxic effect on higher levels of biological organization.

There is a growing number of immunomarkers available for the evaluation of environmental toxic effects. Many of these markers have the advantage of providing an environmentally relevant assessment of a host's immune fitness. Although immunomarkers might not be sufficiently specific to causally identify the source of an immunotoxic effect, they do indicate an impairment of the immune system. While it remains important to identify the source of observed impairments, the immunotoxic effects demonstrated enable us to identify populations at risk regardless of the cause.

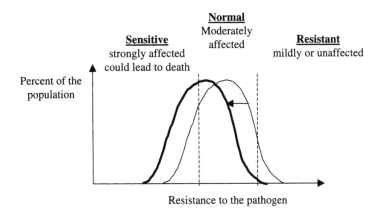

Figure 24.5 Typical shift from normal resistance to a pathogen (indicated by the arrow) following exposure to an immunosuppressor such as an environmental contaminant. The exposed and weaker population (heavy line) will be at higher risk of infection compared to the normal population (fine line).

It is also important to recognize that impairments of the immune system can remain latent. A weakened immune competence will remain undetected until an organism is challenged by a pathogen. A weakened organism or population, as a result of exposure to stressors (including contaminants), is at greater risk of developing an infection (Figure 24.5). In the absence of any immunological monitoring, latent immunotoxic effects would only become perceptible with the occurrence of a disease or epidemic, and quite likely it would be too late to allow for any intervention.

The tier approach described in this chapter would allows us to gather a body of evidence sufficient to establish whether the exposure to one or more toxic contaminants under specific environmental conditions can be associated with an immunological risk. It would also identify specific segments of a population at higher risk (e.g., neonates, juveniles, adults, males, females).

ERA remains a critical step toward better management of environmental problems, and immunomarkers that provide a body of evidence leading to the characterization of risk to the immune fitness of organisms exposed to contaminants should establish a cornerstone of ERA.

ACKNOWLEDGMENTS

The authors thank Darlene Mossman for her editorial comments.

REFERENCES

Aguilar, A., and Borrell, A., Abnormally high polychlorinated biphenyl levels in striped dolphins (*Stenella coeruleoalba*) affected by the 1990–1992 Mediterranean epizootic, *Sci. Total Environ.*, 154, 237–247, 1994.

Beckmen, K.B., et al., Factors affecting organochlorine contaminant concentrations in milk and blood of northern fur seal (*Callorhinus ursinus*) dams and pups from St. George Island, Alaska, *Sci. Total Environ.*, 231, 183–200, 1999.

Beckmen, K.B., et al., Organochlorine contaminant exposure and associations with hematological and humoral immune functional assays with dam age as a factor in free-ranging northern fur seal pups (*Callorhinus ursinus*), *Mar. Pollut. Bull.*, 46, 594–606, 2003.

Bennett, P.M., et al., Exposure to heavy metals and infectious disease mortality in harbour porpoises from England and Wales, *Environ. Pollut.*, 112, 33–40, 2001.

Bernhoft, A., et al., Possible immunotoxic effects of organochlorines in polar bears (*Ursus maritimus*) at Svalbard, *J. Toxicol. Environ. Health A*, 59, 561–574, 2000.

Bernier, J., et al., Immunotoxicity of heavy metals in relation to Great Lakes, *Environ. Health Perspect*, 103, 23–34, 1995.

Betti, C., and Nigro, M., The Comet Assay for the evaluation of the genetic hazard of pollutants in Cetaceans: preliminary results on the genotoxic effects of methyl-mercury on the bottle-nosed dolphin (*Tursiops truncatus*) lymphocytes *in vitro*, *Mar. Pollut. Bull.*, 32, 545–548, 1996.

Boon, J.P., et al., The toxicokinetics of PCBs in marine mammals with special reference to possible interactions of individual congeners with the cytochrome P450-dependent monooxygenase system: an overview, in *Persistent Pollutants in Marine Ecosystems*, Pergamon, Oxford, 1992, pp. 119–159.

Borghesi, L.A., et al., Interactions of metallothionein with murine lymphocytes: plasma membrane binding and proliferation, *Toxicology*, 108, 129–140, 1996.

Brousseau, P., et al., *Manual of Immunological Methods*, CRC Press, Boca Raton, FL, 1999, 141.

Bunn, T.L., et al., Developmental immunotoxicology assessment in the rat: age, gender, and strain comparisons after exposure to lead, *Toxicol. Methods*, 11, 41–58, 2001a.

Bunn, T.L., et al., Gender-based profiles of developmental immunotoxicity to lead in the rat: assessment in juveniles and adults, *J. Toxicol. Environ. Health A*, 64, 223–240, 2001b.

Chandra, R.K., Nutrition and the immune system: an introduction, *Am. J. Clin. Nutr.*, 66, 460S–463S, 1997.

Chandra, R.K., Nutrition and the immune system from birth to old age, *Eur. J. Clin. Nutr.*, 56 Suppl. 3, S73–76, 2002.

Chen, S., et al., Developmental immunotoxicity of lead in the rat: influence of maternal diet, *J. Toxicol. Environ. Health A*, 67, 495–511, 2004.

Crowthers, K.C., et al., Augmented humoral immune function in metallothionein-null mice, *Toxicol. Appl. Pharmacol.*, 166, 161–172, 2000.

Das, K., Debacker, V., and Bouquegneau, J-M., Metallothioneins in marine mammals, *Cell. Mol. Biol.*, 46, 283–294, 2000.

Das, K., et al., Heavy metals in marine mammals, in *Toxicology of Marine Mammals*, Vos, J.G., Bossart, G.D., Fournier, M., and O'Shea, T.J., Eds., Taylor & Francis, New York, 2003, pp. 135–167.

De Guise, S., et al., Immune functions in beluga whales (*Delphinapterus leucas*): evaluation of phagocytosis and respiratory burst with peripheral blood leucocytes using flow cytometry, *Vet. Immunol. Immunopathol.*, 47, 351–362, 1995a.

De Guise, S., et al., Possible mechanisms of action of environmental contaminants on St Lawrence beluga whales (*Delphinapterus leucas*), *Environ. Health Perspect.*, 103, 73–77, 1995b.

De Guise, S., et al., Effects of in vitro exposure of beluga whale splenocytes and thymocytes to heavy metals, *Vet. Immunol. Immunopathol.* 15, 1357–1364, 1996a.

De Guise, S., et al., Immune functions in beluga whales (*Delphinapterus leucas*): Evaluation of mitogen-induced blastic transformation of lymphocytes from peripheral blood, spleen and thymus, *Vet. Immunol. Immunopathol.*, 50, 117–126, 1996b.

De Guise, S., et al., Immune functions in beluga whales (*Delphinapterus leucas*): evaluation of natural killer cell activity, *Vet. Immunol. Immunopathol.*, 58, 345–54, 1997.

De Guise, S., et al., Effects of in vitro exposure of beluga whale leukocytes to selected organochlorines, *J. Toxicol. Environ. Health A*, 55, 479–493, 1998.

De Guise, S., et al., Monoclonal antibodies to lymphocyte surface antigens for cetacean homologues to CD2, CD19 and CD21, *Vet. Immunol. Immunopathol.*, 84, 209–221, 2002.

De Guise, S., et al., Characterization of F21.A, a monoclonal antibody that recognizes a leucocyte surface antigen for killer whale homologue to beta–2 integrin, *Vet. Immunol. Immunopathol.*, 97, 195–206, 2004.

De Swart, R.L., et al., Mitogen and antigen induced B and T cell responses of peripheral blood mononuclear cells from the harbour seal (*Phoca vitulina*), *Vet. Immunol. Immunopathol.*, 37, 217–230, 1993.

De Swart, R.L., et al., Impaired cellular immune response in harbour seals (*Phoca vitulina*) feeding on environmentally contaminated herring, *Clin. Exp. Immunol.*, 101, 480–486, 1995.

Denis, F., and Archambault, D., Molecular cloning and characterization of beluga whale (*Delphinapterus leucas*) interleukin-1beta and tumor necrosis factor-alpha, *Can. J. Vet. Res.*, 65, 233–240, 2001.

Depledge, M.H., and Fossi, M.C., The role of biomarkers in environmental assessment, *Ecotoxicology*, 3, 161–172, 1994.

Feldmann, M., Cytokines, in *Encyclopedia of Immunology*, Roitt, I., and Delves, P., Eds., Academic Press, London, 1992, pp. 438–440.

Fournier, M., et al., Biomarkers of immunotoxicity: an evolutionary perspective, in *Environmental Endocrine Disrupters*, Guillette, L. Jr., and Crain, D.A., Eds., Taylor & Francis, New York, 2000, pp. 182–216.

Galloway, T., and Handy, R., Immunotoxicity of organophosphorous pesticides, *Ecotoxicology*, 12, 345–363, 2003.

Grossman, C.J., Possible underlying mechanisms of sexual dimorphism in the immune response, fact and hypothesis, *J. Steroid Biochem.*, 34, 241–251, 1989.

Habeebu, S.S., et al., Metallothionein-null mice are more sensitive than wild-type mice to liver injury induced by repeated exposure to cadmium, *Toxicol. Sci.*, 55, 223–232, 2000a.

Habeebu, S.S., et al., Metallothionein-null mice are more susceptible than wild-type mice to chronic CdCl(2)-induced bone injury, *Toxicol. Sci.*, 56, 211–219, 2000b.

Hall, A.J., et al., Organochlorine levels in common seals (*Phoca vitulina*) which were victims and survivors of the 1988 phocine distemper epizootic, *Sci. Total Environ.*, 115, 145–162, 1992.

Handy, R.D., et al., Chronic diazinon exposure: pathologies of spleen, thymus, blood cells, and lymph nodes are modulated by dietary protein or lipid in the mouse, *Toxicology*, 172, 13–34, 2002.

Handy, R.H., Galloway, T.S., and Depledge, M.H., A proposal for the use of biomarkers for the assessment of chronic pollution and in regulatory toxicology, *Ecotoxicology*, 12, 331–343, 2003.

Holladay, S.D., and Smialowicz, R.J., Development of the murine and human immune system: differential effects of immunotoxicants depend on time of exposure, *Environ. Health Perspect.*, 108(Suppl. 3), 463–473, 2000.

Jepson, P.D., et al., Investigating potential associations between chronic exposure to polychlorinated biphenyls and infectious disease mortality in harbour porpoises from England and Wales, *Sci. Total Environ.*, 243–244, 339–348, 1999.

Kerkvliet, N. I., et al., Influence of the Ah locus on the humoral immunotoxicity of 2, 3, 7, 8-tetrachlorodibenzo-p-dioxin: evidence for Ah-receptor-dependent and Ah-receptor-independent mechanisms of immuno-suppression, *Toxicol. Appl. Pharmacol.* 105, 26–36, 1990.

Kimber, I., Cytokines and regulation of allergic sensitization to chemicals, *Toxicology*, 93, 1–11, 1994.

Klaassen, C.D., Liu, J., and Choudhuri, S., Metallothionein: An intracellular protein to protect against cadmium toxicity, *Ann. Rev. Pharmacol. Toxicol.*, 39, 267–294, 1999.

Krzystyniak, K., Tryphonas, H., and Fournier, M., Approaches to the evaluation of chemical-induced immuno-toxicity, *Environ. Health Perspect.*, 103, 17–22, 1995.

Kuiken, T., et al., PCBs, cause of death and body condition in harbor porpoises (*Phocoena phocoena*) from British waters, *Aquatic Toxicol.*, 28, 13–28, 1994.

Kwon, O., et al., Expression of cyclooxygenase–2 and pro-inflammatory cytokines induced by 2, 2, 4, 4, 5, 5′-hexachlorobiphenyl (PCB 153) in human mast cells requires NF-kappa B activation, *Biol. Pharm. Bull.*, 25, 1165–1168, 2002.

Lahvis, G.P., et al., Decreased lymphocyte responses in free-ranging bottlenose dolphins (*Tursiops truncatus*) are associated with increased concentrations of PCBs and DDT in peripheral blood, *Environ. Health Perspect.*, 103(Suppl. 4), 67–72, 1995.

Law, R. J., et al., Levels and trends of polybrominated diphenylethers and other brominated flame retardants in wildlife, *Environ. Int.*, 29, 757–770, 2003.

Lefkovits, I., *Immunology Methods*, vols. 1–4, Academic Press, San Diego, 1997.

Leibbrandt, M.E., Khokha, R., and Koropatnick, J., Antisense down-regulation of metallothionein in a human monocytic cell line alters adherence, invasion, and the respiratory burst, *Cell Growth Differ.*, 5, 17–25, 1994.

Luebke, R.W., et al., Aquatic pollution-induced immunotoxicity in wildlife species, *Fundam. Appl. Toxicol.*, 37, 1–15, 1997.

Luster, M.I., et al., Development of a testing battery to assess chemical-induced immunotoxicity: National Toxicology Program's guidelines for immunotoxicity evaluation in mice, *Fundam. Appl. Toxicol.*, 10, 2–19, 1988.

Luster, M.I., et al., Risk assessment in immunotoxicology I. sensitivity and predictability of immune tests, *Fundam. Appl. Toxicol.*, 18, 200–210, 1992.

Luster, M.I., et al., Risk assessment in immunotoxicology. II. Relationships between immune and host resistance tests, *Fundam. Appl. Toxicol.*, 21, 71–82, 1993.

Lynes, M.A., Garvey, J.S., and Lawrence, D.A., Extracellular metallothionein effects on lymphocyte activities, *Mol. Immunol.*, 27, 211–219, 1990.

Lynes, M.A., et al., Immunomodulatory activities of extracellular metallothionein. I. Metallothionein effects on antibody production, *Toxicology*, 85, 161–177, 1993.

Martineau, D., et al., Pathology and toxicology of beluga whales from the St Lawrence Estuary, Quebec, Canada. Past, present and future, *Sci. Total Environ.*, 154, 201–215, 1994.

Miles, A.T., et al., Induction, regulation, degradation, and biological significance of mammalian metallothioneins, *Crit. Rev. Biochem. Mol. Biol.*, 35, 35–70, 2000.

Miller, T.E., et al., Developmental exposure to lead causes persistent immunotoxicity in Fischer 344 rats, *Toxicol. Sci.*, 42, 129–135, 1998.

Moffatt, P., and Denizeau, F., Metallothionein in physiological and physiopathological processes, *Drug Metab. Rev.*, 29, 261–307, 1997.

National Research Council, *Biologic Markers in Reproductive Toxicology*, National Academy Press, Washington, DC, 1989.

National Research Council, Committee on Biological Markers, Biological markers in environmental health research, *Environ. Health Perspect.*, 74, 3–9, 1987.

Navarro, H.A., et al., Neonatal chlorpyrifos administration elicits deficits in immune function in adulthood: a neural effect? *Dev. Brain Res.*, 13, 249–252, 2001.

Norstrom, R.J., et al., Indications of P450 monooxygenase activities in beluga (*Delphinapterus leucas*) and narwhal (*Monodon monoceros*) from patterns of PCB, PCDD and PCDF accumulation, *Mar. Environ. Res.*, 34, 267–272, 1992.

Nyman, M., et al., Contaminant exposure and effects in Baltic ringed and grey seals as assessed by biomarkers, *Mar. Environ. Res.*, 55, 73–99, 2003.

O'Hara, T.M., Woshner, V., and Bratton G., Inorganic pollutants in Arctic marine mammals, in *Toxicology of Marine Mammals*, Vos, J.G., Bossart, G.D., Fournier, M., and O'Shea, T.J., Eds., Taylor & Francis, New York, 2003, pp. 206–246.

O'Hara, T.M., and Becker, P.R., Persistent organic contaminants in Arctic marine mammals, in *Toxicology of Marine Mammals*, Voss, J.G., Bossart, G.D., Fournier, M., et al., Eds., Taylor & Francis, New York, 2003, pp. 168–205.

Olsson, M., Karlsson, B., and Ahnland, E., Diseases and environmental contaminants in seals from the Baltic and the Swedish west coast, *Sci. Total Environ.*, 154, 217–227, 1994.

Pillet, S., et al., In vitro exposure of seal peripheral blood leukocytes to different metals reveal a sex-dependant effect of zinc on phagocytic activity, *Mar. Pollut. Bull.*, 40, 921–927, 2000.

Pillet, S., et al., Presence and regulation of metallothioneins in peripheral blood leukocytes of grey seals, *Toxicol. Appl. Pharmacol.*, 185, 207–217, 2002.

Pulido, M.D., and Parrish, A.R., Metal-induced apoptosis: mechanisms, *Mutation Res.*, 533, 227–241, 2003.

Reijnders, P.J.H., and Simmonds, M.P., Global temporal trends of organochlorines and heavy metals in pinnipeds, in *Toxicology of Marine Mammals*, Vos, J.G., Bossart, G.D., Fournier, M., and O'Shea, T.J., Eds., Taylor & Francis, New York, 2003, pp. 491–506.

Roitt, I.M., Brostoff, J., and Male, D.K., *Immunology*, 3rd ed., Mosby Year, Toronto, 1993.

Rooney, A.A., Matulka, R.A., and Luebke, R.W., Developmental atrazine exposure suppresses immune function in male, but not female Sprague-Dawley rats, *Toxicol. Sci.*, 76, 366–375, 2003.

Ross, P.S., et al., Contaminant-related suppression of delayed-type hypersensitivity and antibody responses in harbor seals fed herring from the Baltic Sea, *Environ. Health Perspect.*, 103, 162–167, 1995.

Ross, P.S., et al., Contaminant-induced immunotoxicity in harbour seals: wildlife at risk? *Toxicology*, 112, 157–169, 1996a.

Ross, P.S., et al., Suppression of naturel killer cell activity in harbour seals (*Phoca vitulina*) fed Baltic Sea herring, *Aquatic Toxicol.*, 34, 71–84, 1996b.

Schuurs, A.H., and Verheul, H.A., Effects of gender and sex steroids on the immune response, *J. Steroid Biochem.*, 35, 157–172, 1990.

Selgrade, M.K., Use of immunotoxicity data in health risk assessments: uncertainties and research to improve the process, *Toxicology*, 133, 59–72, 1999.

Seth, P., et al., Early onset of virus infection and up-regulation of cytokines in mice treated with cadmium and manganese, *Biometals*, 16, 359–368, 2003.

Siebert, U., et al., Potential relation between mercury concentrations and necropsy findings in cetaceans from German waters of the North and Baltic seas, *Mar. Pollut. Bull.*, 38, 285–295, 1999.

St-Laurent, G., Beliveau, C., and Archambault, D., Molecular cloning and phylogenetic analysis of beluga whale (*Delphinapterus leucas*) and grey seal (*Halichoerus grypus*) interleukin-2, *Vet. Immunol. Immunopathol.*, 67, 385–394, 1999.

St-Laurent, G., and Archambault, D., Molecular cloning, phylogenetic analysis and expression of beluga whale (*Delphinapterus leucas*) interleukin 6, *Vet. Immunol. Immunopathol.*, 73, 31–44, 2000.

Taddei, F., et al., Genotoxic hazard of pollutants in cetaceans: DNA damage and repair evaluated in the bottlenose dolphin (*Tursiops truncatus*) by the Comet Assay, *Mar. Pollut. Bull.*, 42, 324–328, 2001.

Tanabe, S., Iwata, H., and Tatsukawa, R., Global contamination by persistent organochlorines and their ecotoxicological impact on marine mammals, *Sci. Total Environ.*, 154, 163–177, 1994.

Theocharis, S., Margeli, A., and Panayiotidis, P., Effects of various metals on DNA synthesis and lymphokines production by human peripheral blood lymphocytes in vitro, *Comp. Biochem. Physiol. C*, 99, 131–133, 1991.

Tian, Y., Rabson, A.B., and Gallo, M.A., Ah receptor and NF-kappaB interactions: mechanisms and physiological implications, *Chem. Biol. Interactions*, 141, 97–115, 2002.

Trinchieri, G., Natural killer (NK) cells, in *Encyclopedia of Immunology*, Roitt, I.M., and Delves, P.J., Eds., Academic Press, London, 1992, pp. 1136–1138.

Tryphonas, H., Immunotoxicology of PBCs (Aroclors) in relation to Great Lakes., *Environ. Health Perspect.*, 103, 35–46, 1995.

Tryphonas, H., Approaches to detecting immunotoxic effects of environmental contaminants in humans, *Environ. Health Perspect.*, 109(Suppl. 6), 877–884, 2001.

U.S. Environmental Protection Agency, Proposed guidelines for ecological risk assessment, *Federal Register*, 61, 47552–47631, 1996.

Van der Oost, R., Beyer, J., and Vermeulen, N.P.E., Fish bioaccumulation and biomarkers in environmental risk assessment: a review, *Environ. Toxicol. Pharmacol.*, 13, 57–149, 2003.

Van Loveren, H., et al., Contaminant-induced immunosuppression and mass mortalities among harbor seals, *Toxicol. Lett.*, 112, 319–324, 2000.

Van Oss, C.J., Phagocytosis: an overview, *Methods Enzymol.*, 132, 3–15, 1986.

Vecchi, A., et al., Immunosuppressive effects of 2, 3, 7, 8-tetrachlorodibenzo-p-dioxin in strains of mice with different susceptibility to induction of aryl hydrocarbon hydroxylase, *Toxicol. Appl. Pharmacol.* 68, 434–441, 1983.

Verhoef, J., Phagocytosis, in *Encyclopedia of Immunology*, Roitt, I., and Delves, P., Eds., Academic Press, London, 1992, pp. 1220–1222.

Viarengo, A., et al., Metallothionein as a tool in biomonitoring programmes, *Biomarkers*, 4, 455–456, 1999.

Vos, J.G., et al., The effects of chemical contaminants on immune functions in harbour seals: results of a semi-field study, in *Toxicology of Marine Mammals*, Vos, J.G., Bossart, G.D., Fournier, M., and O'Shea, T.J., Eds., Taylor & Francis, New York, 2003, pp. 558–570.

Voss, S.D., Hong, R., and Sondel, P.M., Severe combined immunodeficiency, interleukin-2 (IL-2), and the IL-2 receptor: experiments of nature continue to point the way, *Blood*, 83, 626–35, 1994.

World Health Organization International Programme on Chemical Safety (IPCS), Biomarkers and Risk Assessment: Concepts and Principles, World Health Organization, Geneva, 1993.

Yamada, H., and Koizumi, S., Metallothionein induction in human peripheral blood lymphocytes by heavy metals, *Chem. Biol. Interactions*, 78, 347–354, 1991.

Yurkow, E.J., and Makhijani, P.R., Flow cytometric determination of metallothionein levels in human peripheral blood lymphocytes: utility in environmental exposure assessment, *J. Toxicol. Environ. Health*, 54, 445–457, 1998.

CHAPTER 25

Toxicological Considerations: Making the Connection between Toxicologic and Immunotoxicologic Studies as These Relate to Human and Ecosystem Health

Michel Fournier, Barry R. Blakley, Herman J. Boermans, and Pauline Brousseau

CONTENTS

25.1 INTRODUCTION

It is well documented that a broad spectrum of xenobiotics alters immune function. Therefore, immune system endpoints can be useful indicators of toxic effects of chemicals. However, the development of this field has raised basic toxicological considerations related to the selection of immune endpoints. More recently, the use of immunological data in the regulatory process has put greater emphasis on these concerns. For example, the shape of the dose–response curve for immunological responses has become an issue. In the risk assessment process, this factor has influenced strategy selection, particularly the question of threshold or shape of the curve for low-level exposure.

The assessment of chemical toxicity has relied heavily on laboratory animals. The potential of chemicals to alter immune function has traditionally been performed in laboratory rodents (discussed in other chapters of this book). While the immunotoxic effects of many chemicals are well documented in experimental animals as well as in humans, relatively little is known about effects in other vertebrate or invertebrate species (see Section II).

In addition, few attempts have been made to compare the effects of a single chemical on the immune system/immune cells in several species. Comparative studies would provide valuable information on the sensitivity of various immunological endpoints or sensitivity of immune cells from different tissues or species. Such comparative studies would also provide additional information to the existing data derived from rodents about the prediction of potential toxicity of xenobiotics in human, domestic, and wildlife animal immune systems. The possibility of using other animal species in addition to laboratory rodents for toxicity testing of chemicals is of interest. In spite of many similarities among the immune systems or functions of mammals, interspecies differences may exist. Moreover, interspecies differences and extrapolation from *in vitro* and *in vivo* systems have become important issues in the assessment of toxicological risk, including immunotoxicologic risk. In an attempt to fill these gaps, a research program was initiated by the Immunology Theme Team of the Canadian Network of Toxicology Centres. The experimental approach, which involved a multispecies assessment, evaluated the dose–response relationships associated with exposure to heavy metals and immune function.

This comparative immunotoxicologic study was performed with immune cells from a wide variety of invertebrate and vertebrate species. The effects of $CdCl_2$, $CH_3,HgCl$, and $HgCl_2$ were determined over a range of exposures (10^{-9} to 10^{-4} M) on the phagocytic activity of cells collected from the earthworm, selected bivalves, starfish, crab, mummichog, trout, plaice, frog, chicken, kestrel, rat, mouse, sheep, rabbit, horse, cat, dog, squirrel, llama, musk ox, elk, seal, beluga whales, cow, monkey, and human. Phagocytic function was evaluated by measuring the *in vitro* uptake of fluorescent microspheres using flow cytometric methods. In species with circulating lymphocytes, the effect of exposure to heavy metals was also evaluated by measuring the lymphoproliferative response to the concanavalin A (Con-A) mitogen. In selected species, differences of sensitivity between tissues were also monitored using lymphocyte blastogenesis. Studies were also performed to address the question of *in vitro–in vivo* differences. In this chapter, we review the results obtained in this program and discuss the implications with respect to the selection of sentinel species for ecosystem or human health surveillance.

25.2 SHAPE OF THE DOSE–RESPONSE CURVE

As an example of a dose–response curve generated for phagocytic cells exposed to heavy metals, we have selected data generated with bivalve hemocytes, primarily because they produce the response observed in most invertebrate and vertebrate species. Hemocytes from *Mya arenaria* demonstrated three different patterns for *in vitro* metal-specific and concentration-related toxicity within the selected concentration range (Figure 25.1). At the lowest level of exposure (10^{-9} M), stimulation of phagocytic activity without altering viability was evident. Although the mechanisms associated with the stimulation of phagocytosis by hemocytes have not been yet elucidated, this pattern of activity is not unique. It has been reported in several species, including rodents, reptiles, birds, fish, cattle, whales, and primates (see Section II). Based on previous studies, it has been suggested that mercury may stimulate immune function by increasing intracellular calcium levels, enhancing the synthesis or the secretion of soluble factors, and altering the cellular levels of glutathione or tyrosine-phosphorylated proteins, in addition to enhancing kinase and phosphatase activity (reviewed by Bernier et al., 1995). The specific mechanisms involved in this stimulation process, its biological significance, and its consequences on immune function have not been determined. Stimulation of immune function is not necessarily desirable since it may contribute to the development of hypersensitivity and autoimmune diseases.

A second manifestation of toxicity can be observed with methyl mercury at 10^{-6} and 10^{-5} M and mercuric chloride at 10^{-5} M. While the two forms of mercury did not induce evidence of cytotoxicity, phagocytosis was profoundly impaired. This observation has been reported with other

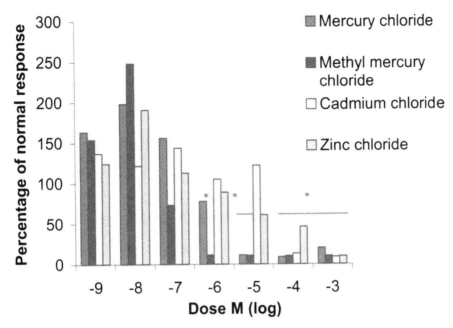

Figure 25.1 Phagocytic activity of hemocytes from *Mya arenaria* measured by flow cytometry after 18 hours of exposure with increasing concentrations of heavy metals. The results are expressed in percent of normal response obtained with nonexposed cultured hemocytes. Hemocytes incubated with hemolymph containing 2.0% sodium azide were used as a negative control for phagocytosis. The phagocytic activity of these cells was 5% of the normal response (*$p \leq 0.05$).

xenobiotics. Potential mechanisms may include inhibition of biochemical pathways, disruption of intracellular integrity, and interference with immunoregulatory networks.

In addition, it was observed that increased cytotoxicity was associated with decreased cell viability with a corresponding impairment of hemocyte phagocytic activity following exposure to cadmium and zinc (10^{-4} M). In this case, immunosuppression was observed only at lethal concentrations.

Mitogenic responses in mammals demonstrated that $CdCl_2$, $HgCl_2$, and CH_3HgCl, produced a biphasic response with Con-A–stimulated blastogenesis. The stimulatory and inhibitory effects were observed at low and high metal concentrations, respectively. This metal-induced biphasic response on the T-lymphocyte mitogenic response was influenced by the metal, concentration of the metal, and test species. Lower concentrations of $CdCl_2$ enhanced the Con-A mitogenic activity in musk ox, elk, and llama lymphocytes. Stimulation of immune function by low concentrations of heavy metals is frequently observed. The biphasic immunomodulatory effects of heavy metals have been reported with leukocytes from human and laboratory animals (Bernier et al., 1995; Zelikoff and Thomas, 1998). The alterations were dependent on the type of heavy metal, concentration of metal, and species of animal or strain of laboratory animal. These observations are consistent with the present study. In general, $HgCl_2$ suppressed the concanavalin A (Con-A) response at higher concentrations ($\geq 10^{-5}$ to 10^{-4} M) compared to CH_3HgCl_2 at $\geq 10^{-7}$ M. At 10^{-4} M, $CdCl_2$ suppressed Con-A responses in all species except llama and sheep, where suppression was also observed at 10^{-5} M $CdCl_2$. The 50% inhibition concentrations (IC_{50}) for CH_3HgCl_2 were usually lower than the corresponding values for $HgCl_2$ and $CdCl_2$.

In summary, when indicators of cell competence were utilized, the dose–response curves generated following exposure to selected xenobiotics exhibit the classical shapes with defined thresholds allowing for determination of the no observed adverse effect level (NOAEL) and lowest observed adverse effect level (LOAEL), which allowed for the use of classical regulatory tools. However, in the absence of biological significance, the low-dose stimulation, particularly in terms of hypersensitivity and autoimmune disease must be investigated since it may shift the NOAEL by several orders of magnitude.

If one considers the clonal nature of the lymphocyte-based immune response, cytotoxicity induced by an immunotoxin on a limited number of lymphocyte clones may represent a severe loss of response potential to a specific antigen. The consequences may be difficult to assess when the immunocompromised organism encounters the specific antigen. The manifestation may not follow a classical dose–response curve. Assessment of these effects experimentally may, in fact, be impossible.

Some chemicals have the potential to induce abnormal immune responses such as hypersensitivity or autoimmune disease. In this instance, the concept of threshold, which triggers the disease or the reoccurrence of disease subsequent to future exposure, remains controversial.

25.3 COMPARISON OF TISSUE SENSITIVITY

Most of the assays used to evaluate the immunotoxic effects or predict the potential health risk associated with exposure to chemicals are frequently performed using rat and mouse splenocytes following *in vitro* or *in vivo* exposure (Lang et al., 1993). In humans, ethical issues limit immunotoxicologic investigations to peripheral blood leukocytes. The use of human immune cells for immunotoxicologic measurement to compare human and animal data is essential to determine human health risk from animal data. It is also important to note that the lymphocyte subpopulations in the spleen, lymph nodes, and peripheral blood are different within the same species (Lebrec et al., 1995). It is critical to determine whether immunotoxicologic endpoints measured with rodent splenocytes are relevant to lymphocytes present in other lymphoid tissues, including the thymus, lymph nodes, and peripheral blood.

In order to evaluate the relevance of immunotoxic effects observed in rodent splenocytes, the intertissue differences associated with immune alteration cells from various lymphoid tissues, including the thymus, lymph nodes, and peripheral blood should be assessed using the same methods of exposure and immunological assessment. The sensitivities of rabbit thymocytes, lymph nodes, and peripheral blood cells were compared using this approach. Heavy metals, which have a well-known suppressive activity on lymphocyte proliferation, were chosen for this investigation. Cells were also collected from seal blood, lymph node, and thymus, and tested using a mixture of polychlorinated biphenyls (PCBs). The proliferation of lymphocytes against the T-lymphocyte mitogen Con-A was employed to assess immune function.

This study demonstrated differences in tissue sensitivity. Methylmercury chloride stimulates rabbit cells (Figure 25.2) at low doses in spleen and blood lymphocytes, but no stimulation was evident in thymocytes. However, lymphocytes from the three tissues exhibited a similar susceptibility to the toxic action of methylmercury. Profound suppression was seen at 10^{-6} M. Similar observations were associated with *in vitro* mercuric chloride.

The effects of mixtures of polychlorinated biphenyls on seal lymphocytes are illustrated in Figure 25.3 for Aroclor 1260. The response pattern is similar for blood and lymph node lymphocytes while thymocytes appear to be more sensitive to Aroclor 1260. Comparable results were obtained for various Aroclors (1232, 1248, 1250).

It is evident from these results that immunotoxicologic evaluation using blood lymphocytes may be useful to predict responses in lymphocytes from other tissues. The differences in sensitivity may be related to differences in lymphocyte subpopulations, stage of maturation, degree of cell activation, and so on. This may be particularly relevant at low-level exposure, where differences in the shape of the dose–response curve are apparent.

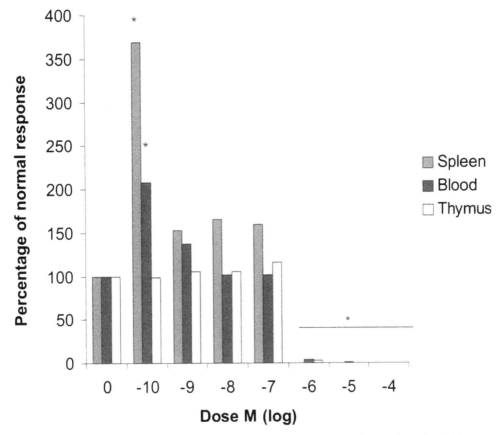

Figure 25.2 Blastogenic activity of spleen, peripheral blood, and thymus lymphocytes from the rabbit measured by tritiated thymidine incorporation after 3 hours of exposure to increasing concentrations of methylmercury chloride. The results are expressed in percent of normal response obtained with non-exposed cultured lymphocytes (*$p \leq 0.05$).

25.4 DIFFERENCES OF SENSITIVITY ASSOCIATED WITH AGE

The establishment of the period of maximal susceptibility is particularly important when the immunotoxic potential of a test compound is determined. The agent may act as an endocrine disruptor during development of the immune system. Most current testing strategies have been designed to assess immune function of a test compound when both exposure and evaluation occur during adulthood. Quantitative and qualitative differences associated with immune function may be encountered if exposure occurs during pregnancy or lactation.

There is extensive evidence suggesting that chemicals known as endocrine disruptors produce detrimental effects on health if exposure occurs at low levels during sensitive periods of embryonic development (Colborn and Clement, 1992; Birnbaum, 1995; Guillette et al., 1995). The developmental stage at the time of exposure is a critical factor, and this is considered in more depth in Chapter 12. Little information is currently available regarding the sensitivity of the immune system to toxicants at different ages. Age-related immune alterations have been identified in fish (Duffy et al., 2002), rodents (Cederbrandt et al., 2003; Dortant et al., 2001; Luebke et al., 2000), and humans (Teig et al., 2002).

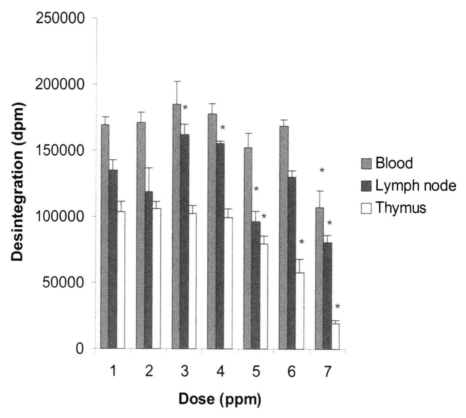

Figure 25.3 Blastogenic activity of lymph node, peripheral blood, and thymus lymphocytes from the grey seal, *Halichoerus grypus*, measured by tritiated thymidine incorporation after 3 hours of exposure with increasing concentrations of the polychlorinated biphenyl, Aroclor 1260. The results are expressed as disintegrations per minute (*$p \leq 0.05$).

This section focuses on differences in sensitivity of the immune system when exposure occurs post-natally, using similar approaches to those described in the previous section. Immune cells collected from animals of different ages were exposed *in vitro* to metals. Figure 25.4 illustrates the effects of methylmercury on lymphocytes and neutrophils collected from seals of different ages. Age-related differences are clearly evident when the responses in the newborn are compared to juvenile or adult seals. The sensitivity of lymphocytes, as compared to neutrophils exposed to mercury, may also be apparent (Lalancette et al., 2003).

Further studies were performed to evaluate the production of coelomocyte immune cells and their phagocytic potential at four different ages in the earthworm, *Eisenia fetida*, following *in vitro* or *in vivo* exposure to methylmercury chloride. The four groups included hatchling, juvenile, and two adult stages as defined by weight. The proportion of phagocytically active cells was lower in the hatchling group, whereas no differences were observed in the other age groups. Following incremental *in vitro* addition of methylmercury chloride or *in vivo* filter paper exposure, no differences in phagocytosis in the adult groups were observed. However, the absolute response of the coelomocytes from the younger worms was significantly reduced, and the estimates of lethal mercury concentrations indicated that the hatchling earthworms were approximately three times more sensitive than the mature earthworms (Sauvé et al., 2004).

In summary, it is evident that immune cells are particularly sensitive to prenatal or neonatal exposure in terms of dose of toxicant or degree of effect. This observation should influence the selection of testing methods and ultimate consequences on human health to be considered in the assessment (Barnett, 1997; Holsapple, 2003). However, once adulthood is attained, there is no clear

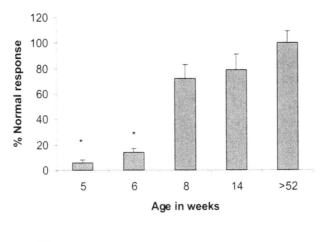

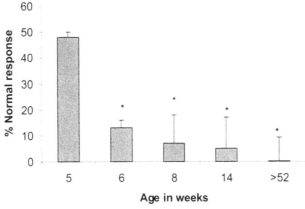

Figure 25.4   Effect on phagocytic activity of neutrophils (upper panel) and blastogenic activity of lymphocytes (lower panel) of methylmercury chloride (10^{-5} M) after 3 hours of exposure. The results are expressed as a percent of the normal response obtained from non-exposed lymphocytes or neutrophils collected from seals of varying age (*$p \leq 0.05$).

indication that increased sensitivity of the immune system to the toxic action of chemicals occurs. Systematic evaluation of the immune system should, therefore, be performed with animal models relevant to human and wildlife species at the appropriate stage of development when exposure is likely to occur.

25.5 DIFFERENCES OF SENSITIVITY ASSOCIATED WITH SEX

Gender differences have been frequently documented in the development of immune response following antigenic challenge. In most species studied, phagocytosis as well as lymphocyte-related responses, such as antibody production, are more efficient in females as compared to males. Studies from our laboratories in brook trout collected in three different lakes in Quebec support these observations as shown in Figure 25.5. However, when cells from these trouts are exposed *in vitro* to methylmercury chloride, female cells expressed a greater sensitivity than cells collected from males when exposed under similar conditions (Figure 25.6).

   In mammals, gender differences in the sensitivity of the immune system to metals are unclear. Peripheral blood leukocytes (PBLs) from females, immature females, male harbor seals (*Phoca vitulina*), and grey seals (*Halichoerus grypus*), were exposed to mercuric and zinc chloride at concentrations ranging from 10^{-9} M to 10^{-3} M. After a 4-hour exposure to 10^{-4} M or 10^{-3} M of

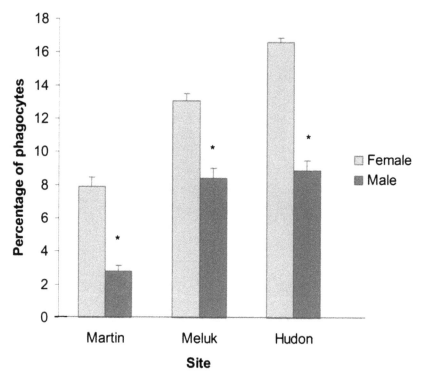

Figure 25.5   Phagocytic activity of head kidney cells from brook trout measured by flow cytometry after 18 hours of exposure with fluorescent beads. The trout were collected from three lakes — Martin, Meluk, and Hudon — in the pristine area of Gaspésie in eastern Quebec. Results are expressed as the percent of phagocytes per kidney (*p≤0.05).

mercuric chloride, cytotoxicity was evident, which resulted in a strong inhibition of the phagocytic activity of cells from all test groups of seals. Zinc chloride was less cytotoxic and induced inhibition of phagocytosis at 10^{-3} M. More interesting was the capacity of zinc to enhance the phagocytic activity of PBLs from mature females (Pillet et al., 2000). This gender difference may be related to the essential biological requirement of zinc.

Sexual differentiation and stages of maturation may represent confounding factors that should be taken into account in any immunotoxicologic evaluation especially when nontraditional species are considered. Avoiding inappropriate interpretation of immunotoxicologic data of chemicals or contaminated sites is recommended by carefully matching sexes and stages of maturation in these studies.

25.6 DIFFERENCES IN SENSITIVITY BETWEEN CELL TYPE AND CELL ACTIVATION STATUS

Evaluation of the effects of a chemical on the immune system for regulatory purposes or for the risk assessment process is based on selected immune endpoints chosen for their biological signif-icance or predictability and their sensitivity to the toxicant (Luster et al., 1992). These endpoints or immune responses are developed through the interaction of many types of cells, including various T-lymphocyte subpopulations, B-lymphocytes, macrophages, natural killer cells, and so on. In order to make the most appropriate selection of cell type for the *in vitro* assessment of chemicals, determining whether there are differences in sensitivity between specific populations is necessary.

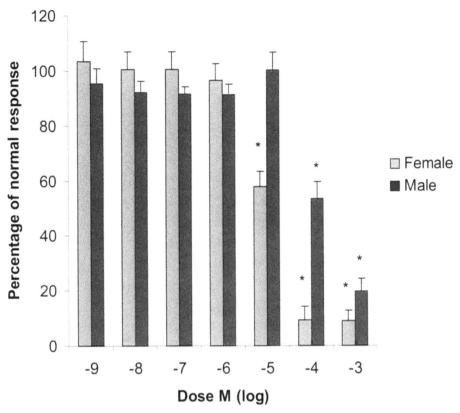

Figure 25.6 Phagocytic activity of head kidney cells from brook trout measured by flow cytometry after 18 hours of exposure to methylmercury chloride. Results are expressed as a percent of the normal unexposed control cells ($^*p \leq 0.05$).

To improve this selection process, the *in vitro* immunotoxic effects of selected heavy metals ($CdCl_2$, CH_3HgCl, and $HgCl_2$) on immunocompetence in several domestic and wildlife animal species (chicken, horse, cat, dog, and musk ox) were assessed using the mitogen-induced proliferation assay as a functional marker for lymphocyte activity and phagocytosis as a marker for macrophage activity.

Differences in sensitivity were evident at several concentrations for some metals. It was decided to perform a statistical analysis to compare the relative phagocytic and blastogenic responses with regression analysis, using as geometrical coordinates the percentage of the normal response of each endpoint (phagocytosis and blastogenesis) for each concentration of metal. The data for this evaluation with methylmercury chloride are presented in Figure 25.7. It is evident that when a significant effect is present, blastogenesis is affected to a greater degree as compared to phagocytosis. The slopes of the regression lines are consistently less than 0.5, suggesting that lymphocytes are more sensitive than macrophages to the toxic action of metals by at least a factor of two. Comparable results were obtained for $HgCl_2$ and $CdCl_2$.

It is well known that the cell activation process induces intracellular changes, including modification of molecules such as thiols. Intracellular thiols such as glutathione play an important role in many physiological processes, including protection from free radical damage and detoxification of chemicals. The nonenzymatic small molecular thiols include non-protein sufhydryls such as glutathione and protein thiols like metallotheionein that scavenge free radicals directly or bind heavy metals (Sugiyama, 1994). As a consequence, activated cells should exhibit less sensitivity to the toxic action of chemicals. To test the hypothesis, differences in susceptibility associated with

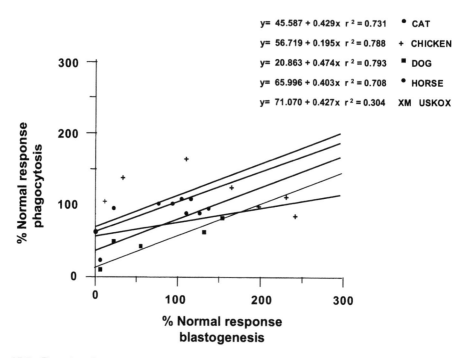

Figure 25.7   The plot of regression lines obtained using as geometrical coordinates, the percentage of the normal response of each endpoint (phagocytosis and blastogenesis) at each level of methylmercury chloride with cells collected from cat, dog, chicken, horse, and musk ox. Phagocytosis and blastogenesis were monitored in a dose–response manner (10^{-9} M to 10^{-3} M).

mercury (organic and inorganic), cadmium, and zinc chloride in both activated and nonactivated rat splenocytes were assessed following heavy metal exposure. Lymphocyte blastogenesis and macrophage phagocytosis were utilized to assess immune function. Cytofluorometric methods were employed to assess intracellular thiol content of lymphocytes and macrophages (Fortier et al., personal communication). Figure 25.8 indicates that macrophages were more resistant than lymphocytes to the toxic effects of cadmium, activated cells are more resistant than resting cells, and levels of thiols in the activated cells are higher than in the nonactivated cells following exposure to cadmium.

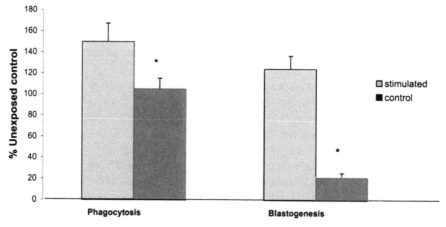

Figure 25.8   Histogram of the effects of *in vitro* exposure to $CdCl_2$ (10^{-5} M) on blastogenic transformation of mouse splenocytes stimulated with concanavalinCon-A. The results are expressed as a percent of the normal response obtained with noncadmium-exposed lymphocytes (*$p \leq 0.05$).

25.7 COMPARISONS OF *IN VITRO–IN VIVO* RESPONSES

The reliability and relevance of *in vitro* chemical exposure to evaluate immunotoxicologic impact have been subjects of controversy for many years, particularly when the complex interactions of cells and soluble factors required for immune function are considered. One can certainly argue that *in vitro* exposure cannot assess possible effects of chemicals on cell interactions, or on immune responses involving messengers from other physiological processes. However, with concerns about the use of animals, *in vitro* approaches remain popular alternatives (Balls and Sabbioni, 2001). The present study was designed to compare the *in vitro* and *in vivo* immunotoxic effects of metals to determine whether the data generated following *in vitro* exposure can be used to predict *in vivo* toxicity. Marine bivalves are aquatic invertebrate organisms that have been used as bioindicators for environmental monitoring. For *in vitro* exposure, hemocytes collected from the clams were exposed to CH_3HgCl or $HgCl_2$ at concentrations ranging from 10^{-9} to 10^{-4} M for 3 hours prior to the assessment of their phagocytic activity. For *in vivo* assessment of metals on phagocytic function of *M. arenaria*'s hemocytes, the clams were exposed to the same metals in water for up to 28 days at concentrations ranging from 10^{-9} to 10^{-5} M. In both instances, the phagocytic activity of hemocytes was determined by measuring the uptake of fluorescent microspheres using flow cytometry (Brousseau et al., 2000; Fournier et al., 2001).

While $HgCl_2$ produced *in vitro* cytolethal effects on hemocytes at 10^{-3} M, the viability was decreased only in hemocytes from clams exposed to 10^{-6} M $HgCl_2$ for 28 days. With respect to hemocyte phagocytic activity (Figure 25.9), *in vitro* exposure at lower concentrations of $HgCl_2$ enhanced phagocytic function, whereas higher concentrations of $HgCl_2$ ($\geq 10^{-5}$ M) inhibited phagocytosis. The suppressive effect of $HgCl_2$ on phagocytic activity appears to be a functional alteration since suppression occurred at noncytotoxic concentrations. Immunostimulatory effects of low-level heavy metal exposure have also been observed in vertebrate species (Cheng and Sullivan, 1984; Bernier et al., 1995; Fournier et al., 2000). As compared to the *in vitro* stimulation of phagocytic activity at lower concentrations of $HgCl_2$, a similar phenomenon was not observed following *in vivo* exposure. In addition, only *in vivo* exposure of *M. arenaria* to 10^{-6} M $HgCl_2$ for 28 days reduced hemocyte phagocytic function. Similar observations were made with CH_3HgCl.

To compare data from *in vitro* and *in vivo* exposures, the 50% inhibitory concentrations (IC_{50}) were determined. These IC_{50} values as well as their ratios are summarized in Table 25.1. Ratios of these IC_{50} for *in vivo/in vitro* exposure demonstrate that *in vivo* toxicity data can be extrapolated,

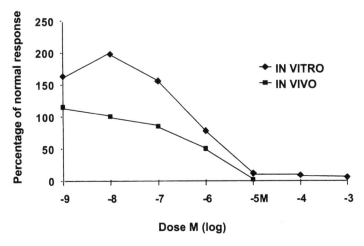

Figure 25.9 Phagocytic activity of hemocytes from *Mya arenaria* measured by flow cytometry after *in vitro* (18 hours) or *in vivo* (28 days) exposure with increasing concentrations of mercury chloride. The results are expressed as a percent of the normal response obtained with nonexposed cultured hemocytes (*$p \leq 0.05$).

Table 25.1 Concentration of Each Metal that Inhibits 50% of Phagocytic Activity (IC_{50}) Following *In Vitro* (18 hours) or *In Vivo* (28 Days) Exposure

	In vivo IC_{50}[a] 10^{-5} M	*In vitro* IC_{50}[a] 10^{-5} M	Ratio
CH_3HgCl	0.17	2.5	0.068
$HgCl_2$	2.8	39.0	0.069
Ratio	15.6	16.4	—

Note: IC_{50}, 50% inhibition concentration.

[a] The IC_{50} was determined graphically on a plot of the percent of phagocytosis versus the concentration of metal. The ratios of the *in vivo–in vitro* and $HgCl_2/CH_3HgCl$ IC_{50} values are also stated.

as least for mercury compounds from *in vitro* generated dose–response information. Moreover, the fact that ratios of IC_{50} values for $HgCl_2/CH_2HgCl$ are similar imply that *in vivo–in vitro* extrapolation is conceivable. It is not possible to generalize these conclusions to all chemicals since differences in bioactivation, absorption, excretion, and so on may alter expression. The use of tissues or tissue extracts to account for these differences may reduce the limitations of *in vitro* exposure models performed with cell suspensions.

In summary, *in vitro* approaches may represent very powerful options to assess immune function especially as our understanding of the immune system expands. In the context of human and ecological risk assessment, *in vitro* approaches allow for the opportunity to use cellular models with more species-specific relevance for the assessment process instead of using palliative rodent models for *in vivo* studies. Recent developments in fields such as quantitative structure activity relationships or microarrays will improve the reliability of these approaches by increasing the robustness of the assessment.

25.8 CONCLUSIONS

The immune system is a structurally and functionally complex system composed of several cell populations and organs strategically placed throughout the host's body. Maturation of the immune system depends largely on its challenge by exogenous agents including microbial infections, while regulation of the immune system is a function of cell–cell interactions mediated by a variety of adhesion molecules and endogenously produced substances, including the cytokines which interact with receptors found on the cell surface. The structural and functional integrity of the immune system is crucial in performing its protective role against pathogenic agents and the development of cancer. Thus, any chemically induced perturbation of the host's immune system can compromise its protective capacity and may lead to adverse health consequences for its host. An array of immunological methods have been developed and validated during the last decade. These methods have been used extensively to study quantitative and functional aspects of the immune system, mostly in adult experimental animals and, to a lesser extent, in humans exposed to potentially immunotoxic chemicals accidentally, occupationally, or via eating contaminated food (Descotes, 2003; Chapman, 2002). More recently, methodologies have been adapted to investigate specific endpoints of the immune system in wildlife species exposed to a wide variety of chemicals in the environment (Arkoosh and Collier, 2002; Fournier et al., 2004). All these studies indicated that chemicals of environmental concern can affect the immune system adversely, and have assisted in unravelling the mechanisms of action for some chemicals. Such studies have contributed to the understanding of the potential risk that such chemicals may pose to human or ecosystem health.

Although impressive progress has been made in the field of immunotoxicology during the last decade, much work remains to be done in the area of *in vitro* immunotoxicology, and in applying the methods developed in experimental animals to the human situation (Luebke, 2002; Barlow et al., 2002; Selgrade, 1999). Such efforts combined with the many established immunotoxicology principles and methods will, undoubtedly, increase the degree of confidence in data extrapolation from experimental animals to human or wildlife species. In this chapter, it was concluded that several parameters should be considered to properly use immunological endpoints in regulatory processes such as quantitative risk assessment. Among these considerations, differences in sensitivity associated with sex, cell type, tissue, endpoint, and cell activation status should be carefully reviewed. Moreover, the impact of endocrine modulators following *in utero* exposure to certain contaminants suggests that the development period represents a key period of high susceptibility. In addition, studies have examined other possible windows of susceptibility (puberty, aging, etc.) when the immune system may be particularly sensitive to hormonal modulation. Other factors must also be considered in any evaluation process including route of administration, nutritional status, stress, season, and so on (Banerjee, 1999; Scapigliati et al., 1999). Since a wide variety of potential consequences of immune alteration may occur, a new battery of tests monitoring a sufficient number of relevant endpoints needs to be validated.

REFERENCES

Arkoosh, M.R., and Collier, T.K, Ecological risk assessment paradigm for salmon: analyzing immune function to evaluate risk, *Hum. Ecol. Risk Assessment*, 8, 265–276, 2002.

Balls, M., and Sabbioni, E., Promotion of research on in vitro immunotoxicology, *Sci. Total Environ.*, 270, 21–25, 2001.

Banerjee, B.D., The influence of various factors on immune toxicity assessment of pesticide chemicals, *Toxicol. Lett.*, 107, 21–31, 1999.

Barlow, S.M., Greig, J.B., Bridges, J.W., Carere, A., Carpy, A.J.M., Galli, C.L., Kleiner, J., Knudsen, L., et al., Hazard identification by methods of animal-based toxicology, *Food Chem. Toxicol.*, 40, 145–191, 2002.

Barnett, J.B., Age-related susceptibility to immunotoxicants: Animal data and human parallels, *Environ. Toxicol. Pharm.*, 4, 315–321, 1997.

Bernier, J., Brouseau, P., Krzystyniak, K., Tryphonas, H., and Fournier, M., Great Lakes health effects — immunotoxicity of heavy metals, *Environ. Health Perspect.*, 103, 23–34, 1995.

Birnbaum, L.S., Developmental effects of dioxins, *Environ. Health Perspect.*, 103, 89–94, 1995.

Brousseau, P., Pellerin, J., Morin, Y., Cyr, D., Blakley, B., Boermans, H., and Fournier, M., Flow cytometry as a tool to monitor the disturbance of phagocytosis in the clam *Mya arenaria* hemocytes following in vitro exposure to heavy metals, *Toxicology*, 142, 145–156, 2000.

Cederbrandt, K., Marcusson-Stahl, M., Condevaux, F., and Descotes, J., NK-cell activity in immunotoxicity drug evaluation, *Toxicology*, 185, 241–250, 2003.

Chapman, P.M., Immunotoxicology and risk assessment — present prospects, future directions, *Hum. Ecol. Risk Assessment*, 8, 251–252, 2002.

Cheng, T.C., and Sullivan, J.T., Effects of heavy metals on phagocytosis by mollusca hemocytes, *Mar. Environ. Res.*, 14, 305–315, 1984.

Colborn, T., and Clement, C., Eds., Chemically-induced alterations in sexual and functional development: the wildlife-human connection, *Advances in Modern Environmental Toxicolovy*, vol. 21, Princeton Scientific Publishing, Princeton, NJ, 1992.

Descotes, J., From clinical to human toxicology: Linking animal research and risk assessment in man, *Toxicol. Lett.*, 11, 140–141, 2003.

Dortant, P.M., Peters-Volleberg, G.W.M., Van Loveren, H., Marquardt, R.R., and Speijers, G.J.A., Age-related differences in the toxicity of ochratoxin A in female rats, *Food Chem. Toxicol.*, 39, 55–65, 2001.

Duffy, J.E., Carlson, E., Li, Y., Prophete, C., and Zelikoff, J.T., Impact of polychlorinated biphenyls (PCBs) on the immune function of fish: age as a variable in determining adverse outcome, *Mar. Environ. Res.*, 54, 559–563, 2002.

Fournier, M., Clermont, Y., Morin, Y., Pellerin, J., and Brousseau, P., Effects of in vivo exposure of Mya arenaria to organic and inorganic mercury on phagocytic activity of hemocytes, *Toxicology*, 161, 201–211, 2001.

Fournier, M., Cyr, D., Blakley, B., Boermans, H., and Brousseau, P., Phagocytosis as a biomarker of immunotoxicity in wildlife species exposed to environmental xenobiotics, *Am. Zoologist*, 40, 412–420, 2000.

Fournier, M., Lalancette, A., Ménard, L., Christin-Piché, M.S., De Guise, S., Brousseau, P., Biomarqueurs immunologiques appliqués a l'écotoxicologie, in *Ecotoxicologie Moléculaire: Principes Fondamentaux et Perspectives de Développement*, Pelletier, E., Campbell, P.G.C., and Denizeau, F., Eds., Presses de l'Université du Québec, Québec, 2004, p. 460.

Guillette, L.J., Crain, D.A., Rooney, A.A., and Pickford, D.B., Organization versus activation: the role of endocrine-disrupting contaminants (EDCs) during embryonic development in wildlife, *Environ. Health Perspect.*, 103, 157–164, 1995.

Hastings, K.L., What are the prospects for regulation in immunotoxicology? *Toxicol. Lett.*, 102–103, 267–270, 1998.

Holsapple, M.P., Developmental immunotoxicity testing: a review, *Toxicology*, 185, 193–203, 2003.

Lalancette A., Morin Y., Measures L., Fortier, M., Brousseau, P., and Fournier M., Contrasting changes of sensitivity by lymphocytes and neutrophils to mercury in developing grey seals, *Dev. Comp. Immunol.*, 27, 735–747, 2003.

Lang, D.S., Meier, K., and Luster, M., Comparative effects of immunotoxic chemicals on in vitro proliferative response of human and rodent lymphocytes, *Fundam. Appl. Toxicol.*, 21, 535–545, 1993.

Lebrec, H., Roger, R., Blot, C., Burleson, G.R., Bhouon, C., and Pallardy, M., Immunotoxicological investigation using pharmaceutical drugs, in vitro evaluation of immune effects using rodent or human immune cells, *Toxicology*, 96, 147–156, 1995.

Luebke, B., Pesticide-induced immunotoxicity: are humans at risk? *Hum. Ecol. Risk Assessment*, 8, 293–303, 2002.

Luebke, R.W., Copeland, C.B., and Andrews, D.L., Aging and resistance to Trichinella spiralis infection following xenobiotic exposure, *Ann. N. Y. Acad. Sci.*, 919, 221–229, 2000.

Luster, M.I., et al., Risk assessment in immunotoxicology. I. Sensitivity and predictability of immune tests, *Fundam. Appl. Toxicol.*, 18, 200–210, 1992.

Pillet, S., et al., In vitro exposures of seals peripheral blood leukocytes to different metals reveal a sex-dependent effect of zinc on phagocytic activity, *Mar. Pollut. Bull.*, 40, 921–927, 2000.

Sauvé, S., Phagocytic activity of marine and freshwater bivalves: In vitro exposure of hemocytes to metals (Ag, Cd, Hg and Zn), *Aquatic Toxicol.*, 58, 189–200, 2002.

Sauvé, S., et al., Phagocytic activity of coelomocytes from terrestrial and aquatic invertebrates following in vitro exposure to trace elements, *Ecotoxicol. Environ. Saf.*, 52, 21–29, 2002.

Sauvé, S., and Fournier, M., Age-specific immunocompetence of the earthworm Eisenia andrei: Exposure to methylmercury chloride, *Ecotoxicol. Environ. Saf.*, 2004.

Scapigliati, G., et al., Immunoglobulin levels in the teleost sea bass Dicentrarchus labrax (L.) in relation to age, season, and water oxygenation, *Aquaculture*, 174, 207–212, 1999.

Selgrade, M.K., Use of immunotoxicity data in health risk assessments: Uncertainties and research to improve the process, *Toxicology*, 133, 59–72, 1999.

Smialowicz, R.J., The rat as a model in developmental immunotoxicology, *Hum. Exp. Toxicol.*, 21, 513–519, 2002.

Sugiyama, M., Role of cellular antioxidants in metal-induced damage, *Cell Biol. Toxicol.*, 10, 1–22, 1994.

Teig, N., Moses, D., Gieseler, S., and Schauer-U., Age-related changes in human blood dendritic cell subpopulations, *Scan. J. Immunol.*, 55, 453–457, 2002.

Zelikoff, J.T., and Thomas, P.T., *Immunotoxicology of Environmental and Occupational Metals*, Taylor & Francis, London, 1998.

CHAPTER 26

Approaches to Immunotoxicity Testing: The Need for National and International Harmonization/Standardization of Immunotoxicity Methods

Pauline Brousseau, Barry R. Blakley, Herman J. Boermans, and Michel Fournier

CONTENTS

26.1 INTRODUCTION

A major drawback to seriously consider immunology as an important component of any hazard identification as well as in risk assessment evaluation, is the observation that many immunological assays, especially functional assays, provide heterogeneous results leading to ambiguous conclusions in a toxicologic study. This is due to a variety of reasons. Among them are factors, such as different test systems, strain and age of animals, and source of reagents. Indeed, even if many of these variables could be made uniform, heterogeneity of results might still be a problem if the methodologies themselves are not well validated and standardized. This book provides numerous examples in various contexts of xenobiotics that produce adverse effects on the immune system, thus having significant health implications for human and wildlife populations as well as for domestic animals. The results presented here are the conclusions of international harmonization experiments performed by various teams of researchers with a special emphasis on the Canadian experience. Researchers unanimously believe in the importance of immunotoxicology, not only as a direct complement to toxicological studies but also for its contribution to understanding the mechanism of immunomodulation. Concurrently, researchers recognize the need for standardized approaches and methodologies in immunotoxicology.

26.2 HARMONIZATION STUDIES

26.2.1 International Collaborative Immunotoxicity Study

Initiated in 1986, the International Collaborative Immunotoxicity Study (ICIS) (see ICIS, 1998) involved 17 laboratories located in 11 countries. It was a joint effort by the International Programme on Chemical Safety, the United Kingdom Department of Health, and the European Union Commission. The aim of this innovative work was to verify through a large interlaboratory collaboration if the classical pathology or enhanced pathology (additional organs, semiquantitative grading of changes in lymphoid organs) performed in a conventional 28-day, repeated-dose oral toxicity test in the rat, as described by the OECD Guideline 407 (1986), would be sufficient to detect the immunotoxic potential of a chemical, or if more specific functional assays would be required.

The three strains of male rats used were Fisher 344, Wistar, and Sprague–Dawley rats. The experiments were conducted initially with azathioprine (AZA) and later with cyclosporin A (CsA) as test articles. Four groups of ten animals per group per compound were used. Three doses for each compound were selected with the high dose being the maximum tolerated one, which was expected to significantly reduce the body weight gain. The vehicles selected were carboxymethyl cellulose and olive oil for AZA and CsA, respectively. The parameters measured were body weights, organ weights, pathology and histopathology of lymphoid organs, heamatology, and bone marrow cell counts. In addition, the study included the functional assays of the antibody-forming cells with sheep red blood cells (SRBC), the lymphoproliferative response of splenocytes to various mitogens, and the natural killer (NK) cell activity of splenocytes.

For the pathology and histopathology, the classical hematoxylin and eosin staining of formaldehyde fixed and paraffin embedded tissues were used. The hematology parameters measured were haemoglobin, hematocrit, erythrocyte, leukocytes, platelet counts, and differential leukocyte counts.

Briefly, for the antibody-forming cells (AFC), each rat was injected once intravenously with 2×10^8 SRBC at the termination of the study. Four days later the animals were sacrificed by CO_2 anoxia, their spleens collected and suspensions of splenocytes prepared. The splenocytes were then mixed with SRBC and guinea pig complement in tubes containing warm agar and diethylaminoethyl dextran. Immediately after the mixing, the content of each tube was poured into a petri dish. These were incubated for 3 hours at 37°C, and the plaques formed in the agar layer (clear area around the antibody forming cell) were counted. The number of AFC per 10^6 splenocytes and per spleen were calculated.

The NK cell activity was performed using the 4-hour ^{51}Chromium (^{51}Cr)-release assay. For this assay, various concentrations of the splenocytes (effector) cells were mixed with a fixed number of ^{51}Cr-labeled YAC-1 (target) cells. After 4 hours of incubation at 37°C, the ^{51}Cr released in the supernatants was determined using a gamma counter.

For the lymphoproliferative response of splenocytes, spleen cell preparations were co-cultured in the presence of *Salmonella typhimirium*, a B-cell mitogen, and concanavalin A, a T-cell mitogen. The cells were incubated at 37°C for 3 days. Eighteen hours before termination of the incubation period, the cells were pulsed with ^{3}H-thymidine. The amount of ^{3}H-thymidine was then determined using a scintillation counter.

For details on the studies, see the International Programme on Chemical Safety (1996a, 1996b), the ICIS (1998). One of the main conclusions drawn from these studies was that use of the classical pathology did not facilitate the identification of immunomodulation by the chemicals used. However, use of the enhanced pathology approach readily detected the expected effects on the immune system of the same chemicals. Another major conclusion was that strict harmonization of methodologies including reagent sources and preparation and distribution of standard operating procedures to all laboratories led to a uniform data agreement across all participating laboratories. One of the strengths of this study remains in the fact that immunomodulation was detected by all laboratories involved at doses below the one that would induce general toxicity. It is also pertinent here to mention that of the immunological assays performed, the humoral immune response against SRBC was the most reliable. Mitogenic assay and NK cell activity were promising a direct consequence of the findings of these studies and was the recommendation for revisions to the existing OECD guideline. The revised OECD guideline 407, adopted in 1995, includes the following parameters: spleen and thymus weight; and histopathology of thymus, spleen, bone marrow, gut-associated lymphoid tissue, Peyer's patches, and lymph nodes (draining and distant).

The ICIS projects and their resulting data were successful in sensitizing regulatory agencies to better define study designs and harmonize methodologies for use in immunotoxicity studies.

26.2.2 Canadian Collaborative Immunotoxicity Study

The harmonization of approaches and methodologies to study the immune system became also a priority in Canada, and a group was created to undertake this task. The Canadian Collaborative Immunotoxicity Study (CCIS) was an important project of the Canadian Network of Toxicology Centres (CNTC). This project, initiated in 1995, involved researchers located in Saskatchewan, Ontario, and Quebec. Because the immune system is a very complex system that involves many organs, cells, and numerous factors, assays to test several aspects of the immune system are required. Therefore, the first task of the study was to select a relevant battery of assays. Various protocols were available in the literature (International Programme on Chemical Safety, 1986; Bloom et al., 1987; Luster et al., 1988, 1992a, 1992b, 1993; Van Loveren and Vos, 1989; Descotes et al., 1993). However, we had to keep in mind that the selected methodologies must be biologically meaningful, sensitive, reproducible, and ideally, that they could be used through the various steps of toxicological assessment (preclinical and clinical phases). The second task was to perform experiments in the spirit of good laboratory practice with a reference compound to test the robustness of the approach and the methodologies.

26.2.2.1 Experimental Design

The aim of the work was to put more emphasis on the reproducibility, reliability, and sensitivity of selected immunological assays in a 14-day, repeated-dose oral toxicity test in the rat. The reference compound selected was CsA. Three doses of CsA — 1.0, 5.0, and 10.0 mg/kg of body weight — were used. Selection of the maximum dose was based on preliminary studies showing that this dose did not produce more than 10% reduction in body weight over a 14-day treatment period. This was a dose well below the maximum tolerated dose of 40 mg/kg per day (ICIS, 1998). This is a key element that needs to be taken into consideration in order to discriminate between direct and indirect immunotoxicity. For the humoral response to a T-dependent antigen, SRBC, a satellite group of animals were required. Animals in these groups were immunized with SRBC. The treatment and control groups are depicted in Table 26.1.

The CsA and vehicle were administered by oral gavage. The volume of the bolus was 0.05 ml/10 grams of body weight. The doses given were corrected weekly to take into account the increase in body weight. The gavage was performed in each laboratory at approximately the same time each day, and always between 8:00 and 11:00 A.M. The animals from all groups were treated for 14 consecutive days. On day 11 of the treatment, all the rats from groups 2, 4, 6, and 8 were challenged with SRBC. On day 15, these animals were sacrificed by CO_2 anoxia and the spleens were collected. For animals in groups 1, 3, 5, and 7, blood was collected from the retro-orbital sinus into EDTA tubes

Table 26.1   Experimental Design of the Harmonization Study

Group	n	Treatment	Dose of CsA	Immunization with SRBC
1	10	Olive oil alone	0	−
2	10	Olive oil alone	0	+
3	10	CsA	1.0 mg/kg	−
4	10	CsA	1.0 mg/kg	+
5	10	CsA	5.0 mg/kg	−
6	10	CsA	5.0 mg/kg	+
7	10	CsA	10.0 mg/kg	−
8	10	CsA	10.0 mg/kg	+

Note: CsA, cyclosporin A; SRBC, sheep red blood cells.

once 3 days prior to the start of the treatment and once on day 15. The blood was used for hematology and phenotyping. The peritoneal macrophages were collected to perform the phagocytosis. At the termination of the study, the rats were sacrificed by CO_2 anoxia and the spleen, thymus, lung, kidneys, and liver were collected. The methodologies were standardized and detailed in standard operating procedures (SOPs) and were distributed to all participating laboratories. Care was taken to ensure that all personnel involved in this collaborative study had the relevant training and experience.

26.2.2.2   Materials and Methods

26.2.2.2.1 Test System

Male Fisher rats (inbred strain) were used in the study. The rats were between 5 to 7 weeks at arrival and they were acclimatized for 1 week before the start of the experiment. At the end of the week, individual body weight measurements were performed for the purpose of randomization. The environmental and husbandry conditions were standard. Rat chow and tap water, suitable for human consumption, were provided *ad libitum*. The food and water were free of contaminants known to interfere with study outcomes. Each laboratory received approval from respective animal care committees before the beginning of the experiment.

26.2.2.2.2 Test Article and Vehicle

CsA was purchased from Sandoz Pharmaceutical Corporation (Broomfield, CO). It was supplied as a solution at a concentration of 50 mg/ml. The vehicle was olive oil (pharmacopoeia-grade, Sigma, St. Louis, MO). It is important to mention that the three laboratories used these two compounds from the same lot numbers. Three dosing solutions of CsA were prepared weekly at concentrations of 0.2 mg/ml, 1.0 mg/ml, and 2.0 mg/ml. The dosing solutions were kept in the dark at room temperature.

26.2.2.2.3 Organ and Body Weight

Body weights were recorded once during the acclimatization, and on days 1, 8, and 15 of the study. Body weight gain was calculated over the 14-day period. Weights were also recorded for the spleen, thymus, lung, kidneys, and liver, and the results were expressed as a ratio of weight of individual organ/body weight.

26.2.2.2.4 Hematology

Prior to the beginning of the treatment and at day 15 into treatment, total erythrocyte counts, total leukocyte counts, hematocrit, hemoglobin, mean corpuscular volume, and mean corpuscular hemoglobin concentration were determined.

26.2.2.2.5 Humoral Response

The agar plaque-forming cell assay, as described prevously (ICIS, 1998), was used in the study.

26.2.2.2.6 Blood Phenotyping

The peripheral blood was collected in EDTA tubes. The expressions of OX-19+–W3/25+ (CD4), OX-19+–OX-8+ (CD8), and OX-6+ (B) were measured by direct immunofluorescence. Briefly, 100 μL of whole blood was mixed with optimal quantities of isotypic controls or specific monoclonal antibodies. Following incubation, the red cells were lysed, the preparation washed thoroughly, and fixed with formalin. The acquisition of 10,000 events was carried out for each sample using a flow cytometer (FACscan, Becton Dickinson, Franklin Lakes, NJ). The lymphocytes were identified by their characteristic appearance on a scattergram of side scatter versus forward scatter, and the fluorescence signals detected within the lymphocyte population were recorded (Brousseau et al., 1999).

26.2.2.2.7 Phagocytosis

Peritoneal macrophages were mixed with fluorescent carboxylate latex beads (1.5 μm). The mixtures were incubated at 37°C. After 60 minutes, free beads were removed with a bovine serum albumin gradient. The acquisition of 10,000 events was carried out for each sample using a flow cytometer (FACscan, Becton Dickinson, Franklin Lakes, NJ). The peritoneal macrophages were identified by their characteristic appearance on a scattergram of side scatter versus forward scatter, and the fluorescence signals detected within the macrophage population were recorded (Brousseau et al., 1999).

26.2.2.2.8 Statistical Analysis

Data were analyzed using a general linear model procedure with a Type III SS with three different effects in the model: (1) laboratory (starting point), (2) CsA, and (3) laboratory x CsA. This analysis, using the SAS statistical package (SAS Institute, Cary, NC), gave us the possibility to detect variations between the laboratories as well as immunotoxicity of CsA.

26.2.2.3 Results

There were no significant reductions of body weight in any of the groups, clearly indicating that indirect systemic toxicity was not involved. Tables were generated from each laboratory and summary tables were prepared. Table 26.2 is an example of the type of data collected on the humoral response expressed in plaque-forming cell (PFC)/spleen. The first part of the analysis with summary tables was performed to determine whether the laboratories were able to detect the expected effect of CsA on the various immune parameters examined. Since the aim of the preliminary studies was to standardize methodologies, the individual laboratory summary tables are not presented here.

Table 26.2 Summary of Data for Humoral Response Expressed in Plaque-Forming Cell/Spleen

Laboratory	Cyclosporin A			
	0.0 mg/kg	1.0 mg/kg	5.0 mg/kg	10.0 mg/kg
1	448,875	371,938	332,500	154,831
2	383,207	220,438	15,819	7,332
3	183,885	136,325	19,764	13,844

Table 26.3   Summary of Results for Differences among Laboratories and Effects of Cyclosporin A

Parameters	Laboratory	Cyclosporin A
Body weight	NSD	NSD
Spleen weight	NSD	NSD
Thymus weight	NSD	SD
Lung weight	SD	NSD
Kidney weight	NSD	NSD
Liver weight	NSD	NSD
White blood cells	NSD	SD
Red blood cells	NSD	NSD
Hemoglobin	NSD	NSD
PFC/10^6 cells	NSD	SD
PFC/spleen	NSD	SD
CD4$^+$ cells	SD	SD
CD8$^+$ cells	NSD	SD
CD5$^+$ cells	NSD	SD
Phagocytosis	NSD	NSD

Note: NSD, no significant differences; PFC, plaque-forming cell; SD, significant differences $p \leq 0.05$.

As expected, a trend toward a statistically significant reduction of the humoral response with increasing doses of CsA was observed. However, certain differences among the three laboratories were observed. First, the level of normal response (vehicle group) ranged from 180,000 PFC up to 440,000. Second, the magnitude of the suppression itself varied from 66% to 98% and 93% for laboratories 1, 2, and 3, respectively.

The second part of the analysis was performed to determine if, although the three laboratories have obtained a significant suppression of the humoral response, they were able to draw the same conclusions in terms of immunotoxicity. Table 26.3 provides a summary of the entire experiment in which discrepancies among laboratories regarding effects of CsA are reported. Three conclusions were drawn from these experiments. First, the expected immunomodulation by CsA was detected by all laboratories. Indeed, the immunosuppressive effects of CsA including reduction in total white blood cell counts, humoral response, and thymus weights, as well as modulations of T-cell subsets are well documented in the literature (Luster et al., 1980; Gunn et al., 1981; Britton and Palacios, 1982; Kahan, 1982; Tosato et al., 1982; Hess and Colombani, 1986; Thomson and Webster, 1988; International Programme on Chemical Safety, 1996b). Second, analysis of the data from the three laboratories demonstrated that all three laboratories agreed in 14 of the 16 parameters used in the study. Third, our data confirmed the conclusions reached by the ICIS studies that the new standardization of methodologies and protocols across laboratories results in meaningful and trustworthy data. Finally, data generated through these collaborative studies indicated that parameters such as peripheral bood phenotyping, NK cell activity, lymphoproliferative response of spleen mononuclear cells to mitogens, and phagocytosis produce comparable results across laboratories and could be good candidate assays for use in preclinical studies.

26.2.3 Other Harmonization Studies

Two more harmonization studies were carried out in Europe using the revised OECD TG 407 protocol with CsA and hexachlorobenzene as the reference compounds with five and nine laboratories, respectively. The novelty of these experiments was in the fact that they have used a classical immunosuppressor compound, CsA, and a second positive control, such as hexachlorobenzene. For detailed results of this study, see Richter-Reichhelm et al. (1995), Richter-Reichhelm and Schulte

(1996), and Schulte et al. (2002). Their conclusions confirmed the appropriateness of the revised protocol that call for harmonization of methodologies and approaches in immunotoxicology across laboratories.

26.2.4 Tangible Outcomes from Harmonization Studies

With the increasing trend in the development of biologics and pharmaceuticals that may target the immune system, it has become clear for regulatory agencies that updated approaches and newer methodologies must be developed to address scientific issues related to immunotoxicology. The U.S. Environmental Protection Agency (1998), U.S. Food and Drug Administration (2002), and European Agency for the Evaluation of Medicinal Products (2000) have already published test guidelines. These documents make recommendations on various immunological parameters that should be investigated in toxicology studies. These recommendations are characterized as flexible allowing for the evaluation of chemical-induced immunotoxicity on a case-by-case approach.

During the Sixth International Conference on Harmonization (ICH), held in Japan in November 2003, it was agreed that immunotoxicology should be among the list of priorities for the ICH. This decision came following a survey in which the relevance of immunoassays used in repeated-dose toxicity studies was questioned. ICH experts have concluded that enough data are available to scientifically support the preparation of a harmonized guideline for immunotoxicology testing.

26.2.5 Validation Issues

One of the key issues to accomplish successful harmonization studies relies on standardized methodologies. However, before proceeding to standardization, validation of the assays is a prerequisite to ensure that assays will be performed under optimal conditions, and consequently, that the results generated by these assays will be reliable.

The evolution of therapeutic products has compelled us to rethink our strategy to validate immunoassays. However, at the present state of knowledge, this task remains still an important scientific challenge. The procedures for the validation of analytical and bioanalytical methods, which methologies such as ELISA, EIA, and RIA, are well defined (Shah et al., 1992, 2000; International Conference on Harmonization, 1996; Findlay et al., 2000; U.S. Food and Drug Administration, 2001; DeSilva et al., 2003). A similar activity regarding the immunoassays used to test immunomodulating effects of chemicals is far from complete. Validation of immunoassays is a mandatory prerequisite for all laboratories engaged in immunotoxicity studies. While the scope of this chapter is not to provide complete guidelines regarding procedures used for validation of immunoassays, we present here pertinent issues that we feel are useful in validation procedures.

Reproducibility of the assays is a critical factor. When all parameters for a specific immunoassay are established, the assay should be run at least three times under the same conditions to determine reproducibility. The difficult task here is to determine the list of relevant criteria. When we look into the guidelines for the validation of bioanalytical assays, the list of potential parameters is well defined. We suggest a rigorous perusal of this list to provide a scientifically based reason for the inclusion or the rejection of each parameter in the list of acceptable criteria. However, due to the diversity of methodologies, the thinking process must be extended further. As an example, when we do flow cytometry testing, criteria such as the minimal number of events within the population of the cells to be analyzed is also a critical point. In summary, the basis to perform acceptable validation relies on sound scientific judgment.

A good example of this type of validation is provided in the work performed with the local lymph node assay (LLNA) (Gerberick et al., 2000; Kimber, 2001; Basketter et al., 2002; Kimber et al., 2002). Historically, to study the sensitizing potential of a compound, assays such as the Buehler and the maximization tests were performed in guinea pigs. However, an important drawback was the fact that the results obtained with these methodologies were very subjective. Following the extensive validation of the LLNA, the U.S. Food and Drug Administration (2002) revised the guidance document, and proposed the LLNA as a good alternative to study chemical-induced sensitising potential.

26.3 SUMMARY

Collectively, research efforts for the harmonization, standardization, and validation of approaches and methodologies applicable to immunotoxicology revealed two major points. First, it was clearly shown that a more in-depth evaluation of the immune system is needed in toxicological assessment. The discipline of immunotoxicology has greatly evolved over the last two decades, and now it is well recognized that classical histopathology must be enhanced to include more tissues of the immune system and complemented with functional assays. Second, the reliability and reproducibility of the various assays used in immunotoxicology increases greatly when proper validation is performed and standard operating procedures are harmonized across laboratories. The result of this exercise is a trustworthy set of data that can be used in risk assessment.

REFERENCES

Basketter, D.A., et al., Local lymph node assay — validation, conduct and use in practice, *Food Chem. Toxicol.*, 40, 593–598, 2002.

Bloom, J.C., Thiem, P.A., Morgan, D.G., The role of conventional pathology and toxicology in evaluating the immunotoxic potential of xenobiotics, *Toxicol. Pathol.*, 15, 283–293, 1987.

Britton, S., Palacios, R., Cyclosporin A — usefulness, risks and mechanism of action, *Immunol. Rev.*, 65, 5–22, 1982.

Brousseau, P., et al., Phenotyping of mononuclear cells, in *Manual of Immunological Methods*, CRC Press, Boca Raton, FL, 1999, pp. 115–125.

Descotes, J., Vial, T., Verdier, F., The how, why and when of immunological testing, *Comp. Haematol. Int.*, 3, 63–66, 1993.

DeSilva, B., et al., Recommendations for the bioanalytical method validation of ligand-binding assays to support pharmacokinetic assessment of macromolecules, *Pharm. Res.*, 20, 1885–1900, 2003.

European Agency for the Evaluation of Medicinal Products, Guidance for Repeated Dose Toxicity Testing, EAEMP, London, England, 2000.

Findlay, J.W.A., et al., Validation of immunoassays for bioanalysis: a pharmaceutical industry perspective, *J. Pharm. Biomed. Anal.*, 21, 1249–1273, 2000.

Gerberick, G.F., et al., Local lymph node assay: validation assessment for regulatory purposes, *Am. J. Contact Dermatitis*, 11, 3–18, 2000.

Gunn, H.C., Varey, A.M., Cooke, A., Effect of cyclosporin A on the function of T cells, *Transplantation*, 32, 338–340, 1981.

Hess, A.D., Colombani, P.M., Mechanism of action of cyclosporine: role of calmodulin, cyclophilin and other cyclosporin-binding proteins, *Transplant. Proc.*, 18, 219–237, 1986.

International Collaborative Immunotoxicity Study, Report of validation study of assessment of direct immunotoxicity in the rat, *Toxicology*, 125, 183–201, 1998.

International Conference on Harmonization, Guideline on Validation of Analytical Procedures: Definitions and Terminology, ICH of Technical Requirements for the Registration of Pharmaceuticals for Human Use, London, England, 1996.

International Programme on Chemical Safety, Immunotoxicology: Development of Predictive Testing for Determining the Immunotoxic Potential of Chemicals, Report of a Technical Review Meeting, World Health Organization, Geneva, 1986.

International Programme on Chemical Safety, International Collaborative Immunotoxicity Study, Phase I. Report of the Interlaboratory Comparison of the Effects of Azathioprine on the Pathology and Functional Tests of Immunotoxicity in the Rat, World Health Organization, Geneva, 1996a.

International Programme on Chemical Safety, International Collaborative Immunotoxicity Study, Phase II. Report of the Interlaboratory Comparison of the Effects of Cyclosporin on the Pathology and Functional Tests of Immunotoxicity in the Rat, World Health Organization, Geneva, 1996b, 245.

Kahan, B.D., Cyclosporin A: a selective anti-T cell agent, *Clin. Haematol.*, 11, 743–761, 1982.

Kimber, I., The local lymph node assay and potential application to the identification of drug allergens, *Toxicology*, 158, 59–64, 2001.

Kimber, I., et al., The local lymph node assay; past, present and future, *Contact Dermatitis*, 47, 315–328, 2002.

Luster, M.I., et al., The effect of adult exposure to diethylstilbestrol in the mouse: alterations of immunological functions, *J. Reticuloendothel. Soc.*, 28, 561–569, 1980.

Luster, M.I., et al., Development of a testing battery to assess chemical-induced immunotoxicity: National Toxicology Program's guidelines for immunotoxicity evaluation in mice, *Fundam. Appl. Toxicol.*, 10, 2–19, 1988.

Luster, M.I., et al., Risk assessment in immunotoxicology I. Sensitivity and predictability of immune tests, *Fundam. Appl. Toxicol.*, 18, 200–210, 1992a.

Luster, M.I., et al., Qualitative and quantitative experimental models to aid in risk assessment for immunotoxicology, *Toxicol. Lett.*, 64/65, 71–78, 1992b.

Luster, M.I., et al., Risk assessment in immunotoxicology II. Relationship between immune and host resistance tests, *Fundam. Appl. Toxicol.*, 21, 71–82, 1993.

Organization of Economic Cooperation and Development, Test Guidelines Programme, Immunotoxocity Testing: Possible Future Work, OECD, Paris, 2000.

Richter-Reichhelm, H.B., et al., Validation of a modified 28-day rat study to evidence effects of test compounds on the immune system, *Regul., Toxicol. Pharmacol.*, 22, 54–56, 1995.

Richter-Reichhelm, H.B., and Schulte, A.E., Evaluation of the Appropriateness of the Extended Subacute (28 Day-) Toxicity Test to Detect Adverse Effects on the Immune System, BgVV-Heft, Berlin, 1996.

Schulte, A., Althoff, J., Ewe, S., and Richter-Reichhelm, H.B., Two immunotoxicity ring studies according to OECD TG 407 comparison data on cyclosporin A and hexachlorobenzene, *Regul., Toxicol. Pharmacol.*, 36, 12–21, 2002.

Shah, V.P., et al., Analytical methods validation: bioavailability, bioequivalence and pharmacokinetic studies, *Int. J. Pharm.*, 82, 1–7, 1992.

Shah, V.P., et al., Bioanalytical method validation: a revisit with a decade of progress, *Pharm. Res.*, 17, 1551–1557, 2000.

Thomson, A.W., and Webster, L.M., The influence of cyclosporin A on cell-mediated immunity, *Clin. Exp. Immunol.*, 71, 369–376, 1988.

Tosato, G., et al., Selective inhibition of immunoregulatory cell functions by cyclosporin, *J. Immunol.*, 128, 1986–1991, 1982.

U.S. Environmental Protection Agency, Health Effects Test Guidelines, OPPTS 870.7800 Immunotoxicity, Environmental Protection Agency, Washington, DC, 1998.

U.S. Food and Drug Administration, Guidance for Industry, Bioanalytical Method Validation, U.S. Department of Health and Human Services, Washington, DC, 2002.

U.S. Food and Drug Administration, Guidance for Industry, Immunotoxicology Evaluation of Investigational New Drugs, FDA, Center for Drug Evaluation and Research, Washington, DC, 2001.

Van Loveren, H., and Vos, J.G., Immunotoxicological considerations: a practical approach to immunotoxicity testing in the rat, in *Advances in Applied Toxicology*, Dayan, A.D., and Paine, A.J., Eds., Taylor & Francis, London, 1989, pp. 143–163.

CHAPTER 27

Approaches to the Statistical Analysis of Immunotoxicologic Data in Support of Risk Assessment

Stephen Hayward

CONTENTS

There once was a Chi-square statistic,
That was somewhat depraved and sadistic,
It gave false hopes of significant slopes,
When an intercept was more realistic.

—*Stephen Hayward*

27.1 INTRODUCTION

Statistical methods provide an essential tool set for use across the field of toxicology, and although many of these have been discussed elsewhere (Maines et al., 1999) and will be discussed here, there are also methods that have particular application in the area of immunotoxicology. Although the primary focus of most of these methods is hypothesis testing, that is, the determination of whether two or more groups are different, there are two other functions that statistical methods serve, either separate from hypothesis testing or as an adjunct to hypothesis testing. The first of these functions is model fitting, which allows researchers to describe a response variable or a set of response variables in terms of another set of variables. The fitting of a variable that depends on time, such as body weight, by a polynomial function, is an example of model fitting. The second function is that of data reduction. The aim of data reduction is to reduce the dimensionality of the data with a minimum of loss in information. The familiar descriptive statistics, means, standard deviations, medians, and so on constitute a subset of this function, but model fitting can also be used to reduce dimensionality. The above example of model fitting, that is, body weight as a function of time, is also an example of data reduction. A large data set consisting of the daily weights for a set of animal subjects can be replaced by the much smaller set of parameters that define the fitted curves for each animal, usually consisting of only a few parameters per subject. This smaller set of parameters can be subsequently used for hypothesis testing, and this will be illustrated with examples of selected immunotoxicologic assays.

Although each of these functions is applied to the results of immunotoxicologic assays, in this discussion model fitting and data reduction will primarily be discussed as a means to perform hypothesis testing.

27.2 EXPERIMENTAL DESIGN

Since immunotoxicity data are usually a subset of data from a larger toxicological study, the experimental design — which includes the selection of dose levels and cofactors, determination of sample sizes, and the allocation of subjects to treatment groups — is determined in the context of this larger study rather than specifically in the context of the immunotoxicity component of the study. Large toxicological studies have been around for some time and the experimental design has been largely standardized. Guidelines for the execution of toxicological studies are available from the Organization for Economic Cooperation and Development (1995, 1997).

27.2.1 Choice of Doses

The selection of doses is outlined in OECD Test Guidelines 451-453, and these require at least three doses in addition to a concurrent control group. For carcinogenicity studies, guideline 451 states that "the highest dose level should be sufficiently high to elicit signs of minimal toxicity without substantially altering the normal life span due to effects other than tumours. The lowest dose should not interfere with normal growth, development, and longevity of the animal; and must not otherwise cause any indication of toxicity and in general should not be lower than 10 percent of the high dose." For chronic toxicity studies, OECD Test Guideline 452 states that "the highest dose level should elicit some signs of toxicity without causing excessive lethality," and "the low dose should not produce toxicity."

Since dose levels are generally chosen on the basis of signs of toxicity other than immunotoxicologic endpoints, there is little else to add with respect to the choice of doses to this discussion.

27.2.2 Sample Size and Randomization

Regarding the sample size, the OECD Test Guidelines suggest, "Statistical advice should be obtained to assure maximum reliability of the study and results amenable to statistical evaluation. The use of adequate randomization procedures for the proper allocation of animals to test and control groups is of particular importance."

Specific recommendations are for at least 50 animals of each sex for each dose group and concurrent control for carcinogenicity studies, and at least 20 animals per group for chronic toxicity studies. The minimum differences that can be detected in a t-test comparison of two groups with an alpha level of 5% with these numbers are shown in Table 27.1. Whether this is adequate for immunological assays will depend on the minimum difference considered important, and this will depend on the health impact of a change in an immune parameter. Sensitivity is also gained by using tests that use all the data, such as an F-test, or take into account the ordered nature of the dose response, so in general the recommended numbers are sufficient for finding meaningful differences. Often the issue is one of more than sufficient power and researchers find themselves explaining differences that are statistically different but not biologically significant.

The use of randomization in the assignment of animal subjects to treatment is now well established in the world of toxicological studies, and several means for accomplishing this from the physical shuffling of a deck of cards with the animal numbers to the use of computer-generated random numbers can be employed. Besides randomizing animals to treatment, the randomization protocol should take into account the assigning of treatments to caging so that the effects of housing are balanced across treatment groups, as well accounting for important cofactors such as initial weight.

Randomization of assay order to subjects or their samples has received little attention in the literature. Assays are subject to several influences including time, instrumentation, reagents, and technician. The effect of time can be seen in the drift of instrumental readings in a single run and in day-to-day variability in assay performance. Changes in reagents, instrumentation, and technician can also contribute to this day-to-day variability. Therefore, in order to make meaningful comparisons between treatment groups, samples from each group, including controls, should be represented in each run of an assay with every effort to achieve balance within each run as well as over time within a run. A common means of achieving this balance is to run samples, selecting from each group in order until the assay series is complete. This achieves some semblance of balance over the course of an assay run, but has an inherent bias built in if there is drift. However, this drift is usually small in comparison with the subject-to-subject variability, and can be determined by running a standard periodically during an assay.

Table 27.1 Power of Statistics at $p < 0.05$ When $n = 20$ (Chronic Toxicity Studies) and $n = 50$ (Carcinogenicity Studies)

n	Power (%)	Minimum Difference (SD)
20	80	0.88
	90	1.02
50	80	0.56
	90	0.65

Note: SD, standard deviation.

27.2.3 Level of Significance

In the literature, significance is declared if the p value for a given test is less than or equal to 5%. Although this is the standard choice of level of significance, being married to this level has its drawbacks. It can steer us away from potentially interesting results, or on the other hand make us pay too much attention to marginal results. The p value is the probability of a test statistic being at least as extreme as the one observed. For the commonly used t-statistic this means the probability of it being higher for a positive difference, lower for a negative difference, or lying outside the range of plus and minus the observed value for a two-sided test. By definition, a lack of significance at the 5% level does not mean that a difference does not exist, only that there is no evidence that a difference exists. A marginally nonsignificant result may actually be indicative of an underlying true difference and should not necessarily be discounted. On the other hand, because the p value is the probability of generating a statistic more extreme than observed, then significance will be declared 5% of the time when no true difference exists. This is particularly a problem when several statistical tests are performed within the context of the same study. The probability of declaring at least one test as significant increases with each additional test, with the result that by time ten independent tests are performed, the probability of declaring at least one result significant is approximately 40%. In order to compensate for this increase in the probability of declaring significance, a lower significance level, $p = 0.05/n$, may be chosen so that the overall probability of at least one significant result is 5%. This is known as Bonferroni's adjustment (Hochberg and Tamhane, 1987). However, this adjustment is conservative if the tests are not independent, which is the case with the repeated measures often seen with immunotoxicologic assays since these measures are correlated. The multivariate techniques as outlined in this chapter adjust for this correlation, and thus do not suffer from being too conservative.

27.3 OUTLIERS AND TRANSFORMATIONS

Before any statistical tests can be applied, two things have to be considered. One is outliers and the other is appropriate transformation of the data. These issues are discussed together because they are somewhat interrelated. In the generation of data from large toxicological studies, suspicious results — much larger or smaller than what would be expected — are common. These are either invalid results derived from technical or data entry errors or valid observations, and it is important to ascertain which as much as possible. Sometimes the underlying technical or data entry problem can be identified, and the outlier can be corrected or omitted from subsequent analyses. Sometimes, extreme observations can be identified as invalid because of context. For example, if an observed weight for an animal is notably heavier one day than the weights for the previous and subsequent days, or if there is a shift in the weights for all the animals weighed on a given day, then it is evident that these are invalid results and can be omitted from subsequent analyses. However, if an observation cannot be clearly identified as an invalid observation, then it must be retained in subsequent analyses; however, results can be given both with and without the outlier included in the analysis. Outliers are often identified by looking at data plots. This introduces a subjective element to the identification of outliers but formal tests for outliers are available and can provide a more rigorous and objective process for detecting outliers (Barnett and Toby, 1984). However, formal methods for outlier detection usually assume that the data are normally distributed, and the data need to be appropriately transformed before such methods are applied.

Whereas the identification of outliers looks for observations that are considered extreme for an observed underlying normal distribution, transformations are often applied to data so that the distribution of the transformed data is made as normal as possible or to stabilize the variance of data, that is, so that the variance of the data is the same across all treatment groups and does not depend on the mean of the data. Usually, an appropriately chosen transformation achieves both normality, and stabilizes the variance for immunotoxicologic data. The most commonly used transformations for immunotoxicologic data are the raw (untransformed), square root, and log transformations. Scientists are most comfortable with the raw and log transformations, but any monotonic function defined on the range of the data can be used as appropriate. One of the advantages of using the log transformation is the interpretability of results. Means of the log-transformed data transform back to geometric means on the raw scale, and differences transform back to ratios. As with the detection of outliers, the appropriate transformation can be chosen using graphical means or formal tests for normality such as the Shapiro–Wilk test (Shapiro and Wilk, 1965). On the other hand, the square root transformation gives results that are more difficult to interpret when transformed back to the raw scale, but is often the appropriate transformation for data that consists of counts, that is, the number of plaques in a plaque-forming cell assay. Although the standard deviation roughly transforms to the coefficient of variation when back-transformed from the log scale to the raw scale, generally measures of variability do not transform back to something interpretable on the raw scale. Confidence intervals, on the other hand, are preserved when transformed back to the raw scale, that is, 95% confidence intervals on a transformed scale, transform to 95% confidence intervals on the raw scale.

Graphs useful for the selection of a transformation include scatter plots of the raw or transformed data versus treatment groups, QQ plots of residuals, and plots of treatment variances versus treatment means. This list is not exhaustive but will serve in most applications. Figure 27.1 illustrates these three types of plots for IgG levels in monkeys challenged with the pneumococcus vaccine. It can be seen in the scatter plot of the raw IgG levels (Figure 27.1a) that the variability is not the same across days postinjection, whereas when the data are log transformed (Figure 27.1b), the variability is quite consistent across both days and treatment. This can also be seen in the plots of the standard deviations versus means, where on the raw scale (Figure 27.1c) the standard deviation is seen to increase linearly with the mean for each day by treatment group, versus the log scale (Figure 27.1d), where the standard deviation is not seen to depend on the mean. For the QQ or quantile plots, the residuals around the means for each day by treatment combination, $y_{ij} - \bar{y}_i$, are ordered and plotted against normal quantiles, which are the inverse of the cumulative normal distribution for n equally spaced values of p from 0 to 1, that is,

$$p_i = \frac{i + 0.5}{n}, \quad i = 1, 2, \ldots, n$$

where n is the number of observations to be plotted against the normal quantiles. When data are normally distributed, then the QQ plots are approximately straight lines. As with the other two types of plots, it can be seen that the log transformation normalizes the data (Figure 27.1e and f). The first two types of plots also indicate that the variances are stabilized by the log transformation.

Since each point or a residual for each point is plotted in the scatter and QQ plots, these plots can also be used to pick out extreme observations. There are four observations and one in particular that can be seen, on the raw scale, to sit apart from the rest of the observations. However, they do not appear as outliers on the log scale, and herein lies the relationship between outliers and transformations. What are selected as outliers depends on the transformation, and the transformation chosen can be influenced by large values. However, in this illustration, even with the removal of the four largest observations, there would still be evidence of a relationship between mean and variance, the QQ plot would not be a straight line on the raw scale, and the log transformation would still be selected as the appropriate transformation.

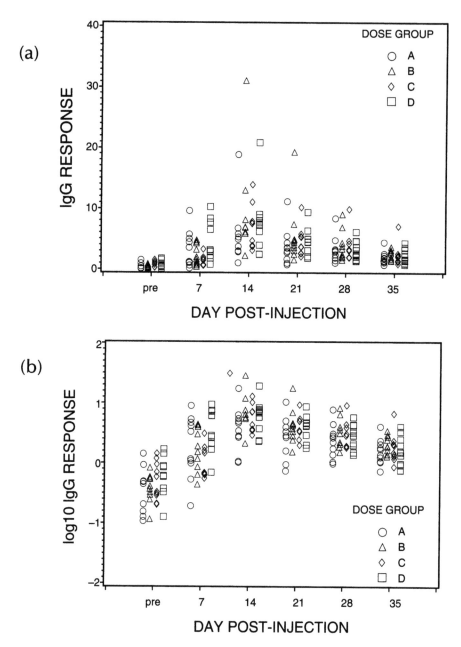

Figure 27.1 IgG levels in monkeys challenged with the pneumococcus vaccine for pre- and five post-immunization serum samples. (a) Scatter plot of IgG levels versus day postimmunization, raw scale. (b)
Scatter plot of IgG levels versus day postimmunization, log scale.

27.4 UNIVARIATE ANALYSIS OF IMMUNOTOXICOLOGIC ASSAYS

For some immunotoxicity assays there may be one observation for each animal in the study. After
removing outliers and choosing the appropriate transformation for analysis, many of the classic
tools of statistical analysis can be applied, such as analysis of variance (ANOVA) and pair-wise
t-tests. There are many good references for such analyses (Snedecor and Cochran, 1980; Steel and
Torrie, 1980); because these methods are widely used, they are not discussed here. However, there

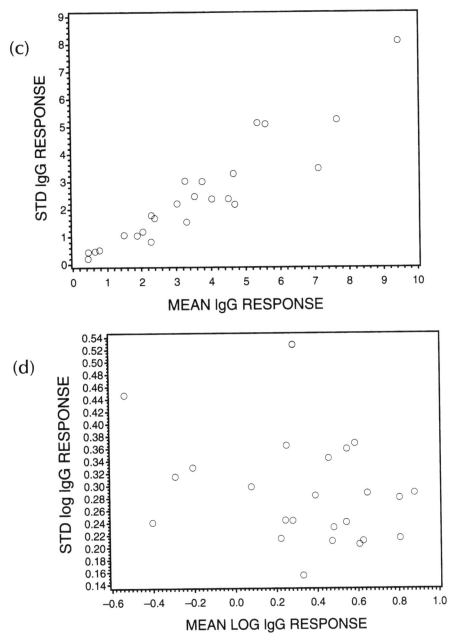

Figure 27.1 (continued) (c) Plot of standard deviations versus mean IgG levels, raw scale. (d) Plot of standard deviations versus mean IgG levels, log scale.

is an extension of ANOVA that is discussed, and some comments are made about the multiple comparison problem inherent in performing pair-wise t-tests.

27.4.1 Test for Trend

As mentioned previously, toxicological studies require at least three doses in addition to a concurrent control group. These doses imply an order to the treatment groups that is not always taken advantage of in the analysis. Sometimes, the effect of treatment is not strong enough to elicit responses for

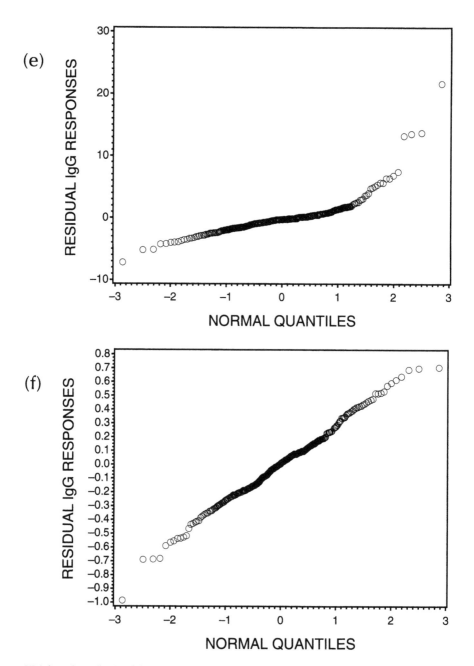

Figure 27.1 (continued) (e) QQ plot of IgG levels, raw scale. (f) QQ plot of IgG levels, log scale.

any of the treatment groups that are significantly different from the control group, but a consistent increase can be seen as the dose increases. The F-test for the effect of treatment that comes from the usual ANOVA gives evidence of any difference between the treatment groups regardless of order and will sometimes not pick up this subtle increase. However, sensitivity to the effect of treatment can be improved by the use of Armitage's trend test (Armitage, 1973). The model used to construct the F-test for the effect of treatment is as follows:

$$y_{ij} = \mu_i + \varepsilon_{ij} \qquad (27.1)$$

where μ_i is the expected response for treatment group i. The estimates of μ_i, $\hat{\mu}_i$ are the means of the responses for each treatment group and the error sums of squares is given by

$$ESS_1 = \sum_{i=1}^{k} \sum_{j=1}^{n_i} \left(y_{ij} - \bar{y}_i \right)^2$$

Alternatively, it can be assumed that the response increases linearly as a function of dose, that is,

$$y_{ij} = \alpha + \beta d_i + \varepsilon_{ij} \qquad (27.2)$$

for which the error sum of squares is

$$ESS_2 = \sum_{i=1}^{k} \sum_{j=1}^{n_i} \left(y_{ij} - \hat{\alpha} - \hat{\beta} d_i \right)^2$$

where $\hat{\alpha}$ and $\hat{\beta}$ are the least squares estimates of α and β.

Presumably the test for the significance of the slope parameter β would provide a more sensitive test for the effect of treatment since the straight-line model in Equation 27.2 takes into account the ordered dose response. When the response is linear with respect to dose $ESS_1 = ESS_2$, that is, there is no loss in sensitivity due to an inflation of the error sums of squares, and the test for a significant slope is indeed more sensitive then the F-test. However when the dose response departs from linearity, the error sum of squares for Equation 27.2 is inflated by this departure and some of the sensitivity gained by considering an ordered dose response is lost. Armitage's test for trend circumvents this problem by using the treatment sum of squares from the linear model in Equation 27.2, and the error sum of squares from the model represented in Equation 27.1 to construct the test for the effect of treatment. This increases sensitivity by taking into account the order inherent in the dose without losing sensitivity because of an inflation of the error term. Armitage's test for trend can be realized by fitting a hybrid model,

$$y_{ij} = \alpha + \beta d_i + \mu_i + \varepsilon_{ij} \qquad (27.3)$$

where α and β are the intercept and slope parameters, respectively, as in Equation 27.2, but the μ_i are the expected residuals responses for each treatment group. This model not only provides Armitage's test for trend, but also a test for the residual effect of treatment after adjusting for the linear fit. This can also be thought of as a test for the nonlinear effect of treatment.

Consider the preimmunization IgG levels in monkeys challenged with the pneumococcus vaccine (Tryphonas et al., 2001). As seen in Figure 27.1, there is an increase in log IgG levels with an increase in toxaphene dose (A = 0, B = 0.1, C = 0.4, D = 0.8 mg/kg of body weight). The standard ANOVA yields a p value for the effect of treatment of 0.18. On the other hand, Armitage's test for trend yields a p value of 0.038. What would otherwise be declared as nonsignificant is found to be significant when the ordering of the effect of dose is taken into account.

27.4.2 Multiple Comparisons

In toxicological comparisons it has been common practice to perform pairwise comparisons of the treated groups with controls to ascertain at which dose there is a significant effect of treatment.

n	$1-0.95^n$
2	0.0975
3	0.143
4	0.185
5	0.226
6	0.265

Table 27.2 Probability of Detecting at Least One Significant Difference for n Independent Tests

This has frequently been realized by performing t-test comparisons between each treatment and control group. However, this practice does not take advantage of the full power available from using all available data to estimate error, since each comparison uses only the data for the controls and that treatment group to which it is being compared. There is also an increased probability of detecting a significant difference when no difference exists. For example, consider two independent comparisons. By definition there is a 5% chance of detecting a difference for each comparison considered separately. Now, the probability of detecting at least one of the comparisons as significant is 1 minus the probability of not finding either significant:

$$1-\left(1-0.05\right)^2 = 1-0.95^2 = 1-0.9025 = 0.0975$$

Thus, the probability of detecting at least one significance difference is substantially higher than 5% and it only gets worse as the number of comparisons increase as Table 27.2 illustrates. One way to compensate for this is to choose a smaller p value at which to declare significance so that the overall probability of detecting a difference is 5%. Bonferroni's adjustment, p/n, accomplishes this and is appropriate when tests are independent, but the comparisons of treatments to control are not independent, because each test involves a comparison with a common control. In this case, Bonferroni's adjustment is conservative, and in any case this still does not take advantage of the power available from all the data. An alternative is to use a multiple comparison test such as Dunnett's test (Hochberg and Tamhane, 1987). This test performs the many-to-one multiple comparisons in the context of an ANOVA, thus using all available data to calculate the error and adjusting appropriately for the multiple comparisons. The test is available in several statistical packages.

27.5 ANALYSIS OF MULTIVARIATE OUTCOMES

For many of the immunotoxicologic assays, there is more than one outcome for each animal. For example, natural killer cell assays are run at different ratios of effector to target cells, and immunization of monkeys with various antigens involves measuring a response for each subject preimmunization and at weekly intervals for several weeks after immunization. The outcomes for separate ratios and times can be considered as separate variables, and the set of all these observations across ratio or time for a subject is referred to as a multivariate outcome. These variables are not independent of each other, and should not be analyzed as though they were. For each of these assays, useful information can be obtained by performing univariate analyses for each variable, that is, each ratio of the natural killer cell assay and at each time point for the immunization assay. However, there are multivariate analysis techniques that can be used with these data that often provide gains in sensitivity and interpretability.

27.5.1 Repeated Measures Data

The two examples listed above, the natural killer cell assay and the immunization assay, are examples of repeated measures data. These are multivariate data for which the same outcome is measured repeatedly for each subject at different levels of one or more within-subject factors, or variables that take on different values within subjects, as opposed to between-subject factors such as treatment or sex, which vary across subjects but not within subjects. The multivariate analysis of these data is referred to as repeated measures, which include, among others, profile analysis and the fitting of polynomials.

27.5.2 Profile Analysis

A good description of profile analysis can be found in Morrison (1976). A profile can be thought of as the shape of the response across the within-subject variable(s) for each subject. In Figure 27.1b, the log IgG levels can be seen to rise from preimmunization levels, peak at 14 days, and decrease until measurements were stopped at 35 days. One such set of observations for a subject constitutes the profile for that subject. A profile analysis answers three questions:

1. Are the levels of the profiles the same or different between treatment groups?
2. Is the response the same or different across the within subject factor(s)?
3. Is the shape of the response different between treatment groups or are they parallel?

Figure 27.2 illustrates three cases of possible mean responses for two treatment groups across some within-subject factor such as time. Case (a) is an example in which there is a difference between the profile levels. However, the profiles are flat, so no differences would be found across the within-subject factor. The profiles are parallel as well. In Figure 27.2b there are also differences across the within-subject factor, but the profiles have the same shape, and thus they are still parallel. In Figure 27.2c, the profiles are not parallel between treatment groups. In this case, the interpretation of the effect of profile level or differences across the within-subject factor is complicated by the lack of parallelism, so care must be taken in interpreting the significance of the test for profile levels and changes across the profiles in the presence of a significant lack of parallelism. In his discussion of parallelism, Morrison (1976) puts the third question about parallelism first, and prefixes the second and third questions with the assumption of parallelism. In any case, plotting the data often aids interpretation. Fortunately, the data analyst does not have to worry about the underlying complex matrix mathematics, since multivariate analyses, including repeated measures analyses, are currently available in any good statistical software package.

One drawback to a profile analysis and other multivariate techniques is that it cannot be applied to the data if there are missing values. Either values have to be imputed for these missing values, which is beyond the scope of this chapter, or all values for the subject with the missing value have to be dropped from the analysis.

As an example, consider the means plus standard error bars of the means of IgG levels in monkeys in response to a repeat challenge with sheep red blood cells (SRBC) plotted in Figure 27.3. IgG titers increase in the first week after the SRBC challenge and decrease thereafter. There is an apparent general decrease with an increase in dose at every time point, both pre- and post-challenge, and some graphical evidence that the profile shapes are different across treatment groups. What evidence can be seen of any of this in a profile analysis? Tables 27.3a, b, and c list the results of the profile analysis and the univariate analyses as well, since these are usually run with a repeated measures analysis. Note that the inclusion of both linear and nonlinear terms for dose is described in the discussion of Armitage's test for trend. There is no significant evidence of a trend in the IgG levels across dose for any of the days sampled or of a nonlinear effect of dose (Table 27.3a). This

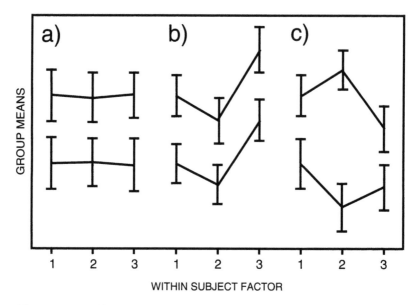

Figure 27.2   Mean responses illustrating three different profile scenarios.

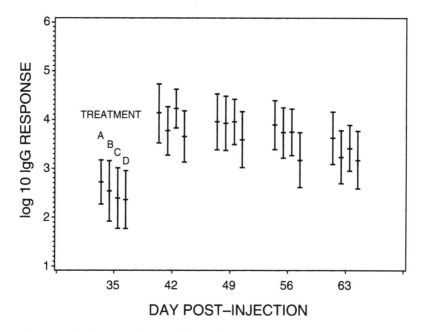

Figure 27.3   Mean IgG levels versus day postinjection plus or minus one standard error.

is also the case for the overall test for profile levels as well, that is, there is no evidence of a trend in the profile levels or of a nonlinear effect of dose. Not surprisingly, day had a highly significant effect. Note that as with the tests for profile levels, there are two tests for parallelism, one for each between-subject effect. One or more significant effects for the parallelism tests would provide evidence of a difference in the profile shapes (Table 27.3b). In this case, the differences in profile shape are the result of differences in the linear effect of dose across days. Inspection of Figure 27.3 would suggest that this is because of a greater decrease in IgG levels on Day 56. In order to find

Table 27.3a   Univariate Analyses and Tests for Profile Levels of Log IgG Titers in Monkey Serum

p Values for the Effect of Dose[a]	Day					Profile Levels
	35	42	49	56	63	
Linear	0.38	0.38	0.34	0.055	0.40	0.21
Nonlinear	0.89	0.27	0.86	0.73	0.61	0.72

[a] Data modified from Tryphonas et al., 2001.

Table 27.3b   Profile Analyses of Log IgG Titers: Test for Differences across Within Subject Factor, Day, and Tests for Parallelism

Within Subject Effect	p Value
Day	<0.0001
Day H dose (linear)	0.0051
Day H dose(nonlinear)	0.11

Table 27.3c   Profile Analyses of Log IgG Titers: Univariate Analyses of Differences between Successive Days

	Day 42 and Day 35	Day 49 and Day 42	Day 56 and Day 49	Day 63 and Day 56
Mean	<0.0001	0.11	0.0003	<0.0001
Dose (linear)	0.94	0.73	0.031	0.0032
Dose(nonlinear)	0.29	0.028	0.78	0.056

out between what days the lack of parallelism is occurring, one can look at univariate analyses of the differences between successive days, which have also been included as Table 27.3c. There is also a line for the mean effect. This is significant when the differences between successive days are different from zero. Thus, the IgG levels are significantly different between days 42 and 35 ($p < 0.0001$), days 56 and 49 ($p = 0.0003$), and days 56 and 63 ($p < 0.0001$), but not between days 42 and 49 ($p = 0.11$). Now, there is no evidence that the linear effect of dose changes between days 35 and 42 ($p = 0.94$) or days 42 and 49 ($p = 0.73$); however, there is evidence that the linear effect of dose changes between days 49 and 56 ($p = 0.031$) and days 56 and 63 ($p = 0.0032$).

27.5.3 Orthogonal Polynomials

Profile analyses cannot be used when the number of measurements per subject is too large, as in the case of a variable that is measured repeatedly over the life of an animal. Body weight is the classic example of such a variable, and the analysis of such data is often referred to as growth curve analysis. One such approach involves the fitting of low-order polynomials to the "growth" curve for each subject and using univariate and multivariate methods to analyze the fitted parameters. As mentioned previously, this serves as a method of data reduction. Instead of weeks worth of observations, the data are described by a small set of parameters for each subject. Often a quadratic or cubic polynomial will suffice to fit the data. These polynomial models can be represented as follows:

$$y_i = a_0 + a_1 x_i + a_2 x_i^2 \quad \left(\text{quadratic}\right)$$

$$y_i = a_0 + a_1 x_i + a_2 x_i^2 + a_3 x_i^3 \quad \left(\text{cubic}\right)$$

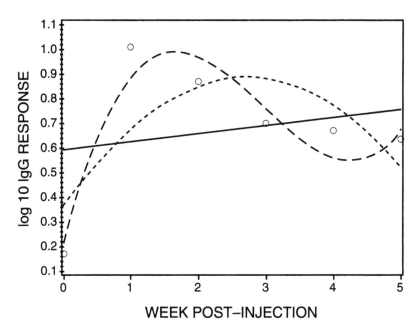

Figure 27.4 Linear, quadratic, and cubic polynomials fitted to the IgG levels of one subject.

where the y_i are the responses to be fitted for each subject, such as body weight, and the x_i are the values of the variable they are to be fitted against, usually time. One of the problems with fitting the above models and the subsequent analyses of the fitted parameters is that the polynomials, x, x^2, x^3, ... are not orthogonal on the set of values x_i. What this implies, without going into a detailed discussion of the geometric meaning of orthogonality, is that the estimates of a_0, a_1, ... will be affected by the estimates of the other parameters. Figure 27.4 illustrates the effect on the intercept of adding higher-order polynomial terms to the model. Linear, quadratic, and cubic polynomials were fitted to the log IgG levels for one monkey in response to pneumococcal antigens. The intercept does not stay put, and small changes in the estimates of higher-order terms can greatly affect the estimate of the intercept, especially as the degree of the polynomial is increased. This is undesirable when comparing estimates of the intercepts across groups of animals and between treatments because the intercept fitted to one subject's data may be affected differently by the higher-order terms than the intercept fitted to another subject's data.

However, if one fits orthogonal polynomials instead of garden-variety polynomials, the estimates of the coefficients for each term are unaffected by the estimates for each of the other terms. They also have straightforward interpretations: The orthogonal intercept is the mean of all the responses for an animal. The first-degree term estimates the average slope across the x_i, the second-degree orthogonal term estimates the average change in the slope or curvature, and the third-degree orthogonal term estimates the average change in that curvature across the x_i. So, how are orthogonal polynomials defined? The kth degree orthogonal polynomial term, p^k, is a linear combination of 1, x^2, x^3, ..., x^k such that the sum of the p^k over the x_i is zero and the sum of the cross-products with every other orthogonal polynomial is zero. Thus,

$$p_{k,i} = \sum_{j=0}^{k} a_{k,i} x_i^j$$

and

$$\sum_{i=1}^{n} p_{k,i} = \sum_{\substack{i=1 \\ k \neq h}}^{n} p_{k,i}\, p_{h,i} = 0$$

where n is the number of "time" points x_i.

In the example given in Figure 27.4, the first three orthogonal polynomials corresponding to weeks 0 to 5 can be defined as follows:

$$p_1 = x - \bar{x} = x - 2.5$$

$$p_2 = x^2 - 5x + \frac{10}{3}$$

$$p_3 = x^3 - \frac{15}{2}x^2 + \frac{137}{10}x - 3$$

As can be seen from this example, except for p_1, which can always be defined as $x - \bar{x}$, the orthogonal polynomials are not straightforward to generate, even for polynomials of degree no greater than three. They depend on the values of the x_i, and are not uniquely defined on any set of x_i. They have been generated for sequences of integers 1, 2, 3, ..., n for several n, and can be found in the statistical tables of texts such as Snedecor and Cochran (1980). However, they have to be built for each set of x_i if the x_i are not evenly spaced. There are algorithms for generating orthogonal polynomials, but these may not be readily available in standard statistical packages. This presents a problem with the practical application of orthogonal polynomials by researchers unless they have a strong background in algebra. A statistician can help if one is available, but if not there is a simple way of generating the appropriate estimates. First, find the mean of the responses for a subject. As mentioned above, this is the orthogonal intercept. Now, fit a straight line. Ignore the estimate of the intercept, and keep the estimate of slope. The estimate of slope is the same estimate as for the first-degree orthogonal polynomial coefficient. Third, fit a quadratic polynomial. Ignore the estimates for intercept and slope, and keep the estimate for the quadratic term. This is the estimate for the second-degree orthogonal polynomial coefficient. Now, fit a cubic polynomial. Ignore the estimates for intercept, slope, and the quadratic term, and keep the estimate for the cubic term. This is the estimate for the third-degree orthogonal polynomial coefficient. This can be repeated up to whatever polynomial degree is desired, but there is likely no reason to go beyond a quadratic or cubic polynomial. Why does this approach work? Since only the estimates from the highest degree term fitted in the model are taken, they cannot be affected by the estimates of any higher term, which implies orthogonality between this term and the higher degree terms, and thus vice versa.

One advantage of orthogonal polynomials over other multivariate techniques is that it can handle missing values as long as there are not too many since the polynomials, as with regular polynomials, can be fitted to the data as long as there are more data points for each subject than degrees in the polynomial. Note, however, that if too many data points are missing, then the orthogonality of the polynomials is compromised.

27.5.4 Other Multivariate Outcomes

Some immunotoxicologic data fall under the category of multivariate data, but are not repeated measures data. In other words, several measures are taken on one subject, but the same outcome is not being measured each time. Flow cytometry data comprise an example of such data. Various cell surface markers are measured in sera that provide counts and relative levels of several types

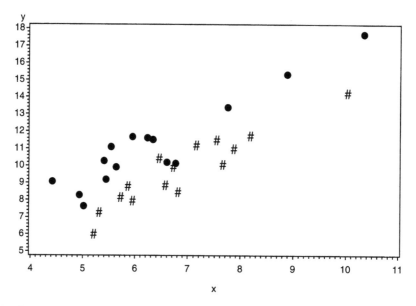

Figure 27.5   Example of highly correlated bivariate data.

Table 27.4   Univariate and Multivariate Analysis of Bivariate Example

Variable	Group 1 Mean ± SD	Group 2 Mean ± SD	P value, Effect of Group
x	6.88 ± 1.27	6.35 ± 1.59	0.32
y	9.70 ± 2.08	11.13 ± 2.65	0.11
Bivariate (multivariate) test for the effect of group			<0.0001

Note: SD, standard deviation.

of cells, including natural killer cells, B-lymphocytes, and T-suppressor/cytotoxic and T-helper/inducer lymphocytes. Although, these levels are often analyzed separately using univariate analyses, the levels of many of the cell surface markers are related, and multivariate methods may be useful in the analysis of these data. Because these are not repeated measures data, profile analysis or the fitting of orthogonal polynomials is not appropriate. However, the extension of ANOVA to the multivariate realm or the multivariate analysis of variance (MANOVA) can be used for these data. Instead of testing for differences in the means of one outcome between treatment groups, the MANOVA looks for differences in the location of a vector of means for several outcomes. This can be thought of as testing for the distance between the centers of clouds of points among treatment groups. The correlation between outcomes comes into play in multivariate analyses, where it does not in univariate analyses. Figure 27.5 provides a simple illustration that highlights one of the advantages of multivariate over univariate analyses. In this figure, there are two cloud points, one for each of two treatments and two outcomes, with one outcome plotted against another. The two outcomes are highly correlated, r = 0.94, and the ellipse of points for each treatment is oriented in such a way that there is no treatment difference when each variable is analyzed separately, but there is a highly significant difference when they are analyzed together in a bivariate analysis. This is summarized in Table 27.4.

When variables are correlated, and the shift in the center of the data due to treatment is in a direction that is, to some degree, at right angles to the direction of the association between the variables, then the MANOVA can find differences that the univariate ANOVAs will not. Yet, the

Table 27.5 Flow Cytometry Data, Test for the Effect of Treatment

Marker	Treatment p Value
Total CD2	0.25
Total CD4	0.34
Total CD8	0.20
CD2:CD8	0.27
Multivariate	0.025

Source: Data modified from Tryphonas, H., Arnold, D.L., Bryce, F., et al., *Food and Chem. Toxicol.*, 39, 947–958, 2001.

MANOVA can lose sensitivity as more variables are included in the analysis. Therefore, rather than applying a MANOVA to all variables indiscriminately from, say, a flow cytometry assay, it would be better to consider how these variables are related to each other and how best to group them for a MANOVA.

Table 27.5 lists the results of analyses of four flow cytometry markers. There was no significant effect of treatment on any of the markers ($p \pm 0.20$), but there was a significant effect of treatment for the multivariate test ($p < 0.05$). It should be noted that this significant effect was found for a small highly correlated subset of all the markers in the flow cytometry data set. A multivariate analysis of the full set of markers did not result in a significant treatment effect. Having found a significant result, one is left with the difficulty of interpreting such a result. It takes a bit of detective work to uncover the nature of the difference, since it is occurring, in this case, in a four-dimensional space. I will leave that for the brave and stout of heart.

Another multivariate technique that can be applied to data such as flow cytometry data is principal components analysis. A linear combination of a set of p variables can be expressed as follows:

$$Y = a_1 X_1 + a_2 X_2 + \cdots + a_p X_p$$

Principal components are linear combinations of the variables that explain, in order, most of the spread in the data. However, this is a complex technique, exploratory in nature, that often as not, does not provide useful information. It is beyond the scope of this chapter, but for those with a masochistic bent, see the principal components section in Morrison (1976).

27.5.5 Application of Statistical Methods to Immunotoxicity Data

Although it may be tempting to go straight to the application of ANOVAs to a set of variables generated by an immunotoxicologic assay, it is a good idea to get to know your data first. Plots, as described previously, will quickly provide you with an idea of the appropriate transformations to use in the ANOVAs, show suspected outliers, and highlight trends in the data if they exist. The results should not be overinterpreted at this point, as an apparent trend may be due to chance or there may not be enough subjects to provide sufficient power to detect a trend seen in the data at this stage. The next step in the analysis of any of the immunotoxicologic assays is to run univariate analyses for each of the variables generated by the assay. As these steps are generally applied to all assays, they are not discussed individually for each assay. More specifically, the appropriate multivariate analyses are discussed for each assay. Some mention is to be made as to the appropriate transformation. However, these should not be applied indiscriminately, without looking at the data first; for instance, I have indicated the log transformation for lymphocyte transformation assays, but have seen data where the quarter root scale performed better at stabilizing the variance.

27.5.6.1 Cell Surface Markers

When used to enumerate cell surface markers, flow cytometry results in data for which there are a large number of variables. As seen in the above examples, many of these are appropriately analyzed on the raw scale, although transformations, usually the log, may be needed for some of the markers. A profile analysis of these data is often not practical because of the large number of variables. It is also not appropriate since there is no natural ordering of the variables. For this reason, fitting orthogonal polynomials is not appropriate either. MANOVAs can be applied to the cell surface marker data in small appropriately selected subsets of the markers. Principal components can be used with this data as well, but this is a somewhat exploratory technique and difficult to apply without experience in working with them.

27.5.6.2 Natural Killer Cell Assay

Since the natural killer cell assay is run at a small selected set of ratios of effector to target cells — usually 100:1, 50:1, and 25:1 — a profile analysis is the appropriate multivariate technique to apply on the raw scale. However, the data for this assay are not always complete, and often suffer from a large number of missing values, particularly at the lowest and highest ratios. This leaves the analyst with the results of the univariate analyses, with the most sensitive analysis being available for the ratio with the least missing results. A transformation is not usually needed for these data.

27.5.6.3 Lymphocyte Transformations

In this assay, mitogen-nonstimulated and mitogen-stimulated lymphocyte proliferation are determined for each subject. Since the primary purpose of this assay is to assess the effect of treatment on lymphocyte proliferation in response to a given mitogen, and the possibility that there may also be an underlying effect of treatment on the mitogen-nonstimulated lymphocyte proliferation, it is recommended that the stimulated lymphocyte proliferation be adjusted by dividing by the nonstimulated levels for each animal. This corresponds to subtracting the nonstimulated lymphocyte proliferation levels from the stimulated levels on the log scale, which is generally the appropriate scale for analysis. An analysis of the nonstimulated levels should also be run so that the background response to treatment can be assessed. The stimulated levels can be easily subjected to a MANOVA, as well as the usual univariate ANOVAs, since only a few mitogens are typically used in this assay.

27.5.6.4 Listeria monocytogenes Assay

For this assay, rats are injected with a predetermined sublethal dose of *Listeria monocytogenes* on the last day of treatment. Colony-forming bacteria are enumerated in the spleen single-cell suspensions of control and treated rats on days 1, 2, and 3. Since the animals are sacrificed to obtain the cells for this assay, separate rats represent days 1, 2, and 3, that is, these data are not multivariate in nature. In this case, day is a between-subject factor, and would be included along with treatment in a univariate analysis.

27.5.6.5 Immunization of Monkeys with SRBC, Tetanus Toxoid, and Pneumococcal Antigens

The data for these assays consist of preimmunization antigen-specific antibody titers (IgG and IgM) and weekly determined titers (IgG and IgM) postimmunization with various antigens (Tryphonas et al., 2001). For the toxaphene data, there were five postimmunization serum samples collected. This is neither too many points for a profile analysis, nor too few for the fitting of orthogonal polynomials. Note, however, that for some profiles there may be a large difference between the preimmunization titers and the first postimmunization titers, followed by a gradual decrease in subsequent postimmunization serum samples. Profiles with this shape would be difficult to fit adequately with low-order orthogonal polynomials. This leaves one with a profile analysis to provide a multivariate analysis. The log scale is the usual transformation for the data from these assays.

27.5.6.6 Plaque-Forming Cell Assay

Plaque-forming cell assay data consist of counts of the number of plaque-forming cells from one or more sources of cells. When there are more than one source of cells (e.g., spleen and thymus), a MANOVA can be applied to the data over and above the usual univariate ANOVAs. Since the data consist of counts, the square root transformation is appropriate.

27.6 SUMMARY

One of the least favorite required college courses for students in nonmathematics disciplines is statistics. It can be intimidating with its plethora of formulae and arcane concepts, but with the availability of modern computing and statistical software, the application of the more commonly used procedures, including multivariate analyses, is relatively straightforward and available to the research scientist. It has been my intention, in this chapter, to discuss most of the issues that must be considered in the statistical analysis, or as an adjunct to the statistical analysis of immunotoxicologic assays. Some of the multivariate techniques may not have been previously considered by the practitioner, and many important results have been discovered, I'm sure, without them. This being said, there can still be some gain in employing these techniques, and nothing to lose. Over and above the multivariate techniques, if there is anything I wish to impress on the researcher is that they look at and become familiar with the idiosyncrasies of their data, and to transform the data if needed. Transformations are often neglected in data analyses and can invalidate results. It is my hope that research scientists move away from the relatively insensitive pairwise t-tests toward more sensitive multiple-comparison techniques and trend analyses. Statistics can be daunting for some, so I hope that I have made the techniques which I have come to generally apply to the data from immunotoxicologic assays, somewhat more accessible to the nonstatistician.

REFERENCES

Armitage, P., *Statistical Methods in Medical Research*, Blackwell Scientific, London, 1973.

Barnett, V., and Toby, L., *Outliers in Statistical Data*, John Wiley & Sons, New York, 1984.

Grubbs, F.E., Procedures for detecting outlying observations in samples, *Technometrics*, 11, 1–21, 1969.

Hochberg, Y., and Tamhane, A.C., *Multiple Comparison Procedures*, John Wiley & Sons, New York, 1987.

Maines, M.D., et al., Eds., *Current Protocols in Toxicology*, John Wiley & Sons, New York, 1999.

Morrison, D.F., *Multivariate Statistical Methods*, 2nd ed., McGraw-Hill, New York, 1976.

Organization of Economic Cooperation and Development, Guideline for the Testing of Chemicals 407: Repeated Dose 28-Day Oral Toxicity Study in Rodents, OECD, Paris, July 27, 1995.

Organization of Economic Cooperation and Development, Guidelines for the Testing of Chemicals, Test No. 453: Combined Chronic Toxicity/Carcinogenicity Studies, vol. 1, no. 4, OECD, Paris, July 1997.

Shapiro, S.S., and Wilk, M.B., An analysis of variance test for normality, *Biometrika*, 52, 591–611, 1965.

Snedecor, G.W., and Cochran, W.G., *Statistical Methods*, Iowa State University Press, Ames, Iowa, 1980, 404–407.

Steel, R.G.D., and Torrie, J.H., *Principles and Procedures of Statistics*, McGraw-Hill, New York, 1980.

Tryphonas, H., Arnold, D.L., Bryce, F., Huang, J., Hodgen, M., Ladouceur, D.T., Fernie, S., Lepage-Parenteau, M., and Hayward, S., Effects of toxaphene on the immune system of cynomolgus (Macaca fascicularis) monkeys, *Food Chem. Toxicol.*, 39, 947–958, 2001.

Index